…OGY FOR HEALTH CARE STUDENTS

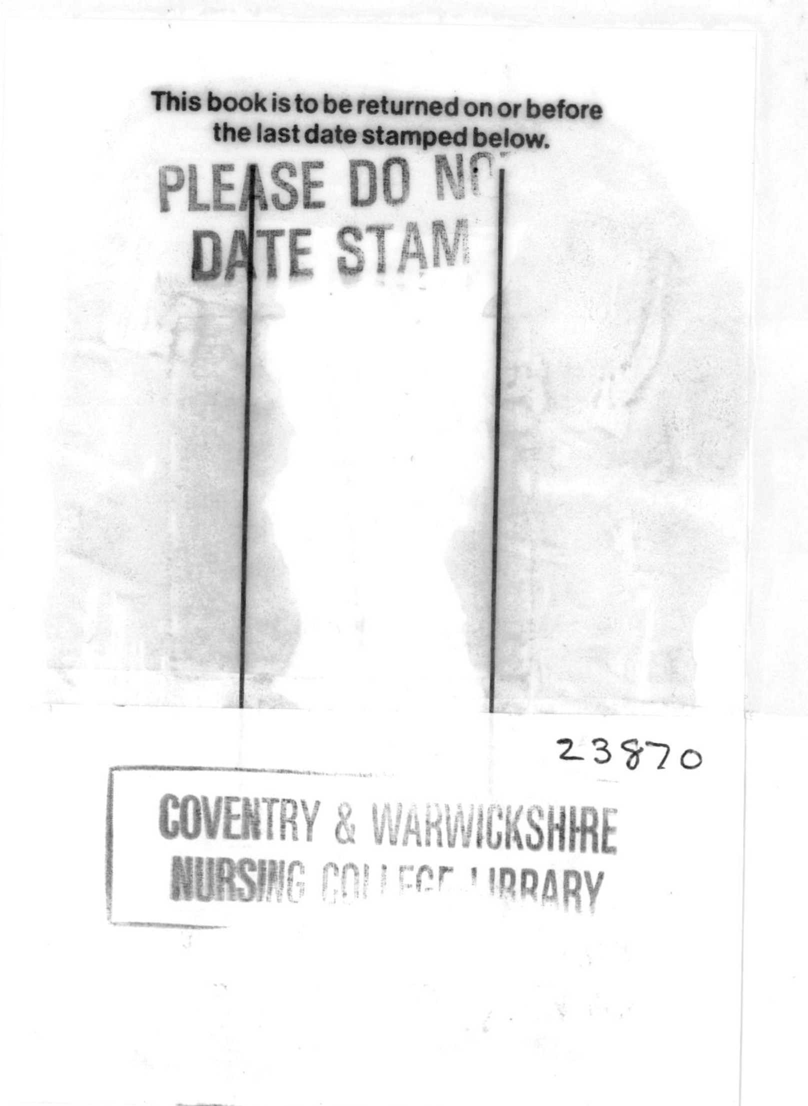

PHYSIOLOGY FOR HEALTH CARE STUDENTS

JUDY L. HUBBARD BSc MPhil

Senior Lecturer in Human Physiology
Department of Health Sciences
City of Birmingham Polytechnic
Birmingham, UK

DEREK J. MECHAN BSc MSc PhD

Co-ordinator of Studies
Institute of Medical and Health Care
Hong Kong Polytechnic
Hong Kong

Illustrated by Derek J. Mechan

Churchill Livingstone
EDINBURGH LONDON MELBOURNE AND NEW YORK 1987

CHURCHILL LIVINGSTONE
Medical Division of Longman Group UK Limited

Distributed in the United States of America by Churchill Livingstone Inc., 1560 Broadway, New York, N.Y. 10036, and by associated companies, branches and representatives throughout the world.

First published 1987

ISBN 0-443-02123-6

British Library Cataloguing in Publication Data
Hubbard, Judy L.
Physiology for health care students.
1. Human physiology
I. Title II. Mechan, Derek J.
612 QP34.5

Library of Congress Cataloging in Publication Data
Hubbard, Judy L.
Physiology for health care students.
Bibliography: p.
Includes index.
1. Human physiology. 2. Allied health personnel.
I. Mechan, Derek J. II. Title. [DNLM: 1. Physiology.
QT 104 H875p]
QP34.5.H83 1987 612 86–23312

Produced by Longman Group (FE) Ltd
Printed in Hong Kong

Preface

This book has been written primarily for students in the various fields allied to medicine, including nurses, physiotherapists, speech and occupational therapists and medical laboratory scientists. We first perceived the need for such a book when we were both lecturers at the City of Birmingham Polytechnic, teaching human physiology to various groups of health care students. Despite the many excellent textbooks then available, we always had great difficulty finding ones that fitted our students' needs precisely. For this reason, we began to produce our own material and several of the chapters in this book began life as student handouts, supplementary to lectures.

We had several aims in writing this textbook. The first aim was to consolidate the information given in lectures but not to add to it substantially. Secondly, we aimed to link the structures of tissues and organs with their functions, and consequently more anatomy is included in this volume than may be found in many textbooks of physiology. Thirdly, although some previous knowledge of biology is assumed, we do not rely on a substantial understanding of physics and chemistry, and appropriate topics from these disciplines are integrated with the physiology in the text. Fourthly, we have tried to make the highly complex subject of physiology accessible to the reader without, we hope, oversimplifying it. We have therefore set out to present the best established theories explaining physiological mechanisms, rather than a series of alternative hypotheses backed up with experimental evidence.

As this book is written for health care students rather than physiology undergraduates, we have chosen not to quote other authors extensively within the text and have only included a short reading list at the end of the book, instead of detailed references at the end of each chapter. Nevertheless, we would like to record our debt to the enormous number of researchers who have built up the present knowledge of anatomy and physiology and to the authors of the many textbooks, monographs and papers from which the information contained within this book has been drawn.

We should like to thank many of our colleagues both in Birmingham and Hong Kong for their useful advice and in particular, David Smith and David Quincey of the City of Birmingham Polytechnic, and Christopher Smith of the University of Aston, who helped us by supplying the photographs in the first two chapters. Last, but not least, we wish to thank Betty So (Hong Kong), who typed the manuscript.

Judy Hubbard
Derek Mechan

Contents

1

Cells and tissues

Optical microscopy
Electron microscopy
***Cell structure and function**, pp 6–22*
Nucleus
Endoplasmic reticulum
Golgi body
Lysosomes
Peroxisomes
Mitochondria
Microtubules and microfilaments
Centrioles and basal bodies
Cilia and flagellae
Membranes
Cell division
Chromosomes
Protein synthesis
***Tissue structure and function**, pp 23–63*
***Epithelia**, pp 23–29*
Simple squamous epithelium
Simple cuboidal epithelium
Simple columnar epithelium
Stratified squamous epithelium
Transitional epithelium
Stratified cuboidal and columnar epithelium
Glands
***Connective tissues**, pp 29–40*
Connective tissue proper
Cartilage
Bone
***Muscle**, pp 40–50*
Smooth muscle
Skeletal muscle
Cardiac muscle
***Nervous tissue**, pp 51–63*
Classification of neurones
Fine structure of neurones
Peripheral nerve fibres
Nerves
Neuroglia
Membrane potential
Origin of the resting potential
Generation of the action potential
Threshold
Refractory periods
Graded potentials
Propagation of the action potential
Synapses
Degeneration and regeneration of neurones

The observation that all living organisms are made of cells was first made by Robert Hooke in 1665. In 1839 the two German scientists Schlieden and Schwann postulated that the cell was the fundamental unit of which all living things are made. A little later, in 1858, Virchow added to this and put forward the idea that all cells arise from other pre-existing ones.

The rate of increase of knowledge of cell structure and function has largely paralleled the technological developments of the past century and both have accelerated in the last twenty five years.

Cells are very small objects, a fact which immediately presents a barrier to their examination. The smallest free-living single-celled organisms are the mycoplasmas which are about 100 nm (nanometres) in diameter (1 nm = 10^{-9} m). The smallest bacteria are about 500 nm diameter, whilst the majority of plant and animal cells vary between about 20 and 30 μm (micrometres) (1 μm = 10^{-6} m) or 20 000 to 30 000 nm diameter.

In the human body the majority of cells are grouped together with cells of similar types to form tissues. Tissues are then grouped to form organs, the functional units of the body.

Before discussing cell and tissue structure and function, it is useful to consider some of the methods used to study them. Since cells are small, in order to visualize them a magnifying system must be used. While the light microscope allows cells and tissues to be observed at relatively low magnification (less than about 2000 diameters), it does not generally permit an examination of the internal structures of cells. This can only be done by using the electron microscope which has far greater magnifying powers (up to several million diameters).

In addition to the different types of microscopes which are available there is also a large battery of other instruments and techniques available to the cell biologist. These include cell fractionation techniques, ultracentrifugation, a vast array of biochemical analyses and many types of chromatographic analysis. While it is generally beyond the scope of this book to include experimental procedures, because of their fundamental importance to an understanding of cell and tissue structure, a brief description of the basic techniques involved in light and electron microscopy is included.

Optical microscopy

In its simplest form the optical or light microscope consists of a light source, and objective and eyepiece lenses. Light passes through the specimen and is collected and focused by the objective lens to create a primary image. The latter is then viewed by the eyepiece lens which produces an enlarged virtual image which is projected on to the retina of the eye (Fig. 1.1).

The modern light microscope normally is much more complicated than this, having a built in light source, condenser lenses which focus the light on to the specimen, and compound objective and eyepiece lenses which eliminate the chromatic and spherical aberrations produced by single lenses.

It is evident that if light is to pass through the specimen, then the specimen must be transparent or at least translucent. For this reason only thin sections of tissues and organs and in some cases smears or squashes, can be studied. In addition since most tissues, in section, are not only transparent but also colourless they must be stained in order that they can be seen at all.

A wide variety of techniques exists for the preparation of materials for optical microscopy. Tissues may be examined fresh, frozen or fixed. They may be sectioned, smeared, squashed or teased. Probably the most commonly used technique is that where tissue is fixed and then sectioned, and this is the only one which will be described here.

It is important if an accurate picture of tissue structure is to be achieved that tissue is removed from the body, be it animal or human, as soon as possible after death. A small piece of tissue is taken and placed in a fixative solution (e.g. a solution which includes formaldehyde). This not only kills the tissue, but

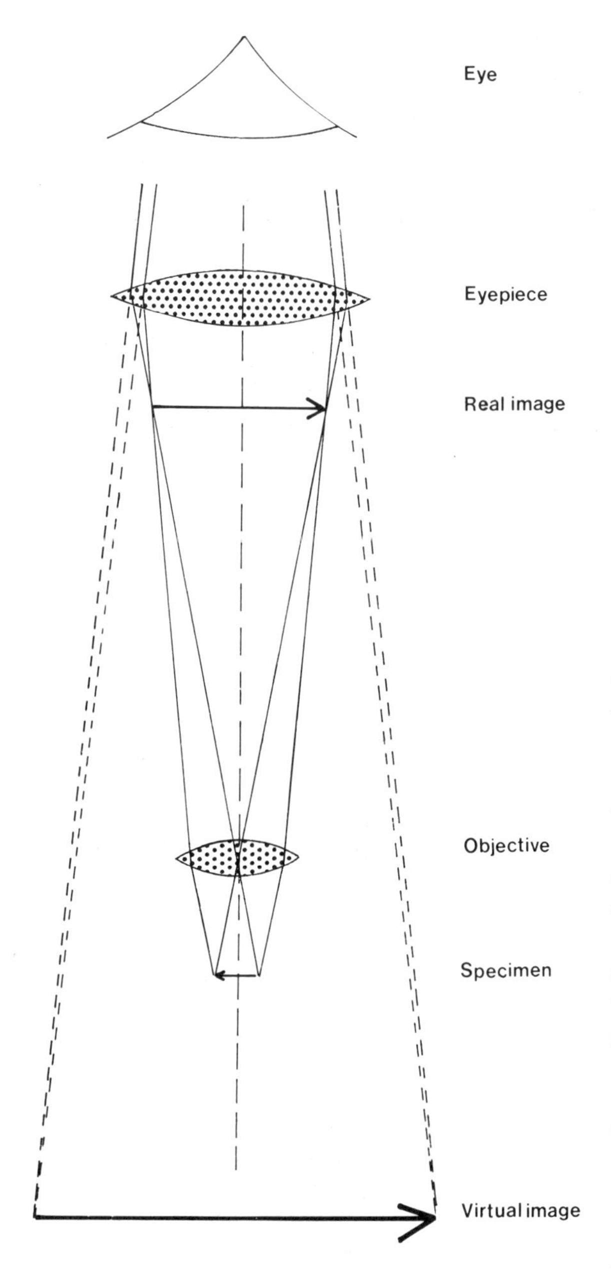

Fig. 1.1 Principle of the optical microscope

also preserves it. After several hours, or in some cases days, the fixative is washed away and the tissue is dehydrated by replacing the water held within its structure by an organic agent (e.g. alcohol). The tissue is then impregnated with a substance which can be made to set hard. Molten paraffin wax may be used for this, but since it is not miscible with alcohol, some intermediate or 'clearing' agent must be employed prior to wax impregnation.

Once the tissue is fully impregnated with wax, it is allowed to cool and harden. The tissue is now fully embedded within a wax block and is hard enough that thin sections (usually less than 60 μm) can be cut with a microtome.

The tissue sections are floated out on to the surface of a warm water bath from which they are collected on to glass slides. After drying, the wax is removed with an organic solvent and the section is hydrated again, by reversing the dehydration procedure. This is followed by a staining procedure which may be simple, involving one stain only, or complex involving several different stains. The tissue will now be coloured and will probably have been treated so that different elements appear to be different colours (differential staining).

Once again the tissue is dehydrated and then cleared by treating it with a solvent which is miscible with the medium used to mount the final product.

The prepared section is now ready to be viewed through the microscope. The tissue is no longer alive and its structure is incomplete since the section only represents a thin slice of the original (Fig. 1.2). Nevertheless it is possible, by cutting many such sections, to assemble an overall picture of the tissue's three-dimensional structure; this is called micro-reconstruction.

Paraffin wax embedding techniques are now considered by many to be rather old-fashioned and many new methods exist which involve the use of hard-setting plastics and resins.

In addition, other types of microscope are also available for specific purposes, e.g. phase contrast for examining living cells, the interference microscope which shows up masses of individual organelles and the polarizing microscope which allows tissues with a particular in-built structural regularity to be viewed.

Electron microscopy

Generally speaking, the minute structures

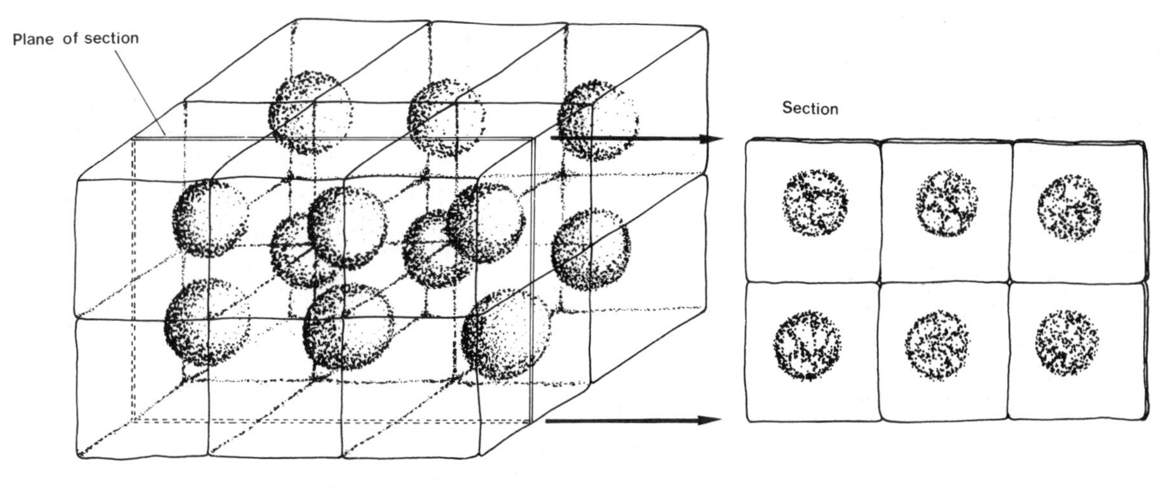

Fig. 1.2 Removal of a thin section from a block of tissue.

within cells are too small to be visualized with the long wavelengths employed by the optical microscope. For this reason the electron microscope is used, since it uses a beam of electrons of very short wavelength. This beam is fired down a column and through a thin section before striking a fluorescent plate upon which an image is produced. Glass lenses would absorb the electrons and prevent them from reaching the specimen and therefore the electron beam has to be focused by a series of magnets.

Superficially the electron microscope is very unlike the optical microscope; however, fundamentally their designs are very similar (see Figs. 1.1 and 1.3).

The preparative techniques are also similar to those employed by light microscopists. Fresh tissue is taken and fixed, then embedded, sectioned and stained. Normally very small pieces are taken to facilitate the rapid penetration of fixative. This would not normally be a formalin solution but might include substances such as gluteraldehyde or osmium tetroxide.

After fixation the tissue is dehydrated and then mounted in a plastic or resin (not wax) which is hardened by the addition of chemical hardeners and the application of heat. The block, which is much harder than the wax blocks employed by optical microscopists is

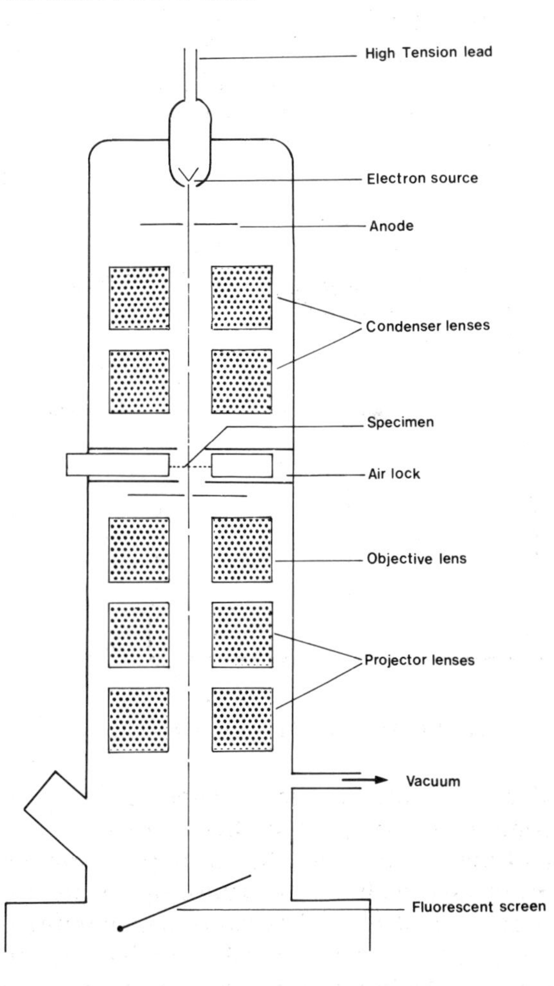

Fig. 1.3 Structure of the electron microscope.

sectioned on an ultramicrotome which includes an extremely sharp glass or diamond knife. Sections of less than 150 nm are normally required.

The sections are stained by the addition of heavy metal ions, not chemical dyes since these would not show up in the electron beam. The stained sections, each of which is probably less than 1 mm square, are introduced into the electron beam. The latter passes down to the specimen which freely transmits the electrons through unstained areas while heavy metal ions which have been deposited in the section absorb electrons; the amount of absorption being proportional to the amount of stain. At the bottom of the column is a fluorescent screen which is excited by the transmitted electrons such that a 'shadow picture' of the section is produced. Thus, the images produced by the electron microscope are very different from those produced by the light microscope (Fig. 1.4).

Again, electron microscopists recognize that the internal structure of the cell does not have the appearance of an electron micrograph when it is in the living state, but they conventionally describe these structures as though they did.

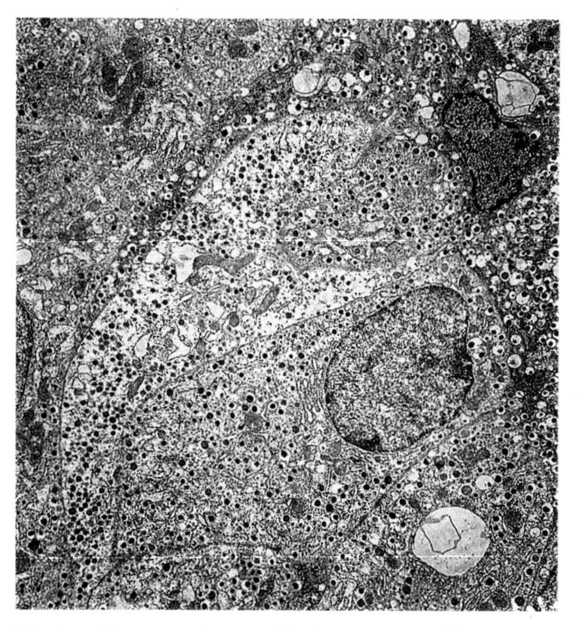

Fig. 1.4 Electron micrograph of pancreatic cells (mouse) × 3000. (Courtesy of C. U. M. Smith.)

CELL STRUCTURE AND FUNCTION

With the exception of red blood cells all human cells contain at least one nucleus and a mass of surrounding cytoplasm containing a variety of small components or organelles. In addition the cell may also contain temporary structures such as secretory vesicles which are known as inclusions. No cells contain all of the cytoplasmic organelles or inclusions which are known to exist and some cells contain very few. Nucleated cells are described as eukaryotic as opposed to bacterial cells which are prokaryotic, since they contain their genetic material scattered in the cytoplasm and have no discrete nuclear structure.

Nucleus

The contents of the nucleus are separated from the cytoplasm by a double nuclear membrane. The two layers are fused at some points to produce nuclear pores which are thought to allow molecules to pass between nucleus and cytoplasm (Figs. 1.5 and 1.6).

Electron micrographs reveal irregular masses of a heavily stained material in the nucleus. This is called chromatin and is a mixture of protein, deoxyribonucleic acid (DNA) and some ribonucleic acid (RNA). When the cell reproduces itself the chromatin condenses to form 46 small threads, the chromosomes.

Normally one or two areas appear to be especially condensed, these are the nucleoli which are rich in RNA. The nucleoli tend to disappear during cell replication and then reappear in the daughter cells.

The nucleus is the repository of genetic information for the whole body. However, while each cell contains the same information, in any one cell most of it will be 'switched off'. In cells of the same type the same part of the message will be 'switched on' but in different types of cells, different parts will be 'switched

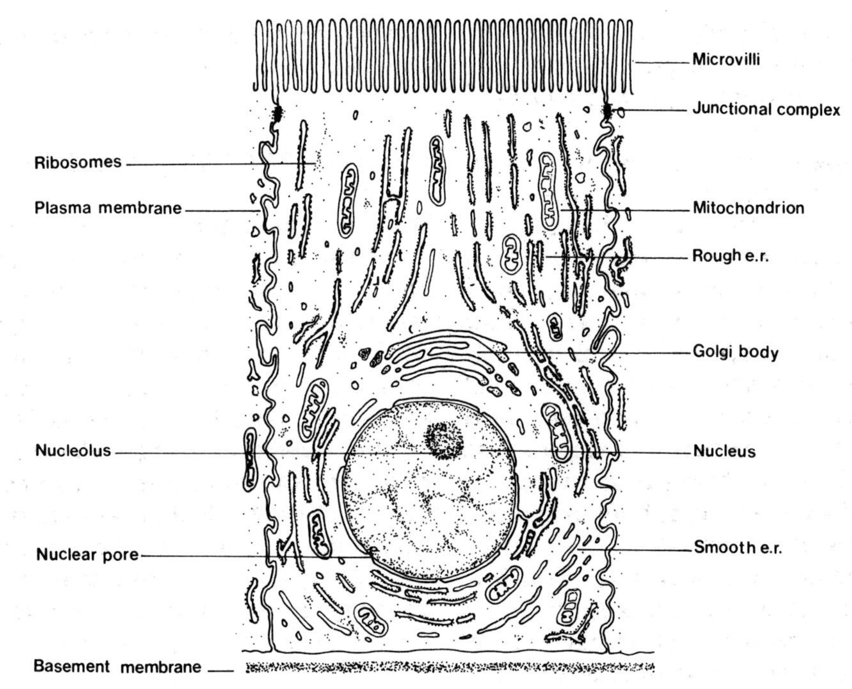

Fig. 1.5 Section through a cell of striated epithelium to show some of the common intracellular organelles (e.r. = endoplasmic reticulum).

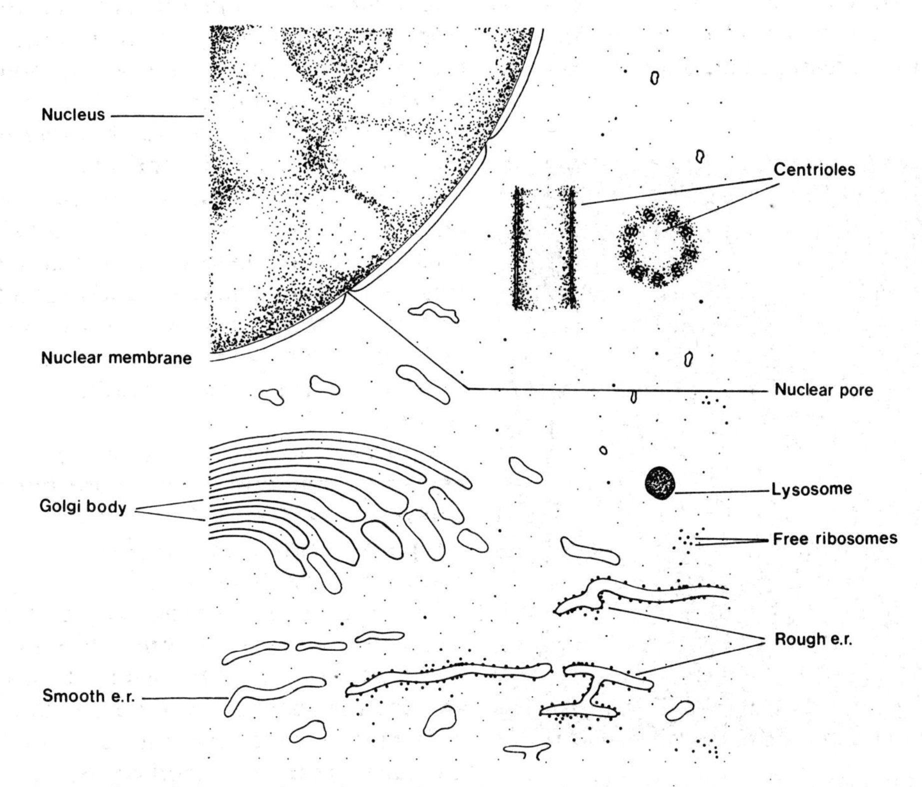

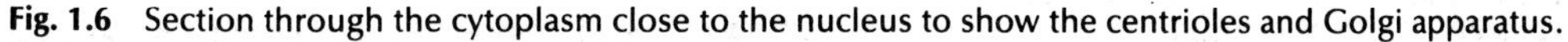

Fig. 1.6 Section through the cytoplasm close to the nucleus to show the centrioles and Golgi apparatus.

on'. The nature of this genetic information and the way in which it functions will be dealt with in a later section.

Endoplasmic reticulum

The endoplasmic reticulum is usually one of the more obvious cytoplasmic organelles. It is composed of a series of interconnecting membrane-bound tubules or channels and is found in the cytoplasm of all nucleated cells (Figs. 1.5 and 1.6). Very small amounts of endoplasmic reticulum may be present or the cytoplasm may be almost completely filled as for example in pancreatic cells. Endoplasmic reticulum is always being destroyed and reformed and the amount is not constant in any one particular cell.

The membranes of the endoplasmic reticulum may have a smooth or a granular appearance (smooth e.r. and rough e.r.). The latter is due to the presence of minute RNA-rich particles called ribosomes which are attached to the membrane surface.

Smooth endoplasmic reticulum is involved in the synthesis of lipid and, in liver cells, contains the enzymes which catalyse glycogen breakdown. It may also be associated with the detoxification of drugs.

Rough endoplasmic reticulum synthesizes protein molecules, most of which are to be secreted by the cell. For example the cells of the pancreas synthesize digestive enzymes. Molecules pass into the cavities of the e.r. and travel to the Golgi body (see below). Protein for intracellular use is probably assembled on the free ribosomes in the cell.

Golgi body

The Golgi body (Golgi apparatus or Golgi complex) is similar in structure to smooth endoplasmic reticulum. It consists of a small number of flattened sacs (often less than seven), which are usually arranged in a cup-like configuration. The 'cup' may face towards the nucleus but more often it faces away.

The Golgi body receives membrane-bound protein from the rough endoplasmic reticulum on its convex surface. Vesicles containing the stored protein are pinched off from the concave surface and pass into the cytoplasm. These vesicles may be of the secretory type, in which case they will pass to the edge of the cell and the contents will be discharged, or they may form lysosomes (see below).

Carbohydrate may be added to the protein within the Golgi body to form glycoproteins and mucus is also formed in this area.

Lysosomes

Lysosomes are small membrane-bound bodies about 0.5 μm in diameter which are formed from vesicles separated off from the Golgi body. Typically they appear darkly-stained in electron micrographs, and they are present in varying numbers in all the nucleated cells of the human body.

Lysosomes contain a number of powerful enzymes, including hydrolases which can break down all types of macromolecules and lysozyme which digests the cell walls of bacteria.

Their function is primarily digestive. When the body is starved they may break open and cause the digestion of whole cells, while in more normal times they may simply cause the breakdown of individual organelles. In this case the organelle becomes surrounded by a membrane, forming an isolation body. The lysosome then combines with this body and the contents are broken down (Fig. 1.10).

In phagocytic cells the lysosome combines with the phagocytic vesicle and useful molecules are absorbed into the cytoplasm while debris is left in what is now termed a secondary lysosome. Pinocytic vesicles also may be digested by lysosomes (see also *Exocytosis and Endocytosis*.)

Peroxisomes

Peroxisomes are also small membrane-bound bodies which are similar in appearance to lysosomes. They are different in function, however, in that they contain catalase which causes the breakdown of hydrogen peroxide.

The latter is formed by some of the metabolic reactions of the cell and is extremely toxic. The peroxisomes remove the molecules as they are formed and thus alleviate any danger to the cell.

Mitochondria

Mitochondria are found in varying numbers in all nucleated cells. They may be distributed evenly throughout the cytoplasm or concentrated in areas of high energy requirement; for example they lie between the fibrils of muscle fibres where they produce energy for contraction.

Mitochondria are small bodies which may be round or filamentous and may be less than 1 μm wide and several μm long; they are sometimes found to be branched. Mitochondrial form is not necessarily fixed and may change depending upon external factors acting upon the cell. Each mitochondrion is bounded by a smooth outer membrane which is separated by a small space of about 8 nm from a folded inner membrane. These folds are called cristae (Fig. 1.7) and they are studded with minute, mushroom-like particles.

Both inner and outer membranes, the space between them, the membrane bound particles and the inner matrix all contain enzymes. All of the enzymes which break down nutrient substances into carbon dioxide and water, together with the enzymes which enable the transfer of released energy to stable high-energy compounds are present within the mitochondrial structure. Virtually all of the cell's high-energy compounds are synthesized within the mitochondria.

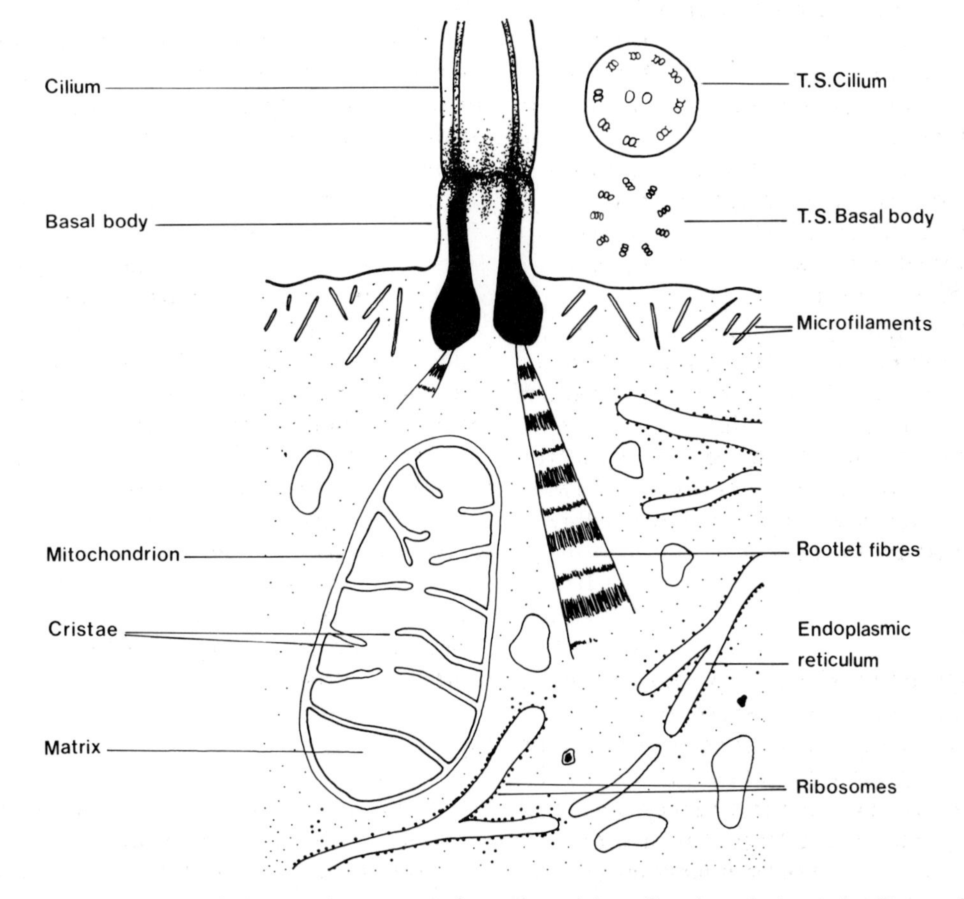

Fig. 1.7 Section through the cytoplasm near to the surface of the cell to show the base of a cilium and a mitochondrion.

An interesting point about mitochondria is that they contain a minute amount of DNA which can direct the synthesis of small quantities of protein. In addition mitochondria are able to replicate themselves. When cells divide, the original number of mitochondria distribute themselves between the daughter cells and then later replicate themselves so that each daughter will have roughly the same number as the parent cell.

Microtubules and microfilaments

Microfilaments are thin filaments 4 or 5 nm in diameter but of indefinite length, which form a lacework (the cytoskeleton) within the cytoplasm of most cells, helping to maintain the shape of the cell. The filaments are composed of actin and a small amount of myosin and are capable of a limited amount of contraction. It is suggested that they may bring about the movement of organelles within the cell.

Microtubules are somewhat larger, having a diameter of about 25 nm, and are more highly organized, being composed of 13 protofibrils which are arranged so as to form a hollow cylinder. Like microfilaments they contribute to the maintenance of cell shape and are found as neurofibrils in the axons of nerve fibres. They also provide the motile structures of cilia and flagellae and form the spindle of dividing cells (see below).

Centrioles and basal bodies

Two short cylindrical bodies (approximately 0.4 μm long and 0.15 μm diameter) lie close to the nucleus, usually at right angles to one another. Each of these bodies or centrioles is composed of nine sets of triple microtubules (Fig. 1.6).

The centrioles assist in cell division since they give rise to the spindle fibres (which are, in fact, microtubules) which pull the daughter chromatids apart.

Centrioles are self-replicating and in the early stages of cell division they separate slightly and reproduce themselves. Thus, for a while, the dividing cell contains two sets of centrioles.

In addition to their role in cell division, centrioles, or structures which appear remarkably like them, form the basal bodies of cilia and flagellae. Their function is to organize the microtubules which are always present in those organelles. Basal bodies also are able to replicate themselves.

Cilia and flagellae

Both cilia and flagellae are thread-like processes which protrude from the apical surfaces of some cells. Flagellae are normally quite long, up to 200 μm, while cilia are much shorter, being less than 10 μm in length. Flagellae are normally present in small numbers. For example, the male human spermatozoon has only one. Cilia, on the other hand, are more numerous and the epithelial cells which line the upper respiratory tract are covered by them.

Both types of thread have the same internal structure, having a core of microtubules (the axoneme) which extends up from the basal body and are arranged with nine double units (doublets) around the outside and two single ones in the centre. This 'nine plus two' organization is slightly different from the 'nine plus zero' structure of centrioles and basal bodies (Fig. 1.7).

Flagellae tend to move the cell to which they are attached through the surrounding liquid medium. Cilia on the other hand tend to move the medium over the cell surface (see *Epithelia*).

Both types of organelle are able to move due to the sliding of adjacent doublets against one another. This process is ATP-driven and is similar to the contractile activity of myosin in muscle.

Membranes

In an electron micrograph of an animal cell, membrane structures usually form the most obvious features. The cell is bounded by the plasma membrane, the nucleus by the nuclear membrane and the cytoplasmic organelles, e.g. endoplasmic reticulum, Golgi body and mitochondria are all membrane bound structures. Membranes regulate the entry and exit of molecules between the cytoplasm and the surrounding medium and perform vital functions within the cell.

Cell membranes are composed of protein, lipids and small amounts of carbohydrates. A study of the penetration of various substances into cells indicates that the lipid extends over all or almost all of the cell surface. Much of the lipid is in the form of phospholipid molecules, which have a charged 'head' region and a pair of hydrocarbon 'tails'. Thus one end of the molecule, the 'head', is hydrophilic and the other end hydrophobic.

Under appropriate conditions phospholipids can be demonstrated to form sheets, one molecule thick, e.g. at an air/water interface the molecules form a layer, one molecule thick, with the 'heads' in the water, the 'tails' in the air.

Gortner and Grendel, as long ago as 1925, suggested that membranes must consist of *two* layers of lipids and in 1935 Davson and Danielli carried this a little further. They proposed that membranes consist of two layers of phospholipids arranged with their 'tails' pointing towards one another, each free surface

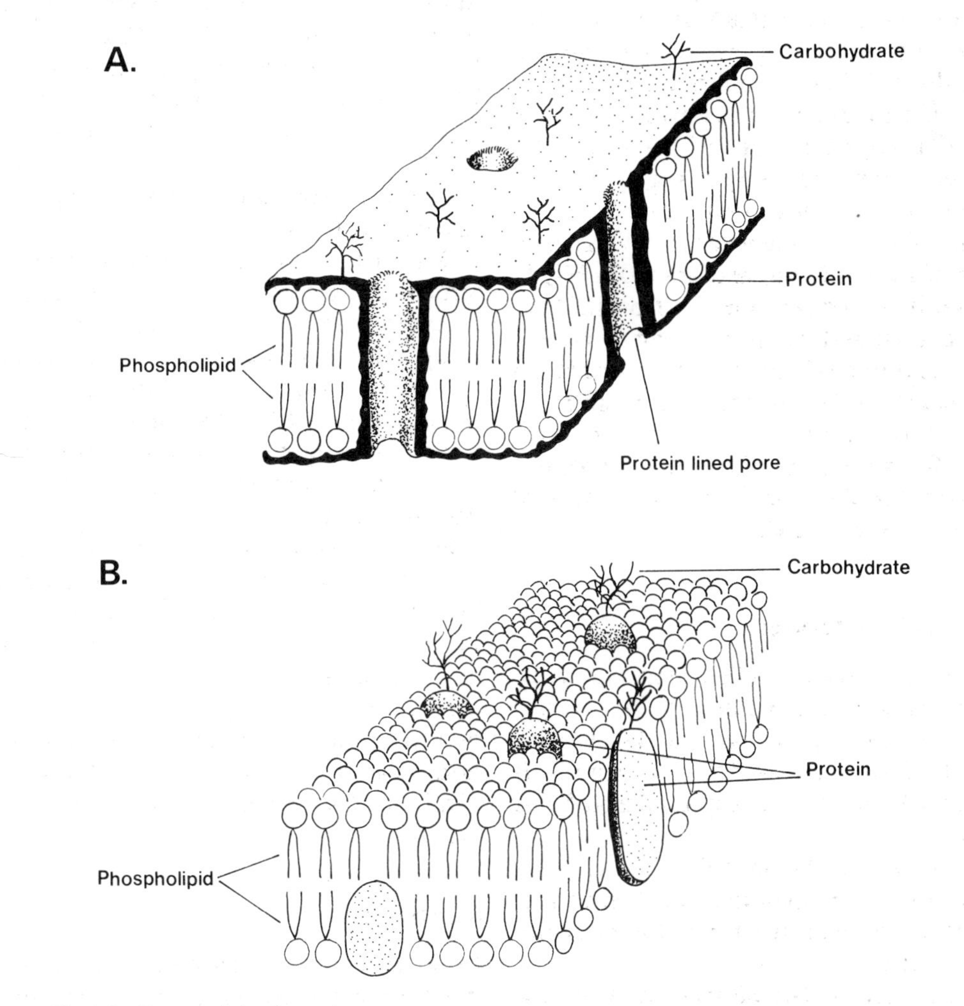

Fig. 1.8 Two models of membrane structure.
(A) Davson-Danielli model — protein sheets cover the two surfaces of the phospholipid layers.
(B) Singer-Nicholson model — protein molecules float in a 'sea' of phospholipid.

being covered by a layer of protein (Fig. 1.8A). This model has survived almost to the present day.

In more recent years Singer and Nicholson have put forward the theory that the protein does not actually coat the surfaces of the lipid but rather individual molecules 'float' in a 'sea' of phospholipid. Some proteins lie in one lipid layer, some in the other, while others extend right through the membrane. Some of these are known as permeases and are thought to act as channels or 'pores' for the passage of water and small solutes. This is known as the 'fluid mosaic' model (Fig. 1.8B). According to this proposal, proteins can move about within the lipid layer, although it is suggested that cholesterol (which is also a part of the membrane) is able to restrict this movement. Furthermore, some proteins may be anchored to underlying microtubules and/or microfilaments which would severely restrict their movement.

Membrane structure appears to be asymmetrical, and carbohydrate (in plasma membranes at least) is only found on the outer surface. Lipids too may be assymetrically placed within the membrane so that molecules which are found in the outer layer do not migrate to the inner layer and vice versa.

The plasma membrane acts as a permeability barrier, regulating the entry and exit of material into and out of the cell.

There are several mechanisms by which substances gain entry to (or leave) a cell. Firstly, they may pass along a concentration gradient (i.e. from a region of high concentration to a region of low concentration) using their own inherent kinetic energy. This process is known as simple diffusion and is generally employed by small ions or molecules and lipids. Large molecules tend to become attached to carrier molecules which themselves are an integral part of the membrane structure. When such transport is from high to low concentration then it is known as facilitated diffusion. The number of carrier molecules present in the membrane sets a limit to the number of molecules (e.g. glucose) that can cross a cell membrane at one time.

A different transport mechanism employs energy derived from the metabolic activities of the cell so that molecules can be taken against the concentration gradient. This is carrier mediated active transport.

Molecules may also enter or leave the cell by a mechanism which involves a physical change in the structure of the membrane. Phagocytosis or pinocytosis literally mean cell eating or cell drinking and involve the uptake of relatively large amounts of particulate matter or fluid. Exocytosis is the reverse of this and involves the elimination of material from the cell, as in secretion. This method of transport also requires energy from the cell.

Simple diffusion

There are several rules which appear to govern the diffusion of molecules across the cell membrane. Lipid solubility seems to be the most important factor and molecules of high lipid solubility tend to pass through more easily than those with low lipid solubility. This is logical if the structure of the membrane is considered. Small molecules pass more easily than large ones when lipid solubilities are the same and presumably this is related to variations in the sizes of pores in the membrane. Electrolytes, which tend to be repelled by the charge on the membrane surface, pass through less easily than non-electrolytes. The greater the charge the more difficult it becomes for molecules to pass; thus Ca^{++} will enter a cell less easily than Na^{+}. Ionic size is a further regulating factor in that small ions will pass more easily than large ions of the same charge. Size in this case, however, does not necessarily just mean ionic size, for ions tend to collect water molecules which travel with them. Sodium is a smaller ion than potassium, but sodium gathers more water molecules around itself than does potassium. Therefore the 'hydrated radius' of sodium is larger and it passes less easily than potassium.

The permeability of the membrane to various substance is not necessarily fixed and for example in the activated nerve axon the permeability for sodium is markedly raised above its resting value.

Water is able to move through the membrane by osmosis when water molecules pass from a region of low to a region of higher solute concentration. Since the water is actually passing from a region where there is more water to a region of less water, this is really another type of passive diffusion.

In addition to this, water and solutes can also pass through the membrane by ultrafiltration when there is a hydrostatic pressure difference between the two compartments. Water will pass from a region of high pressure to one of low, and the membrane will act like a sieve allowing solute molecules to pass through according to size, small ones passing easily and larger ones with difficulty, or not at all.

Active transport and facilitated diffusion

Active transport and facilitated diffusion both depend upon the presence of carrier molecules in the membrane. Since there will be a finite number of such molecules, then both processes are liable to saturation, i.e. a maximum rate of transfer may be achieved.

The two processes differ in that active transport requires metabolic energy while facilitated diffusion is still a passive process and requires only the kinetic energy of the transported molecules.

Carrier molecules are proteins and may be enzymes or molecules which are enzyme-linked. ATP is used to drive the process which requires that a molecule is taken up on one side of the membrane by the carrier which then undergoes some sort of structural change or may even physically move and carry the molecule to the other side.

Figure 1.9 shows two possible mechanisms for the active transport of materials through a membrane. In (A), a molecule becomes attached to an active site on the carrier molecule (1). The latter then moves through the membrane (2, 3) and gives up the molecule into the cytoplasm (4) and then returns to the outer surface (5). Such a mechanism would require a great deal of metabolic energy. In (B), a molecule passes into the space between two carrier molecules within the membrane (1), and attaches to a binding site (2). The carrier molecules then alter their configuration so that the binding site will release the transported molecule into the cytoplasm (3). This is

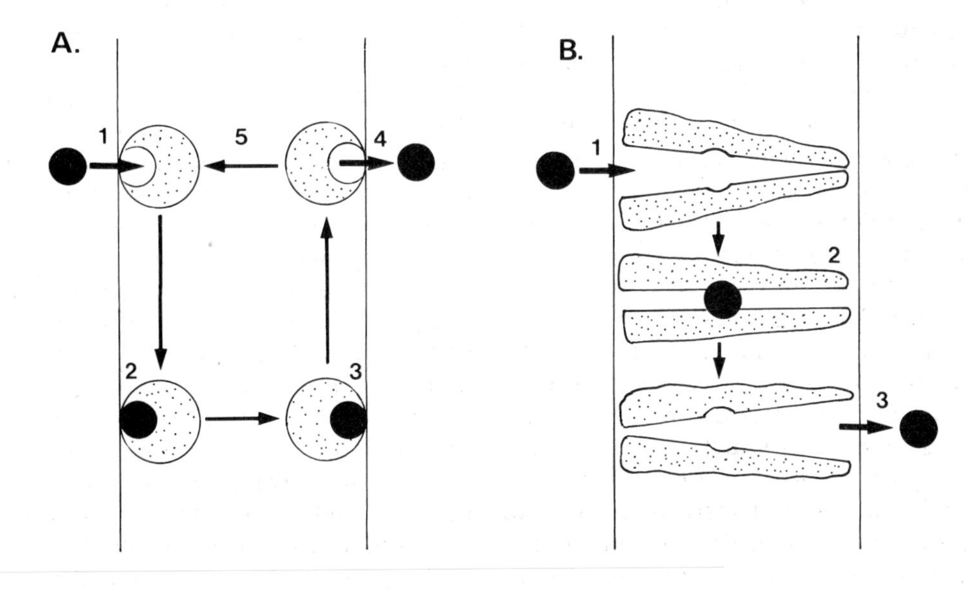

Fig. 1.9 Two possible mechanisms of active transport through a membrane. (A) High energy system. (B) Low energy system (see text).

known as a 'fixed-pore' mechanism and requires much less energy than a moving carrier molecule.

Endocytosis and exocytosis

Endocytosis involves the uptake of particles or fluid by the cell membrane with the formation of a vacuole (Fig. 1.10).

Small amounts of materials, usually macromolecules such as proteins, are normally taken up by pinocytosis. Minute channels form in the surface membrane and then pinch off as pinosomes which are then thought to be broken down by lysosomes.

Large particles such as bacteria and cellular debris are absorbed by phagocytosis. This is a much more spectacular process since the cell normally puts out large cytoplasmic extensions or pseudopodia which surround and engulf the particle, forming a phagosome, which is then digested by lysosomal enzymes. In the human body there are several types of cells whose primary role appears to be that of a phagocyte. There are 'free' or mobile cells, e.g. the neutrophils of the blood and 'fixed' cells such as those lining the cavities of the liver, spleen, lymphatic tissue and bone marrow.

If particles or macromolecules are passed out of the cell it is called exocytosis, a process by which, for example, pancreatic cells secrete their enzymes.

Cell division

As you may recall, Virchow proposed that all cells are derived from pre-existing cells. Since his time, of course, it has been discovered how this process is achieved.

With the exception of some cells in the ovaries and testes all of the nucleated cells of the human body contain 46 chromosomes; 23 of these would have originally been derived from the individual's mother and 23 from the father. Therefore, in reality, the cells contain 23 pairs of chromosomes rather than 46 different ones.

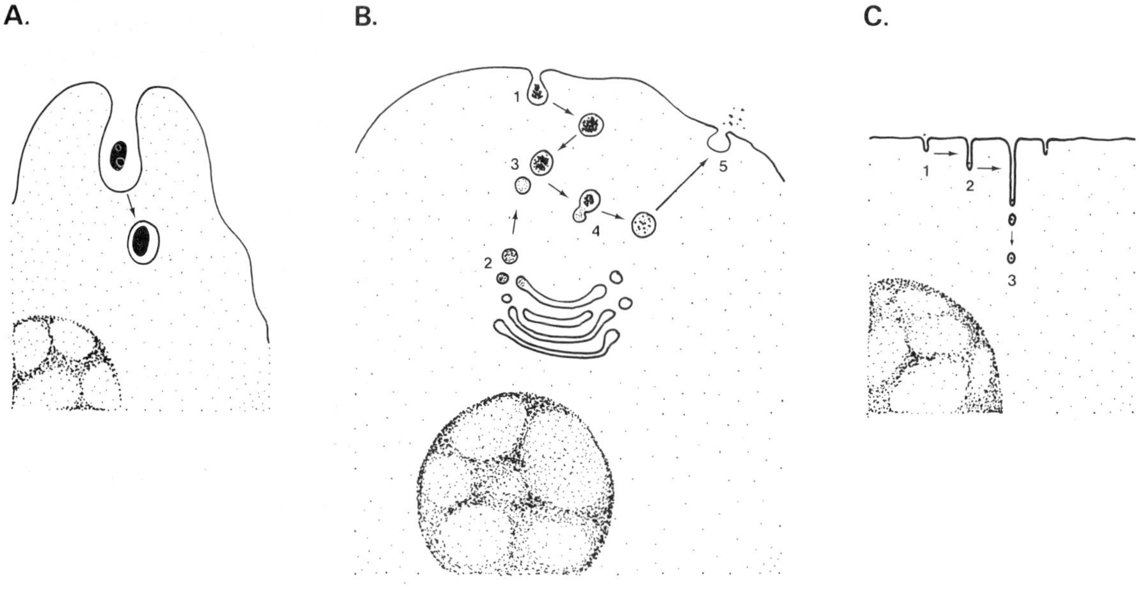

A. B. C.

Fig. 1.10 Exocytosis and endocytosis.
(A) Phagocytosis — pseudopodia engulf the particle and fuse to enclose the particle in a vesicle.
(B) The fate of particulate matter. The particle is taken into the cytoplasm (1) and is carried to a lysosome (3) with which it fuses (4). The lysosome has been formed on the Golgi body (2). The particle is digested by the lysosomal enzymes and the waste is ejected (5).
(C) Pinocytosis. A small channel forms in the surface membrane (1), it deepens (2) and pinches off at the bottom to form a vesicle (3).

The chromosomes of a pair are said to be homologous. They are alike in size and shape but since one is maternal and the other paternal they differ in the expression of their genetic information.

It is the chromosomes which play the leading part in the division, or more accurately the replication of cells. The nature and importance of chromosomes is considered in a later section; here only the process itself will be dealt with.

The two sex chromosomes of a male are exceptional in that they are not homologous. One chromosome is identified as the X-chromosome, which is maternal in origin, whilst the other is known as the Y-chromosome and derives from the father. The cells of females contain two X-chromosomes.

It is conventional to describe 22 pairs of 'autosomes' and one pair of 'sex' chromosomes.

There are two mechanisms by which eukaryotic cells may replicate themselves. Most cells of the body replicate by mitosis which gives rise to two daughter cells identical to their parent and to each other. Reproductive cells undergo meiosis where four dissimilar cells are formed, each containing only half the parental chromosome number. Cells containing 23 pairs of chromosomes are said to be diploid while cells containing half this number, i.e. 23 single chromosomes are haploid.

Chromosomes are not visible, even with a microscope, in a non-dividing ('resting') cell. Instead the nuclear material is in the form of chromatin, long extremely fine threads of DNA and protein. At the commencement of both mitosis and meiosis these threads condense, coiling up tightly to become visible (with appropriate stains) as chromosomes.

Mitosis

Immediately before a cell undergoes mitotic division, before condensation of the DNA threads, each chromosome replicates itself. Therefore when the chromosomes appear each one is in the form of two chromatids, each of which will, at the end of mitosis, give rise to a new chromosome.

For convenience mitosis may be divided into four phases even though, in reality, it is a continuous process (Fig. 1.11).

The initial appearance of the chromatids signals the beginning of prophase. The replicated chromosomes appear as fine threads which become thicker as the genetic material condenses more and more. Each pair of chromatids are attached only at one point, the centromere.

Outside the nuclear membrane, the centrioles replicate and begin to migrate to the opposite poles of the cell.

The end of prophase and the beginning of metaphase occurs when the nuclear membrane disappears. The chromosomes are cast free in the cytoplasm and migrate towards the equator of the cell. The centrioles which have, by now, moved to opposite poles of the cell, give rise to a number of radiating microfilaments, the other ends of which are attached to the centromeres of the chromosomes lying at the equator of the cell. These microfilaments form what is known as the spindle.

The spindle filaments begin to contract so that the two chromatids derived from each of the original chromosomes are pulled apart as the centromere breaks. This phase, which involves the physical movement of the chromatids towards the poles of the cell is called anaphase.

Finally, nuclear membranes form around the chromosomes so that for a short period the cell contains two nuclei. This is telophase and it ends when the new chromosomes disappear and the cytoplasm becomes pinched off to form two new cells.

This type of division occurs when tissues grow and also when they repair themselves following injury.

Meiosis

It is obvious that if, as a part of the reproductive process, a male cell (spermatozoon) is to fertilize a female cell (ovum) some reduction of the chromosome number must occur,

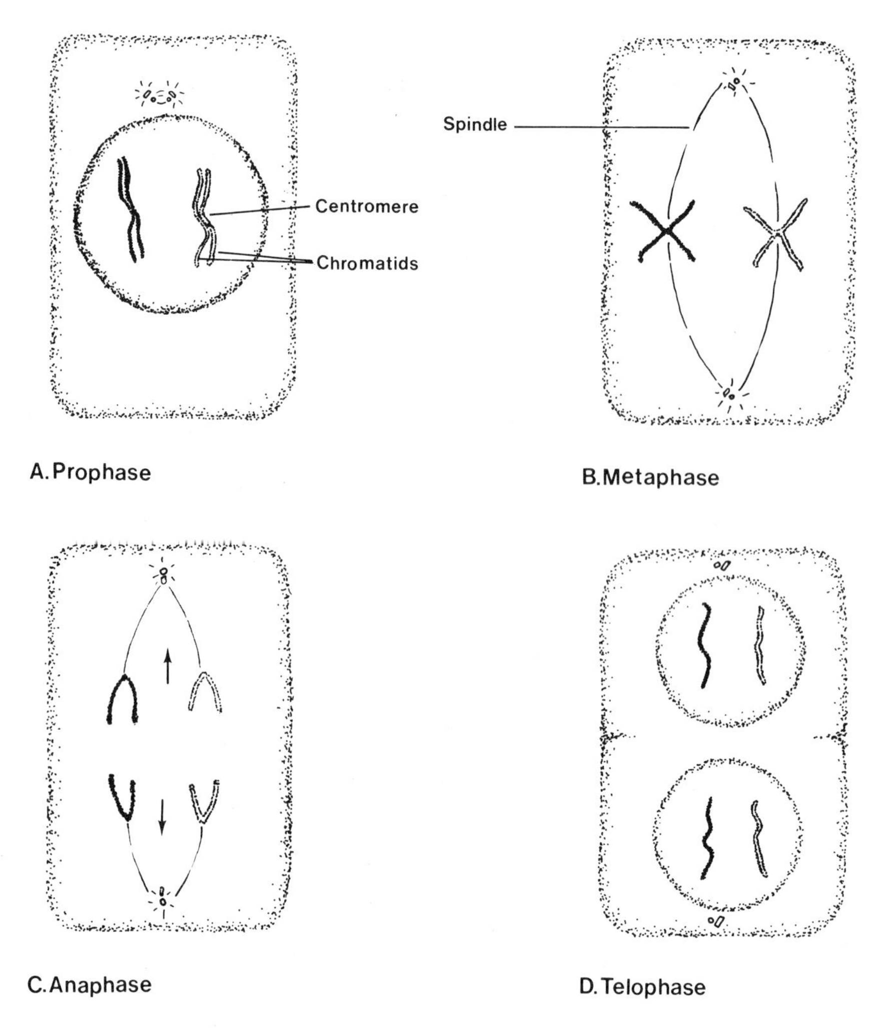

Fig. 1.11 Mitosis. For simplicity only two chromosomes are shown (see text).

otherwise there would be a doubling of the chromosome number at each generation.

Meiosis, or reduction division, solves this problem since it consists of two consecutive cell divisions but only one chromosomal replication. Thus four daughter cells are formed, each containing half the original chromosome number i.e. four haploid cells (Fig. 1.12).

Each division may be separated into the same four phases as mitosis. However prophase of the first division is extended to include the reassortment of genetic material between homologous chromosomes. For this reason it is conventional to recognize five subphases in this first phase of the first division.

The first of these subphases is leptotene when the chromosomal threads first become visible within the nucleus. Replication has already occurred so that each consists of two chromatids. However these chromatids are so tightly wrapped around one another that the chromosomes have the appearance of single threads.

In the second subphase, zygotene, homologous chromosomes pair up and form bivalents. At this stage the nucleus contains 23 pairs of chromosomes, each pair consisting of four chromatids. The chromosomes shorten and thicken during the third subphase, pachytene, and in the fourth, diplotene, they

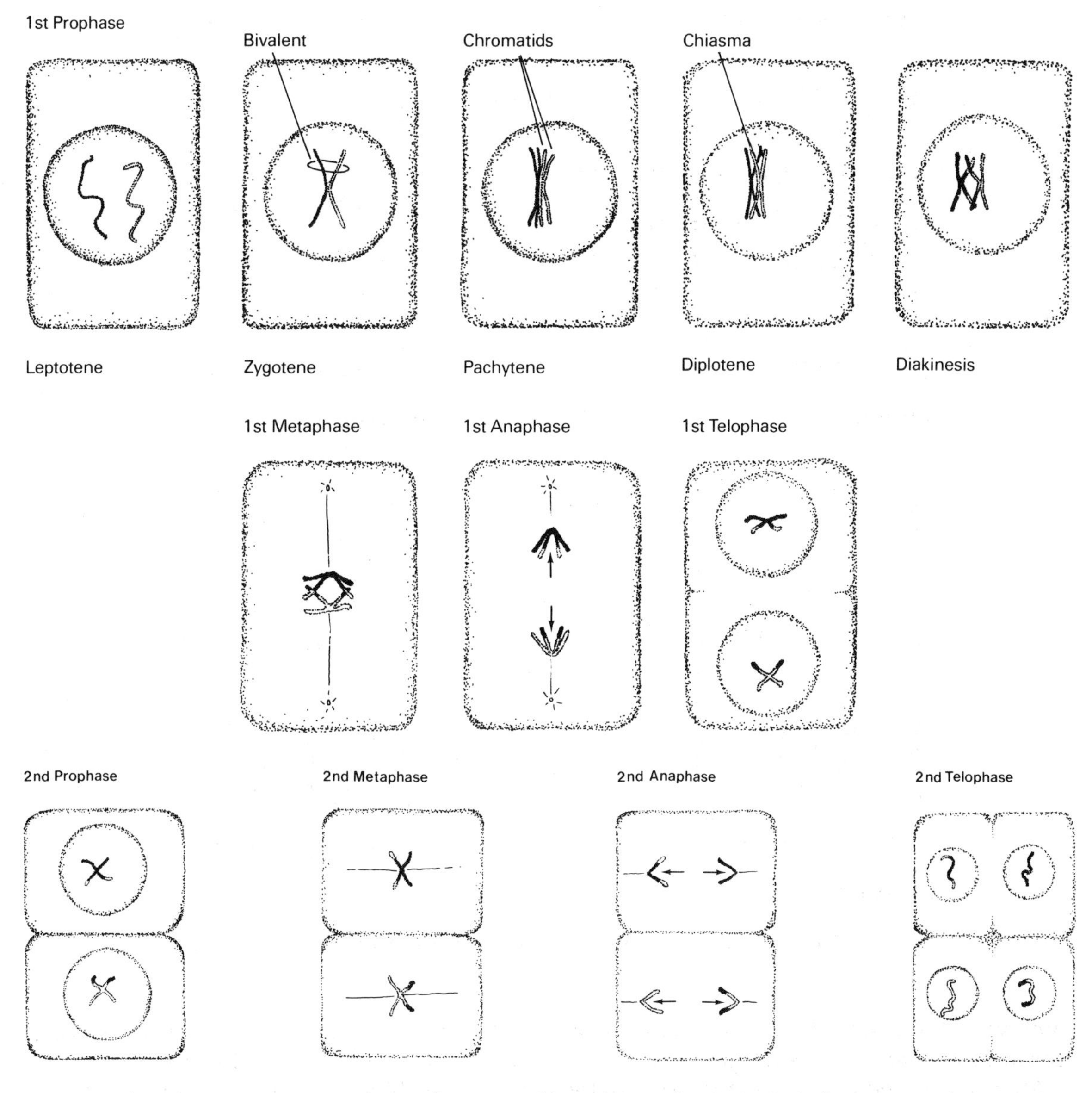

Fig. 1.12 Meiosis. For simplicity only one set of homologous chromosomes is shown (see text).

begin to pull apart from one another. At this point individual chromatids are visible, still held together at the centromere. The chromatids of homologous chromosomes are also held together at a variable number of other points or chiasmata (see Fig. 1.12) where non-sister chromatids have exchanged genetic material. This process is called crossing over and it allows the reciprocal exchange of genetic information between maternal and paternal chromosomes. Although no new genetic information appears at this time, the remixing of information from two different sources is a very important method of ensuring genetic variability in each generation.

The final stage, diakinesis, occurs when the centromeres of each pair of chromosomes start to pull apart.

The end of prophase and beginning of metaphase is marked by the disappearance of the nuclear membrane. The chromosomes then move to the centre of the cell and arrange themselves at the equator so that their centromeres lie equidistant on either side.

In anaphase, homologous chromosomes move apart; a very different situation from that in mitosis when the two chromatids of a chromosome are separated.

Telophase follows anaphase and is then followed immediately by a second prophase. Each nucleus at this stage contains 23 chromosomes, each one composed of two chromatids. In the second division these chromatids separate, without further replication, giving rise to four nuclei, each of which consists of 23 single-thread chromosomes. The four nuclei are not identical to one another, nor are they identical to that of the parent cell.

These haploid cells will give rise to spermatozoa in the male and ova in the female and may then be fertilized by the reproductive cells of the opposite sex. It is evident from a study of meiosis that the parental genetic information will be randomly distributed in the gametes. In addition to this, fertilization is a random event so that it is inevitable that a child will never be identical to his parents or brothers or sisters (unless he has an identical twin).

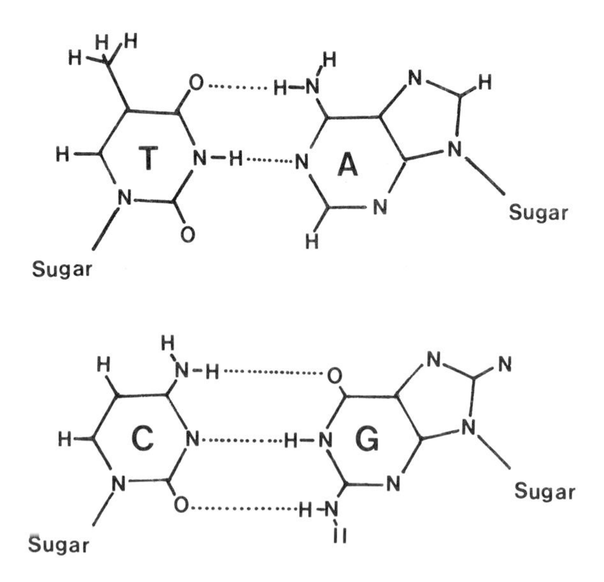

Fig. 1.13 The four bases of DNA. Adenine (A) always joins to Thymine (T) and Cytosine (C) always joins to Guanine (G).

Chromosomes

The chromosomes of eukaryotic cells are made up of a mixture of proteins and nucleic acid, mostly DNA (deoxyribonucleic acid) with small amounts of RNA (ribonucleic acid).

About half of the protein is a mixture of five different types of histones while the other half is a heterogeneous collection of non-histones. It is, however, the DNA which is of prime genetic importance, since it is this which regulates the function of the cell.

DNA is a deceptively simple molecule consisting of four different nucleotides, each of which contains a simple sugar (deoxyribose), a phosphate group and one of four different nitrogenous bases. Two of these bases, adenine and guanine are purines and have structures that are similar. The other two, thymine and cytosine are pyrimidines (Fig. 1.13). Again, the two pyrimidines are very similar to each other but differ from the purines. In any DNA molecule the concentration of adenine is equal to that of thymine, while the amount of guanine present equals the amount of cytosine.

For many years scientists attempted to understand how the nucleotides could be arranged in such a way as to produce a molecule complicated enough to determine the structure of all of the proteins in a cell. Furthermore, the molecule had to be capable of replicating itself so accurately that a daughter cell would be identical to its parent.

In 1953, James Watson and Francis Crick proposed a model for DNA which fitted most of the facts known about the molecule at that time.

The Watson-Crick model shows a molecule with two back-bones, each consisting of alternating sugar and phosphate groups. Each backbone is twisted around the other to produce a double helix (Fig. 1.14).

To each sugar is bonded a purine or a pyrimidine which projects inwards between the backbones and links to the opposite unit

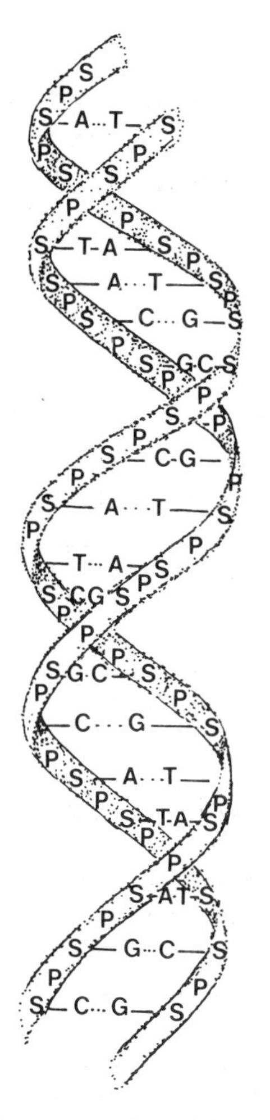

Fig. 1.14 The Watson-Crick model of DNA.
P = Phosphate, S = Sugar (deoxyribose), A = Adenine, G = Guanine, T = Thymine, C = Cytosine.

of the other chain. Thus a purine is linked to a pyrimidine, adenine with thymine and cytosine with guanine, which explains why the amount of adenine in the molecule is always equal to the amount of thymine and guanine equal to cytosine.

The exact arrangement of the DNA and protein components of chromosomes is not fully understood. It is suggested that the histones are grouped into a globular complex around which is wrapped a DNA double helix. These 'beads' are then linked together by a 'string' of DNA (Fig. 1.15).

The DNA is able to replicate itself by splitting the parent molecule so that the two back-

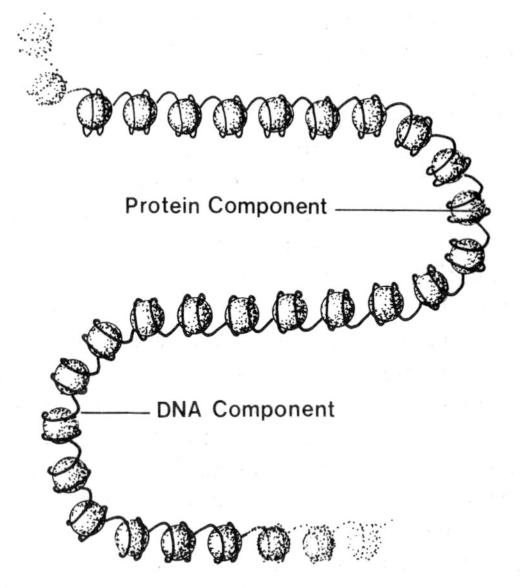

Fig. 1.15 Possible arrangement of DNA and protein in chromosomes (see text).

bones separate. On to each backbone, nucleotides are then built up in the correct sequence i.e. adenine linking with thymine and guanine with cytosine (Fig. 1.16). Replication occurs during interphase prior to mitosis or meiosis. The daughter DNA strands correspond to the chromatids which will separate in cell division.

Enzymes determine the size, shape and physiological functions of a cell by regulating its internal biochemical activities. It is believed that each gene, which is the functional unit of the chromosome, is responsible for the formation of a single enzyme. Since an organism is a collection of cells, it follows that the genetic makeup of the organism (the genotype) is responsible for its physical and physiolgical makeup (the phenotype).

It has long been established that the genetic code is a triplet code, i.e. each group of three nucleotides on one chain is responsible for the placement of one amino acid into a protein chain.

Protein synthesis is carried out in the cytoplasm and while DNA governs the process it is not itself directly involved. Therefore the genetic message must be copied and taken into the cytoplasm. This copy is in the form of

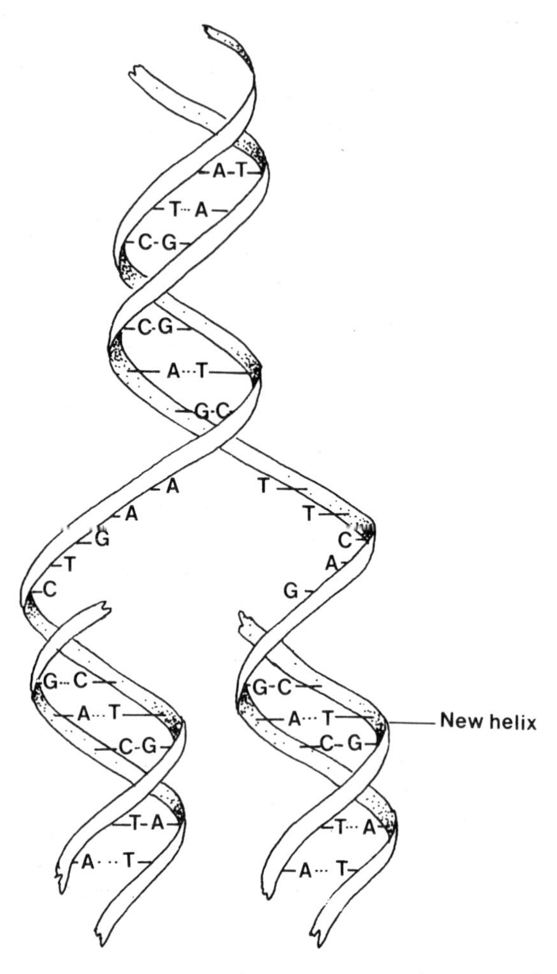

Fig. 1.16 Replication of DNA. The DNA splits open and new helices are built up on the original ones.

an RNA (ribonucleic acid) molecule. RNA has a similar structure to DNA: the sugar ribose replaces deoxyribose and the pyrimidine thymine is replaced by uracil. Several different forms of RNA exist within the nucleus and cytoplasm.

RNA is found within the ribosomes which cover the surfaces of the rough endoplasmic reticulum. Each ribosome is made of two subunits, the smaller one consisting of a single molecule of RNA and two to three dozen polypeptides. The larger unit contains three different RNAs and about 50 polypeptides. These RNA molecules are formed in the nucleus on a DNA template.

In addition to ribosomal RNA, two other forms are found in the cell. The original DNA message which determines the structure of a protein is transferred to a messenger RNA molecule which then passes out into the cytoplasm where it associates with one or more ribosomes. Amino acids are transported to the mRNA on the ribosomes by another type of RNA, known as transfer or tRNA.

The latter are able to form bridges between the amino acids and mRNA.

Protein synthesis

The process of protein synthesis begins when a portion of DNA splits open and one strand acts as a template for mRNA formation. Nucleotides are then assembled on this template with cytosine on the DNA linking with guanine on the RNA and adenine linking with uracil and so on until a complete mirror image of the DNA base sequence is created. This process is called transcription.

The messenger RNA then passes out into

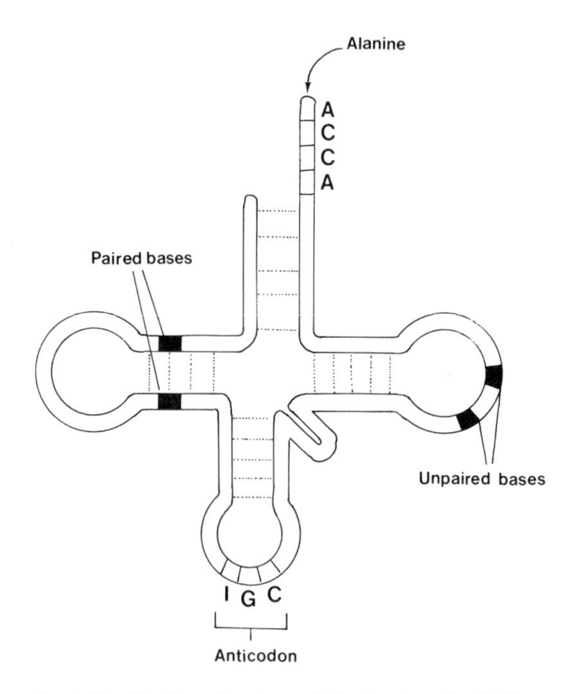

Fig. 1.17 Outline structure of alanine transfer RNA. Paired bases are joined by dotted lines. I = Inosine a base varient found only in transfer RNA.

the cytoplasm where the base sequence is translated into an amino acid sequence, with the assistance of transfer RNA.

Transfer RNA is a short chain molecule of between about 75 and 90 nucleotides (Fig. 1.17), synthesized in the nucleus. The molecule is only partially coiled and some of the free nucleotides at the end are able to form attachments with amino acids. There are many more different types of tRNA than amino acids and it appears therefore that there may be several for each amino acid. A triplet of unpaired bases also exists in each tRNA which is able to 'recognize' a corresponding triplet on the mRNA. Again, there are several mRNA triplets known to code for each amino acid.

Each triplet of nucleotides on the mRNA which encodes a single amino acid is called a 'codon' and the mirror image on the tRNA is known as an 'anticodon'. The nucleotides which make up each codon (64 are known, representing all of the possible combinations of the four bases) are known and appear to be universal in all living organisms (Table 1.1).

Each amino acid is able to link, with the assistance of appropriate enzymes, with one of a small group of tRNAs, each of these is then able to recognize one of a small group of mRNA codons. Thus while the code is not strictly amino acid-codon specific, it is specific enough to ensure that the correct amino acid is placed in the correct position within a protein molecule.

Once the mRNA is in position threaded through a ribosome, translation can begin (Fig. 1.18). The ribosome appears to move along the mRNA 'reading' the message as it progresses. As a particular codon passes through the ribosome, the appropriate tRNA, with an amino acid in tow, is picked up and the mRNA codon and tRNA anticodon link up. A ribosome has two positions into which tRNAs can be slotted. One of these is occupied by a newly arrived tRNA while the other contains a tRNA which anchors the growing polypeptide chain. The chain is then attached to the newly arrived amino acid and the now freed tRNA passes out of the ribosome which moves along to occupy the next position. A new codon is now available for the next tRNA-amino acid complex to slot into.

Thus as the ribosome moves along the length of the mRNA a polypeptide chain is formed, its structure being determined by the RNA base sequence. Sometimes an mRNA may be threaded through several ribosomes

Table 1.1 The Genetic Code. Sequences of three bases on mRNA which are read by the ribosome and bring about the incorporation of the corresponding amino acid into a protein molecule.
TERM — terminating codon, signals the end of an amino acid sequence.

	U		C		A		G		
	UUU	Phe	UCU		UAU	Tyr	UGU	Cys	U
U	UUC		UCC	Ser	UAC		UGC		C
	UUA	Leu	UCA		UAA	TERM	UGA	TERM	A
	UUG		UCG		UAG		UGG	Trp	G
	CUU		CCU		CAU	His	CGU		U
C	CUC	Leu	CCC	Pro	CAC		CGC	Arg	C
	CUA		CCA		CAA	GlN	CGA		A
	CUG		CCG		CAG		CGC		G
	AUU		ACU		AAU	AsN	AGU	Ser	U
A	AUC	Ile	ACC	Thr	AAC		AGC		C
	AUA		ACA		AAA	Lys	AGA	Arg	A
	AUG	Met	ACG		AAG		AGG		G
	GUU		GCU		GAU	Asp	GGU		U
G	GUC	Val	GCC	Ala	GAC		GGC	Gly	C
	GUA		GCA		GAA	Glu	GGA		A
	GUG		GCG		GAG		GGG		G

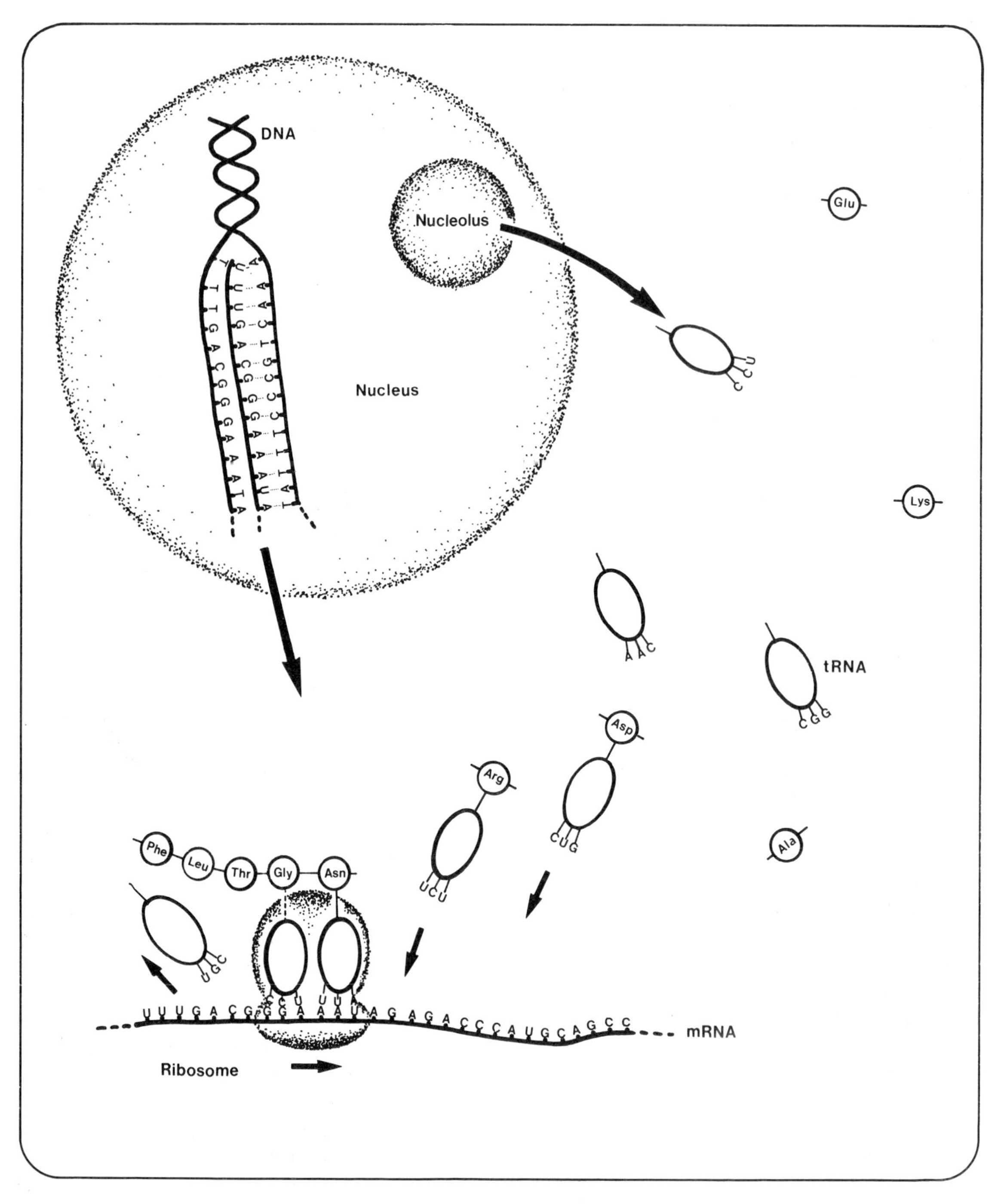

Fig. 1.18 Protein synthesis. For explanation, see text.

and each individual ribosome will still produce a protein. In addition to this a single mRNA can be 'read' several times by a single ribosome. It is evident that a single messenger RNA molecule can be responsible for the formation of many protein molecules, although exactly how this facet of protein synthesis is governed is not fully understood.

TISSUE STRUCTURE AND FUNCTION

A tissue is a collection of structurally similar cells together with intercellular substance (also known as matrix or ground substance) which may contain fibres.

It is usual to classify tissues into four primary groups:

1. epithelia
2. connective tissues (including blood, cartilage and bone)
3. muscle
4. nervous tissue.

The principal structural features of these tissues together with their functions are given in the succeeding sections. Blood is considered separately in Chapter 2.

EPITHELIA

Epithelial tissues are composed of cells which are tightly packed together forming a continuous sheet and are found covering or lining organs. Secretory epithelium is found in some glands.

Simple epithelium consists of a single layer of cells which are variously specialized for secretion, absorption or transport of substances over their surface.

Stratified epithelium on the other hand is many cells thick and acts to protect underlying structures against chemical or mechanical damage or dehydration.

All epithelial tissues are supported by a basement membrane which comprises a basal lamina consisting of a network of fine filaments in a matrix, with reticular fibres beneath.

There are several ways in which epithelial cells are held together on their lateral borders. Some cells (such as those lining the alimentary tract) have interlocking projections forming a jigsaw-like structure. Columnar cells frequently have a collection of attachments called a junctional complex just below their free surfaces. These are illustrated in Figure 1.19. The uppermost region is called the zonula occludens; here, the adjacent membranes appear to fuse. The middle section is known as the zonula adhaerens, in which the membranes are reinforced by fine filaments. The lower region is known as the macula adhaerens or desmosome; the membranes are attached to a feltwork of fine filaments in a matrix, and filaments from the cytoplasm anchor to the structure.

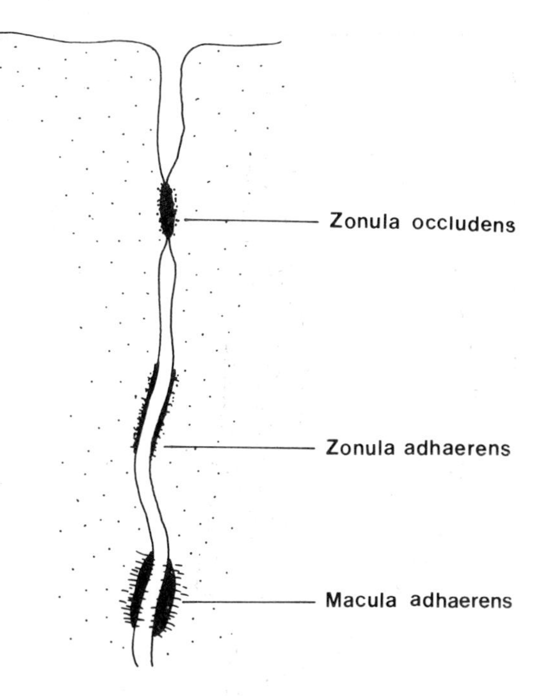

Fig. 1.19 A junctional complex.

Epithelial cells are capable of regenerating damaged tissue by mitosis. In striated epithelium, the basal layers of cells divide and replace the cells which are constantly desquamated from the superficial layer. Cells deep in the gastric and intestinal glands divide and replace the more differentiated cells on the surface.

The respiratory, alimentary and urinogenital tracts are all lined by a mucous membrane. This comprises a layer of epithelium and the connective tissue (lamina propria) beneath.

Serous membranes include the peritoneum, pericardium and pleurae. The simple squamous epthelium (see below) which is found at these sites is known as mesothelium.

Epithelia do not have a blood supply as such, the cells exchange water and solutes with capillaries in adjacent tissue, by diffusion.

The types of epithelia are named according to the shape of the cells in longitudinal section i.e. squamous, cuboidal or columnar.

Simple squamous epithelium

In longitudinal section, the cells appear to be extremely thin, down to 0.1 μm thickness, except where the nucleus causes a bulge (see Fig. 1.20). In transverse section (or surface view), the cells have an irregular wavy outline which has been likened to crazy paving. The cells form a continuous sheet and are anchored together by zonulae occludentes.

This extremely thin tissue is appropriately found in the alveoli in the lungs, where rapid diffusion of oxygen and carbon dioxide takes place across the cells. Simple squamous epithelium lines the whole of the cardiovascular and lymphatic systems, where it is called endothelium. Capillaries are composed largely of endothelial cells with occasional pericytes and therefore only a very thin barrier exists between blood and interstitial fluid. The paucity of organelles is thought to limit cellular activity to passive transport in the main, although the large number of flask shaped invaginations of the membrane (caveolae) suggest that pinocytosis is common. Some capillaries, such as those in the renal glomeruli, endocrine glands and intestinal villi are fenestrated. That is they have what appear as pores in them which may or may not be closed by a very thin membrane. The smoothness of endothelium facilitates blood flow and reduces clotting.

Other sites where this tissue is found include parts of the kidney nephron (the outer layer of Bowman's capsule and the thin part of the loop of Henle); in the ear (covering the tympanum on the middle ear side and lining the membranous labyrinth); the rete testis; small ducts in glands; lining cavities as mesothelium.

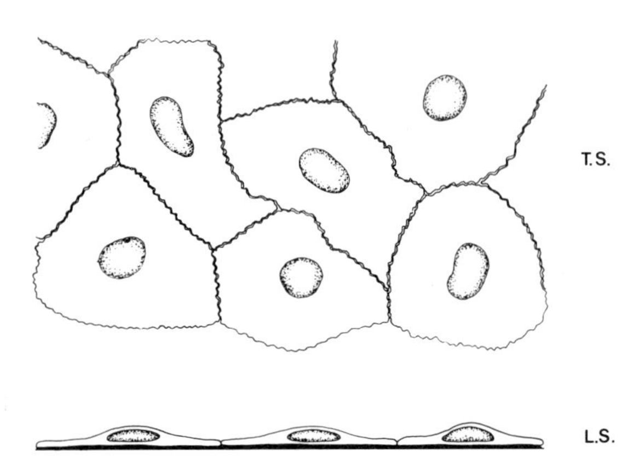

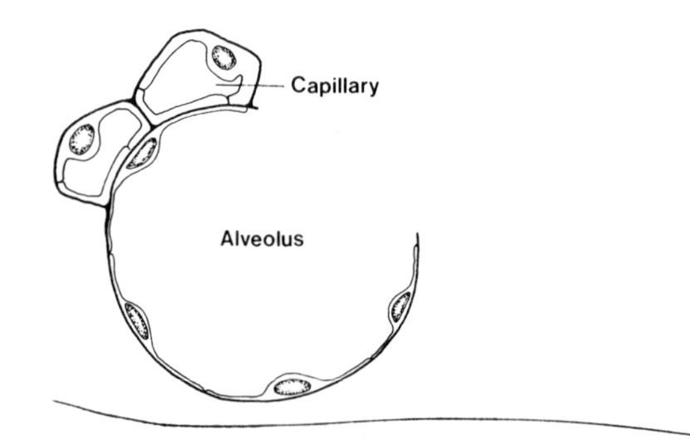

Fig 1.20 Simple squamous (pavement) epithelium in transverse and longitudinal section and as the principal component of capillaries and alveoli in the lungs.

Simple cuboidal epithelium

These cells appear square in longitudinal section, but in transverse section they are polygonal, usually hexagonal (see Fig. 1.21A).

Much of the kidney nephron is composed of simple cuboidal epithelium. The proximal convoluted tubule is primarily engaged in reabsorption of solute and water from the tubular lumen. The relative thinness of the tissue is appropriate for the rapid transfer of materials across it, but compared with squamous cells, there are many more mitochondria and therefore a greater capacity for active transport. The luminal surfaces of the cells project to form microvilli which increase the surface area available for absorption.

Cuboidal epithelium is found covering the ovaries, where it is known as germinal epithelium while the pigment layer of the retina is

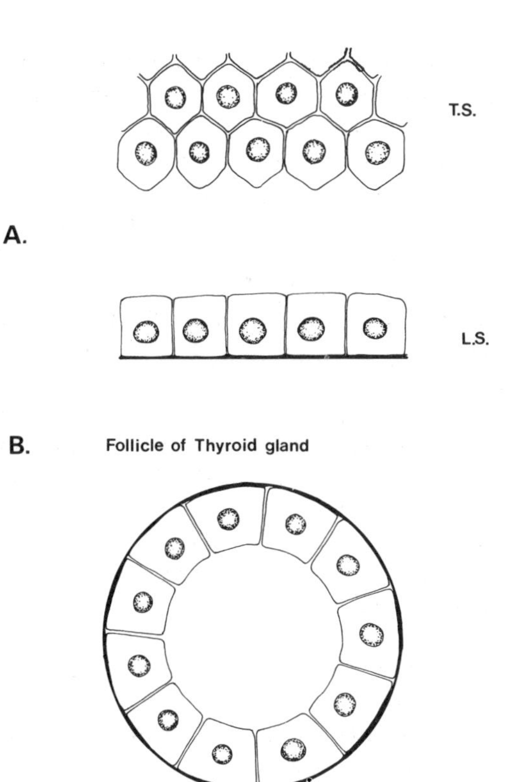

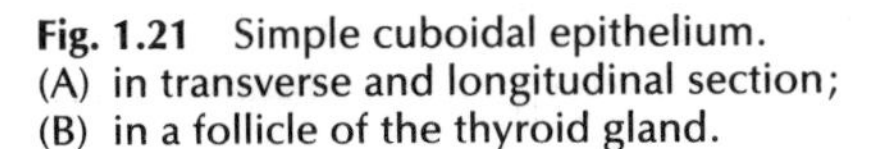

Fig. 1.21 Simple cuboidal epithelium.
(A) in transverse and longitudinal section;
(B) in a follicle of the thyroid gland.

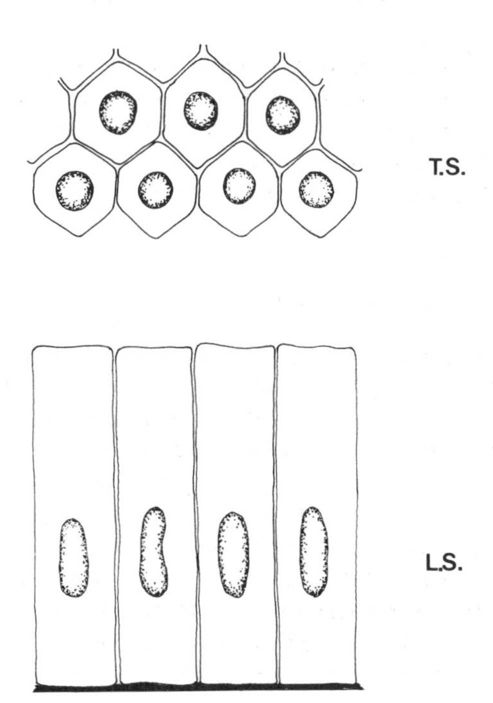

Fig. 1.22 Simple columnar epithelium in transverse and longitudinal section.

also composed of cuboidal cells. The excretory ducts as well as the secretory parts of many glands are composed of this type of tissue. The cuboidal cells found in the follicles of the thyroid gland are shown in Figure 1.21B.

Simple columnar epithelium

This type of epithelium consists of cells which are taller and thinner than the cuboidal group. Each nucleus is also elongated and tends to be located towards the base of the cells (see Fig. 1.22). The cells are hexagonal in cross section, like those of cuboidal epithelium. Columnar epithelium can be subdivided into several specialized types (see Fig. 1.23).

Striated or brush border cells

These cells are adapted for absorption by the presence of microvilli which consist of finger-like projections of the cell membrane between 0.5 and 1 μm in length. The majority of cells comprising the epithelial covering of the intestinal villi are of this type. Another site where these cells are found is the gall bladder in which bile is stored and concentrated by the absorption of salts and water.

Ciliated cells

The cilia are larger than microvilli (7–10 μm long), and have a characteristic arrangement of nine paired peripheral microtubules and two central ones, and each has a basal body at its base. Each cilium has a stiff, rapid effective stroke and a slower, flexible recovery stroke. All the cilia in a given site beat in the same direction, and commonly in epithelial tissues the beat is metachronal. This means that the cilia in each row start their cycle at slightly different times, so that a wave sweeps over the surface. The function of this activity in the respiratory tract is that a sheet of mucus is propelled towards the pharynx.

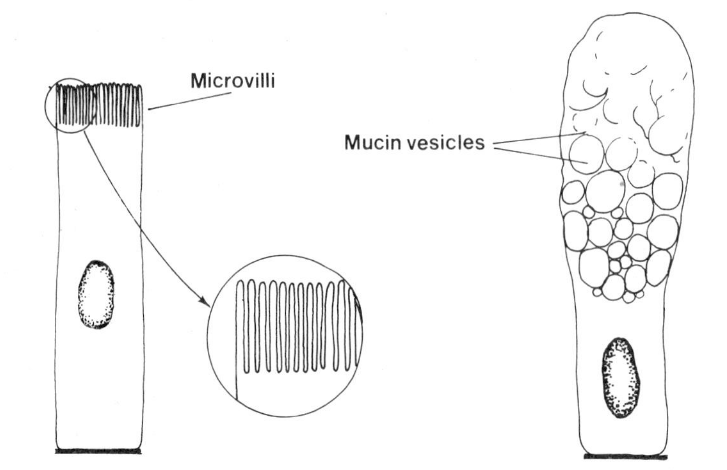

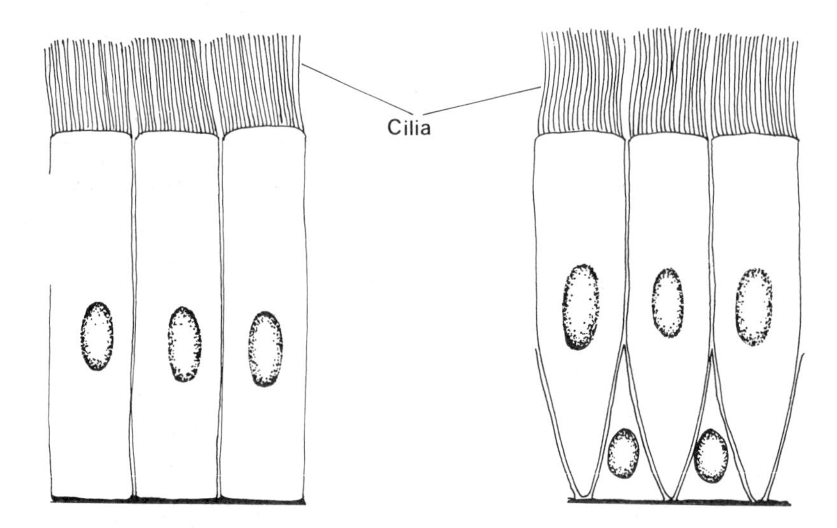

Fig. 1.23 Specializations of columnar epithelium.

Ciliated columnar cells are found lining part of the uterus and oviducts, the paranasal sinuses, the central canal of the spinal cord and the ventricles of the brain.

Glandular cells

The epithelial lining of the stomach is made up of mucous cells which secrete mucin into the lumen of the stomach. The chief cells of the gastric glands are also a secretory form of columnar epithelium. In this case, the secretory product is pepsinogen. The mucous cells, interspersed with the brush border cells covering the intestinal villi, have expanded apices, distended with mucin. These cells are known as goblet or chalice cells, and they are also found lining the respiratory tract.

Pseudostratified epithelium

In this tissue, a false impression of more than one layer of cells is given by the nuclei which lie at more than one level. Although all the bases of the cells lie on the basement membrane, their heights vary, and so do the positions of the nuclei. Such a tissue is afforded strength by the packing of the cells, and is found in the male urethra and some large excretory ducts.

In other sites, the taller cells are ciliated. This tissue is found lining the trachea and large bronchi, the Eustachian tube, lacrimal sac and part of the tympanic cavity.

Some of the smaller cells divide and replenish the other cells, and some grow into the taller cells.

Stratified squamous epithelium

In this type of tissue it is the superficial layers of cells that are squamous in longitudinal section, whereas the basal cells are relatively longer. Intermediate cells show progressive flattening towards the surface (see Fig. 1.24).

Stratified squamous epithelium may be keratinized or non-keratinized. The former type constitutes the epidermis of the skin (Fig. 1.25). The epidermis is composed of two zones, the deeper zona germinativa and the superficial zona cornea. Within each zone there are two or three layers of cells.

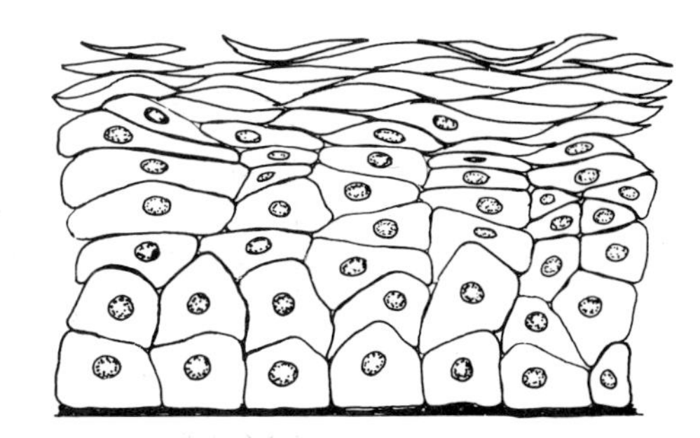

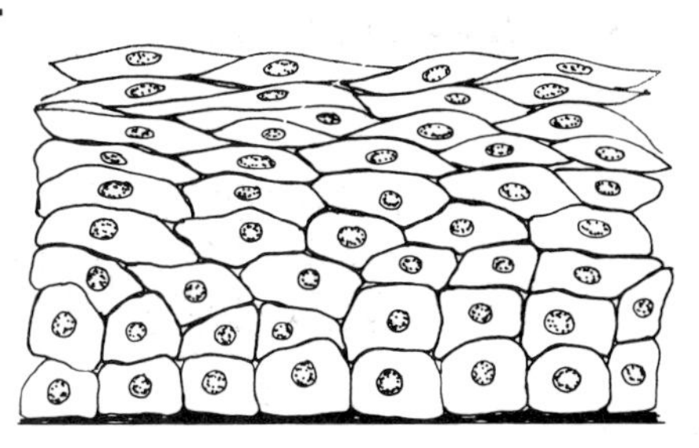

Fig. 1.24 Stratified squamous epithelium.
(A) keratinized;
(B) non-keratinized.

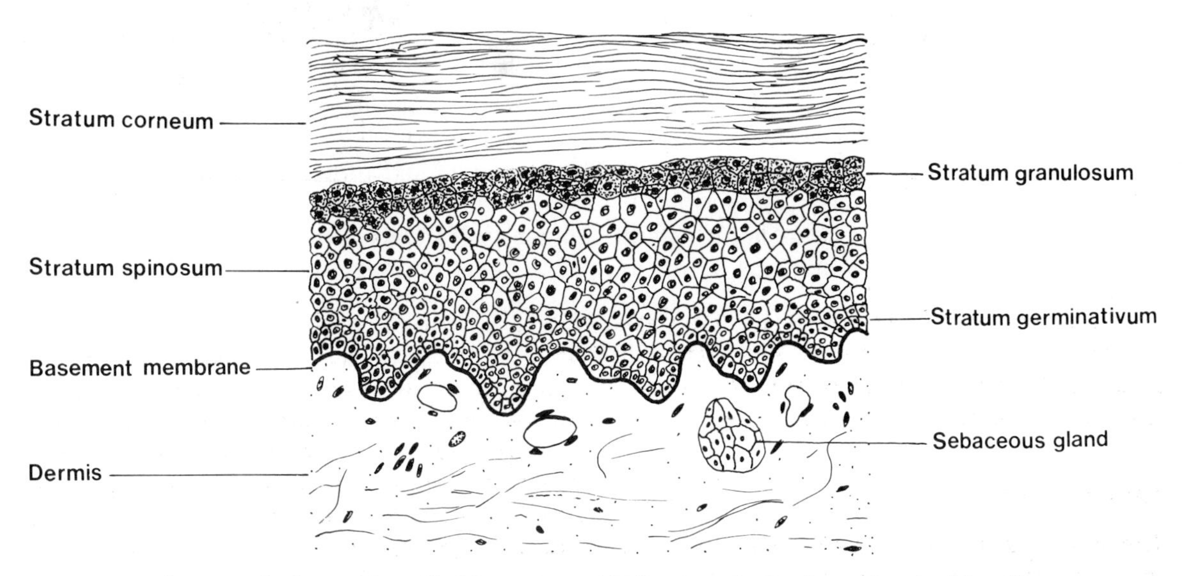

Fig. 1.25 The layers of stratified squamous epithelium constituting the epidermis of the skin.

The zona germinativa consists of a layer of columnar cells on the basement membrane, the stratum germinativum; and a layer of variable thickness above, the stratum spinosum. The zona germinativa is so named because it is in this region that the cells divide and replenish those sloughed off from the surface. Cell division occurs principally in the stratum germinativum, but also in the stratum spinosum. The cells in this layer have been named 'prickle' cells because they appear to have spines all over their surfaces. The spines are actually anchorage points (desmosomes) between the cells. There are also chromatocytes in the stratum germinativum. These are pigment cells derived from nervous tissue.

The deepest layer of the zona cornea is called the stratum granulosum because the flattened cells synthesize granules of the protein keratohyalin. As the cells move nearer to the surface they lose their nuclei. In the epidermis of the palms of the hands and the soles of the feet, a clear layer, the stratum lucidum may be seen. The superficial layer of cells, the stratum corneum, consists of dead, enucleate cells full of the protein keratin.

The thickness of the epidermis is generally between 0.07 to 0.12 mm, but reaches a thickness of 1.4 mm on the soles of the feet. The transit time from a cell starting in the deepest layer to being rubbed off the surface as lifeless squame is between 15 and 30 days.

The keratin imparts many properties to the skin: toughness, flexibility, resistance to bacterial penetration and water loss.

Non-keratinized stratified squamous epithelium is found on wet surfaces subject to a lot of wear and tear such as the lining of the mouth, pharynx and oesophagus, and the covering of the cornea.

Transitional epithelium

This type of tissue is found lining the ureters, the bladder and the proximal urethra. The superficial cells are large, often binucleate and umbrella-shaped. Beneath this layer are pear-shaped cells which have their apices pointing downwards, and the rounded part fits into the concavity of the outer layer of cells. Between the tapering cells are a third type of smaller cells (see Fig. 1.26). A characteristic of this tissue is that it will withstand stretching. The superficial cells flatten and the pear-shaped cells shorten and thicken on stretch.

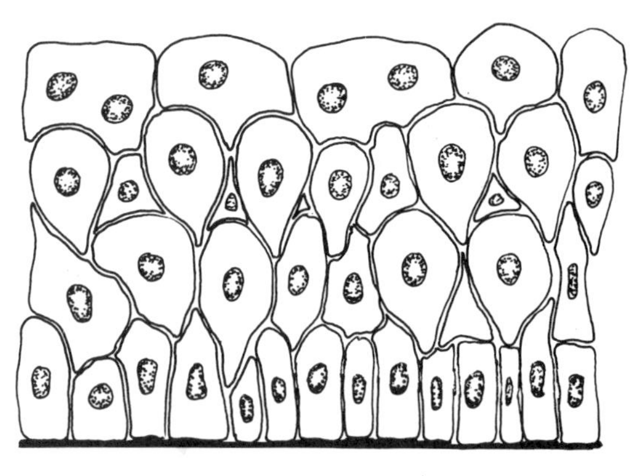

Fig. 1.26 Transitional epithelium.

Stratified cuboidal and columnar epithelium

There are relatively few sites where stratified cuboidal or columnar epithelium is found. Such sites include the larger ducts of glands such as the salivary and pancreas, fornix of the conjunctiva, in the cavernous urethra, some parts of the anal mucous membrane, in the pharynx and on the epiglottis.

Glands

Glands of epithelial origin are usually found in connective tissue underlying the surface membrane which has grown down. Exocrine glands retain their connection with the surface by means of a duct, whereas endocrine glands lose their connection and are therefore an island of cells from which the secretion passes into the blood, rather than a duct.

Exocrine glands may be unicellular, such as goblet cells, or more usually multicellular. If the secretory part of the gland opens directly onto the surface, or indirectly via a single duct, then the gland is designated simple; whereas if the duct is branched, then the gland is a compound one.

The shape of the secretory units may be

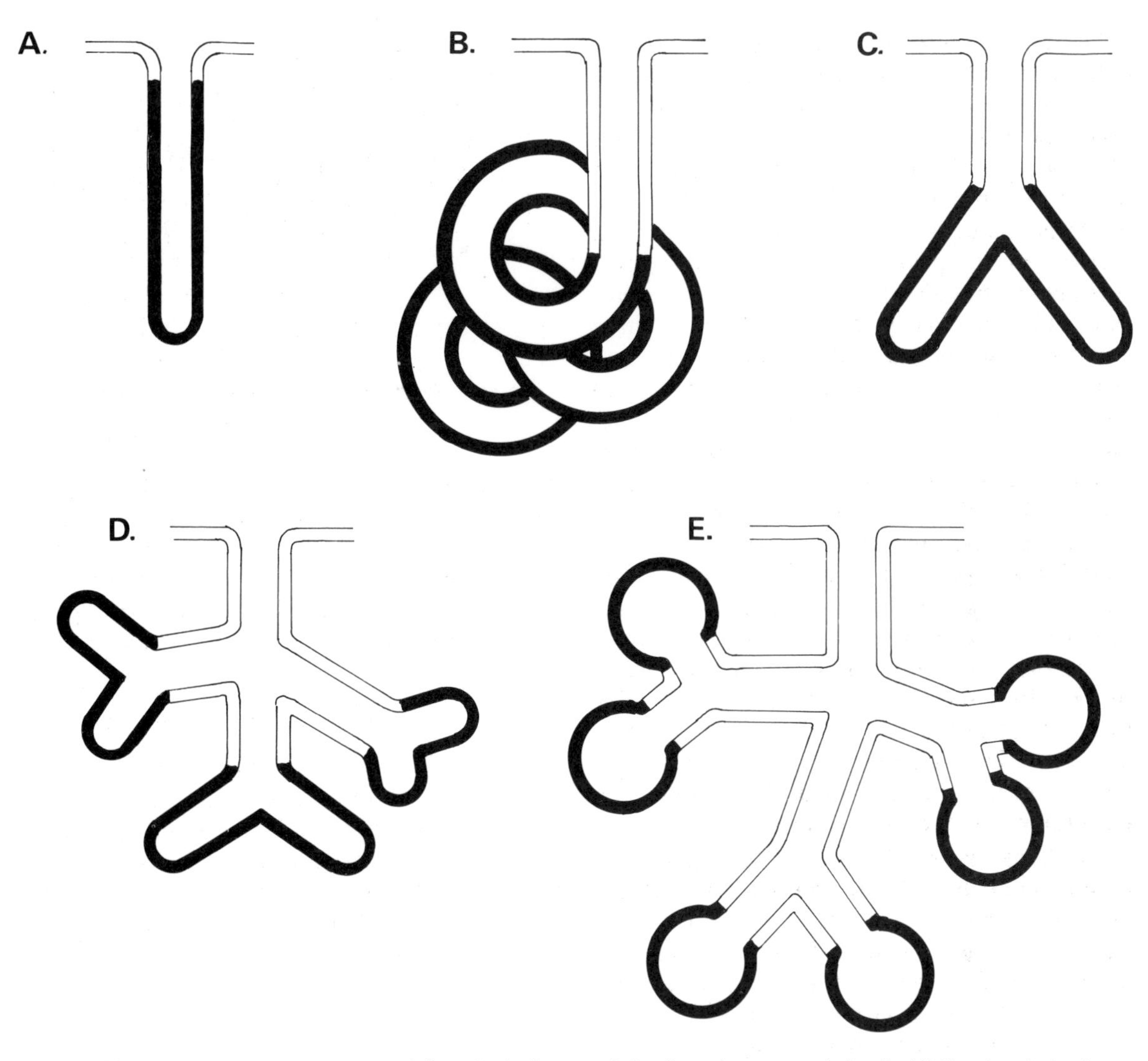

Fig. 1.27 Diagrammatic representation of the principal types of simple and compound glands. (A) Simple tubular (e.g., intestinal gland of Lieberkuhn); (B) Simple coiled tubular (e.g., sweat gland); (C) Simple branched tubular (e.g., gastric gland); (D) Compound tubular (e.g., mucous gland in oral cavity); (E) compound acinar (e.g., salivary gland).

tubular, acinar (spherical) or alveolar (flask-shaped). Figure 1.27 illustrates various types of gland. In large glands containing connective tissue and secretory cells, the former tissue is known as stroma, whilst the glandular tissue is called parenchyma.

CONNECTIVE TISSUES

The characteristic feature of connective tissues is that they have a high proportion of intercellular substance so that the cells are relatively widely spaced. It is usual to subdivide the many diverse tissues included in this group into four types:

1. connective tissue proper
2. cartilage
3. bone
4. blood and lymph.

With the exception of blood and lymph, the intercellular material consists of an amorphous ground substance which contains extracellular fluid, and fibres. Connective tissues contain three types of fibres called elastic, collagen and reticular. The quantity and arrangement of these fibres gives rise to a further classification. Thus loosely woven fibres are found in loose connective tissues and dense connective tissue contains densely packed fibres. The structure and properties of the intercellular material largely determine the particular function of the tissue. The intercellular material of loose connective tissue, for example, is the site of various defence mechanisms involved in combating infection or tissue damage. The closely packed fibres found in the intercellular material of dense connective tissue, on the other hand, imparts the property of mechanical strength.

The blood supply to connective tissue itself is not extensive, though many vessels may be seen en route to adjacent tissues. Lymphatic vessels on the other hand are frequently abundant.

Connective tissue proper

The principal types of connective tissue proper include areolar (loose connective tissue); two types of dense connective tissue — elastic and white fibrous; and adipose tissue (although some classifications have this type as a separate tissue group).

Areolar tissue (loose connective tissue)

This type of tissue has a soft, transparent appearance and contains a variety of cell types, some of which derive from the blood, together with collagen, elastic and reticular fibres (Fig. 1.28). The tissue is found both between organs and also around the blood vessels and nerves which lie within them. The open nature of the tissue enables the various reactions constituting inflammation to take place; loose connective tissue is synthesized in the repair of wounds.

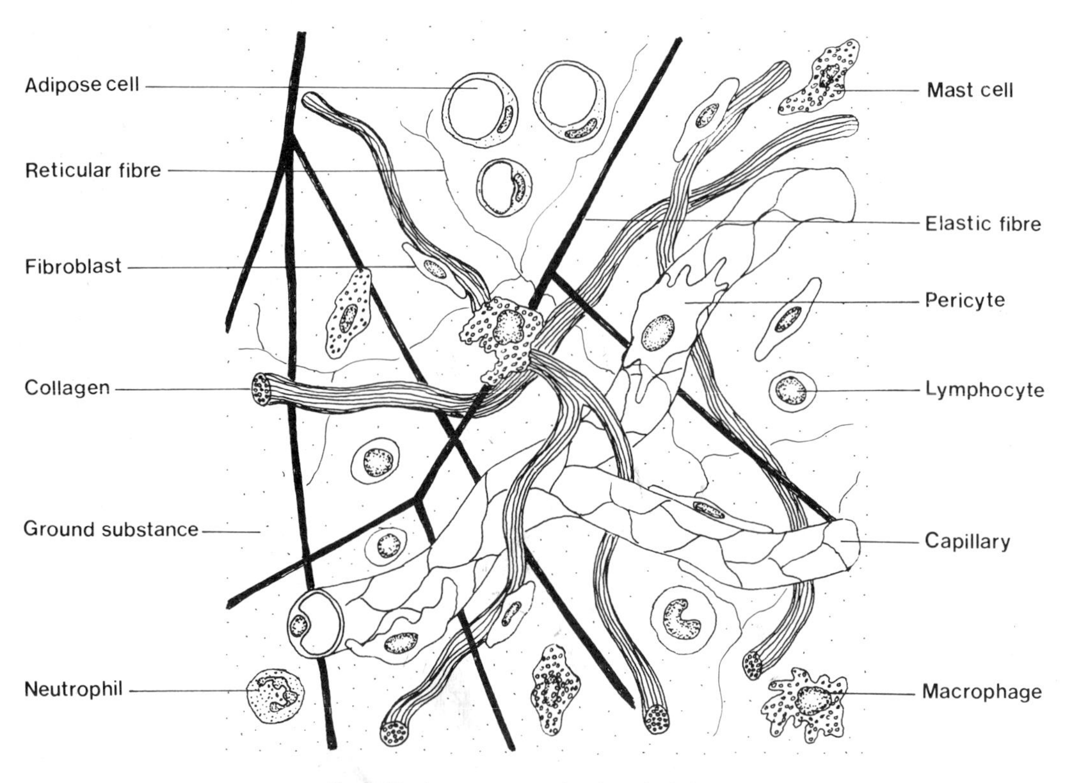

Fig. 1.28 Loose connective (areolar) tissue.

The ground substance, which appears structureless under the light microscope, contains extracellular fluid and several protein polysaccharide complexes. The two principal ones are chondroitin sulphate and hyaluronic acid. The presence of these substances imparts a degree of viscosity to the medium, which is thought to limit the movement of microorganisms through the tissue. Some bacteria, however, secrete the enzyme hyaluronidase which changes the structure of the hyaluronic acid and reduces the viscosity of the ground substance around it. Such bacteria are therefore able to move more easily through the tissue. The protein polysaccharides also have the property of binding water which is important for the diffusion of materials between the capillaries and connective tissue cells.

The collagen or white fibres are usually found in bundles running in various directions (see Fig. 1.28). Each fibre is made up of individual fibrils which show cross striations in electron micrographs. The striations result from the way the molecules of the constituent protein tropocollagen are aligned (see Fig. 1.29).

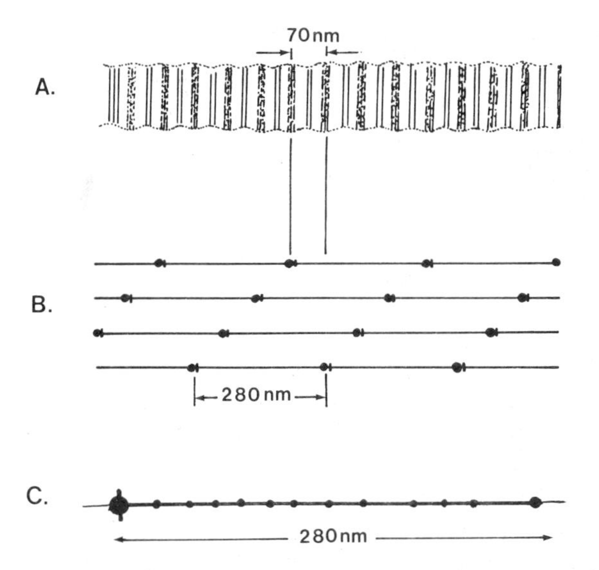

Fig. 1.29 (A) Collagen fibre; (B) Collagen fibre with tropocollagen components separated to show their overlapping arrangement; (C) Single tropocollagen molecule.

Reticular fibres are finer than the collagen type and they stain black with silver salts. They are usually found arranged as a network supporting fine structures such as the basement membranes of epithelial tissues. They are now known to be composed of tropocollagen molecules and although they are still classified separately are essentially small collagen fibres.

Elastic fibres are thinner than collagen fibres and they are not arranged in bundles. As their name suggests, their main property is elasticity. There are fewer elastic than collagen fibres in loose connective tissue and they are not arranged in bundles, rather they form an open network of branching fibres (see Fig. 1.28). They are composed of the protein elastin and the fibres can be stretched up to about 150% of their original length and return to their resting length without damage. Electron micrographs show that the fibres have an amorphous core surrounded by microfibrils.

There are a variety of cells present in loose connective tissue, fibroblasts which may be regarded as the cells characteristic of connective tissue, together with mast cells and fat cells and a variable population of macrophages, lymphocytes, neutrophils, eosinophils and plasma cells which have migrated from the blood and whose numbers may dramatically increase in pathological states.

Fibroblasts are the most abundant cells in areolar tissue. They are flattened, spindle shaped, and may have several processs. They are usually found in close proximity to the fibres which they synthesize. These cells both produce and maintain the extracellular material. In the repair of wounds, fibroblasts produce new connective tissue (granulation tissue and scar tissue).

The appearance of macrophages (histiocytes) varies according to whether they are 'fixed' (stationary) or 'free' (mobile). When stationary, they appear irregular, whereas they adopt a rounded shape as they move through the matrix. The nucleus is indented and the cytoplasm contains granules and lysosomes. The name macrophage reflects the highly developed phagocytic ability of the cells which is used to ingest cell debris, bacteria or inert foreign matter. The lysosomes then break down the ingested material. When tissue is damaged or invaded by micro-organisms the 'fixed' macrophages are stimulated to move and their numbers are supplemented by monocytes leaving the blood and transforming into macrophages in the connective tissue. In some instances where there is a large amount of material to be phagocytosed, several macphages fuse together to form giant cells around the object. Macrophages play an important role in the immune response whereby they present antigen to lymphocytes which then transform and produce antibodies against the antigen. (This aspect is covered more fully in Chapter 2.)

Mast cells are ovoid, with a small, round nucleus. They contain many granules which stain blue with basic dye. The granules contain the physiologically active substances histamine and heparin. The cells are disrupted by tissue damage or infection, thus releasing the active agents, thereby increasing the blood supply and facilitating the migration of white cells from the blood. Mast cells are implicated in the tissue changes occurring in allergic and hypersensitivity reactions. It is possible that heparin prevents extravascular clotting of protein which leaks out of the capillaries in inflammatory states. There is a great similarity between mast cells in the tissue and basophils in the blood, but it is probable that they are independent cell types.

Some cells in the connective tissue are specialized for the synthesis and storage of fat. These adipose cells may be present singly or in groups. Their appearance varies according to the amount of fat present. When empty, the cells resemble fibroblasts, whereas when full of fat they become spherical with the centre filled with fat and only a thin rim of cytoplasm visible around the periphery. The nucleus is flattened and displaced.

Neutrophils, eosinophils and lymphocytes derive from the blood and are all involved in different ways in combating infection and tissue damage. The connective tissue population of these cells increase in pathological states.

Neutrophils have a lobed nucleus and fine, purple staining granules. They are quickly attracted to the site of tissue damage and are phagocytic.

Eosinophils are not very numerous in most connective tissue, but they are found particularly in the lamina propria of the small intestine, in the interstitial connective tissue of the lungs, in the omentum, and in the stroma of mammary glands. The cells have a bilobed nucleus and coarse granules containing hydrolytic enzymes, which stain red with the dye eosin. Eosinophils have been shown to phagocytose antigen-antibody complexes and they have a probable role in combating parasitic infections. Their numbers are increased in some hypersensitivity states.

Lymphocytes have a large nucleus and normally appear rounded, and although they are capable of amoeboid movement, they are not phagocytic. These cells are concerned with the production of antibodies. When stimulated by specific antigens they are capable of transforming into larger cells which divide and some of which differentiate to form plasma cells. It is the plasma cells which synthesize antibody against the antigen. Plasma cells have an eccentrically placed nucleus and they are found in greater numbers underneath the epithelial membranes lining the respiratory and alimentary tracts, where micro-organisms are liable to gain entry to the body.

A more detailed account of the cells which derive from the blood is given in Chapter 2.

Elastic tissue (dense connective tissue)

This tissue is a type of dense connective tissue which contains a high proportion of elastic fibres (Fig. 1.30). This type of tissue is not widely distributed, it is found in sites where its ability to be stretched and then regain its original configuration is important. In small arteries, for example, the elastic tissue forms a sheet of tissue called the internal elastic lamina. The vocal cords are composed of elastic tissue as are the ligamenta flava of the vertebral column.

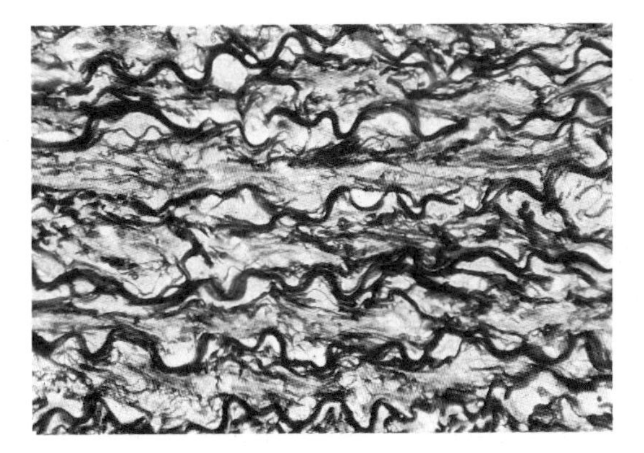

Fig. 1.30 Photomicrograph of elastic tissue (carotid artery). Weigert's Resorcin Fuchsin Stain × 128. (Courtesy of D. R. Smith.)

White fibrous tissue (dense connective tisue)

The name of this tissue derives from the preponderance of bundles of collagen fibres which have fibroblasts lying between them. This tissue has great strength and its locations reflect this function.

In the dense regular connective tissue of tendons, the bundles of collagen run parallel. In ligaments, the arrangement is similar, but slightly less regular. Other sites where regular connective tissue is found include fasciae and aponeuroses and the cornea. In these sites the tissue is built up of sheets, and in each sheet the fibres run parallel; however, the orientation of the sheets may be different.

In other sites, the arrangement of collagen bundles is irregular and has more of a woven appearance (see Fig. 1.31). Such fibrous membranes are found surrounding some organs: as periosteum surrounding bones and perichondrium surrounding cartilage; as the dura mater, one of the meninges surrounding the central nervous system; as the fibrous pericardium surrounding the heart; and the sclera of the eye.

Adipose tissue

This tissue may warrant a category of its own as more discoveries are made about it. The cells are specialized for the synthesis, storage and release of fat, and they are under hormonal and nervous control. The most widespread type of adipose tissue is known as white adipose tissue.

The cells can be very large (up to 120 μm diameter) and are basically spherical, but may appear polyhedral because they are squashed by their neighbours (Fig. 1.32). The fat is usually stored as a single droplet, in which case the cells are described as unilocular.

This type of tissue is present in the superficial fasciae under the skin. The distribution is different in the two sexes, so that in men the main sites are overlying the cervical vertebrae and the deltoid and triceps muscles, the lumbosacral region and the buttocks. In women, adipose tissue is principally found in the breasts, thighs and the buttocks. Both sexes have stores of fat in the omentum and mesenteries.

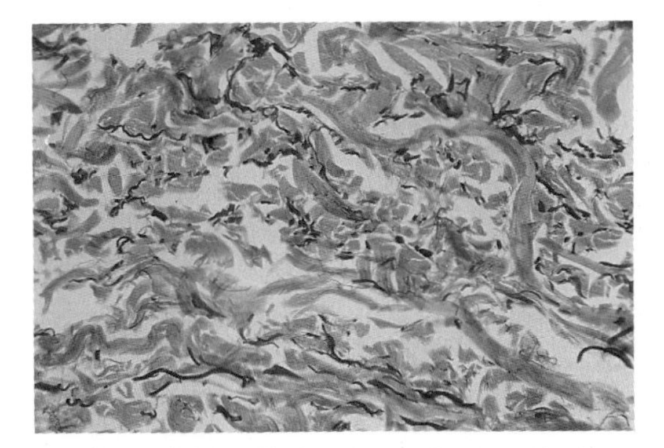

Fig. 1.31 Photomicrograph of white fibrous tissue, dense irregular (skin). Weigert's Resorcin Fuchsin, Van Gieson Stain × 128. (Courtesy of D. R. Smith.)

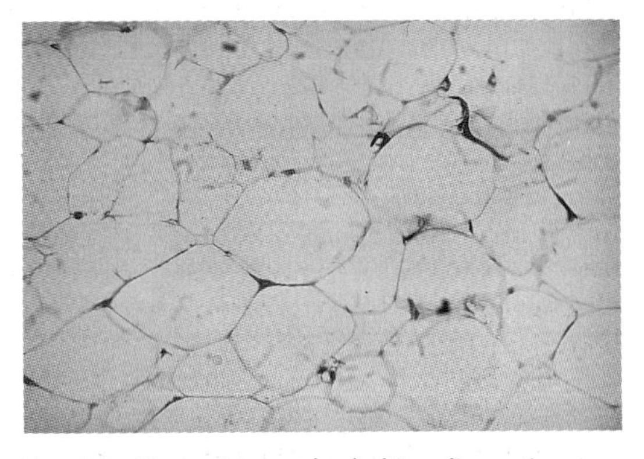

Fig. 1.32 Photomicrograph of white adipose tissue. Stained with toluidine blue × 200 (Courtesy of D. R. Smith.)

The quantity of fat from all these areas will diminish with prolonged fasting when the fat is released and used by other cells. In some sites, however, the fat has primarily a protective function and is not released so readily for metabolic functions. Such sites include the tissue in the orbit of the eye, in the joints and on the palms of the hands and the soles of the feet.

The cells of brown adipose tissue contain large numbers of mitochondria. The cells are described as multilocular because the fat is present as many droplets rather than one large one. The cells are smaller than those found in the white fat (Fig. 1.33).

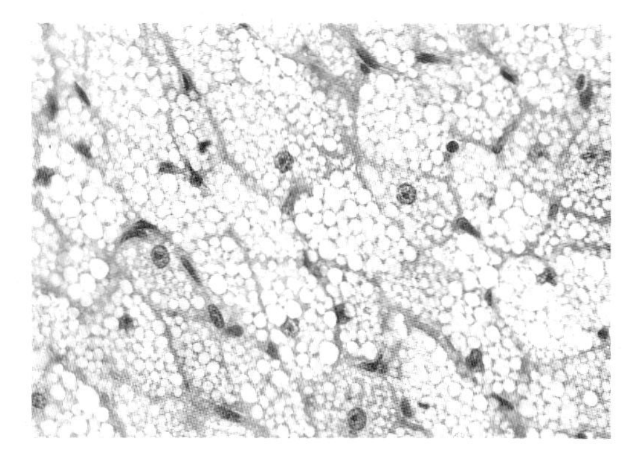

Fig. 1.33 Photomicrograph of brown adipose tissue (monkey). Haematoxylin & Eosin Stain × 320. (Courtesy of D. R. Smith.)

The function of brown adipose tissue appears to be primarily one of heat generation. Mitochondrial oxidation of the fat results in the the energy being released as heat, rather than partly as ATP which is the usual pattern in cells. This oxidation is stimulated by the sympathetic nervous system and the hormone noradrenaline. Brown fat is found between the shoulder blades, around the kidneys, in the axillae, mediastinum and along the aorta in the thorax. The quantity of brown fat diminishes with increasing age.

Cartilage

The cells of cartilage are called chondrocytes and they are present in spaces or lacunae in the intercellular matrix (see Fig. 1.34). The matrix is firm and solid and can therefore withstand pressure, tension and torsion. The type and quantity of fibres present affects the nature of the intercellular material and gives rise to three types of cartilage: hyaline (glassy) cartilage, fibrocartilage and elastic cartilage.

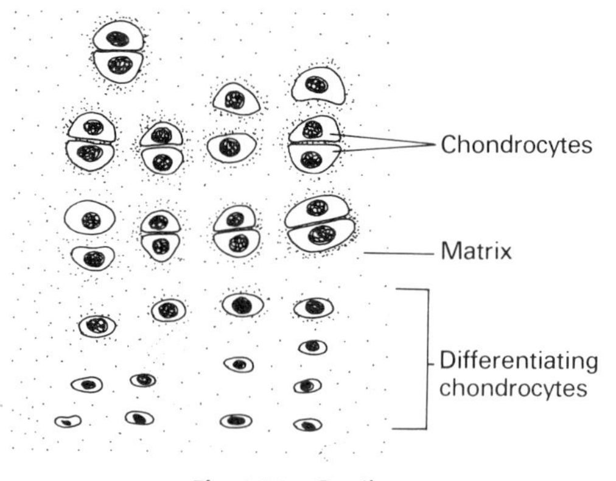

Fig. 1.34 Cartilage.

The ground substance and the fibres are produced by immature cells which are small and irregular in shape. They have a round or oval nucleus and are usually flattened. As the matrix is formed, the cells are separated from each other by it, although cell division often results in several isogenous cells lying together in a lacuna. The mature chondrocytes have a more rounded shape.

Growth of cartilage occurs in two ways, from within (interstitial) or by the addition of new tissue to the outer edges (appositional). The latter occurs by chondrocytes differentiating from connective tissue cells in the perichondrium which surrounds cartilage except at articular surfaces.

If cartilage is damaged, repair can take place by fibroblasts in the perichondrium invading the damaged area to form granulation tissue. The fibroblasts may then transform into cartilage cells.

The matrix contains chondromucoprotein, which is a copolymer of protein and chondroitin sulphate. Nutrition of chondrocytes occurs largely, if not entirely, by diffusion of substances from blood vessels in adjacent tissue, or from synovial fluid in the case of articular cartilage.

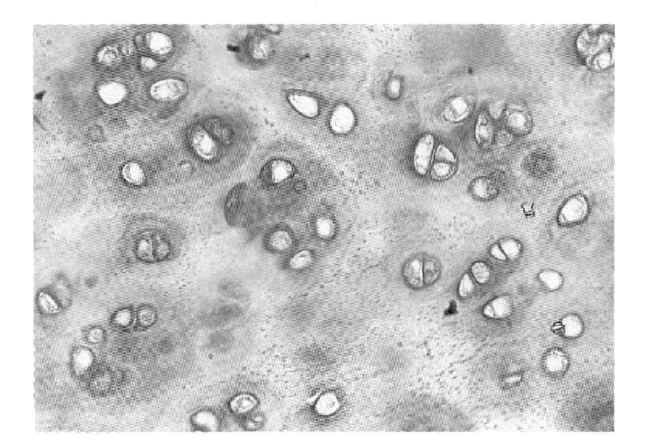

Fig. 1.35 Photomicrograph of hyaline cartilage (trachea). Gomori's Aldehyde Fuchsin Stain × 128. (Courtesy of D. R. Smith.)

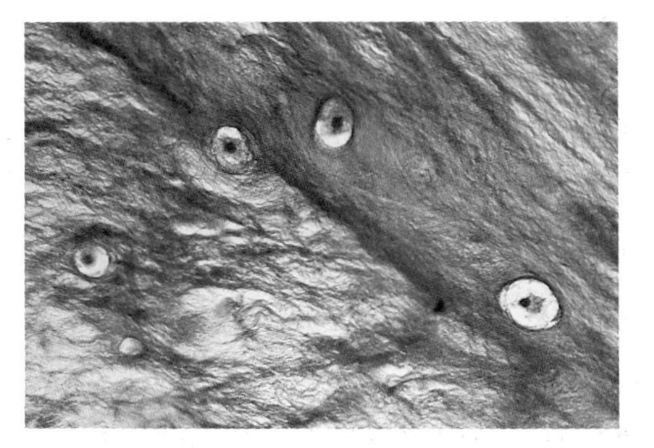

Fig. 1.36 Photomicrograph of fibrocartilage (intervertebral disc). Stained with toluidine blue × 160. (Courtesy of D. R. Smith.)

Hyaline cartilage

The matrix of hyaline cartilage appears pearly and translucent and has some flexibility. Collagen fibres are present in the form of fine, interlacing fibrils which cannot be seen with the light microscope (Fig. 1.35).

The fetal skeleton is laid down as hyaline cartilage, most of which is subsequently replaced by bone. The cartilage persists in the costal cartilages of the ribs, in the nose, larynx, trachea and bronchi. In long bones, hyaline cartilage persists until adolescence as epiphyseal cartilage plates, which, because of their capacity for interstitial growth, enable the bones to increase in length (see Fig. 1.38). Hyaline cartilage is found covering articular surfaces where the smoothness of the tissue enables the bones to move easily against each other.

Fibrocartilage

The principal structural feature of fibrocartilage is thick bundles of collagen fibres running parallel with each other. The chondrocytes lie in lacunae between the fibres, and there is little matrix present (Fig. 1.36). This structure renders the tissue rigid and very strong, although it still has some flexibility.

This tissue is found in the intervertebral discs, the pubic symphysis, the atttachments of some tendons and the linings of tendon grooves in long bones. Fibrocartilage is continuous with the dense connective tissue in joint capsules and ligaments and has no perichondrium around it.

Elastic cartilage

This tissue is yellow, opaque and elastic. It contains branching and anastomosing elastic fibres which are continuous with those in the surrounding perichondrium. There are some fine collagen fibres present as well. The chondrocytes in lacunae are present singly or in isogenous groups of two or four cells (Fig. 1.37).

Elastic cartilage will withstand deformation and return to its original shape. It is found in

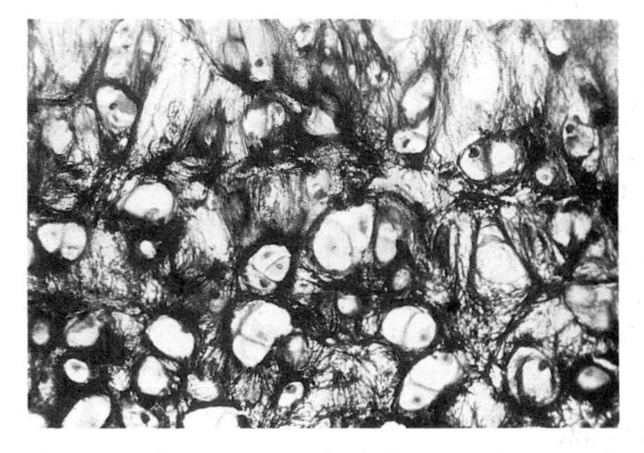

Fig. 1.37 Photomicrograph of elastic cartilage (epiglottis). Weigert's Resorcin Fuchsin Stain × 128. (Courtesy of D. R. Smith.)

the external ear, external auditory meatus and eustachian tube; in the epiglottis and some laryngeal cartilages.

Bone

Bone has the characteristic components of connective tissues, that is cells, ground substance and fibres, but its principal feature is that the intercellular material is calcified. The tissue has, therefore, high tensile and compressive strength. Bone has some elasticity and the structure of bones renders them remarkably lightweight in view of the composition of the tissue.

The skeleton gives support to the body, as well as protection to soft organs such as the brain, heart and lungs. Muscles are attached to bone, and the cells of the blood are produced in the marrow cavities of some bones. The tissue is a dynamic one, in that it is constantly being renewed (accretion) and removed (resorption). Blood calcium concentration is kept relatively constant partly by the maintenance of an equilibrium with calcium in the bone. If blood calcium level falls, for example, then calcium is added from the bone.

Bone, in contrast to the other connective tissues, has an extensive blood supply. The hard nature of the matrix would not enable the bone cells to rely on diffusion for nutrition and removal of waste. There are fine channels or canaliculi present in the matrix which connect the cells to the blood vessels.

Gross structure of bones

Complete bones are composed of two types of bone tissue which can be distinguished by the naked eye, that is, compact and cancellous bone. The former, as its name suggests, is dense and appears like ivory, whereas cancellous bone has a honeycomb-like structure with spicules or trabeculae of bone enclosing gaps in the matrix containing bone marrow. The parts of a long bone (e.g. the humerus and tibia of the limbs) are shown in Figure 1.38. The rounded ends or epiphyses are composed of cancellous bone with a thin peripheral layer of compact bone. Hyaline cartilage covers the bone at the articular surfaces.

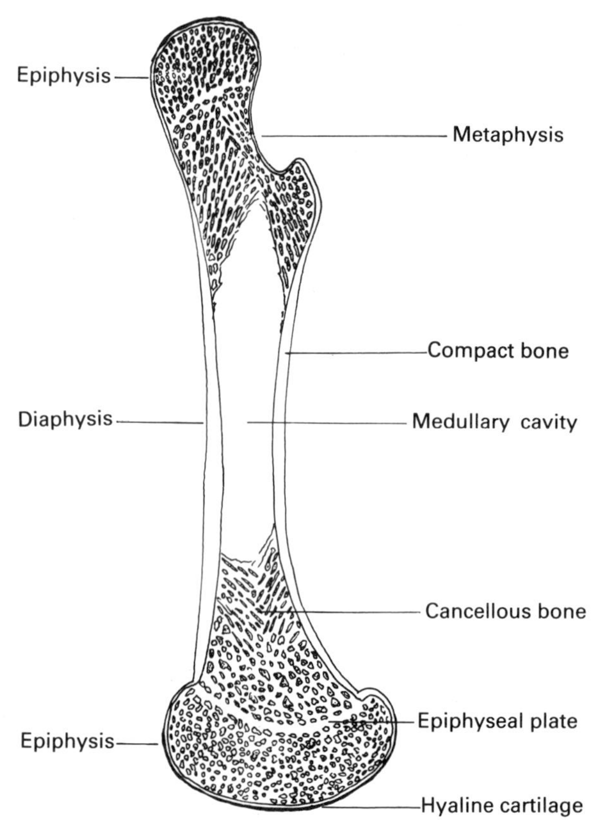

Fig. 1.38 Diagrammatic longitudinal section through a long bone.

In children and young adults who are still growing, a cartilage disc called the epiphyseal plate is present next to each epiphysis. This is connected to the shaft of the bone (diaphysis) by columns of spongy bone in a region called the metaphysis. The diaphysis has a central medullary cavity containing marrow which is continuous with cancellous bone and surrounded by a thick, peripheral layer of compact bone.

The skull is composed of flat bones which consist of a layer of cancellous bone (diplöe) sandwiched between the two layers of compact bone.

Bones are surrounded by a layer of connective tissue called periosteum (except at the articular surfaces), which has the potential to form bone-producing cells (osteogenic potency). The bone cavities are lined by endosteum, which consists of a thin layer of squamous cells, which also have osteogenic potency.

Fine structure of bone

Compact bone is constructed in lamellae or layers which are arranged in concentric rings to form cylindrical structures known as Haversian systems or osteons (see Fig. 1.39). The lamellae comprise the ground substance and collagen fibres which are orientated parallel with each other within each lamella, but the direction changes in adjacent lamellae. The bone cells or osteocytes lie in cavities (lacunae) in the matrix which connect with each other and with the central Haversian canals, each of which contains one or two blood vessels, usually capillaries. The canaliculi are effectively an extension of the circulatory system in that they contain fluid filtered from the blood vessels, which conveys nutrients to the

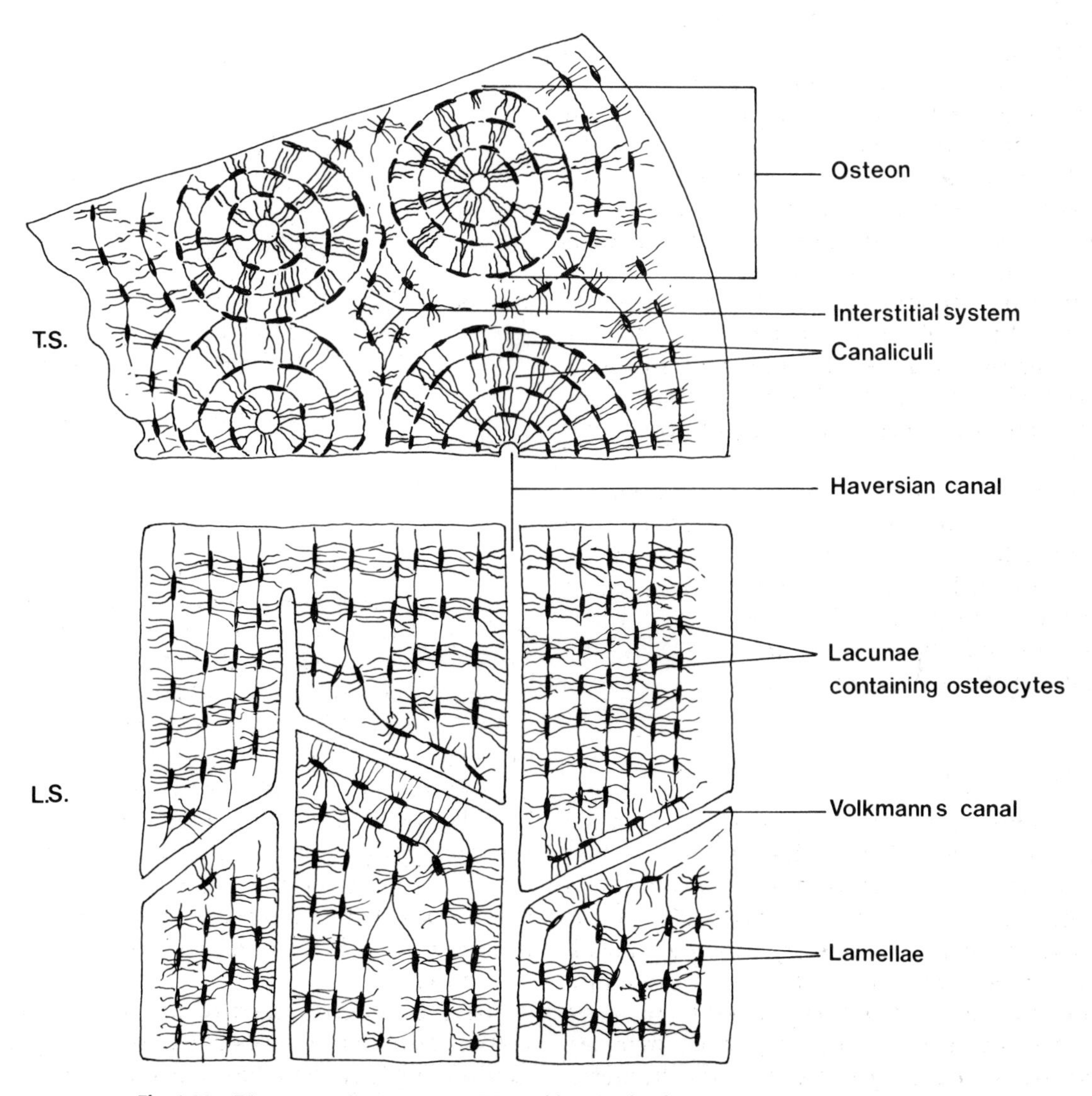

Fig. 1.39 Diagrammatic transverse (TS) and longitudinal (LS) sections of compact bone.

bone cells and waste products and metabolites back to the blood. The lamellae are 3–7 μm thick, and each osteon is made up of four to twenty lamellae, so that the size of the diameter of individual osteons varies. Some osteons are simple cylindrical structures, whilst others branch and anastomose with nearby osteons. The Haversian canals are interconnected by channels called Volkmann's canals.

Between the osteons, there are lamellae filling in the spaces. These are called interstitial systems, and on the surfaces of bones beneath the periosteum and endosteum, there are usually a few circumferential lamellae which extend around the periphery of the bone.

Cancellous bone also has a lamellar structure, but it does not have Haversian canals containing blood vessels, and the cells are connected to the blood vessels in the endosteum by the canaliculi.

Bone salt is mainly the complex molecule hydroxyapatite (Ca_{10} $(PO_4)_6$ OH_2) and it is found on and within the collagen fibres. Other ions may either associate with the crystals of apatite (e.g. carbonate and citrate) or be substituted for ions within the molecule (e.g. fluoride may substitute for the hydroxyl group OH^-). Sodium and magnesium ions are also present. Some radioactive isotopes released in the fission of uranium or plutonium may become incorporated into the hydroxyapatite molecules. The most hazardous of these bone-seeking isotopes is strontium-90 (Sr^{90}).

Bone cells

Bone is laid down by cells called osteoblasts, which are present on the growing surfaces of bone. These cells are ovoid in shape with several cytoplasmic processes radiating outwards to connect with those of adjacent cells.

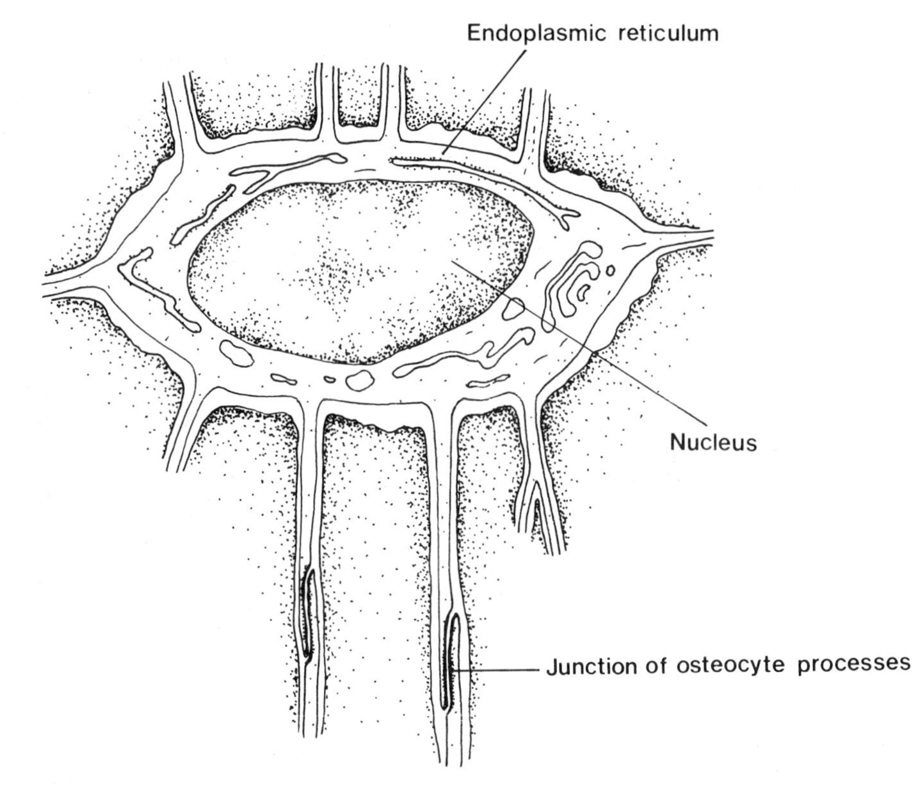

Fig. 1.40 Osteocyte within its lacuna. The cell's processes are in contact with those of adjacent cells. The removal of bone salt by the cell leaves an adjacent clear area of unmineralized matrix.

As the bone matrix is laid down around the cells, it leaves the lacunae and canaliculi surrounding the cells. The 'trapped' cells are the mature bone cells, the osteocytes (see Fig. 1.40), which maintain the matrix and also actively remove bone salt from the bone to the extracellular fluid filling the canaliculi and thence to the blood. This transfer of bone salt from bone to blood by the osteocytes is called osteolysis and is thought to be stimulated by parathyroid hormone.

Bone surfaces which are undergoing resorption contain large, multinucleate cells called osteoclasts. These cells are often seen in shallow depressions on the bone surface called Howship's lacunae. The surface of the cell next to an area of bone undergoing resorption appears 'ruffled' because the cell membrane is deeply folded. The precise mechanisms of bone resorption are unknown, but it is probable that the osteoclasts secrete proteolytic enzymes into the matrix, and then the digestion products are taken into the cells by pinocytosis. Parathyroid hormone stimulates osteoclast activity.

Ossification

Bone is produced by osteoblasts, and the events preceding its formation during development and growth are described as either intramembranous or endochondral ossification.

Most of the skull bones and the clavicles are formed by intramembranous ossification, that is, from primitive connective tissue (mesenchyme) which undergoes structural changes including vascularization and the development of osteoblasts from the connective tissue cells.

The first bone to be laid down is called woven bone, which has collagen fibres running in all directions, rather than in the regular lamellae which are laid down subsequently. Woven bone has large channels running through it containing blood vessels.

Most bones, however, are formed by endochondral ossification. In this process, a template of cartilage is laid down which replaces the embryonic mesenchyme, and then the cartilage is in turn replaced by bone. The details of endochondral ossification are given here in connection with the ossification of a long bone (see Fig. 1.41). The first or primary centre of ossification occurs in the diaphysis by the third month of fetal life. The chondrocytes enlarge and the matrix diminishes. Calcium phosphate crystals are deposited in the matrix, which causes the cells to die. Concurrently, there are changes in the perichondrium resulting in the production of osteoblasts, which produce a periosteal band or collar of bone. Blood vessels grow into the diaphysis

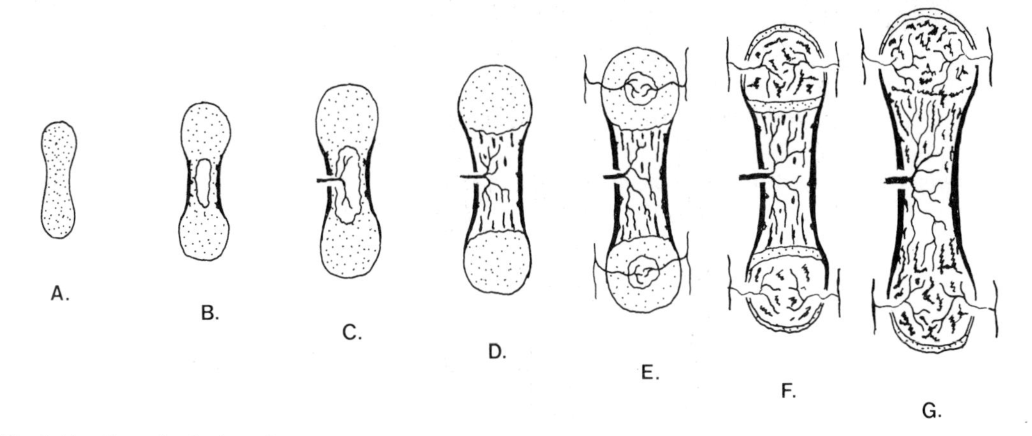

Fig. 1.41 Growth of a long bone. (A) Cartilage template of long bone in fetus; (B) Primary centre of ossification and periosteal collar (3rd month of fetal life); (C) Invasion of diaphysis by blood vessels and connective tissue; (D) Trabeculae formation: (E) Postnatal formation of secondary centres of ossification; (F) Ossified epiphyses, leaving cartilaginous epiphyseal plates; (G) Closure of epiphyses.

and carry with them, in the perivascular tissue, primitive connective tissue cells which subsequently develop into haemopoietic cells and osteoblasts. These osteoblasts congregate to form a layer on the spicules of calcified cartilage, which is disintegrating. The cells deposit bone matrix and form trabeculae.

Further increase in length occurs by interstitial growth of the cartilage, particularly at the junctions of the diaphysis and the epiphyses of the bone. The diaphysis thickens by appositional growth, which increases the thickness of the periosteal band. Ossification progresses from the centre of the shaft towards the epiphyses, with cartilage being replaced by bone in the metaphyseal region. The bones are modelled as they grow by resorption by osteoclasts on the inner (endosteal) surface, which enlarges the marrow cavity, and also by osteoclasts on the periosteal surface, which shape the bone surface.

The primary bone is subsequently replaced by ordered secondary bone from about one year onwards and the replacement of bone continues throughout life. The primary bone is eroded by osteoclasts and then replaced by secondary bone arranged in osteons.

After birth, secondary centres of ossification occur in the epiphyses. In this case, there is no periosteal band, and the ossification gradually replaces all but the articular cartilage, and the cartilage between the epiphysis and the diaphysis. This cartilage becomes known as the epiphyseal plate, and because of its ability to divide, it is responsible for the increase in length of long bones during childhood and adolescence. The thickness of the plate remains relatively constant, as ossification keeps pace with the additional cartilage produced. Eventually, the cartilage is completely replaced by bone, a process called closure of the epiphysis.

MUSCLE

Muscle tissues are capable of contraction and thereby cause movement of the whole body or a part of it.

The cells are elongated in the direction of contraction and contain the proteins actin and myosin in much larger amounts than other types of cell. These proteins are arranged to form myofilaments in the cytoplasm (sarcoplasm) of the cells and it is here that the contractile machinery may be said to reside. Details of how the process is thought to take place are given in *Skeletal muscle*.

When viewed microscopically under polarized light or after staining, muscle cells appear either homogenous (smooth), or show alternate light and dark cross banding (striated).

Smooth muscle is also known as involuntary muscle as it is found in those structures not generally associated with voluntary activity, such as the walls of blood vessels and the gastro-intestinal tract. Contraction is controlled by the autonomic nervous system.

There are two types of striated muscle: skeletal and cardiac (Fig. 1.42). The former is usually, though not always, found in muscles attached to bone and is therefore responsible for such movements as walking and raising the rib cage during respiration. This tissue is also called voluntary muscle, although it can contract involuntarily by reflex action. Skeletal muscle is controlled by somatic motor neurones.

A.

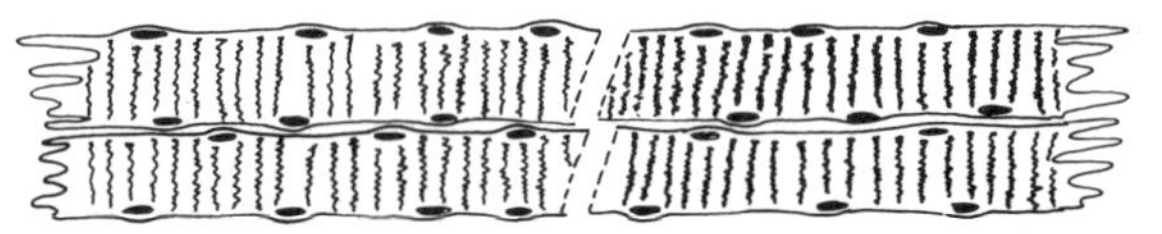

B.

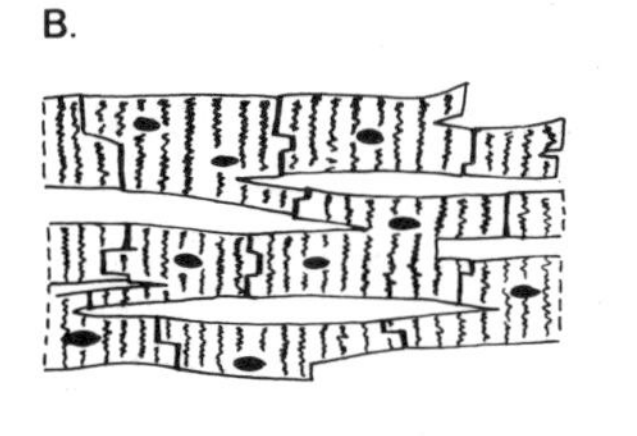

C.

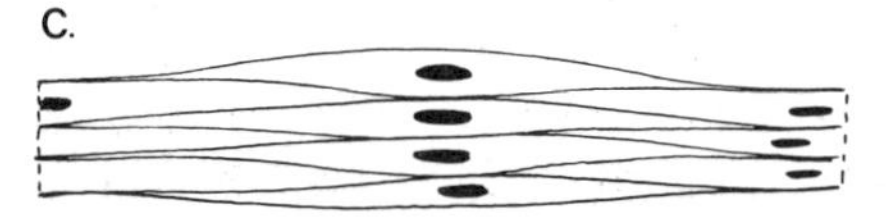

Fig. 1.42 Types of muscle. (A) Skeletal; (B) Cardiac; (C) Smooth.

Cardiac muscle is found in the myocardium of the heart and is controlled by the autonomic nervous system.

Smooth muscle

Smooth muscle cells are long (15–500 μm) and spindle shaped, with an ovoid central nucleus (Fig. 1.42). Each smooth muscle cell is surrounded by a basal lamina (like that seen in epithelial tissues) and there are gap junctions or nexuses linking the cells together. They are usually arranged in small bundles or fasciculi which are themselves arranged as sheets.

The orientation of the cells determines the type of movement effected, so that the concentric arrangement in the gastro-intestinal tract and blood vessels will cause constriction, whilst the longitudinal cells will cause local shortening.

Connective tissue surrounds the fasciculi and carries blood and lymphatic vessels and nerves. The collagen, reticular and elastic fibres in the connective tissue are continuous with those forming a sheath around individual cells, so that when a cell contracts, the force is distributed via the fibres to the surrounding connective tissue.

The autonomic motor nerve fibres which supply the cells, are unmyelinated and branch at their terminals. Each axon terminal ending lies in a groove on the surface of a muscle cell. Although only a few of the cells actually lie next to a nerve ending, the spread of excitation is via the muscle cell membranes and the nexuses which join adjacent cells.

Some smooth muscle appears to contract spontaneously i.e. it is myogenic, or is stimulated by stretching the muscle as for example in the gastro-intestinal tract, the uterus and ureters. In these cases the nervous control serves to modify the rate or force of the pre-existing rhythmic contraction. Smooth muscle also exhibits a sustained or tonic contraction.

Repair of damaged smooth muscle is thought to be achieved largely by the formation of scar tissue by the fibroblasts in the connective tissue, although some regeneration of muscle cells may also take place.

Smooth muscle is found in the iris and ciliary body of the eye, in the walls of the respiratory tract, in the urinary and genital tracts, in the duct walls of glands, as arrector pili muscles attached to hairs in the skin, as well as in blood vessels and the gastro-intestinal tract.

Skeletal muscle

There are some 600 whole muscles in the body composed of skeletal (voluntary) muscle tissue. They are mostly attached to the skeleton, usually by tendons.

Muscles are surrounded by a dense connective tissue sheath called the epimysium, from which loose connective tissue fasciae extend into the muscle and divide it into primary bundles or fasciculi, containing 20–40 fibres (Fig. 1.43). The fasciculi may be arranged parallel, oblique or spiralized in relation to the direction of pull of the muscle.

The connective tissue surrounding each fasciculus is known as the perimysium, and this in turn is continuous with the connective tissue surrounding each individual muscle fibre, the endomysium.

Arterial, venous and lymphatic vessels lie in the perimysium and epimysium whilst the endomysium contains an extensive capillary network, supplying each muscle fibre.

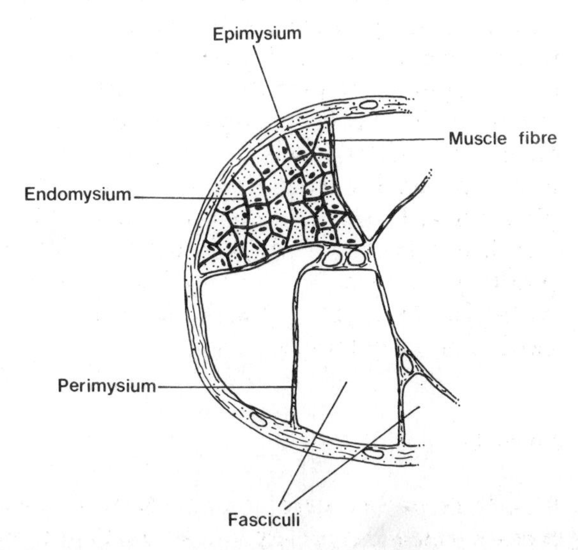

Fig. 1.43 Arrangement of connective tissue in a muscle.

The muscle fibres are 10–60 μm in diameter and their length varies from a few mm to about 30 cm (in the sartorius muscle of the thigh). Usually, a fibre does not extend from one end of the muscle to the other, but is connected to a tendon at one end and to connective tissue within the muscle at the other.

Muscle fibres are cylindrical in shape and contain many hundreds of peripherally situated nuclei. Each fibre develops from a number of myeloblasts which fuse to form myotubes which then grow to become fetal muscle fibres.

Under the light microscope, a characteristic banding pattern can be seen across a muscle cell (see Fig. 1.44). There are alternate dark and light bands, the dark ones are known as A bands, because they are anisotropic (strongly rotate the plane of polarized light), whilst the light bands are isotropic (weakly rotate the plane of polarized light). Each of these bands has additional, thinner bands bisecting them. The A bands have a pale central region known as the H band or zone (after Hensen), which itself has a dark M band crossing it; and the I bands are bisected by a dark Z band or disc (zwischensheibe).

The muscle fibres contain cylindrical struc-

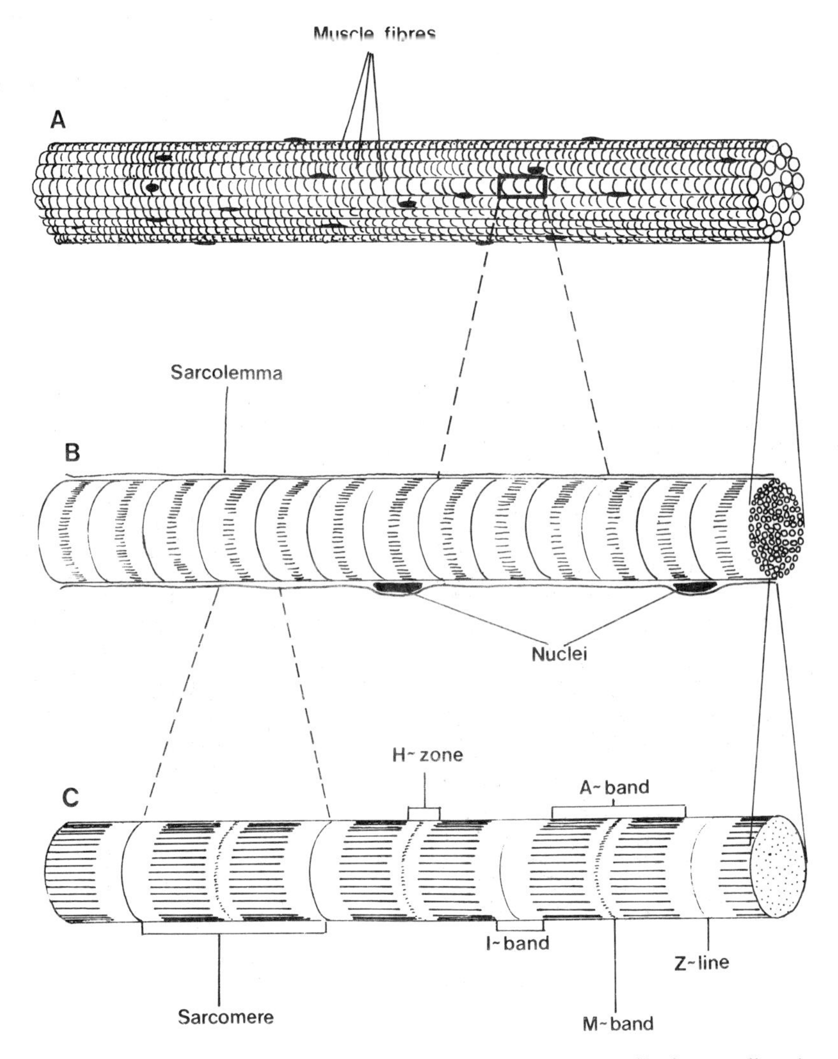

Fig. 1.44 Skeletal muscle. (A) Fasciculus; (B) Individual cell (fibre); (C) Myofibril. (Not all to the same scale.)

tures called myofibrils which run parallel to the long axis of the cell (Fig. 1.44). The banding pattern is present on the myofibrils within the cell, and as the bands of adjacent myofibrils are in register across the cell, then the whole cell appears cross striated. The repeating units from Z line to Z line are called sarcomeres and each myofibril may be regarded as a column of disc-like sarcomeres.

The cross striations of the myofibrils result from the very regular arrangement of the myofilaments which they contain. There are two types of myofilament, thick (mainly myosin) and thin (mainly actin).

The large number of mitochondria present between the myofibrils and under the sarcolemma, reflects the high production rate and usage of ATP.

The sarcolemma is invaginated into the cell to form structures called T-tubules which run transversely across the cell and around each fibril. As these tubules are an extension of the sarcolemma, they contain extracellular fluid. Each sarcomere contains two such tubules which cross the cell at each A-I band junction (see Fig. 1.45).

The smooth endoplasmic reticulum in skeletal muscle cells is called sarcoplasmic reticulum and it runs longitudinally around each myofibril. A T-tubule runs between adjacent sections of sarcoplasmic reticulum and either side of it there are dilated tubules known as terminal cisternae. In section a triad can be seen, consisting of two terminal cisternae and a T-tubule in between (see Fig. 1.45).

Myoneural junction

The somatic motor neurones which innervate skeletal muscle cells terminate as structures called motor end-plates. Each axon branches towards its end and each branch supplies a separate muscle fibre. When the neurone is stimulated, the impulses pass along each of the branches to the muscle fibres, which then contract in unison. All the branches of a single

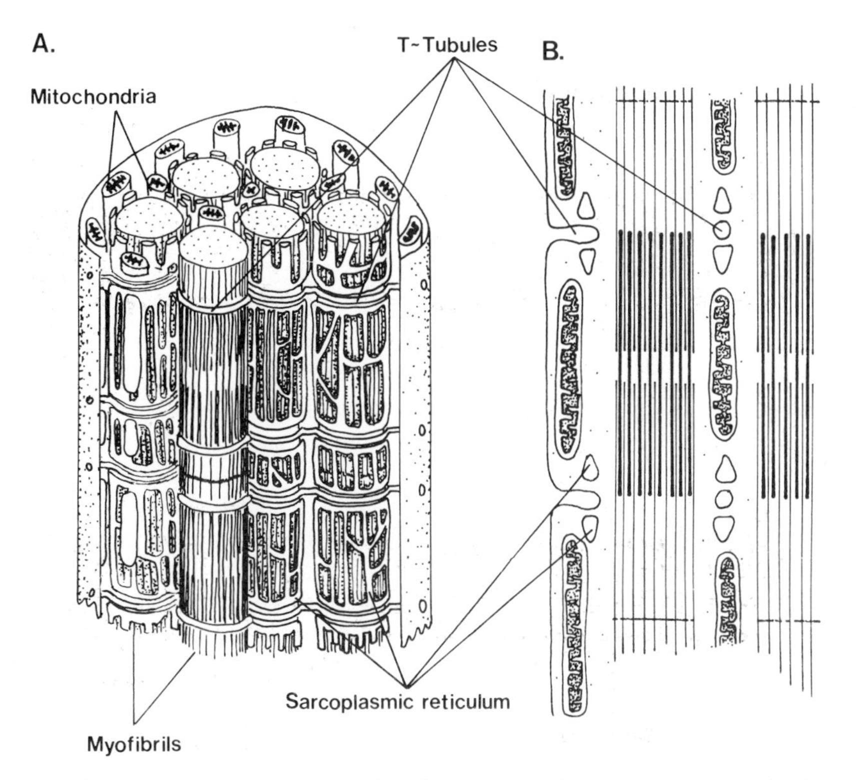

Fig. 1.45 Ultrastructure of a skeletal muscle cell. (A) Longitudinal view; (B) Longitudinal section

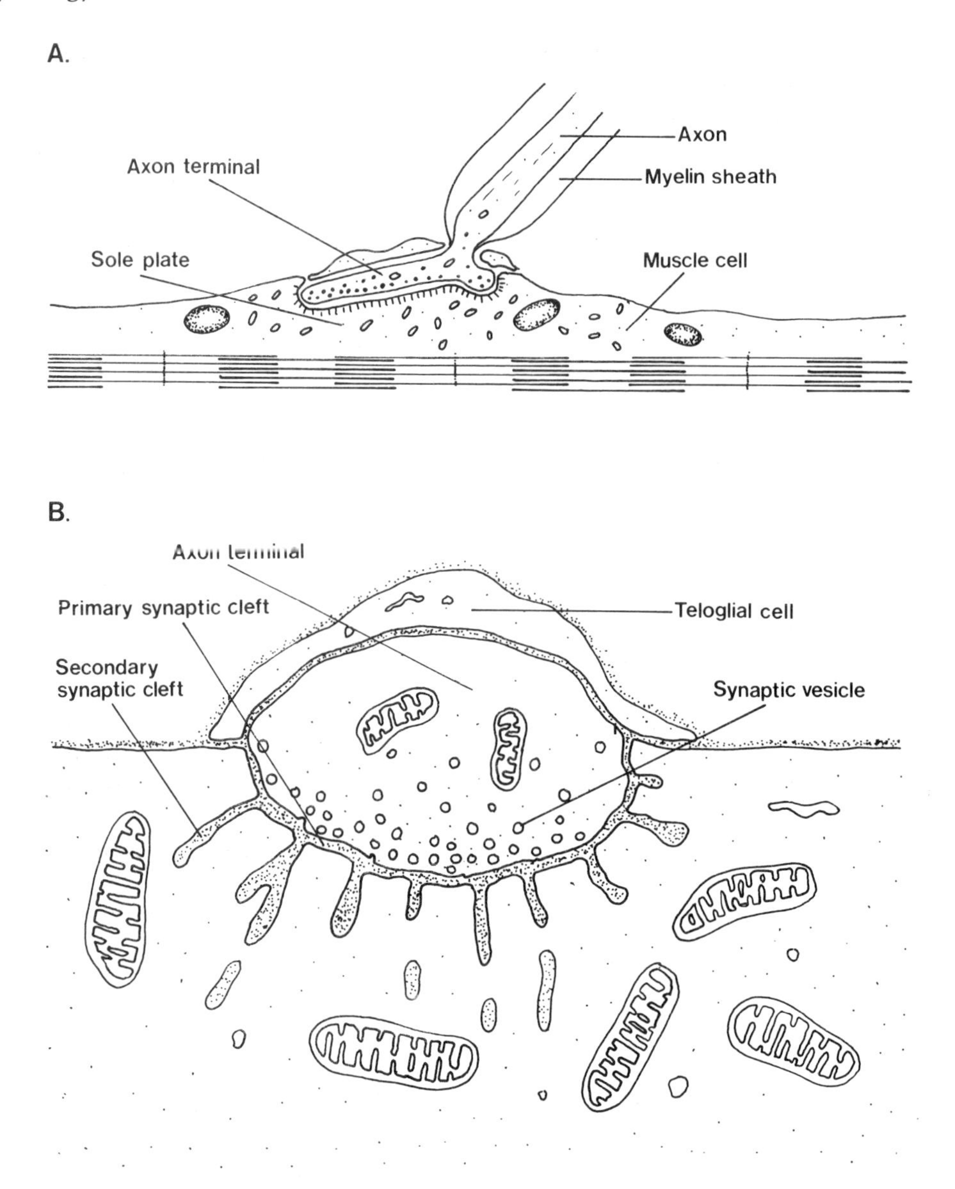

Fig. 1.46 Motor end-plate. (A) Axon terminal and sole plate. (B) Section through the motor end-plate to show the synaptic clefts.

axon together with the muscle cells they innervate is called a motor unit.

The motor end-plate consists of the terminal branches of each axon branch supplying an individual cell together with specialized areas of muscle cell underneath. Each axon terminal is unmyelinated, lies in a trough (primary synaptic cleft) formed by an invagination of the sarcolemma and is covered by Schwann cells (teloglia).

The cleft has secondary folds or clefts in it, and the gap between the axon ending and the sarcolemma is filled with a protein-carbohydrate substances like that found in basal laminae (see Fig. 1.46).

There is an accumulation of sarcoplasm beneath the sarcolemma called a sole plate. It may be noted that some authorities confine the term motor end-plate to this section of the myoneural junction.

The axon terminals themselves contain large numbers of vesicles containing the neurotransmitter acetylcholine. There are also large numbers of mitochondria which provide the ATP required for the release of acetylcholine into the synaptic clefts. Acetylcholinesterase is present in the cleft, and the molecules are probably attached to the sarcolemma.

Fine structure of the myofilaments

If a myofibril is cut through at an A band and the cut surface examined, it can be seen that the myofilaments are arranged in a very regular pattern. Each thick filament is surrounded by six thin filaments and each thin filament is surrounded by three thick ones (see Fig. 1.47).

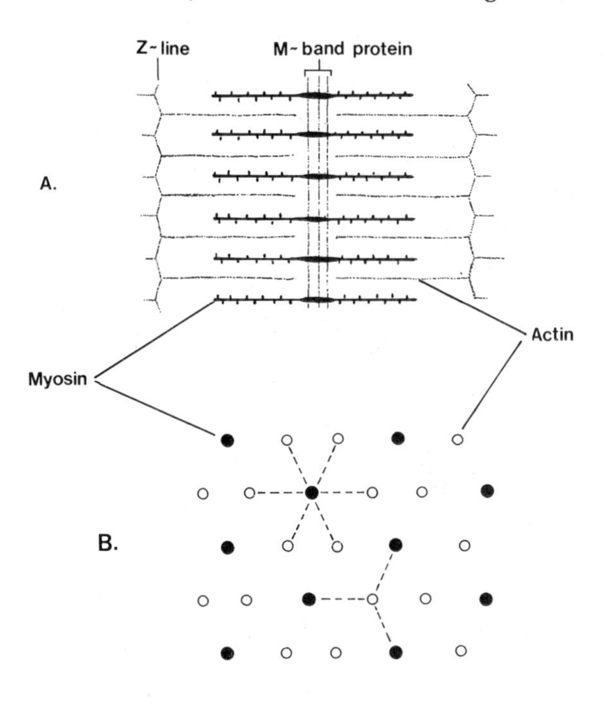

Fig. 1.47 Arrangement of myofibrils within a sarcomere. (A) Longitudinal section; (B) Transverse section through the A band.

The principal protein comprising the thick filaments is myosin. Each myosin molecule has a long tail region composed of light meromyosin (LMM) which consists of two helical polypeptide chains wound round each other and a neck and head region composed of heavy meromyosin (HMM). The neck region is also helical and the head is double and globular. Each head has a binding site for ATP and a separate site for combining with actin (thus each HMM molecule has two binding sites for ATP and two binding sites for actin, see Fig. 1.48). The myosin molecules are arranged head to tail, with the tails forming the thick filament and the head and neck parts of the molecule sticking out from the main axis, and forming cross bridges. The orientation of these cross bridges is very regular. Each filament has cross bridges sticking out in pairs opposite each other and each pair is rotated 120° with respect to adjacent pairs. Every fourth pair is thus orientated in the same direction. This arrangement means that there are cross bridge positions all along the thick filament, which correspond with the arrangement of the six thin filaments which surround it (Figs. 1.47B and 1.49). At the M band, the myosin molecules are oriented tail to tail and anchored in position by the M band protein.

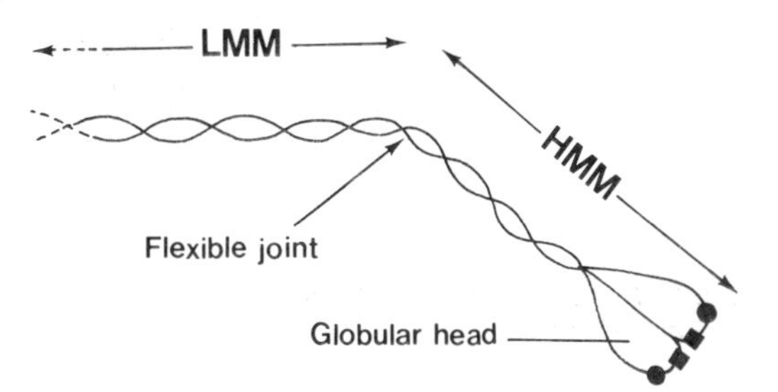

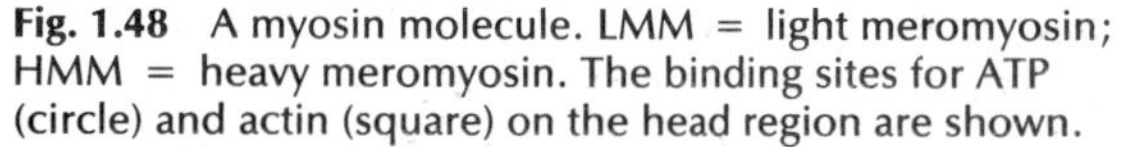

Fig. 1.48 A myosin molecule. LMM = light meromyosin; HMM = heavy meromyosin. The binding sites for ATP (circle) and actin (square) on the head region are shown.

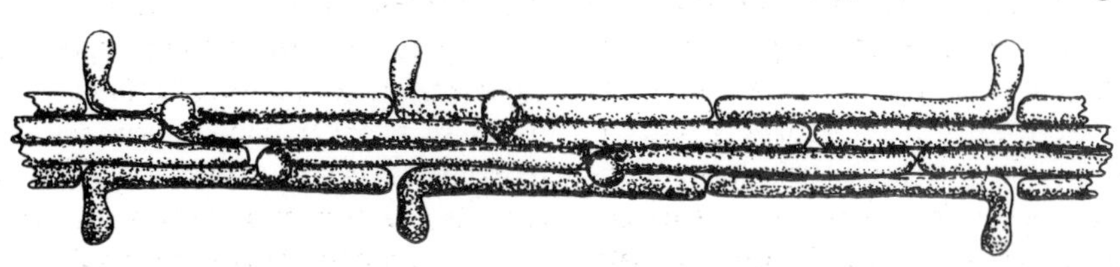

Fig. 1.49 Arrangement of myosin molecules to form a thick myofilament. Consecutive molecules are rotated 120°.

The thin filaments consist mainly of the protein actin. The globular molecules (G-actin) form a polymer of fibrous actin (F-actin) which resembles two strings of beads twisted round each other to form a double helix. There are seven G-actin molecules in each chain per twist (see Fig. 1.50).

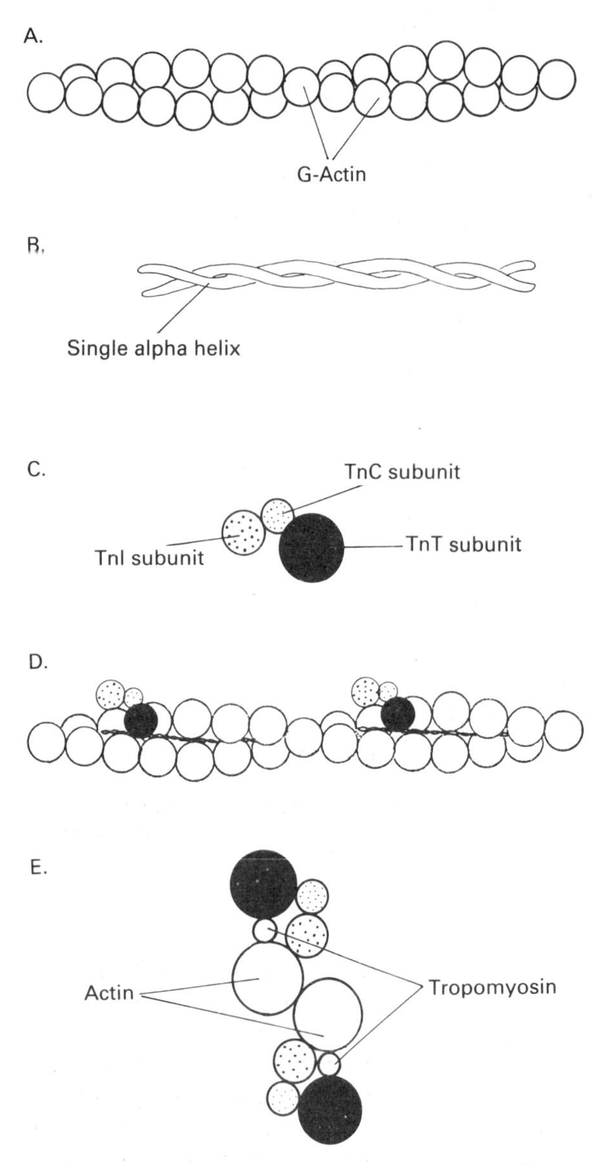

Fig. 1.50 Components of a thin myofilament. (A) Fibrous actin formed from a double helix of individual globular molecules; (B) Tropomyosin; (C) Troponin complex; (D) Thin filament assembled; (E) Cross section of filament to show the location of two troponin complexes and two tropomyosin helices on either side of the actin helix.

There is a second protein, tropomyosin, running alongside the groove between the two strands of F-actin. Tropomyosin consists of two α-helices wound round each other, spanning seven G-actin molecules.

A third protein complex, troponin, is positioned about one third of the distance along from the end of each tropomyosin molecule (see Fig. 1.50). Troponin has three subunits, TnT (binds to tropomyosin); TnC (binds to calcium ions), and TnI (binds to actin and inhibits the interaction between actin and myosin).

Sliding filament theory

The sliding filament theory of muscle contraction dates from around 1950 and is attributed primarily to the scientists H E Huxley, A F Huxley & J Hanson. It was observed that when a muscle contracts and the sarcomeres reduce in length, the I bands become narrower, but the A bands do not, and that if a resting muscle is stretched, then the H zones increase in size. These observations can be explained in terms of a varying degree of overlap between the thick and the thin filaments. When a muscle contracts, the thin filaments slide in between the thick ones towards the M band and, conversely, if the muscle is stretched, the thin filaments slide away from the M band and increase the length of the H zone (see Fig. 1.51). The force of contraction is developed by the cross bridges in the overlap region. The thin filaments are pushed towards the M band by the repeated movement of the cross bridges making and breaking contact with them, like oars in a boat with water.

Excitation-contraction coupling

Skeletal muscle contraction is controlled by the somatic motor neurones which terminate in motor end plates on the muscle cells. Each muscle fibre has its own end plate, usually located centrally along the fibre. An active neurone transmits impulses (action potentials) along the axon to the terminals. Each action potential causes some of the vesicles in the

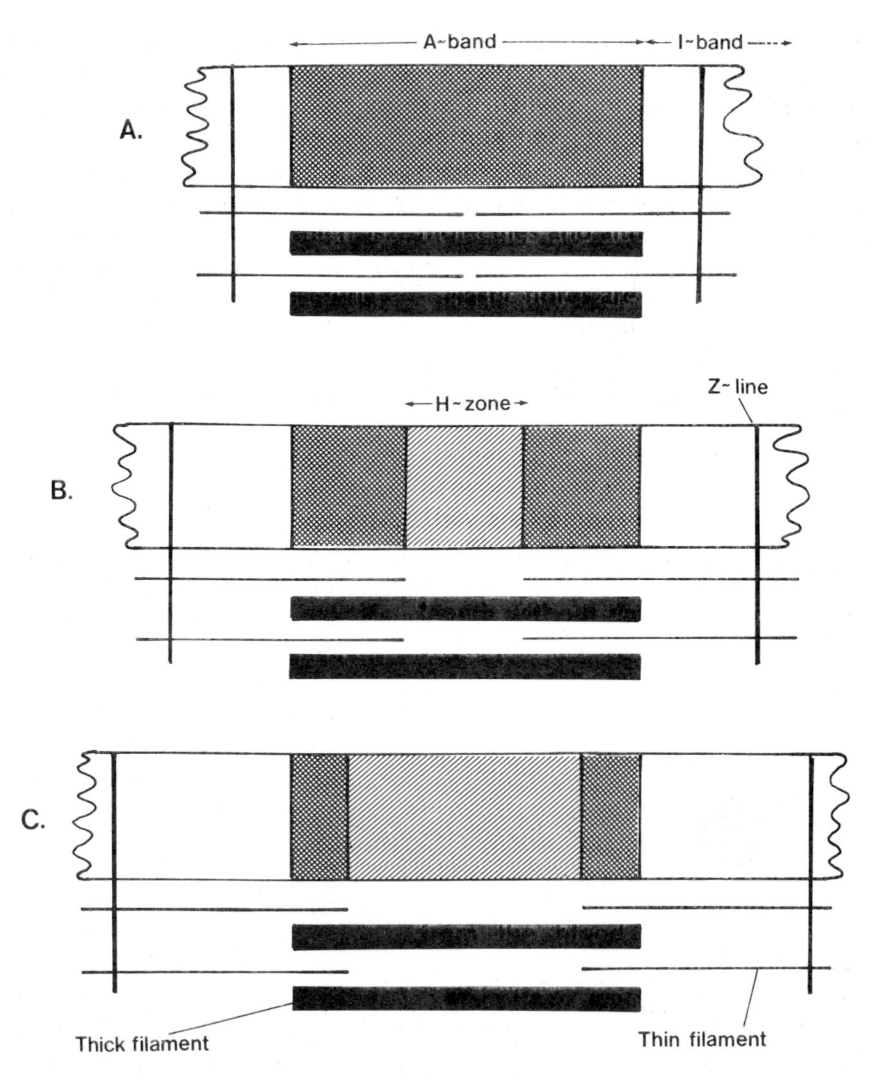

Fig. 1.51 Changes in the banding pattern with contraction or relaxation of the muscle. (A) Muscle contracted, thin myofilaments overlapping thick myofilaments, narrow I band and no H band; (B) Muscle partially contracted, some overlap between myofilaments, I band wider and H zone apparent; (C) Muscle relaxed, very little overlap between myofilaments, wide I band and H zone.

axon terminal to fuse with the neurone membrane and release a quantity of acetylcholine into the synaptic cleft. The neurotransmitter diffuses across the cleft, combines with receptor sites on the sarcolemma, and increases the sarcolemmal permeability to ions. This results in sodium ions entering the muscle cell and initiating an action potential in the sarcolemma. The impulses are transmitted rapidly over the sarcolemma and into the muscle fibres by means of the T-tubules. The presence of action potentials in the T-tubules cause the adjacent terminal cisternae to release calcium ions into the sarcoplasm.

Troponin and tropomyosin are known as regulatory proteins because in resting muscle they prevent the combination of actin with myosin to form actomyosin, that is they prevent the cross bridges of the thick filaments from attaching to the thin filaments surrounding them.

When calcium ions are released into the sar-

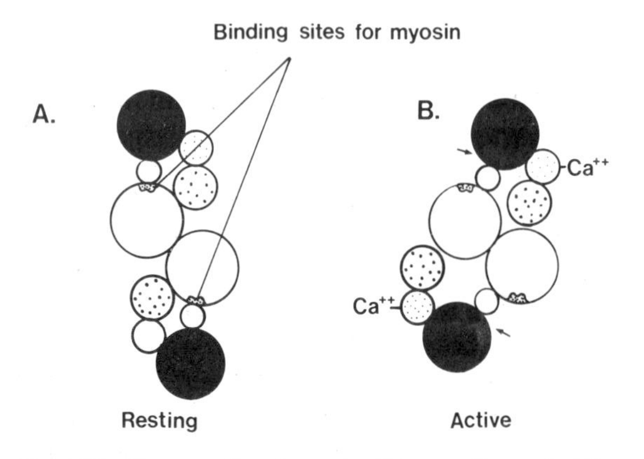

Fig. 1.52 Cross section through a thin myofilament: (A) Resting muscle, tropomyosin is covering the binding sites for myosin on the actin molecules; (B) Active muscle, calcium binds to TnC and the troponin subunits draw away from the filament, tropomyosin rolls further into the groove thereby exposing the myosin binding site.

coplasm they bind to the troponin subunits TnC and the inhibitory action of the regulatory proteins is lifted so that actin and myosin combine. It is thought that the inhibitory action of the regulatory proteins is effected by tropomyosin physically covering the myosin binding sites on the actin molecules, and that the troponin complex holds the tropomyosin in position.

When calcium is bound to TnC, the troponin subunits are drawn closer together and slightly away from the filament so that tropomyosin rolls further into the groove between the two F-actin strands and exposes the binding site for myosin (see Fig. 1.52). The cross bridge (myosin head) then makes contact with the two nearby actin molecules.

When the muscle is relaxed and the cross bridges are not making contact with the thin filaments they are thought to be bent at about 45° with the tip pointing away from the Z line. When calcium binds to troponin, the cross bridges make contact with the nearby actin filaments. The sequence of events constituting the contraction cycle is shown in Figure 1.53. ATP binds to myosin and causes dissociation of actomyosin. Myosin has ATPase activity, so the ATP is broken down to ADP and phosphate. The energy released during the breakdown of ATP is used to move the cross bridges

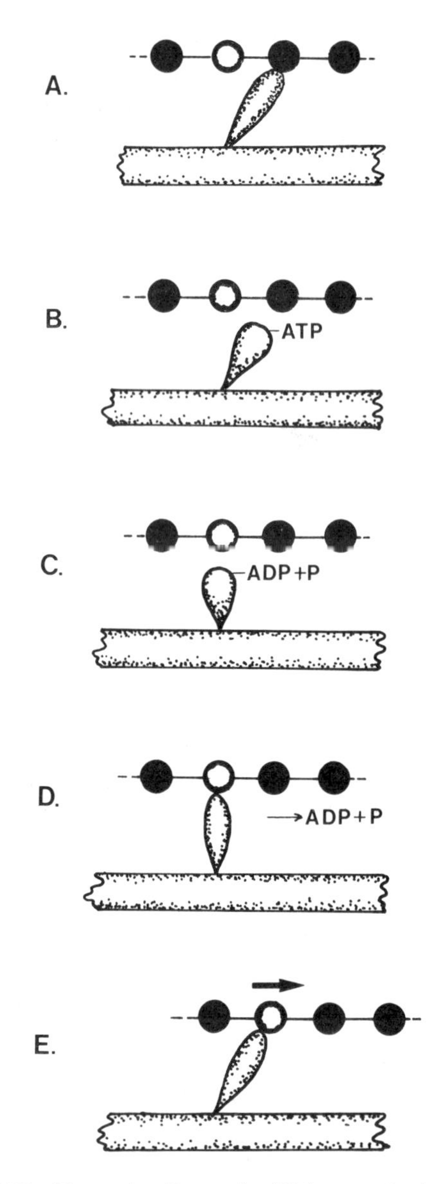

Fig. 1.53 The contraction cycle. (A) As a result of calcium binding to the thin myofilament, contact is made between myosin and actin; (B) ATP binds to the myosin head and causes dissociation of the actomyosin; (C) ATP is broken down to ADP and phosphate and the energy released straightens the cross bridge; (D) Myosin now makes contact with a different actin, ADP and P dissociate from the myosin; (E) Myosin head returns to its 'bent' configuration and drives the thin filament towards the M-band.

to a straight configuration. The cross bridges are now adjacent to actin molecules further along the filament, they make contact and this promotes the dissociation of ADP and phosphate from the myosin heads. The cross bridges then return to the unstrained shape and thereby push the thin filaments away from the Z lines, towards the M band. Each stroke moves the thin filament along about 12 nm which is equivalent to two actin diameters.

The precision of neural control is dependent on its being of short duration, so that a single nerve impulse does not result in an inappropriately long muscle contraction. Two mechanisms serve to keep the response to each impulse short. Acetylcholinesterase rapidly destroys acetylcholine in the synaptic cleft, and calcium ions are 'pumped' back into the sarcoplasmic reticulum. Muscle contraction, therefore, only takes place as long as action potentials continue to arrive at the motor end plate and maintain a raised intracellular calcium concentration.

All skeletal muscle fibres are not identical and they may be classified according to their 'twitch' characteristics into slow or fast fibres. These in fact are two extremes of a range of fibre types. Slow, red fibres contain the protein myoglobin, which provides an intracellular oxygen store, have a high density of mitochondria and are therefore suited for oxidative metabolism and sustained muscular contraction. Fast, white fibres are larger than the red ones, have fewer mitochondria, but are rich in glycogen and glycolytic enzymes. They are capable of anaerobic glycolysis (breakdown of glycogen to pyruvate and lactate) which gives a fast, but small supply of ATP. Whole muscles contain a mixture of these fibre types, but postural muscles which exhibit slow, tonic contraction have a higher proportion of red fibres, whilst muscles which bring about rapid movements of parts of the body have a higher proportion of white fibres.

Muscle contraction may be either isotonic, in which case external work is done as the fibres shorten under a constant load (e.g. lifting a weight or pedalling a bicycle) or isometric, in which case no external work is done and the muscle is generating tension at constant length (e.g. pressing two hands together).

The strength of muscle contraction may be varied in two ways, by summation of action potentials and by the recruitment of additional motor units. When an action potential causes a motor unit to twitch, the duration of the action potential (about 1 msec) is much shorter than the duration of the twitch (25 to 75 msec). It is, therefore, possible for a second action potential to arrive at the myoneural junction before the contraction caused by the previous action potential is over. In this case, the contraction initiated by the second action potential is added to that initiated by the first and therefore the tension developed is higher. This effect is called summation (see Fig. 1.54). If the frequency of stimulation is increased to about 50/sec, then the muscle tension does not fall between successive stimuli, so that a sustained contraction known as tetanus is achieved.

The overall tension developed by a whole muscle depends on the number of fibres which are contracting. The greater the number of motor units which are stimulated simultaneously then the greater the tension will be.

Muscle cells have only a limited capacity for repair. If cells are completely destroyed, then new muscle cells will not be formed. If a damaged cell has an intact sarcoplasmic reticulum, then repair can be effected. The nerve supply

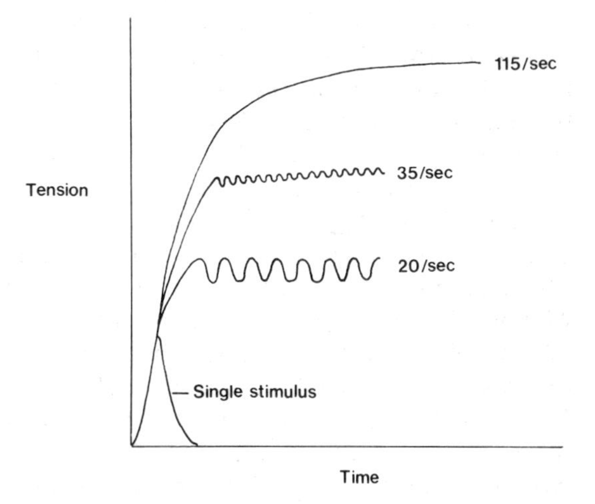

Fig. 1.54 Summation of action potentials, showing the effect of a single impulse and increasing the frequency of impulses arriving at the muscle. The very high frequency of 115/sec induces a sustained contraction known as tetanus. (Modified from Astrand & Rodahl, 1970.)

to muscle is essential for its maintenance and denervated muscle atrophies.

It is frequently observed that repeated use of particular muscles (e.g. weight training) causes them to increase in size. This is due to hypertrophy of the existing cells and not to an increase in number.

Cardiac muscle

Cardiac muscle fibres are faintly striated, branched cylinders, with periodic, darkly staining transverse lines called intercalated discs running across them. The fibres consist of many individual cells about 80 μm in length and 15–20 μm diameter, arranged end to end. Each cell has one or more centrally placed nuclei, and one or two branches, so that adjacent fibres are joined (see Fig. 1.42).

The cells are surrounded by connective tissue endomysium (like the skeletal muscle cells), which contains lymphatic as well as blood capillaries, and autonomic nerves. There are extensive nerve plexuses around the sinoatrial node and the conducting system of the heart (see Ch. 4), but the cardiac muscle is generally supplied with parasympathetic and sympathetic fibres, which have no specialized junctions with the muscle cells. The fibres are grouped in bundles of between a few hundred and a few thousand, surrounded by perimysium.

The striations of cardiac muscle arise from the same arrangement of myofilaments as in skeletal muscle. There are, however, several intracellular features which are different from skeletal muscle cells.

Cardiac muscle cells have abundant sarcoplasm and larger, more numerous mitochondria (usually about the length of one sarcomere, but sometimes more). The myofilaments are not grouped in regular myofibrils, rather the filaments form a continuous mass, incompletely subdivided into fibrils. The T-tubules are larger, and cross the cells at the level of the Z band, although in the smaller cells found in the atria of the heart, the T-tubules are frequently absent. The longitudinal sarcoplasmic reticulum is less extensive than in skeletal muscle, and instead of terminal cisternae, one of the longitudinal tubules next to each T-tubule has a smaller expansion, so that the tubules form a dyad instead of a triad.

The cardiac cells have stores of glycogen and lipid and oxygen attached to myoglobin. Oxidative enzymes are abundant. The intercalated discs have various types of attachments joining adjacent cells together, maculae adhaerentes (desmosomes), gap junctions or nexuses (like those in smooth muscle) and atypical zonulae adhaerentes called fasciae adhaerentes. The cytoplasm immediately either side of the cell membranes serves as the attachment point for the thin myofilaments (Fig. 1.55).

Although cardiac muscle cells are myogenic, the rate of contraction is dominated by the frequency of impulses arising from the sinoatrial node or pacemaker. The impulses are transmitted along the cell membranes of the branching fibres and through the gap junctions very rapidly, so that the interconnected cells function as a unit or syncytium. There are in fact two syncytia in the heart, the atrial and ventricular syncytia, separated by the fibrous connective tissue surrounding the heart valves.

Repair of cardiac muscle is effected by fibroblasts which form scar tissue, rather than by regeneration of muscle cells.

As in skeletal muscle, hypertrophy of heart muscle involves an increase in the sizes of individual fibres rather than an increase in the number of cells.

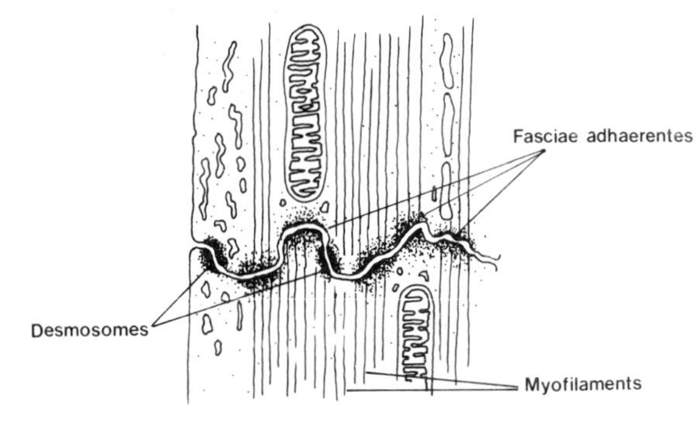

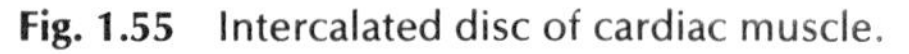
Fig. 1.55 Intercalated disc of cardiac muscle.

NERVOUS TISSUE

Ten per cent of the total number of cells comprising nervous tissue are neurones (although they occupy 50% of the volume) which are cells capable of receiving, generating and transmitting impulses at great speed from one part of the body to another. The remaining 90% of the cells are neuroglia, which have a secondary, supportive role to the neurones.

The nervous system can be divided into two primary sections, the central nervous system (CNS) which comprises the brain and spinal cord, and the peripheral nerves which connect the CNS with the tissues.

Classification of neurones

The neurones which are found in the peripheral nerves are either sensory (afferent), in which case they transmit impulses from the tissues to the CNS; or motor (efferent), in which case they transmit impulses from the CNS to the tissues (effectors).

Motor neurones are further subdivided according to the type of effector innervated. Somatic motor neurones innervate skeletal muscle, whilst autonomic motor neurones innervate smooth muscle, cardiac muscle and glandular tissue. Furthermore there are two types of autonomic neurones, preganglionic and post-ganglionic, because the autonomic pathways have two neurones between the CNS and the effector, in contrast to the sensory and somatic motor pathways which only have one (see Fig. 1.56).

There are many other types of neurones which are confined exclusively to the CNS and these may be collectively classified as interneurones. 99% of all neurones fall into this category.

Neurones consist of a cell body, soma, or perikaryon, which is the nucleus and surrounding cytoplasm, and branched processes or neurites. The processes are of two kinds, one or more dendrites, which conduct impulses towards the cell body, and a single axon, which conducts impulses away from the cell body.

Neurones may be classified according to the number of processes joining the cell body (i.e. the number of poles) as multipolar, bipolar

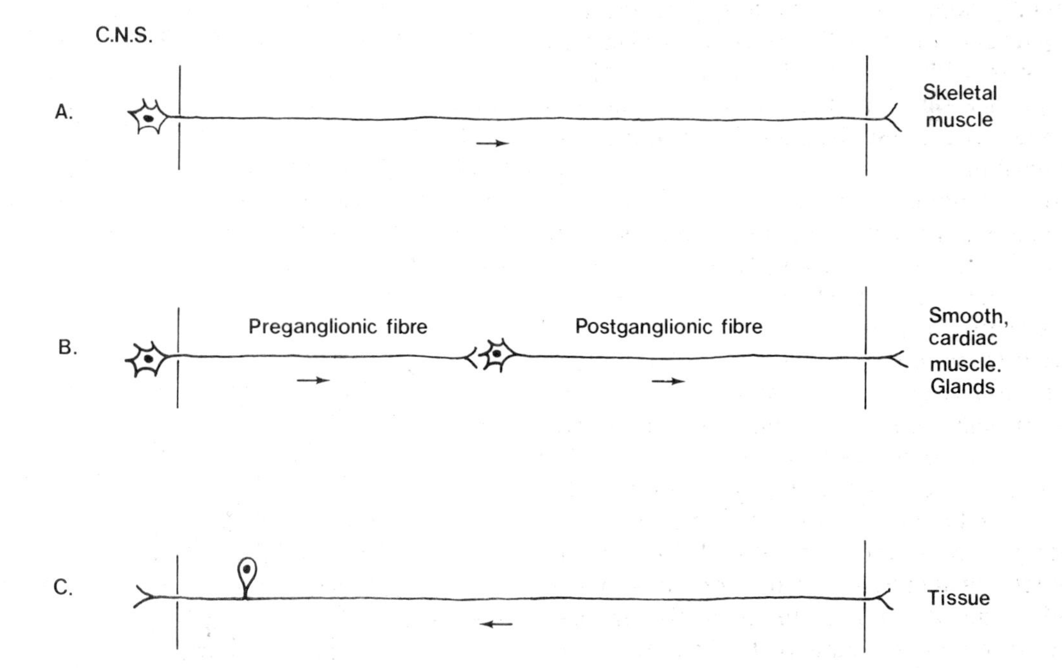

Fig. 1.56 Nerve pathways. (A) Somatic motor pathway; (B) Autonomic pathway with two neurones; (C) Sensory pathway.

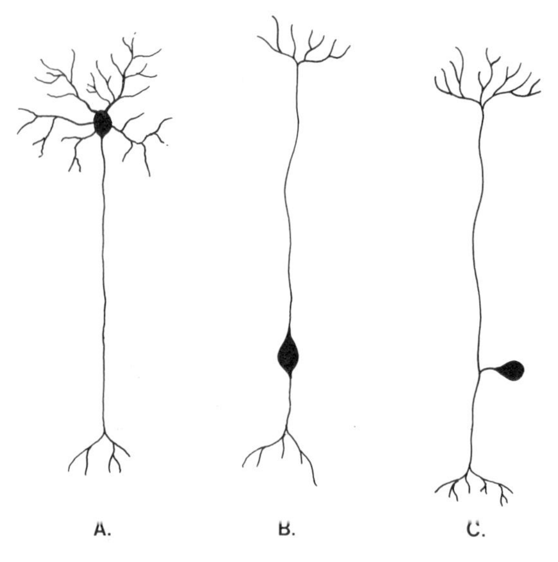

Fig. 1.57 (A) Multipolar neurone; (B) Bipolar neurone; (C) Pseudounipolar neurone.

and unipolar (pseudounipolar) (see Fig. 1.57). Multipolar cells have many dendrites and one axon connected to the cell body, which may have a variety of shapes. Cells in this category include somatic motor neurones, the Purkinje cells in the cerebellar cortices and the pyramidal cells in the motor cortices of the cerebrum.

Bipolar cells have a spindle-shaped cell body with one axon and one dendrite. Most sensory cells pass through this stage before becoming pseudounipolar. Cells which remain bipolar include those in the retina, in the sensory ganglia of the cochlea and vestibular apparatus and cells in the olfactory epithelium.

Most sensory neurones are classified as pseudounipolar, because after the bipolar stage, the two processes move together and combine, so that the cell has a single process attached to a pear-shaped cell body. The process divides into a dendrite and an axon.

Another way of classifying neurones is by the length of the axon. Short ones are classified as Golgi type II, and these are found in the retina and cerebellar and cerebral cortices, whereas neurones with a long axon are Golgi type I neurones and they are found in peripheral nerves and fibre tracts in the CNS.

Fine structure of neurones

The cell body is concerned with the maintenance and repair of the cell and has the appropriate organelles present including mitochondria, Golgi apparatus, lysosomes and yellow lipofuscin granules (which are probably a product of lysosomal activity, and they accumulate with age), microfilaments, Nissl bodies (rough e.r.) and centrioles.

Dendrites are the receptive surface of the neurone. They are branched processes. The branches leave at an acute angle, and the dendrites are covered in thorn-like spines or gemmules (see Fig. 1.58). In sensory neurones the dendritic tree is remote from the cell body and the process that connects this with the cell body is structurally an axon, despite the fact that it transmits impulses towards the cell

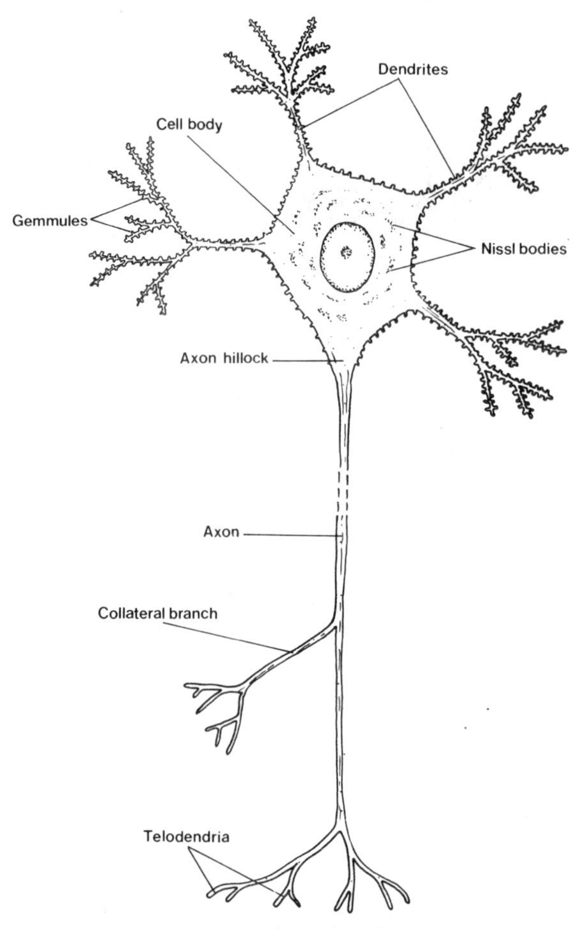

Fig. 1.58 Structure of a multipolar neurone.

body. Normally, though, dendrites connect directly with the cell body.

Dendrites contain microfilaments and microtubules, Nissl bodies and rod-shaped or spherical mitochondria.

The axon usually arises from a conical extension of the cell body, the axon hillock. The first section is bare and is known as the initial segment of the axon. Branches along the axon are known as collaterals, and they leave at right angles. The terminal branches of the axon are known as telodendria.

Axons contain long, slender mitochondria, which are particularly abundant in the terminals. Thick and thin microfilaments run parallel to the long axis of the axon, although they interlace with each other. There are few microtubules present compared with dendrites. There is no endoplasmic reticulum present. Proteins are transported either by bulk flow, which involves axoplasmic streaming at the rate of 1–3 mm per day; or by rapid flow, which involves the transport of selected proteins by the microtubules at about 100 mm per day, although in the hypothalamo-hypophyseal tract, speeds of 2800 mm per day have been recorded. Axons, in contrast to dendrites, may be surrounded by a myelin sheath.

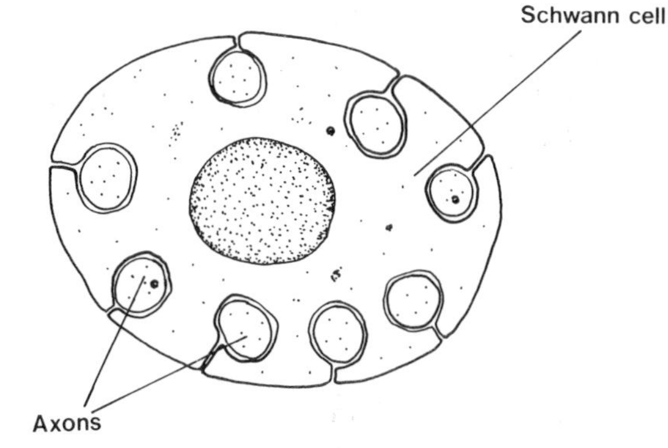

Fig. 1.59 An unmyelinated peripheral nerve fibre consisting of several axons embedded in a single Schwann cell.

Peripheral nerve fibres

The white colour of cranial and spinal nerves is due to the presence of myelin sheaths surrounding the axons of some of the neurones. An axon together with its myelin sheath is the nerve fibre. In peripheral nerves, axons are associated with Schwann cells (similar to neuroglia). Several unmyelinated fibres share a single Schwann cell (see Fig. 1.59) whereas myelinated fibres have a much more intimate association with Schwann cell (see Fig. 1.60).

The process of myelination starts before birth and is completed after birth at various times in different sites.

The Schwann cells develop alongside the axons and each section of myelin is formed from a single Schwann cell wrapping itself around the axon up to as many as 50 times forming layer upon layer of Schwann cell membrane. The myelin sheath so formed

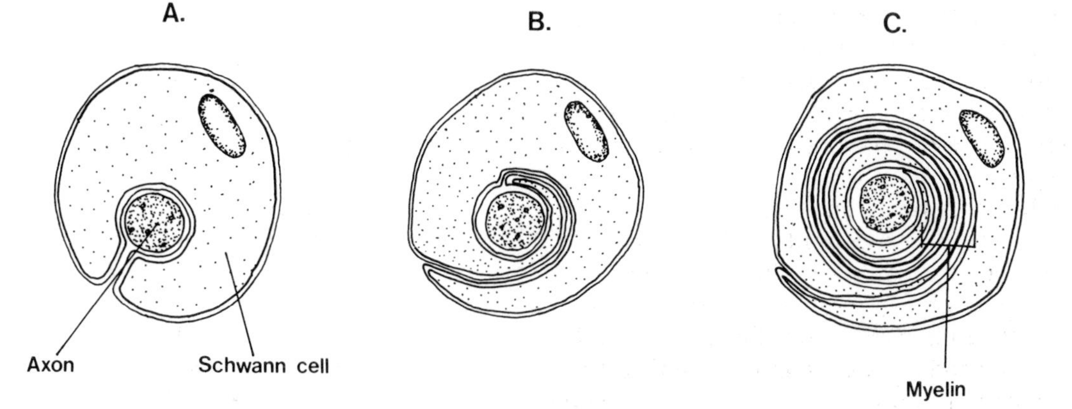

Fig. 1.60 Formation of a myelin sheath in a peripheral nerve. (A) Schwann cell wrapped around an axon; (B) Schwann cell overlaps itself around the axon; (C) Myelin formed by many layers of Schwann cell membrane.

appears striated, the major dense lines are due to the apposed inner surfaces of the membrane and these alternate with the dark intraperiod lines formed from the outer surfaces (see Fig. 1.60).

The section of myelin formed from a single Schwann cell is known as an internode, and the junction between adjacent cells where the axon is partially uncovered, is called the node of Ranvier. The nuclei of the Schwann cells are found peripherally near the outer edge of the myelin. Within each internode there are several oblique structures called clefts of Schmidt-Lantermann. These may act as channels to convey nutrients and metabolites to and from the axon (see Fig. 1.61).

The white matter in the CNS is due to myelin, but in this site, the myelin sheath is formed from a type of neuroglial cell, the oligodendrocyte (see below). The clefts of Schmidt-Lantermann are absent from myelin in the CNS.

Peripheral nerve fibres are classified according to their conduction velocities and fibre diameters. Erlanger and Gasser in the 1930s divided fibres into groups A, B or C. Group A are the large diameter myelinated fibres (up to 22 μm), with velocities between 30 and 120 m/sec; group B are intermediate diameter, with velocities between 4 and 30 m/sec; group C are small diameter, unmyeliated fibres (down to 0.1 μm), with velocities of 0.5–4 m/sec. Group A fibres may be further subdivided into α β and γ, again according to conduction speed and fibre diameter.

Another classification system for sensory neurones was introduced in the 1940s by Lloyd. This system has four groups, I and II (equivalent to group A), group III (equivalent to group B) and group IV (equivalent to group C).

Nerves

A nerve (such as the vagus) consists of bundles of nerve fibres, both afferent and efferent, in which case it is described as being a 'mixed' nerve. All spinal nerves are mixed, but some cranial nerves, such as the vestibulocochlear nerves are entirely sensory and others such as the oculomotor are entirely motor.

The fibres within a nerve are grouped in bundles or fasciculi, containing a variable

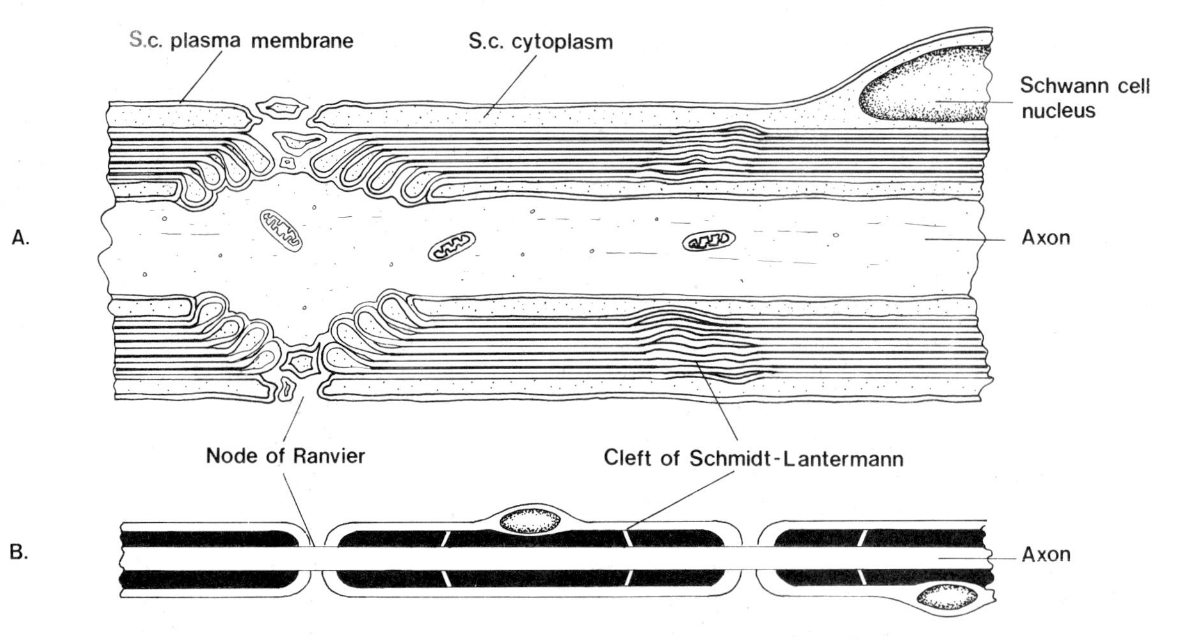

Fig. 1.61 Longitudinal section through a myelinated axon as it would appear in (A) An electron micrograph; (B) A light micrograph.

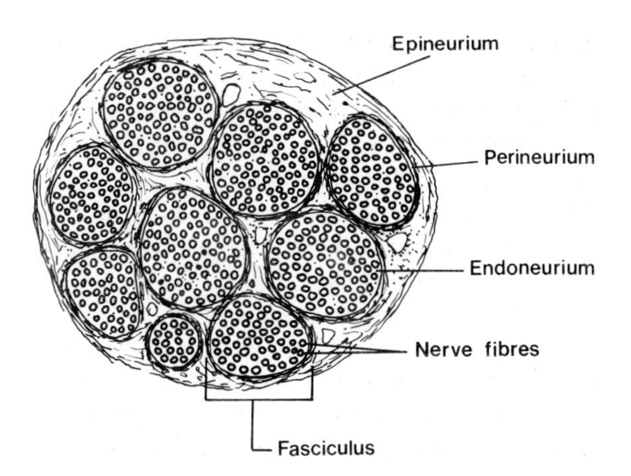

Fig. 1.62 Arrangement of connective tissue and nerve fibres within a nerve.

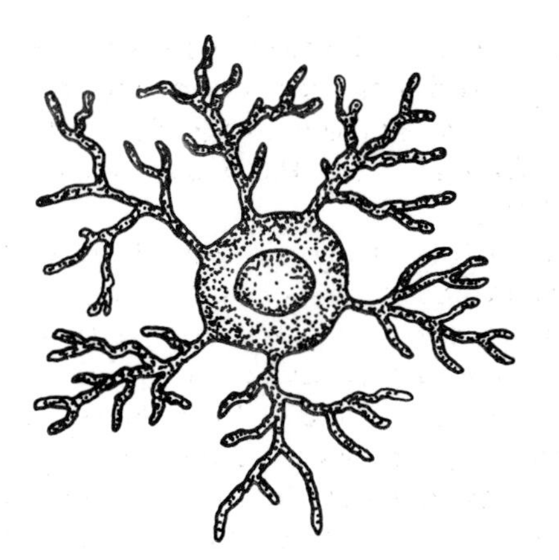

Fig. 1.63 Astrocyte.

number of fibres (from a few to several hundred) (see Fig. 1.62). The arrangement of connective tissue is analagous to that in skeletal muscle. Epineurium, a dense, irregular connective tissue sheath, surrounds the whole nerve; perineurium, less fibrous connective tissue, surrounds each fasciculus, and endoneurium, which is a loose, delicate connective tissue, surrounds each nerve fibre. The connective tissue is the distribution route for blood and lymphatic vessels within the nerve.

Neuroglia

The principal types of neuroglia can be subdivided into macroglia and microglia. The latter are found between neurones or outside capillaries in the CNS and they are now thought to be the same as connective tissue macrophages (that is they start life as monocytes in the blood). The macroglia comprise astrocytes, oligodendrocytes and glioblasts.

Astrocytes are cells with small cell bodies and cytoplasmic processes, like dendrites, which have leaf-like structures on them (see Fig. 1.63). They are found in contact with blood vessels, the ependyma (lining the ventricles of the brain) and in the grey and white matter generally. They divide if brain tissue is damaged and phagocytose cell debris.

Oligodendrocytes are so named because they have a few cell processes (see Fig. 1.64). They produce the myelin sheaths within the CNS, and are therefore found in the white matter. Unlike Schwann cells, one oligodendrocyte can enclose several adjacent axons with separate myelin sheaths.

The glioblasts are found particularly beneath the ependyma, and they are capable of differentiating into macroglial cell types.

Overall, the glial cells offer mechanical support and insulation to the neurones, they act as phagocytes and can form scar tissue. They play a regulatory role in neurone activity by affecting the ionic environment and also by taking up neurotransmitters released from synapses. They may play a role in providing nutrients to the neurones.

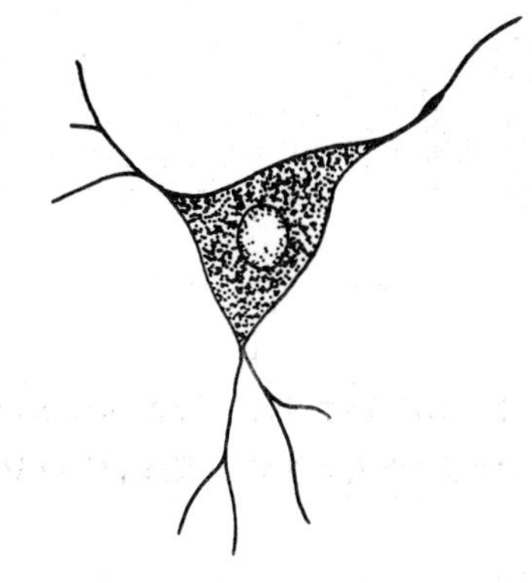

Fig 1.64 Oligodendrocyte.

Membrane potential

If a microelectrode is inserted into a cell, connected to a voltmeter and to a second electrode placed outside the cell, then a voltage difference across the membrane can be measured (see Fig. 1.65). This voltage or potential difference (the measure of the potential of separated charges to do work) is very small, varying between 5 and 100 millivolts (mV) in different cells, with the inside of cells normally having a net negative charge with respect to the outside.

The membrane potential at a particular site on a neurone varies according to whether the cell is in a resting state (in which case it has a resting potential) or whether it is in an active or excited state (action potential). In the latter the charge is reversed so that the inside of the cell becomes positive with respect to the outside.

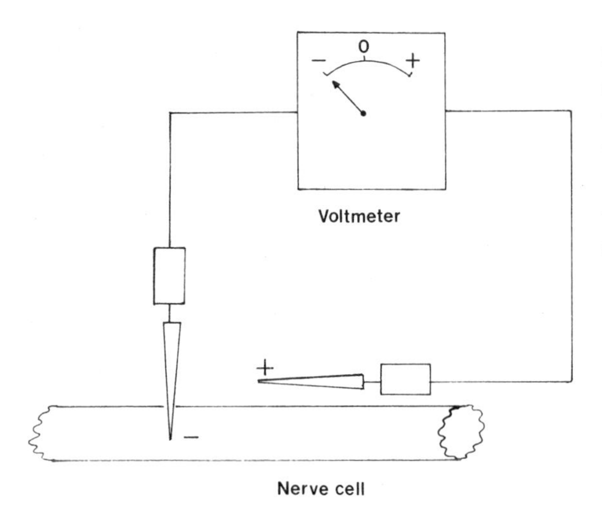

Fig. 1.65 Measurement of the potential difference across the membrane of a nerve cell.

Origin of the resting potential

Intracellular and extracellur fluids are very different in composition (see Ch. 2). Intracellular fluid has a relatively high concentration of potassium ions and a relatively low concentration of sodium ions, whereas extracellular fluid is rich is sodium ions and low in potassium ions.

The unequal distribution of these ions is dependent upon the presence of an active transport system within the cell membrane which uses ATPase as a carrier to transport potassium ions into the cell and sodium ions out of the cell (the sodium-potassium pump). The exchange of ions is thought to be unequal so that, generally, three sodium ions are exchanged for only two potassium ions. Such a pump is described as being electrogenic because it results in the separation of charge across the cell membrane, in this case a net negative charge inside the cell.

As concentration differences are established between the inside and outside of the cell, then the gradient for each ion acts as a force promoting diffusion of ions across the membrane. In the case of potassium ions, the concentration force acts outwards across the membrane, whereas for sodium ions the concentration force acts inwards (see Fig. 1.66).

The extent to which ions diffuse across the membrane is limited by the membrane permeability. The permeability of the membrane to potassium ions is 50 to 100 times higher than the permeability to sodium ions.

Potassium ions diffuse out of the cell down their concentration gradient, but movement is opposed by the build up of positive charge

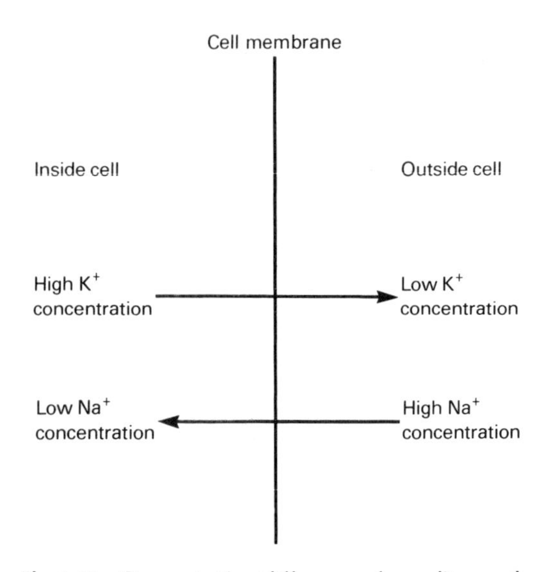

Fig. 1.66 Concentration differences for sodium and potassium across the nerve cell membrane.

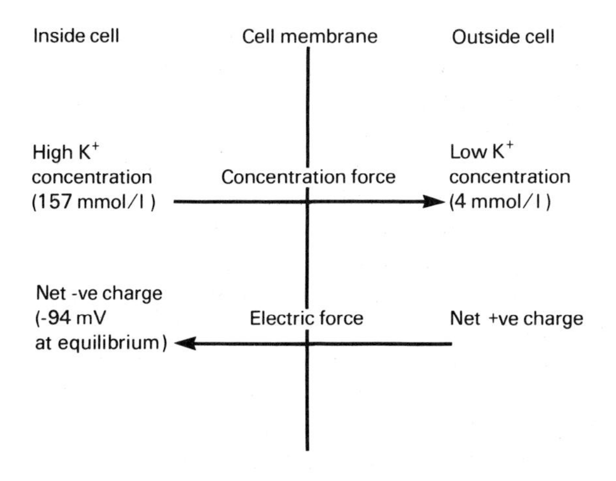

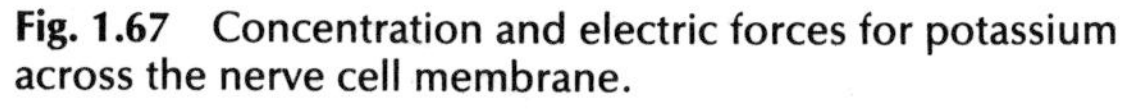

Fig. 1.67 Concentration and electric forces for potassium across the nerve cell membrane.

outside the membrane which will act as a repellent force.

If potassium was the only cation present, then an equilibrium would be established when the concentration force driving potassium out of the cell is equalled by the electric force repelling such movement. At this point there would be no net movement of potassium ions and the membrane potential would be −94 mV. This is known as the equilibrium or Nernst potential for potassium (see Fig. 1.67). For sodium ions alone, the equivalent potential would be +61 mV.

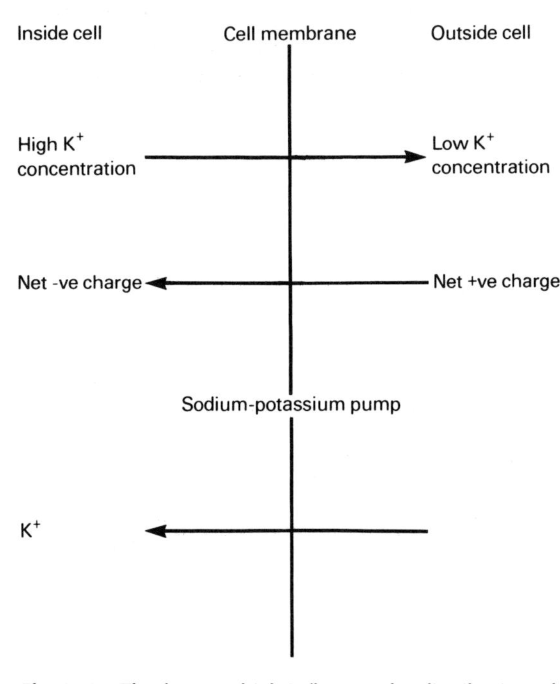

Fig. 1.68 The forces which influence the distribution of potassium across the nerve cell membrane.

In vivo, the equilibrium point for potassium is also influenced by the activity of the sodium-potassium pump, so that the two forces promoting inward movement of potassium ions are balanced by the single force for outward movement (see Fig. 1.68).

So the resting potential, which is between −70 and −90 mV, represents an equilibrium point where there is no net movement of potassium ions nor sodium ions, as the number of sodium ions diffusing in down the concentration gradient is balanced by the number of ions being pumped out.

Generation of the action potential

When a neurone is stimulated, the cell membrane becomes more permeable to sodium ions. For most neurones the stimulus will be chemical, a neurontransmitter released from a previous neurone or neurones in a nerve pathway. Sensory receptors, however, may be adapted to respond to other stimuli such as temperature, pressure, pain or light.

The action potential or nerve impulse is the change in membrane potential from its resting

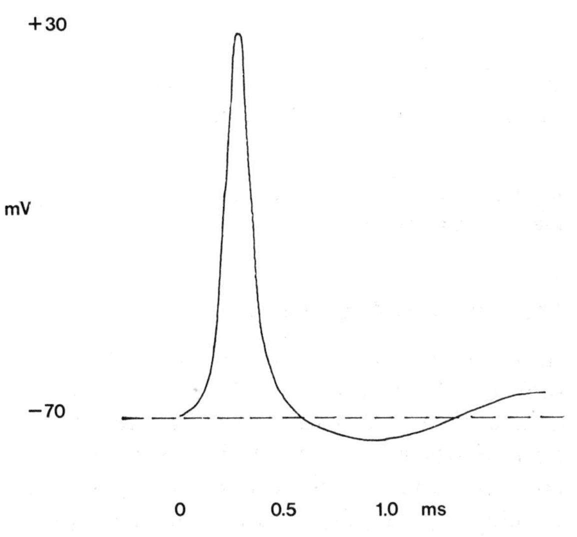

Fig. 1.69 The action potential.

value (around –70 mV) to its peak (around +30 mV) and back again (see Fig. 1.69).

When the neurone is stimulated and becomes more permeable to Na^+, these ions diffuse into the cell down their electrochemical gradient and thereby reduce the resting potential towards zero (depolarization). The equilibrium potential for Na^+ alone is +61 mV and so the ions continue to diffuse in, changing the membrane potential to positive (reversal potential). The reversal potential does not however reach 61 mV because the sodium permeability is not high enough. The increased permeability only lasts for a fraction of a millisecond and is then suddenly reduced again (see Fig. 1.70). At this point the cell membrane permeability to K^+ increases and the downward phase of the action potential is caused by a loss of K^+ from the cell by diffusion down its concentration gradient, the electrical gradient and enhanced by the increased permeability. This phase is called repolarization.

The membrane potential actually becomes more negative than the resting value (hyperpolarization) because the permeability to potassium remains higher than normal. The original ionic composition is finally restored by the sodium-potassium pump.

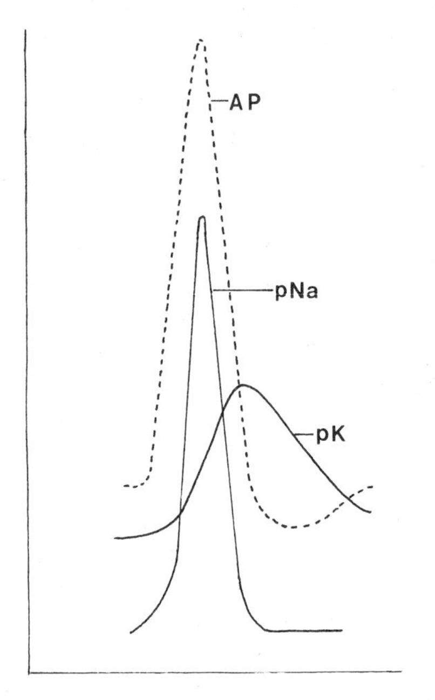

Fig. 1.70 Diagram to show the changes in membrane permeability to sodium ions (pNa) and potassium ions (pK) during the action potential (AP).

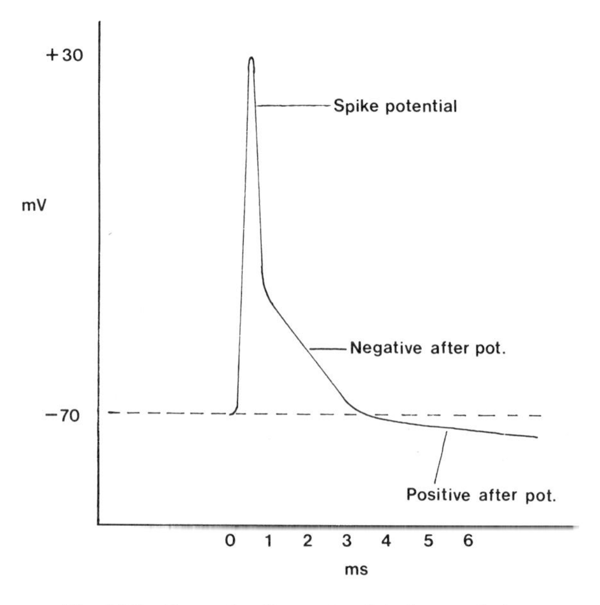

Fig. 1.71 Type of action potential observed after generating many such potentials.

The sharp change in membrane potential is called the spike potential and in type A fibres, this lasts for about 0.4 msec. Following the spike potential, several msec may elapse before the potential returns to its resting value. This effect is seen particularly after a series of action potentials, and is thought to be due to a build up of K^+ outside the membrane which reduces the rate of diffusion out of the cell. This phase of the action potential is called the negative after-potential (see Fig. 1.71). The subsequent hyperpolarization phase or positive after-potential lasts between 50 msec and many seconds.

Mechanism of permeability changes

Sodium and potassium ions are thought to diffuse through the cell membrane by means of 'pores' which are guarded by 'gates' (probably cations). The change in permeability is thought to be achieved by a variation in the degree to which the gates are opened. In the resting state then, the gates to sodium pores are opened a little, whereas the gates to potas-

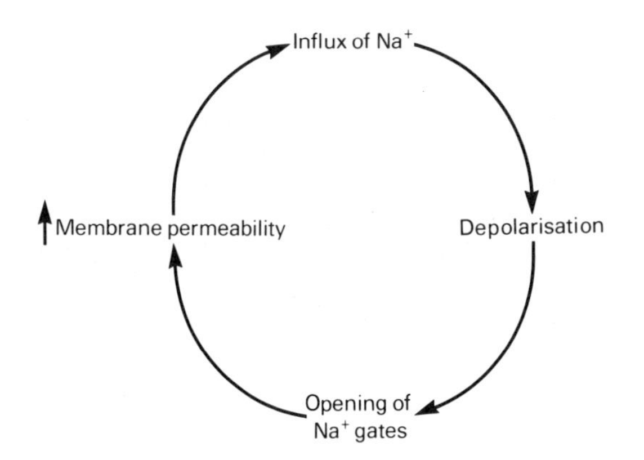

Fig. 1.72 Role of positive feedback in the generation of the action potential

sium pores are opened wider. In the active state, first the sodium and subsequently the potassium gates open wider.

Depolarization of the membrane has been shown experimentally to increase the permeability to sodium ions, which will cause further depolarization because of the sodium influx, which will then increase the permeability further (see Fig. 1.72). This relationship is an example of positive feedback and it occurs once the threshold potential is reached, causing the upward phase of the action potential.

Threshold

For each neurone, there is a particular membrane potential (some 5 to 15 mV less negative than the resting potential) which, if reached, will result in the generation of an action potential. The potential is known as the threshold for that particular neurone. The size (in millivolts) of the action potential is always constant for a particular neurone. Thus either an action potential is initiated by a threshold or suprathreshold stimulus, or it is not, by a subthreshold stimulus. This effect is known as the 'all-or-nothing' principle.

Refractory periods

It is not possible to restimulate a neurone while the action potential is in progress. This absolute refractory period lasts for about 0.4 msec in type A fibres. During the hyperpolarization phase the neurone can be restimulated if a suprathreshold stimulus is applied so that the threshold potential is reached. This is the relative refractory period and lasts 10–15 msec or more.

Graded potentials

In spite of the all-or-nothing principle described above, it is possible for small changes in membrane potential to occur, either depolarizations or hyperpolarizations, which vary in size according to the degree of stimulation. These occur at sensory receptors or at synapses. Such potentials are not action potentials because they vary in size and they do not reach the threshold potential of the neurone.

In the case of sensory receptor activation, the stimulus (mechanical deformation, heat, etc) causes an increase in cell membrane permeability with the result that there is a net influx of sodium ions down the electrochemical gradient and the membrane is depolarized. This is called a generator potential and its size varies according to the strength of the stimulus.

Graded potentials of this type can be transmitted by local current flow although the conduction is decremental and therefore will die out after a few millimetres.

If the generator potential reaches threshold, then action potentials will be generated instead. The lowest threshold of a neurone is found at the beginning of the axon (the initial segment, or the first node of a myelinated axon), so that is where the action potential first appears. Further increases in stimulation (suprathreshold stimuli) cause an increase in the frequency of action potentials (but not an increase in their size). In addition, with increasing levels of stimulation, adjacent neurones will be stimulated. Thus the intensity of a stimulus is conveyed to the central nervous system by the frequency of impulses travelling along a neurone, and by the number of neurones firing.

Propagation of the action potential

In unmyelinated neurones the action potential is transmitted to adjacent sections of the cell membrane by local current flow. The active region of the nerve cell has the opposite charge inside and outside the membrane compared with adjacent areas (see Fig. 1.73). Ions are attracted to an area of opposite charge. The positive ions on the inside of the cell therefore diffuse to the adjacent section which has a resting negative charge and depolarizes the membrane to the threshold value. The positive feedback cycle between depolarization and sodium permeability is initiated and an action potential of the same value is thereby generated. The conduction of action potentials, therefore, is not decremental in contrast to the conduction of graded potentials, and can therefore cover relatively long distances.

If an isolated neurone is experimentally stimulated midway along its axon, then action potentials will be propagated towards each end. Similarly, action potentials are propagated in vivo along the sarcolemma of muscle cells from the central motor end-plate. In neurones in vivo, however, action potentials travel in one direction only, along dendrites towards the cell body, and along the axon away from the cell body. This is because the cell is stimulated at the synaptic connections around the cell body and dendrites (i.e. at one end of the cell) and the action potentials then travel from the beginning of the axon to its terminals.

The speed of conduction along axons depends upon their diameter. The larger the diameter the lower the resistance to ion flow and therefore the faster the conduction velocity. The major influence on conduction velocity, however, is myelination. The myelin sheath acts as an insulator; it offers some 5000 fold increase in resistance to ion flow. The nodes of Ranvier, which interrupt the myelin sheath, on the other hand, offer little resistance to ion flow.

Action potentials are only generated at the nodes of Ranvier in myelinated neurones and therefore the speed of transmission is greatly increased, since only a fraction of the total membrane is involved. Conduction in myelinated axons is described as saltatory because the impulse appears to jump from node to node.

The presence of a net positive charge inside the axon at an active node repels adjacent positive ions in the axoplasm, which in turn

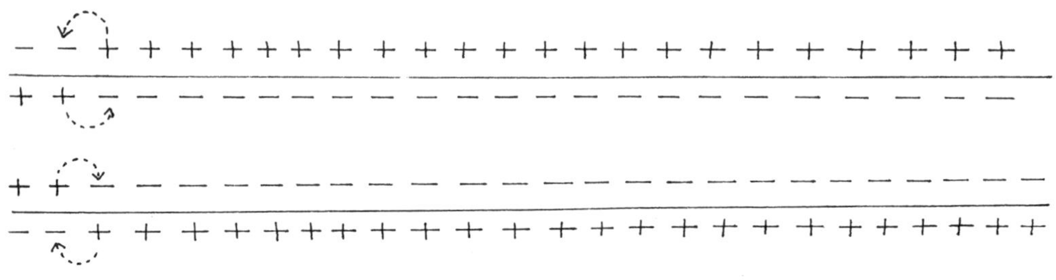

Fig. 1.73 Local current flow in a non-myelinated fibre.

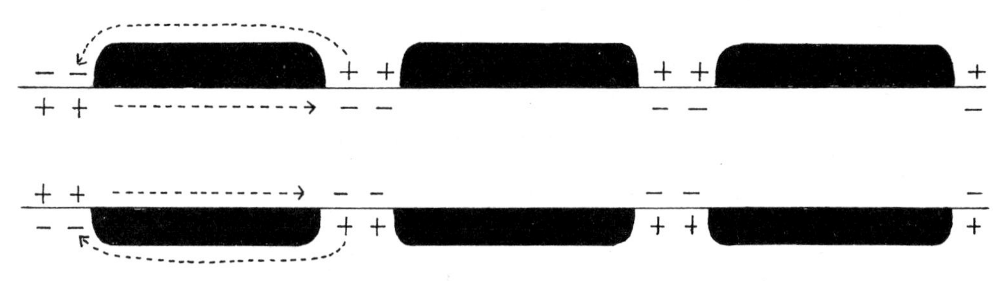

Fig. 1.74 Saltatory conduction in a myelinated fibre.

repel those adjacent to them. Effectively, a column of positive ions drifts towards the next node, so that the ions that depolarize it are in fact local ones. Ions do not jump from node to node! The mechanism can be likened to a tube filled with a column of ball bearings. The net positive charge at one node is like an extra ball bearing being pushed into one end of the tube causing a ball bearing to fall out of the other end, or in this case depolarizing the next node (see Fig. 1.74).

Synapses

The junctions between two or more neurones in a pathway are known as synapses. The commonest synapse is between an axon and a dendrite or soma, in which the axon branch ending is a spherical structure known as a bouton. There may be one or a row of several boutons at each axon ending.

A synapse between an axon and a dendrite (either a spine or the flat surface) is known as an axodendritic synapse, whilst a synapse between an axon and the soma is termed an axosomatic synapse. Synapses are less commonly found between other parts of neurones (dendrodendritic, axoaxonic, somatodendritic, somatosomatic, dendroaxonic).

An axodendritic synapse consists of the bouton of the presynaptic cell, the synaptic cleft of 20 or 30 nm width, and the postsynaptic membrane (see Fig. 1.75).

The bouton contains zones of dense cytoplasm containing microtubules and filaments and a protein lattice next to the cleft, as well as mitochondria and numerous small synaptic vesicles containing neurotransmitter. Some filaments extend across the cleft. The postsynaptic side of the synapse has a filamentous mesh (the subsynaptic web) which appears dark. There are frequently glial cells wrapping round the synaptic junctions.

Neurochemical transmission

The influence of presynaptic over postsynaptic neurones is exerted chemically, by the release of neurotransmitters contained within the synaptic vesicles. These are released into the synaptic cleft and become physically attached to receptor sites on the postsynaptic cell membrane.

The mechanism of release of the neurotransmitters involves the action potentials arriving at the bouton which depolarize the membrane thereby causing an increase in permeability to calcium ions. These enter the presynaptic terminal and cause the migration of a number of synaptic vesicles to the presynaptic cell membrane. It is likely that the complex arrangement of microtubules, fila-

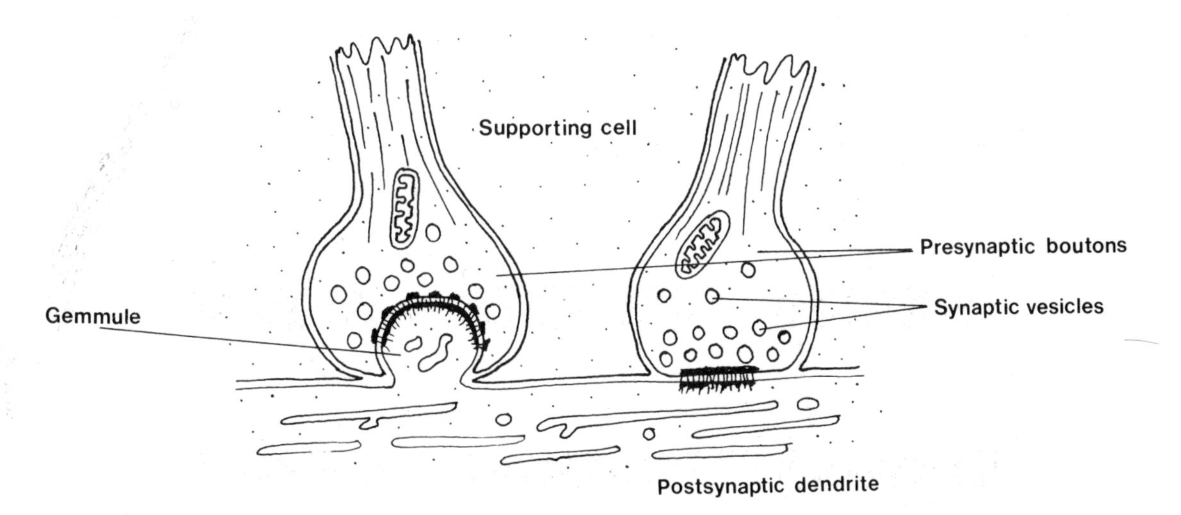

Fig. 1.75 Two presynaptic boutons associated with the postsynaptic membrane in an axodendritic synapse.

ments and protein lattice are concerned with guiding the vesicles to specific sites on the presynaptic cell membrane (synaptopores) where exocytosis occurs and the contents are liberated into the synaptic cleft. Diffusion across the cleft is followed by binding of the neurotransmitter to receptor sites on the postsynaptic cell membrane.

The effect of transmitters on the postsynaptic cell depends on both the nature of the neurotransmitter involved and the nature of the postsynaptic cell membrane. Each neurone has traditionally been thought to produce only one transmitter, (although this has more recently been questioned) and each synapse can be classified as either excitatory or inhibitory.

Excitatory synapses

At these junctions, the neurotransmitter-receptor combination causes a general increase in permeability to ions in the postsynaptic cell membrane. The principal consequence is that there is an influx of sodium ions (because the electrochemical gradient favours its movement), so that the postsynaptic cell membrane is depolarized. An individual synapse will depolarize the cell slightly, and the difference in voltage between the resting and depolarized levels is called the excitatory postsynaptic potential (EPSP). This is a type of graded potential.

In some cases, e.g. acetylcholine, the depolarisation is caused by a decrease in permeability of the postsynaptic neurone to potassium ions.

Inhibitory synapses

In this case, the neurotransmitter-receptor combination has the effect of reducing transmission between neurones. This is achieved by the postsynaptic cell membrane becoming hyperpolarized and therefore refractory to stimulation.

The permeability to smaller ions (potassium and chloride) is increased so that they diffuse down their concentration gradients and increase the negativity of the membrane potential. The difference between the hyperpolarized potential and the resting potential is known as the inhibitory postsynaptic potential (IPSP).

Integration of synapses

A given postsynaptic neurone may have thousands of synaptic connections and the net effect of these will depend on the number of active boutons and the frequency of their firing. If the excitatory effect dominates and there is net depolarization, then the neurone is said to be facilitated, as it is nearer to the threshold potential. If the inhibitory effect dominates, then the neurone becomes hyperpolarized.

For an impulse to be transmitted from one cell to another, there must be summation of the effects of individual action potentials. Summation may be spatial or temporal.

Spatial summation involves several excitatory synapses firing simultaneously, the EPSPs being transmitted to the initial segment of the axon, and if they collectively reach threshold, then the neurone will fire.

The duration of the EPSP is of the order of 15 msec. This is much longer than an action potential and if a bouton receives a volley of impulses within 15 msec, then the effects will be additive and the initial segment depolarized to the threshold value. This is temporal summation of action potentials.

Neurotransmitters

Within the central nervous system, the number of possible candidates for neurotransmitters is very high. The best known to date include acetylcholine, noradrenaline, serotonin, histamine, glutamic and aspartic acids, all regarded as excitatory transmitters. Substance P is the transmitter at sensory neurone endings, and γ-amino butyric acid (GABA), glycine and dopamine are likely inhibitory transmitters. The more recently discovered enkephalins and endorphins are also thought to act as inhibitory transmitters which bind to opiate receptors in the CNS thereby reducing pain. Some trans-

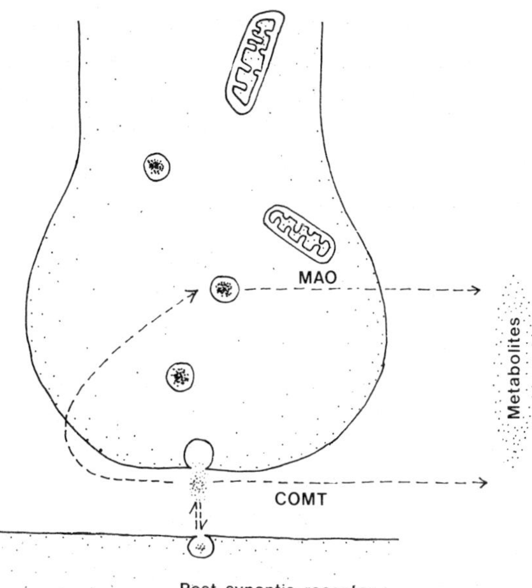

Fig. 1.76 Chemical activity at an adrenergic synapse. The transmitter (noradrenaline) is metabolized with the aid of the enzymes monomine oxidase (MAO) in the mitochondria and catechol-O-methyl transferase (COMT) in the synapse. Alternatively, noradrenaline may be taken back into the bouton.

mitters may be excitatory in some sites and inhibitory in others.

The duration of action of neurotransmitters is necessarily short, so that the action of neural pathways is short lived and therefore appropriate for a precise control system. The transmitters may be enzymatically destroyed in the synapse, as in the case of acetylcholine, where the enzyme cholinesterase is present in the postsynaptic cell membrane where it splits the molecule into choline and acetate. The choline can be actively transported into the nerve terminal and reutilized by the cell.

The catecholamines (noradrenaline, adrenaline and dopamine) are broken down in the synapse by the enzyme catechol-O-methyl transferase (COMT) or removed by reuptake into the axon terminal where the transmitter may be recycled or broken down by another enzyme MAO (monoamine oxidase) in the mitochondria (see Fig. 1.76).

Degeneration and regeneration of neurones

Neurones cannot divide and replace others that are destroyed or degenerate. Degeneration of neurones starts during fetal development and continues throughout life, so that in old age the number is reduced to 80% of the original.

If a neurone is damaged then its recovery is dependent on an intact cell body. If an axon is severed or crushed, then the section distal to the injury degenerates and the debris are removed by tissue macrophages. The section proximal to the damage may gradually recover over some weeks.

Within the CNS, the axon segment distal to the injury is not replaced. In peripheral nerves, however, contact may be re-established with the end-organ, provided that the endoneurium is intact.

Schwann cells divide and fill the space within the endoneurial tube and connect with the cut axon by cords. The proximal part of the axon develops a terminal swelling and several sprouts develop from this, but only one persists and grows into the endoneurial tube. After the new axon tip makes contact with the end-organ, the Schwann cells start to form myelin.

2

Body fluids, lymphatic system and blood cells

BODY FLUIDS

Water constitutes about 60% of the total body mass of an average young adult male. This water is distributed between two major compartments with about two-thirds being found inside the cells as intracellular fluid and the other one-third outside, as extracellular fluid.

Lean body tissue contains 71–72 ml of water per 100 g of tissue. Fatty tissue, on the other hand, contains much less water. Thus total water depends upon the proportion of adipose tissue which is present. The figure of 60% for the total water content of the body quoted above represents that of a thin young person.

Intracellular fluid provides a medium within which biochemical reactions can take place, while extracellular fluid supports the cells and allows transport of nutrients and waste molecules. Extracellular fluid can be subdivided into two components, interstitial fluid and plasma. Interstitial fluid occupies the spaces between cells, whereas plasma which comprises some 55% of the blood, circulates within the cardiovascular system.

Units of concentration and osmotic pressure

The concentration of a solute is expressed as moles (gram molecular mass) or millimoles (mmol) per unit volume (kilogram of water or litre of solution). In body fluids, however, it is common for equivalents to be used. One equivalent (Eq) is one mole of an ionized substance divided by its valency, so that the equivalent of Ca^{++} is 40/2 g, whereas the equivalent of Na^{+} is 23 g. By using these units, the chemical combining power of ions is taken into account.

The osmotic pressure of a solution is proportional to the number of particles present, irrespective of their molecular weight. The contribution made by protein molecules to the total osmotic pressure is small because, although they have high molecular weight, they are present in relatively small numbers.

Osmotic pressure may be measured in osmoles. One osmole is defined as the number of particles (molecules or ions) in one mole of undissociated solute. If one mole of glucose (180 g) is dissolved in a kilogram of water, then it will have an osmolality of one osmole per kilogram. If, however, the solute dissociates in water (e.g. sodium chloride), then if all the molecules dissociated into two ions, the osmolality of a solution containing one mole in a kilogram of water would be two osmoles per kilogram.

Note that the term osmolality has been introduced and that it has a different meaning from the term osmolarity. Osmolality is measured as osmoles per kilogram of water. Osmolarity on the other hand is expressed as osmoles per litre of solution. The latter is more convenient to use, and the quantitative difference is, in practice, very small.

The osmolality of both intracellular and extracellular fluids is about 300 milliosmoles per kilogram (the osmolarity is about 300 milliosmoles per litre (mosmol/l)).

COMPOSITION AND EXCHANGE OF CONSTITUENTS BETWEEN INTRACELLULAR, INTERSTITIAL AND EXTRACELLULAR FLUIDS

The composition of intracellular fluid varies in different cells, but there are consistent qualitative differences between it and extracellular fluid (see Table 2.1).

Intracellular fluid contains a large amount of protein and the negative charges on the protein molecules make a significant contribution

Table 2.1 Electrolyte composition of body fluids (mEq/l).

Substance	Intracellular	Interstitial	Serum
Cations			
Na^{+}	10	140	142
K^{+}	150	4	4
Ca^{++}	2	5	5
Mg^{++}	30	2	2
Anions			
Cl^{-}	10	110	100
HCO_3^{-}	10	28	27
$HPO_4^{=}/H_2PO_4^{-}$	100	2	2
$SO_4^{=}$	20	1	1
Protein	60	1	16

to the total amount of intracellular anion. Chloride and bicarbonate ions are present in small quantities compared with interstitial fluid, but the amount of phosphate is relatively high. Potassium is the most abundant cation in intracellular fluid, followed by magnesium ions, whereas sodium ions predominate in interstitial fluid.

Differences in composition between intracellular and interstitial fluid are maintained by active transport systems operating across the cell membranes.

Interstitial fluid is the largest component of extracellular fluid, that is about 11 litres in a 70 kg man. This fluid lies between cells, usually held in a gel of polymerized hyaluronic acid. Dissolved materials pass between the cells and through this gel by diffusion.

Plasma volume averages about three litres (with about two litres of cells, total blood volume is around five litres). Plasma is a watery fluid, rich in protein. It acts as a carrying and supporting medium for the cells; it also carries molecules which have been absorbed through the gut wall, waste products of metabolism and enzymes and hormones. Oxygen and carbon dioxide are also carried in plasma (details of the modes of transport of these gases are given in Ch. 5). Plasma is also the medium by which heat is transferred to the surface of the body.

Plasma and interstitial fluid have a similar ionic composition (see Table 2.1), but plasma has a much higher protein content. As the movement of constituents between plasma and interstitial fluid is by diffusion across the capillary walls, then it is understandable that the composition of the two fluids is similar. Protein is largely retained within the blood vessels because of its large molecular size.

Table 2.2 shows the composition of blood plasma. Protein is the single largest organic component and sodium and chloride are the most abundant mineral ions.

Twenty or more litres of fluid per day filters out from the blood capillaries into the tissue spaces and while about 90% of this fluid is normally reabsorbed back into the blood (see Ch. 4) some drains away through the lymphatic system. Lymph, therefore, is derived from blood and is also ultimately returned to it.

Table 2.2 Composition of blood plasma

Substance	Amount/100 ml
Total protein	6.0–8.0 g
Albumin	3.5–5.5 g
Globulin	1.5–3.0 g
Fibrinogen	0.2–0.6 g
Total lipid	450–850 mg
Glucose	60–120 mg
Na^+	310–335 mg
K^+	14–20 mg
Ca^{++}	8.5–10.5 mg
Fe^{++}	65–175 μg
Mg^{++}	1.8–3.0 mg
$C1^-$	340–375 mg
Inorganic phosphate	3.0–4.5 mg

Homeostasis

The mineral and organic composition of the plasma and interstitial fluid is maintained within very narrow limits by a variety of physiological mechanisms. This 'constant internal environment' is a necessary condition of life, an observation first made by Claude Bernard in 1857. 'It is the fixity of the internal environment which is the condition of free and independent life.' Later, in 1929, W.B. Cannon coined the term 'homeostasis' to denote the stable environment resulting from the many varied regulatory mechanisms involved. Physiology is largely concerned with how these mechanisms may operate within the body.

Plasma protein

Plasma proteins are classified as albumin and several types of globulin (α1, α2, β, (including fibrinogen) and γ). Proteins are large molecules and generally do not cross capillary walls in large numbers to enter the tissue spaces. They therefore exert an osmotic pressure across the capillary wall which draws water into the blood (see Ch. 4).

Plasma proteins also contribute to the buffering capacity of the blood since at the normal blood pH of 7.4 they are in an anionic form and are able to take up hydrogen ions (see Ch. 3).

Some protein molecules are involved in blood clotting while others (primarily the γ globulins) exhibit antibody activity (see below). Others (α and β globulins) act as carriers for minerals, hormones and many other organic substances.

The characteristic ability of plasma to clot on standing is dependent upon the presence of fibrinogen, calcium ions and a group of proteins known as the coagulation or clotting factors. The fluid component of blood after it has clotted is known as serum. The ability of plasma, and therefore whole blood, to clot is very important in that it prevents loss of fluid through the walls of damaged vessels (haemostasis).

Table 2.3 Blood clotting factors

I	Fibrinogen
II	Prothrombin
III	Tissue factor
IV	Calcium ion
V	Proaccelarin
VII	Proconvertin
VIII	AHF — Anti-haemophilic factor
IX	Christmas factor
X	Stuart-Prower factor
XI	PTA — Plasma thromboplastin antecedent
XII	Hageman factor
XIII	Fibrin stabilizing factor

Haemostasis

Haemostasis may be divided into three phases:

1. vasoconstriction
2. formation of a temporary platelet plug
3. formation of a clot.

When a small vessel becomes damaged in some way, there is initially a local constriction which reduces blood flow and therefore blood loss. Damage to the endothelium exposes the underlying collagen to which blood platelets become stuck. Platelets are small disc shaped elements, 2–4 μm diameter, numbering 250 000–500 000/mm^3. They are enucleate because they are fragments of large cells called megakaryocytes which are formed in the bone marrow (see Fig. 2.10 and the section on *leucocytes*). The platelets liberate serotonin and adenosine diphosphate (ADP) which attract other platelets, leading to the formation of a platelet plug which will temporarily prevent blood loss.

The temporary platelet plug will be converted into a clot by the deposition of fibrin which is formed from fibrinogen. Although the clotting mechanism involves a whole series of biochemical reactions, the conversion of fibrinogen to fibrin is comparatively simple. Fibrinogen is converted to a fibrin monomer which is then itself polymerized to form a loose gel of fibrin molecules. The latter is then enzymatically converted to a dense network which forms the basis of the clot (see Fig. 2.1).

The mechanism by which fibrinogen is converted to fibrin involves a 'cascade' of reactions which require a number of plasma factors (numbered I to XIII, listed in Table 2.3). Two pathways, the extrinsic and intrinsic, have been identified as leading to the formation of fibrin.

The initial reaction of the intrinsic pathway (Fig. 2.2) is the activation of factor XII by collagen exposed by damage to the wall of the vessel. This triggers a series of reactions culminating in the conversion of prothrombin to thrombin by activated factor X (Xa). This conversion also requires the presence of platelets, factor V and calcium ions.

The extrinsic pathway (Fig. 2.2) is initiated by a lipoprotein complex known as tissue factor which is released from damaged tissue.

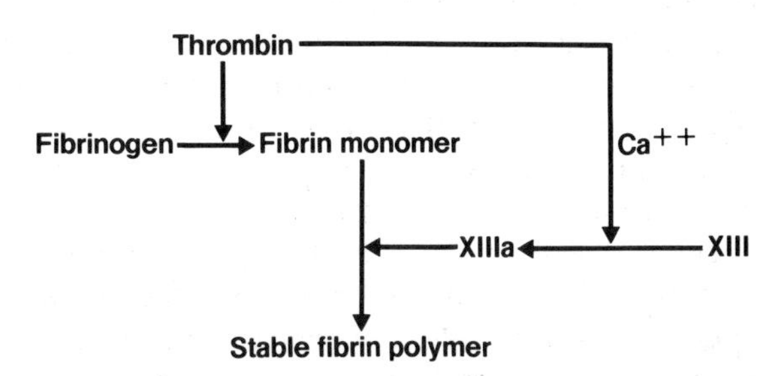

Fig. 2.1 The final stage of clot formation.

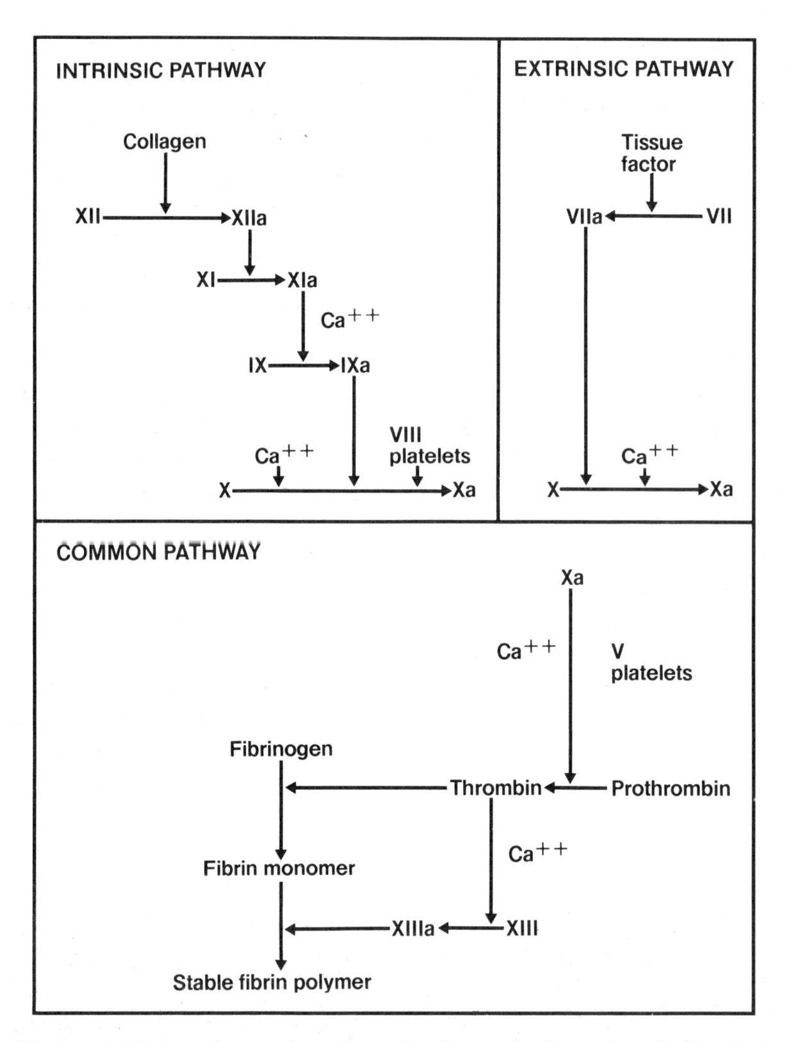

Fig. 2.2 Intrinsic and extrinsic pathways leading to the formation of a blood clot.

This activates factor VII which, in turn, brings about the activation of factor X and leads to the conversion of prothrombin to thrombin.

The formation of a clot is important in the prevention of blood loss from the body but the formation of clots within healthy blood vessels would be extremely dangerous. There are a number of reactions which prevent spontaneous clotting within vessels and destroy any clots which do form. Plasmin (fibrinolysin) brings about the breakdown of fibrin and fibrinogen with the production of degradation products which inhibit thrombin activity. Plasmin itself is formed from an inactive precursor in plasma called plasminogen, by the action of a tissue activator released from the blood vessel wall (see Fig. 2.3).

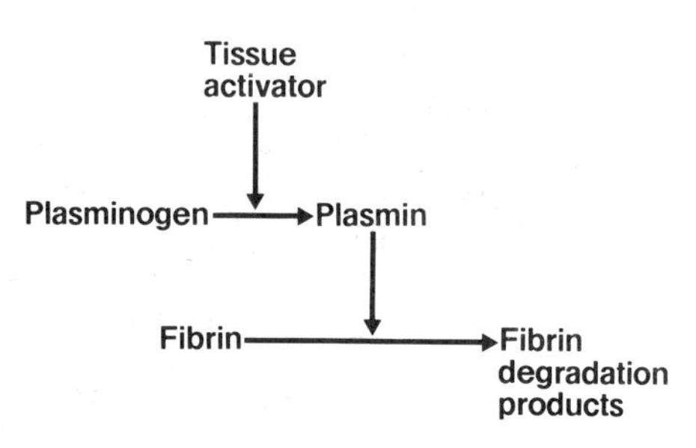

Fig. 2.3 Formation and action of plasmin.

Heparin, which is found in basophils and mast cells, combines with the plasma globulin, antithrombin III (ATIII) and this complex inhibits the actions of thrombin, Xa and IXa. The levels of activated clotting factors are also reduced as the liver removes them from the circulation.

LYMPHATIC SYSTEM

The lymphatic system consists of networks of minute lymph capillaries which originate as blind-ended vessels in many of the tissues of the body. These capillaries drain into larger vessels which ultimately carry lymph to the cardiovascular system. Small solid masses of lymphoid tissue known as lymph nodes are distributed throughout the system in such a manner that lymph will normally pass through one or more before reaching the blood stream. The system is considered to include aggregations of lymphoid tissue in the walls of the alimentary tract and in the spleen and thymus.

The lymphatic capillaries are blind-ended vessels with extremely thin walls consisting of a single layer of endothelium. The cells overlap at their edges and since they are not firmly joined together, molecules can pass between them. The arrangement of the cells creates a series of 'valves' which allow one-way flow of large molecules such as proteins and debris from the tissue spaces into the capillary (Fig. 2.4).

The capillaries give rise to larger thicker-walled vessels which eventually take the lymph to the blood stream. Lymph from the lower part of the body and upper left hand side of the body and head drains through the thoracic duct into the left brachiocephalic vein. Lymph from the upper right hand side of the body and head drains through the right lymph duct into the right brachiocephalic vein (see Fig. 2.5).

The larger lymphatics have valves in their walls rather like those found in the veins. These valves have an important function since

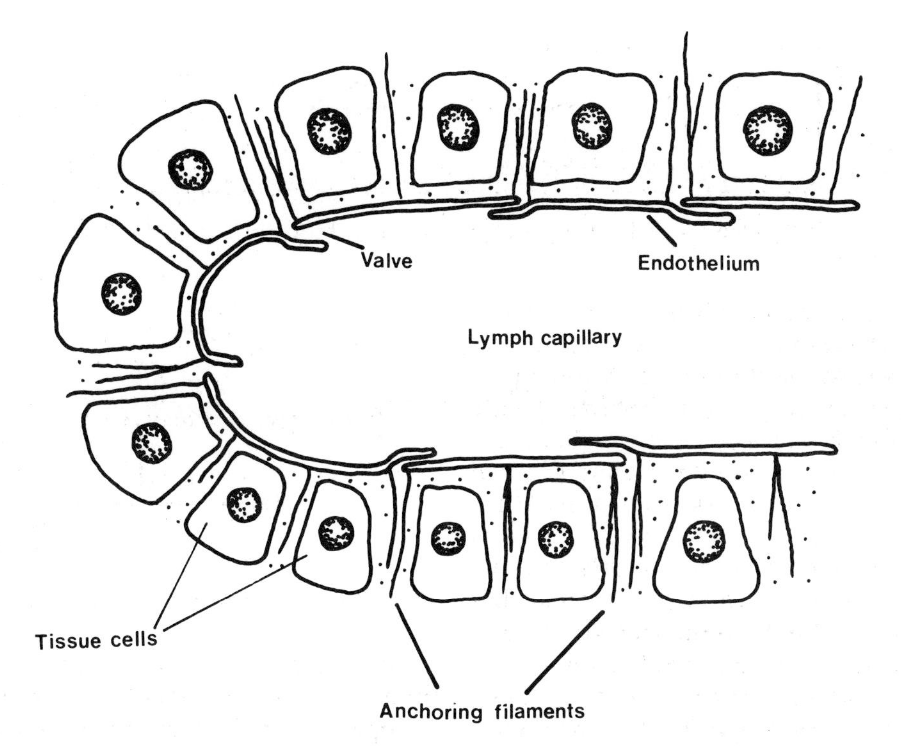

Fig. 2.4 Lymphatic capillary (redrawn from Guyton, 1981).

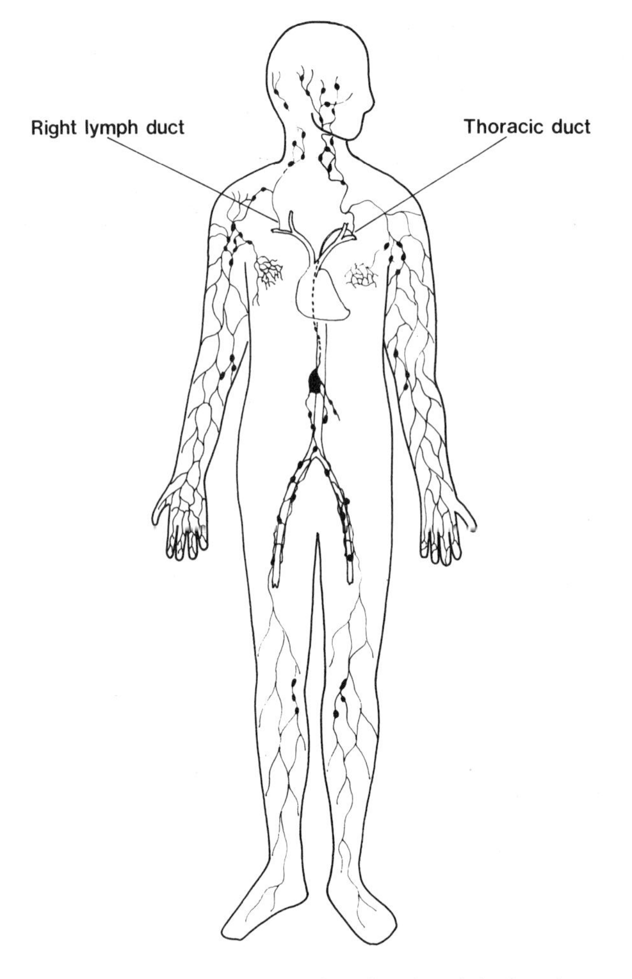

Fig. 2.5 Distribution of the major vessels and nodes of the lymphatic system.

movement of lymph is largely dependent upon external forces, e.g. skeletal muscular activity.

The lymph nodes are small bean-shaped structures of variable size (0.1–2.5 cm long) (Fig. 2.6). Each consists of a collagenous capsule from which arise inwardly-directed partitions or trabeculae. A fine network of reticular fibres occupies the space inside the capsule. This network supports a large mass of lymphocytes which fills much of the lymph node; they are absent from the area immediately next to the capsule and the trabeculae. Other cells, e.g. macrophages are also found within the cell mass. The lymphocytes form particularly dense aggregations called lymphatic follicles in the cortex of the node. The centre of each lymphatic follicle is a germinal centre, where new lymphocytes are formed.

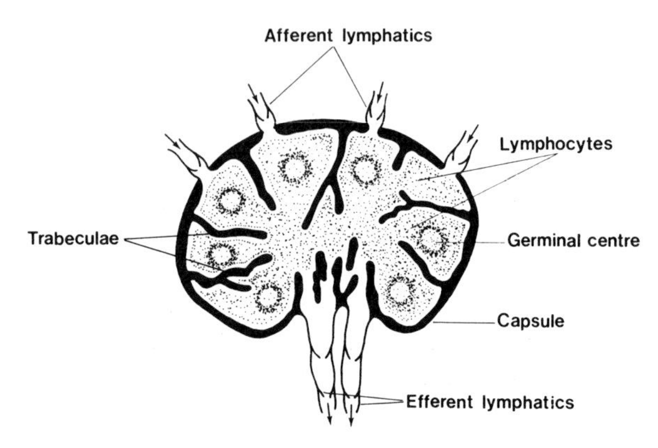

Fig. 2.6 Section through a lymph node.

Lymph enters the node through several afferent vessels on the convex surface and a smaller number of efferent vessels takes lymph away from the node at the concave surface (hilus). Blood vessels also enter and leave through the hilus. Lymphocytes are exchanged between the blood and lymph within the lymph nodes. Lymphocytes can be brought from other parts of the body via the blood and deposited in the lymph node cortex from post-capillary venules. Lymphocytes leave the node in the lymph and are subsequently returned to the blood via the thoracic and right lymph ducts.

The lymph nodes act as filters, removing cellular debris and micro-organisms from the lymph. Antigens within macrophages initiate the immune reponse and lead to the formation of more lymphocytes (see *Acquired (specific) immunity*).

LYMPH

One function of the lymphatic system is to drain the tissue spaces of excess fluid and other materials and return them to the blood. Lymph, therefore, is formed from interstitial fluid. As fluid leaks out of the capillaries of the cardiovascular system, minute quantities of protein are carried out which accumulate in the tissue spaces. The presence of this protein increases the tonicity of the interstitial fluid which leads to the osmotic retention of water. Excess fluid and protein are forced into the blind-ended lymph capillaries through the 'valves' in their walls.

Lymph has a similar composition to blood plasma, although the amount of protein present is variable since it depends upon the amount which leaks out of the blood and this varies in different part of the body. Thus lymph originating in skeletal muscle only contains about 15% of the amount of protein found in the blood, whilst that from the liver contains nearly as much as plasma. Lymph originating in the specialized lymphatics (lacteals) in the wall of the small intestine contains large amounts of lipid and indeed the lymphatic system is the route through which much of the digested dietary fat is taken to the blood. Foreign materials, bacteria etc. are also found in the lymph and are normally removed at the lymph nodes by macrophages. Lymph travels slowly through the lymphatic system and total lymph flow may only be 2–4 litres a day. Lymph pressure is also very low and in peripheral vessels may be less than 3 mmHg.

BLOOD CELLS

The formed elements in blood comprise red cells (erythrocytes), white cells (leucocytes) and platelets (thrombocytes), which are fragments of large cells called megakaryocytes.

If whole blood is centrifuged, then the formed elements can be separated off from the supernatant fluid (plasma). The percentage of formed elements in blood is normally around 45% (about 42% in woman, about 47% in men), which leaves about 55% plasma. Red cells comprise 99% of all the formed elements and the fraction of blood consisting of red cells is known as the haematocrit.

ERYTHROCYTES

Red cells measure about 7.5 μm in diameter. They are biconcave discs, with no nucleus,

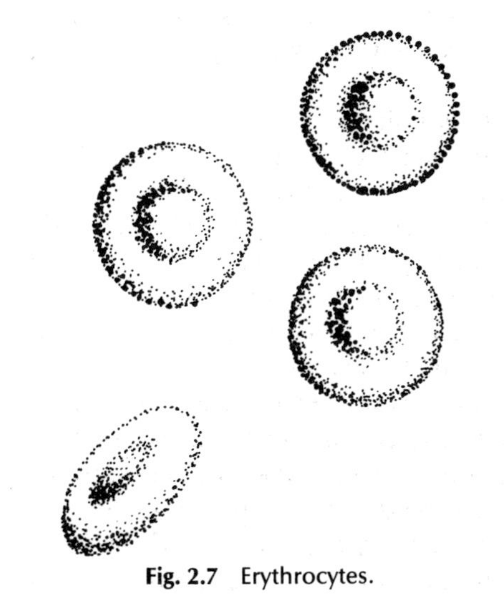

Fig. 2.7 Erythrocytes.

measuring about 1 μm thick in the centre and 2 μm at the edges (see Fig. 2.7). The red cell count is around 5 million per cubic millimetre of blood (4.8 million/mm^3 in women (4.8×10^{12}/litre), 5.4 million/mm^3 (5.4×10^{12}/litre) in men).

Erythrocytes contain the protein haemoglobin which contributes to the carriage of both oxygen and carbon dioxide, and also acts as a buffer. The quantity of haemoglobin is about 15 g/100 ml blood (14 g/100 ml in women and 16 g/100 ml in men).

Haemoglobin consists of one molecule of globin and four molecules of haem (see Fig. 2.8). Globin is made up of four amino acid chains and the haem comprises iron and protoporphyrin.

When oxygen combines with haemoglobin (in the lungs), it becomes loosely attached to the iron which is in the ferrous form. Oxygen can easily be freed from the iron in the tissues.

Carbon dioxide combines with amino groups in the globin part of the molecule. Further details are given in Chapter 5.

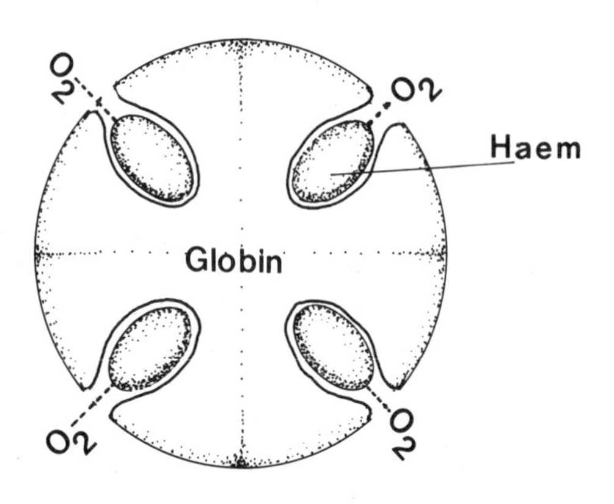

Fig. 2.8 Haemoglobin.

Erythropoiesis

In the fetus, red cells are firstly produced by the yolk sac, later the liver becomes the predominant site, and the spleen also contributes. By the end of gestation, the site of red cell production has changed to the bone marrow. During childhood, most bones contribute, but the number of sites reduces so that in the adult, red cells are produced only in the marrow of skull bones, vertebrae, ribs, sternum, pelvis and the proximal ends of the femur and humerus.

Erythrocytes, like other blood cells, originate from stem cells in the marrow which are capable of becoming any kind of blood cell. These are called 'uncommitted' stem cells and they are able to divide to produce daughter cells, some of which will differentiate to become stem cells 'committed' to a particular cell line. Pools of uncommitted stem cells are present in the marrow together with committed stem cells for the various types of blood cells.

The morphological identity of the uncommitted stem cell is uncertain. The committed stem cell normally has a diameter of about 18 μm, with a large nucleus and dark blue staining cytoplasm; this cell is called a haemocytoblast (Fig. 2.9). Each haemocytoblast divides and one daughter cell differentiates to form a proerythroblast which is smaller than its precursor, and its nucleus occupies a smaller proportion of the total cell volume. These trends towards smaller cells and nuclei continue with successive cell types in the series, the early, intermediate and late erythroblast. Haemoglobin synthesis begins in the early erythroblast and is complete by the late erythroblast stage and the cytoplasm consequently stains red.

When erythroblasts divide, both daughter cells differentiate into the next cell type, so that a greater number of cells is produced by each division.

The late erythroblast loses its nucleus and thereby becomes a reticulocyte. These cells are so named because they have fragments of endoplasmic reticulum (the site of haemoglobin synthesis) remaining.

The reticulocytes enter the blood by squeezing between the endothelial cells (diapedesis) of the sinusoids in the bone marrow. The cells lose the endoplasmic reticulum and mature into erythrocytes in one or two days. From proerythroblast to erythrocyte takes three to four days.

For the red cell count to remain constant, the rate of production of red cells must equal the rate of destruction of old cells. The rate of production is indirectly controlled by the oxygen content of the blood.

If the red cell count falls, then the oxygen content of the blood also falls and this acts as a stimulus for the increased rate of release of erythropoietic factor from the kidneys. This factor acts on a plasma globulin (erythropoietinogen) and splits off erythropoietin which stimulates the conversion of committed stem cells to proerythroblasts, thereby increasing the rate of erythropoiesis. After a few days, the oxygen content of the blood is restored and the rate of erythropoiesis falls again.

Normal development of erythrocytes is dependent upon an adequate supply of various growth factors.

Vitamin B_{12} is required for the synthesis of DNA, and if insufficient is present, then inadequate numbers of red cells are produced, resulting in anaemia. The cells produced in these conditions are abnormally large (megaloblasts instead of erythroblasts), and the mature cells, macrocytes, are fragile and oval instead of the usual biconcave shape, and their survival time is reduced. This type of anaemia is called megaloblastic anaemia.

Lack of vitamin B_{12} may be due to poor absorption or rarely an inadequate dietary supply. Absorption of the vitamin requires 'intrinsic factor' which is produced by the gastric glands and combines with B_{12} when it is in the stomach.The complex is resistant to digestion and travels through the small intestine as far as the ileum where it is absorbed. In the absence of intrinsic factor, therefore, B_{12} is liable to be digested and not absorbed, giving rise to a condition called pernicious anaemia. The liver stores relatively large amounts of B_{12} and so a deficiency may not be manifested for some months.

Another B vitamin, folic acid, is also required for normal DNA synthesis, and a severe dietary lack will also result in megaloblastic anaemia. Folic acid deficiency is common in pregnant women and old people.

Iron is required for the synthesis of haemoglobin, because although the iron released from worn-out red cells is largely recycled, there is a small loss in the faeces, urine, sweat and sloughed off skin cells (about 0.6 mg per day), and menstrual losses in women account for a further 0.7 mg per day.

Iron is actively absorbed from the duodenum in the ferrous form. Although there may be a large amount of iron in the diet, only a few milligrams are absorbed daily. The rate of iron absorption appears to be controlled by the intestinal epithelium itself, and depends upon the total iron balance in the body. For example, if there is an excess of iron in the body, then there is a build-up in the mucosal cells and this inhibits further uptake.

Following absorption from the intestine, iron combines with one of the plasma proteins, transferrin (a β globulin), from which it can easily be split for uptake by developing red cells. The principal store of iron is in the liver as ferritin, which is formed by the combination of iron with the protein apoferritin.

The average lifespan of a red cell is 120 days. The lack of a nucleus means that replacement of essentail protein, such as the enzymes, cannot occur. Old erythrocytes may be engulfed intact by phagocytes lining some blood vessels, mainly in the spleen, or it may be that they rupture in 'tight spots' of the circulation. In any event, the haemoglobin is broken down by the phagocytes into haem and globin. The iron from the haem, and the globin are used again in the marrow, whereas the porphyrin part of the haem is broken down to bilirubin. This is taken up by the liver cells and eventually excreted in the bile.

Blood groups

Blood groups are due to substances present in blood, which are capable of inducing the production of antibodies. They are therefore types of antigens, although they are also known as agglutinogens. There are a great number of such antigens present in blood, but the two most important systems, because of

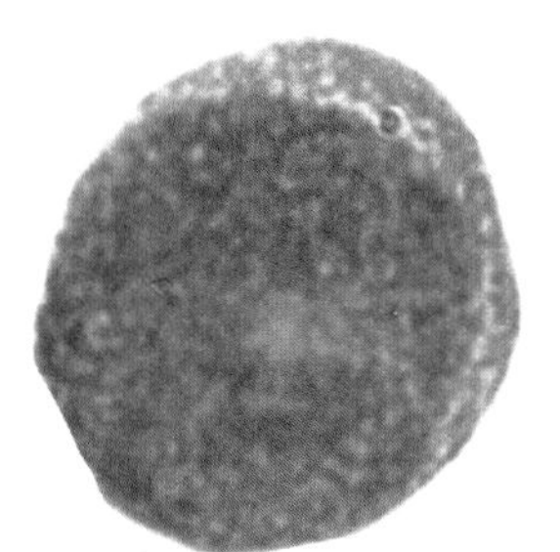

A. Haemocytoblast

E. Late erythroblast

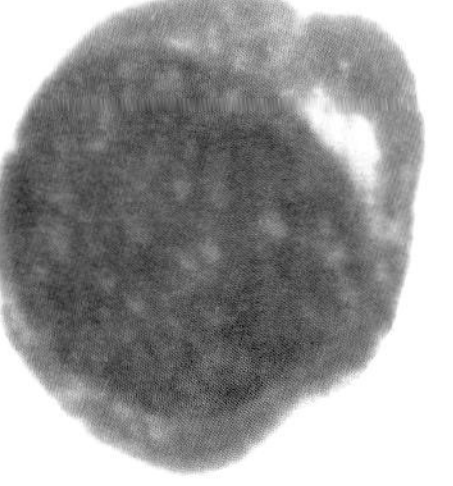

B. Proerythroblast

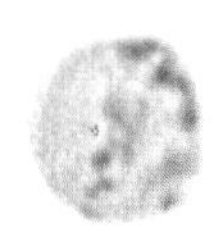

F. Reticulocyte

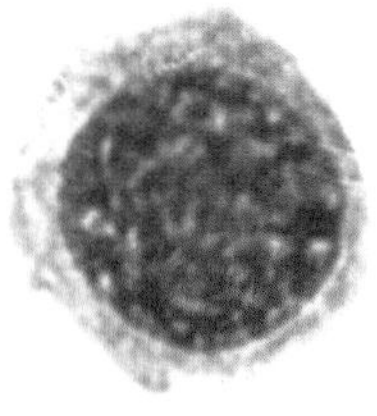

C. Early erythroblast

G. Mature red cell

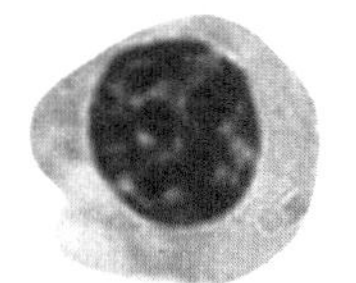

D. Intermediate erythroblast

Fig. 2.9 The precursors of erythrocytes in bone marrow smears (× 2000). (Courtesy of D. M. Quincey.)

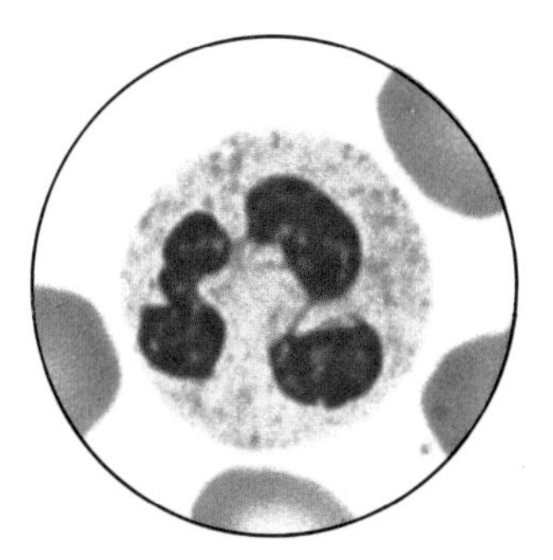

A.. Neutrophil

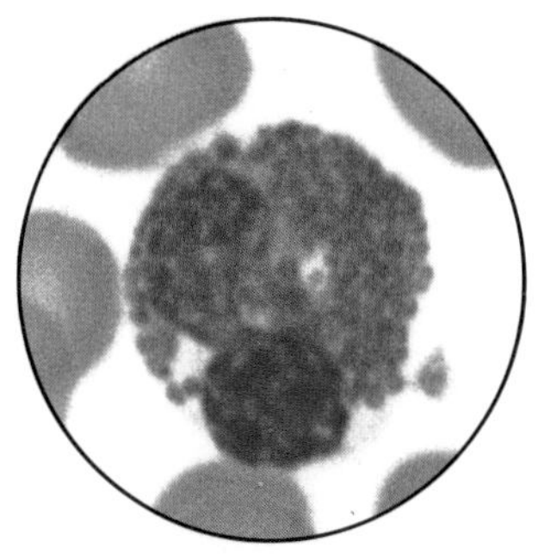

B. Eosinophil

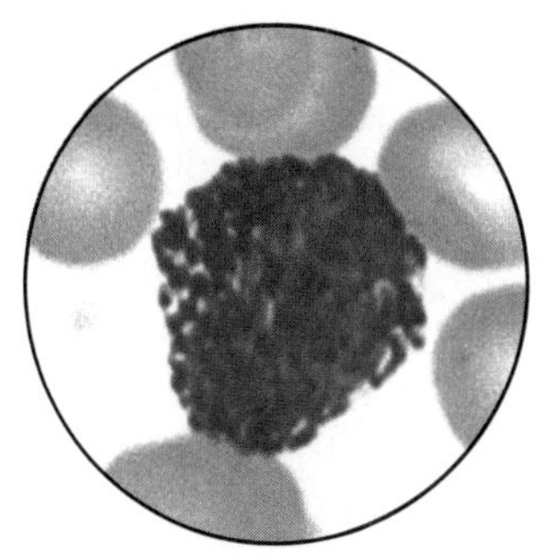

C. Basophil

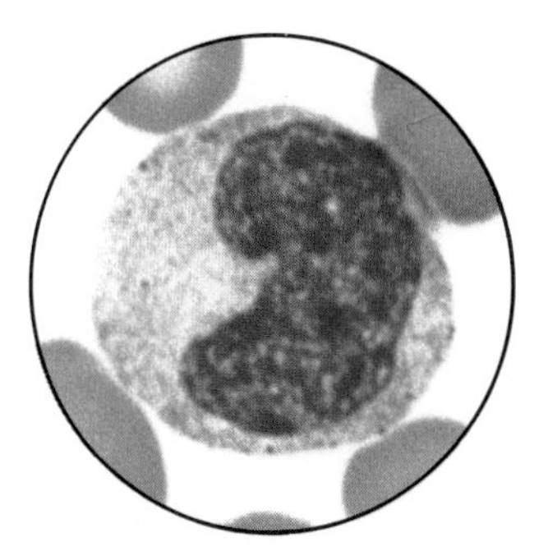

D. Monocyte

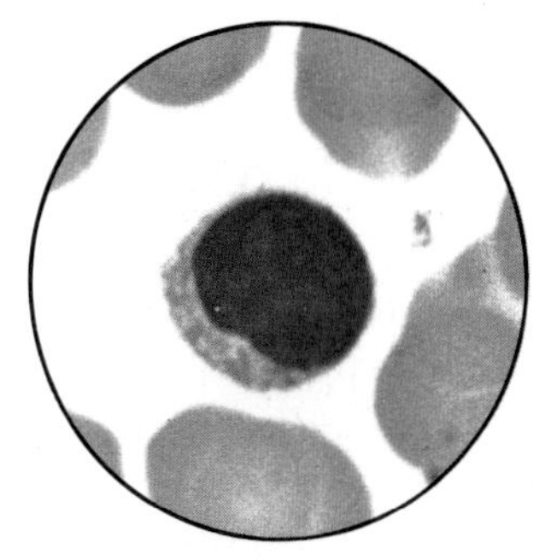

E. Small lymphocyte

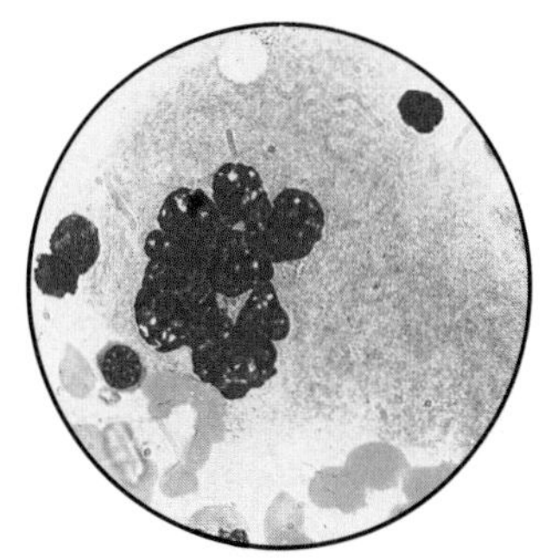

F. Megakaryocyte

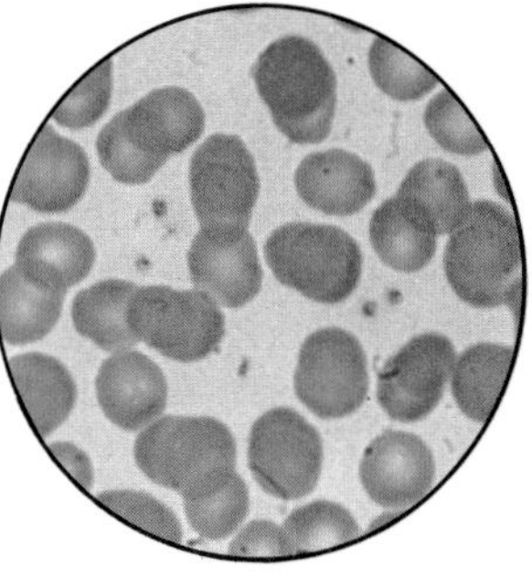

G. Platelets

Fig. 2.10 White cells (× 2000) and platelets (× 800) in peripheral blood. A megakaryocyte (F.), the precursor of platelets from the bone marrow, is also shown (× 500). (Courtesy of D. M. Quincey.)

the potential transfusion hazards, are the ABO and Rh (rhesus) systems. The antigens of these systems are found on red cell membranes.

ABO system

ABO antigens are glycoproteins and glycolipids with a slight difference in the carbohydrate component in each group. Individuals may be type A, B, O or AB. These groups are inherited by one gene on each of two homologous chromosomes (one from each parent). The A and B genes are dominant, so that a blood group A person may be homozygous (AA) or heterozygous (OA). The genetic make-up (AA or OA) is known as the genotype, whereas the blood group A is the phenotype. Similarly there are genotypes OB and BB for phenotype B, but phenotype O is always genotype OO, and phenotype AB is always genotype AB.

During fetal life, no antibodies to the A or B agglutinogens are found but they start to appear after two or three months of post-natal life, probably in response to A and B antigens which are found in bacteria in the respiratory tract and alimentary tract.

Antibodies are not formed against the individual's own red cell antigens, rather to the antigens he or she lacks, so that blood group A individuals produce anti-B antibodies (agglutinins) in the plasma. Blood group B individuals produce anti-A antibodies, blood group O individuals have both anti-A and anti-B antibodies, whereas blood group AB individuals have neither antibody (see Table 2.4).

Three of the four main groups were discovered by Landsteiner in 1901, the fourth (AB) a year later; since then group A has been subdivided into group A_1 (80%) and group A_2 (20%).

Blood group A_1 individuals have both A and A_1 antigens, whereas A_2 individuals have only A antigen (see Table 2.5). In some cases, A_2 individuals have appreciable amounts of antibody to the A_1 antigen.

The anti-A antibodies found in blood group B and O individuals actually consist of anti-A and anti-A_1 components.

If erythrocytes are mixed with serum containing the antibody to them, the cells agglutinate, or clump together, as a result of the antigen-antibody reaction. Blood groups may be tested in this way, by the cells being mixed, in turn, with serum containing anti-A antibodies and anti-B antibodies. If the cells agglutinate only with the anti-A serum, then the blood must be group A. If the cells also agglutinate with the anti-B serum, then the blood is group AB. If the cells only agglutinate with anti-B serum, then the blood must be group B. If neither antiserum causes agglutination, then the blood is group O.

The property of agglutination means that if blood is to be transfused, then it is vital that the donor and recipient's bloods are compatible, that is do not agglutinate. Table 2.6 shows which transfusions would cause agglutination and which would not. It can be seen that

Table 2.4 ABO blood groups

Genotype	A and B antigens on cells	Antibodies in serum	Percentage distribution (England)
AA or OA	A	anti-B	41
BB or OB	B	anti-A	9
AB	AB	–	3
OO	–	anti-A anti-B	47

Table 2.5 A_1 and A_2 blood groups

Phenotype	Genotype	Antigens on cells	Antibodies in serum
A_1	A_1A_1, A_1A_2 A_1O	A A_1	anti-B
A_2	A_2A_2, A_2O	A	anti-B anti-A_1 (weak)

Table 2.6 Transfusion reactions

Recipient	Incompatible donor (agglutination)	Compatible donor
O	A B AB	O
A	B AB	O A
B	A AB	O B
AB		O A B AB

theoretically, blood group O can be transfused into a patient with any blood group and is therefore called the universal donor, whereas blood group AB will not cause agglutination of any of the other types and is therefore called the universal recipient.

If incompatible blood is transfused, then antibodies in the recipient's blood cause agglutination of the donor's cells. Antibodies from the donor's blood would be so diluted that they would be unlikely to cause agglutination of the recipient's cells.

Agglutinated cells block small vessels in the circulation and are then destroyed by phagocytes which liberate haemoglobin into the blood stream. This in turn leads to a rise in plasma bilirubin concentration, which, if it becomes high enough constitutes jaundice. Fatal damage may be caused to renal tubules.

In practice, prior to a transfusion, the donor and recipient's bloods are compatibility tested, that is, the donor's cells are mixed with the recipient's serum under a range of conditions. If no agglutination occurs with any test, then the two bloods are compatible.

Rh system

A great number of antigens are included in the rhesus system, named after the rhesus monkey in which the system was first studied. The principal antigens of the system are identified as C, c, D, E and e and the most potent is D. Individuals are classified, as either 'Rh positive' or 'Rh negative', but these terms should always be qualified by reference to the particular antigens involved, e.g. RhD positive. Usually the term 'Rh positive' is taken to mean 'D positive', but for transfusion purposes a donor is Rh positive if he possesses any of the antigens C, D or E and Rh negative if he possesses only c and e. Some 85% of Caucasians are RhD positive and 15% are RhD negative.

Rhesus factors, like the ABO system are inherited. They differ, however, in that they are inherited by genes on three separate loci as opposed to one locus in the ABO system. The principal alternate genes (alleles) of the first locus are identified as C and c, of the second as D and d, and of the third as E and e. From each parent an individual inherits any one allele at each locus, e.g. CDe, cDe, cde, etc. The Rh genotype consists of two such 'sets', one from each parent, e.g. CDe/cde — this individual will possess red cells having surface antigens C, c, D and e, (the d allele does not describe an antigen). A rhesus negative donor thus always has the genotype cde/cde.

In contrast to the ABO system, antibodies are not automatically produced against the antigens an individual lacks, but only when exposure to 'foreign' cells occurs.

The most likely situation in which this may occur (apart from a rhesus incompatible transfusion) is when an RhD negative mother carrying an RhD positive fetus suffers a transplacental haemorrhage (usually during parturition). In this case some fetal red cells may leak across the placenta and induce the production of anti-D antibodies by the mother.

Nowadays RhD negative mothers are routinely given 100 μg of anti-D within 60 hours of delivery to prevent the formation of the mother's own anti-D. If this were not done, a second pregnancy with an RhD positive fetus would result in the transplacental transport of the antibody to the fetus.

The resulting antigen-antibody reaction causes agglutination and haemolysis of the fetal red cells which in turn can cause jaundice. The bilirubin can pass through the fetal blood-brain barrier and can be deposited in the brain (kernicterus), which causes neurological damage. These changes constitute haemolytic disease of the newborn or erythroblastosis fetalis. Erythropoiesis is stimulated by the reduced red cell count following haemolysis and an increased number of erythroblasts are found in the circulation. Rh antigens other than D can occasionally cause haemolytic disease of the newborn.

LEUCOCYTES

White cells collectively function as the body's major defence against micro-organisms, as well as removing the body's own old or defective cells. The blood is simply a means of conveying these cells to their sites of action in the tissues. The total number of leucocytes varies with different states of health, such as the presence of bacterial or viral infection, allergic reactions etc, but normally the total count is between 5000 and 10 000 per cubic millimetre. Table 2.7 shows the normal percentage distribution of leucocytes in the blood.

Table 2.7 Percentage distribution of leucocytes normally found in blood

Neutrophils	50–70%
Eosinophils	1– 4%
Basophils	1%
Lymphocytes	20–40%
Monocytes	2– 8%

Structure

Leucocytes can be classified as granulocytes, (which have granules in the cytoplasm), and agranulocytes. The cells are shown in Figure 2.10.

There are three types of granulocyte, polymorphonuclear neutrophils, which are the most common white cells, polymorphonuclear eosinophils and polymorphonuclear basophils. The polymorphs are so named because of the lobed nature of the nucleus. The three types are distinguished by Romanowsky stains.

Neutrophils, which have up to six lobes in the nucleus, have granules which take up either red (eosin an acid dye) or blue (methylene blue a basic dye) stain so that the small granules (lysosomes) appear purple. Eosinophils have two lobes in the nucleus and their granules which are oval and crystalline take up eosin. Basophils do not always have a lobed nucleus and it is often obscured by the granules which take up methylene blue dye and thus appear blue-black. These granules contain histamine, heparin and serotonin. Figure 2.10 shows the appearance of these cells after staining with a Romanowsky stain. Neutrophils and eosinophils are usually 12–14 μm in diameter, whereas basophils are a little smaller (10–12 μm).

The agranulocytes comprise lymphocytes and monocytes. The principal circulating lymphocytes are small, about 10 μm in diameter, whereas large lymphocytes, which are mainly found in lymphoid tissue, are up to 20 μm in diameter. The large lymphocytes are a transition stage between actively dividing lymphoblasts and dormant small lymphocytes (see Fig. 2.11). The principal distinguishing feature of lymphocytes is their large, spherical nucleus, leaving a small rim of cytoplasm which stains light blue.

Monocytes are the largest of the circulating cells, 16–22 μm in diameter, with a nucleus indented on one side and grey-blue cytoplasm. These cells migrate into the tissues and become macrophages (see *Areolar Tissue* in Ch. 1).

Formation

Like the red blood cells, white cells originate from stem cells in the bone marrow, although in the case of lymphocytes, subsequent production occurs in lymphoid tissue such as the lymph nodes, spleen and thymus. Precise details of the sequence of cell types in the white cell series are still uncertain. Figure 2.11 gives a probable sequence of development for each cell type. It will be noted that megakaryocytes from which platelets are derived, are also derived from the same uncommitted stem cell.

Lifespan

The marrow (and lymphoid tissue) has a large pool of leucocytes which spend a relatively short time in the blood when they are 'mobilized' to combat infection or tissue damage.

Granulocytes typically spend only a few hours in the circulation and a few days in the tissues. In times of infection their time in the blood and the tissues may be much shorter than this.

Monocytes also have a short time in the blood, but their survival time in the tissues (as macrophages) is much longer than granulocytes. They are powerful phagocytic cells, which are capable of amoeboid movement

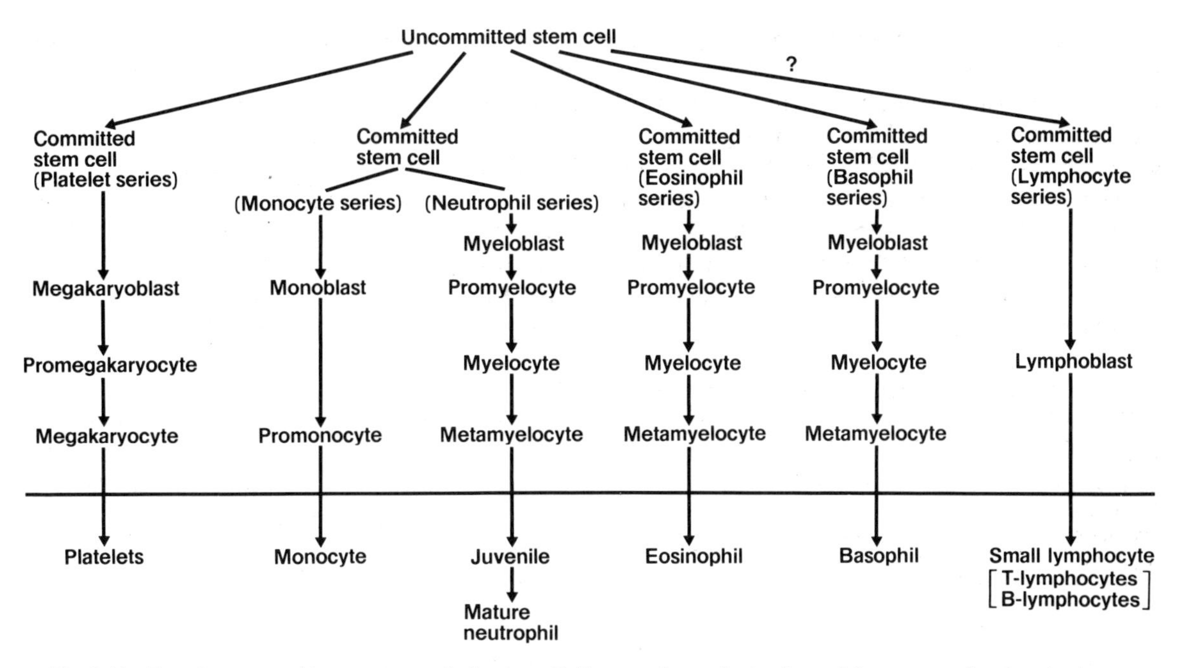

Fig. 2.11 Development of leucocytes and platelets. Cell types above the horizontal line are confined to the bone marrow (and lymphoid tissue in the case of lymphocytes).

and may stay in the tissues for months or years.

Lymphocytes are continually added to the blood from the lymph and usually only spend a few hours in the blood. They leave the blood and enter the tissues (including lymphoid tissue) and then are drained by the lymph and are eventually returned to the blood again. The lifespan of lymphocytes is very variable, some lasting about one hundred days, others surviving for years.

Role of leucocytes in immune responses

Innate (non specific) immunity

There are several important defence mechanisms which prevent micro-organisms from gaining entry to the body. The skin offers an external barrier, and the mucous membranes lining the respiratory and alimentary tracts prevent entry by offering a physical barrier; by removal of trapped organisms in the mucus by ciliary action; and by coughing and sneezing.

Various secretions have antimicrobial properties by virtue of their acidity, including sweat, sebum and gastric juice. Lysozyme (which breaks down bacterial walls) is found in lacrymal fluid, nasal secretions and saliva.

The presence of microbial flora on the skin and internal linings of the body inhibits growth of more harmful micro-organisms.

When micro-organisms gain entry to the body, then the mechanisms available for their destruction are either direct chemical destruction (principally by the complement system, see below), or by phagocytosis (mainly by neutrophils and macrophages). The exit of phagocytic cells from the blood is greatly aided by the inflammatory reaction.

Inflammation was described by Celsus in the first century as comprising rubor (redness), tumor (swelling) calor (heat) and dolor (pain). The redness and heat are caused by local vasodilation thereby increasing blood flow to the area. There is also an increase in permeability of the capillaries to protein with a consequent filtration of fluid into the tissues causing swelling. The pain is caused by activation of the kinin system (see below) which stimulates sensory nerve endings.

The improved blood flow brings phagocytes

to the area at an increased rate. Neutrophils begin to stick to the sides of the capillaries within one minute of microbial invasion and then begin to migrate out of the blood by squeezing between the endothelial cells (diapedesis). They are attracted to the infected or damaged tissue by chemical changes induced by the kinins and complement (see below) and this migration of neutrophils is known as chemotaxis.

The macrophages already present in the tissues multiply and become mobile and they are joined by additional cells which migrate from the blood (as monocytes) an hour or so after the neutrophils.

The neutrophils, then, are the first phagocytes to be mobilized and they engulf infectious material. In many cases they are short lived in the tissues. If vast numbers of them are involved they contribute to the formation of pus. The more powerful macrophages arrive later and they are capable of phagocytosing the neutrophils as well as the micro-organisms, particularly those which can live within cells.

Eosinophils are also phagocytic and they are mobilized particularly in parasitic infections as well as allergic reactions.

The processes involved in the inflammatory reaction are mediated by some non-specific defence mechanisms (as well as some specific mechanisms covered in the next section).

The vasodilation and increased capillary permeability are mediated in three different ways, by histamine, the kinin system and the complement system.

Histamine is found in the basophils of the blood, the mast cells in the tissues, as well as in platelets. The mast cells are probably the major source of histamine and it is released by mechanical disruption of the cells (along with tissue damage generally) as well as by chemicals released from neutrophils and by complement (see Fig. 2.12).

Histamine is a vasodilator, that is it causes relaxation of the smooth muscle in the walls of arterioles so that their diameter is increased and blood flow increases. It also loosens the junctions between the endothelial cells of

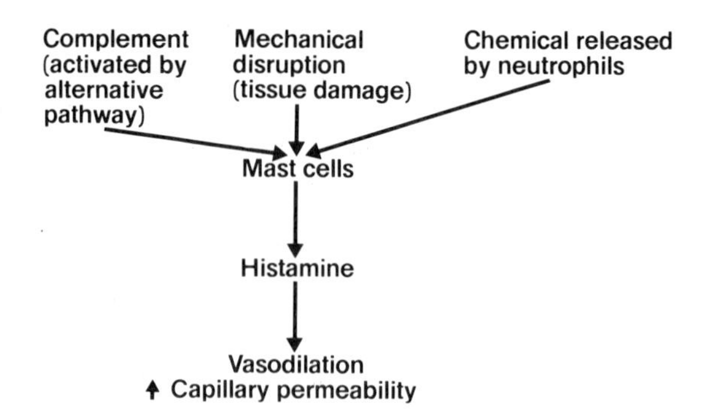

Fig. 2.12 Action and release of histamine in non-specific inflammation.

capillaries thereby facilitating the transudation of protein and cells from the blood into the surrounding tissue.

The kinins are polypeptides which are normally present in an inactive form as the protein kininogen. They are split off from kininogen under the action of the activated plasma enzyme kallikrein. Kallikrein, in turn, is activated by activated factor XII, which is also involved in haemostasis. Factor XII is activated by contact with damaged tissue (see Fig. 2.13). Active kallikrein is also found outside the blood, in tissue fluid, and when capillary permeability is increased, the plasma kininogen leaks out and becomes activated in the tissue.

The kinins enhance phagocytosis and cause

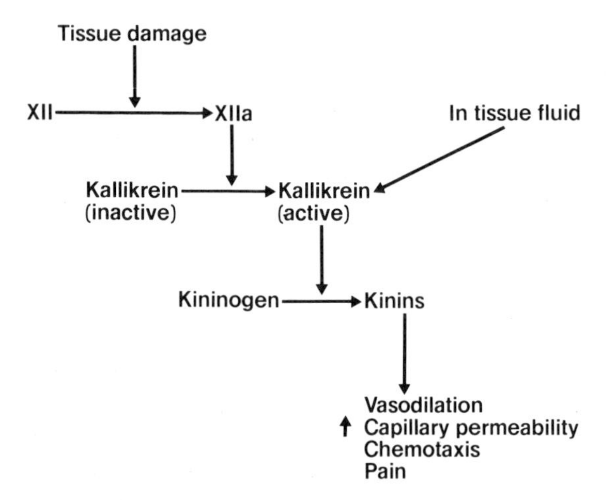

Fig. 2.13 Action and activation of the Kinin system in non-specific inflammation.

pain, as well as enhancing vasodilation and increased capillary permeability along with histamine.

The complement system consists of nine plasma proteins named C1 to C9, but C1 has three subunits C1q, C1r and C1s which makes a total of eleven elements. They are normally present in an inactive form, and activation may be initiated by an antigen-antibody complex (classical pathway) or by an alternative pathway, where the initiation process involves only non specific immune mechanisms (e.g. bacterial endotoxins).

In the classical pathway of complement activation, C1 is activated by the antigen-antibody complex and then a cascade of activation of the other proteins ensues. In the alternative pathway, activation bypasses the initial stage and starts with the activation of C3.

Collectively the complement proteins amplify the inflammatory response by stimulating histamine release, (mediated by the anaphylatoxins C3a and C5a), chemotaxis, and enhanced phagocytosis (C3b coats the micro-organisms, a process called opsonization). The terminal components of complement are cytolytic, that is they directly damage the outer membrane of the microbe. Inflammation is further enhanced by the direct effect of complement proteins causing vasodilation and increased capillary permeability, as well as the activation of plasma kallikrein.

Viral invasion of the body causes the production of a protein, interferon, particularly in lymphocytes. This can migrate and bind to the cell membrane of other cells where it stimulates the synthesis of antiviral proteins. Interferon represents another form of non specific immunity.

Acquired (specific) immunity

In specific defence mechanisms, the body reacts to 'foreign' antigens (usually proteins) in one of two ways. Firstly it may be by the production of circulating antibodies (gamma globulins); this is known as humoral immunity and is the method by which the majority of bacterial infections are counteracted. Alternatively 'sensitized' lymphocytes, with antibody-like molecules on their surface combat 'foreign' cells such as transplants, cancer cells and cells containing viruses, fungi and bacteria. This is known as cell-mediated immunity.

Both humoral and cell-mediated immunity are brought about by the actions of lymphocytes. The stem cells from which lymphocytes are derived, start life in the bone marrow, differentiate to lymphoblasts and then small lymphocytes and shortly before birth the cells circulate and become 'processed' by different areas into two different populations, T lymphocytes and B lymphocytes. The former are processed by the thymus gland and then circulate and are deposited in other lymphoid tissue, early in life. In birds, B lymphocytes are processed by a gland deriving from the gut called the bursa of Fabricius, before populating lymphoid tissue. In man it is uncertain which structure is equivalent to the bursa, but it may be the bone marrow itself, or the liver (see Fig. 2.14).

Antibodies are proteins which are capable of combining with a specific antigen. They have two longer heavy chains and two shorter lighter chains, joined together by disulphide links (see Fig. 2.15). Part of the antibody molecule is different in structure from any other antibody (variable portion Fig. 2.15) and this makes it able to combine with only one particular antigen in most cases.

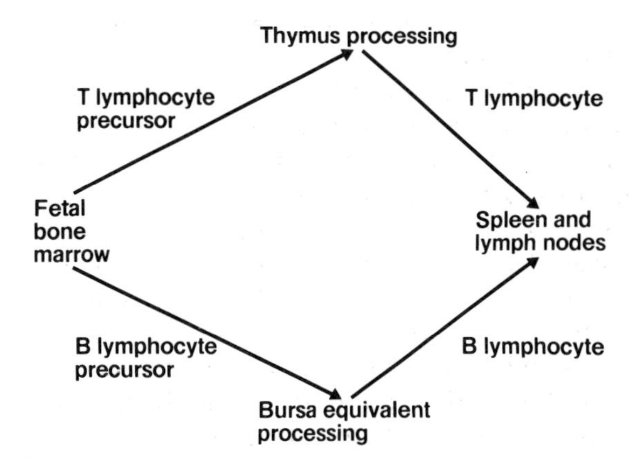

Fig. 2.14 Origin of T and B lymphocytes in the spleen and lymph nodes.

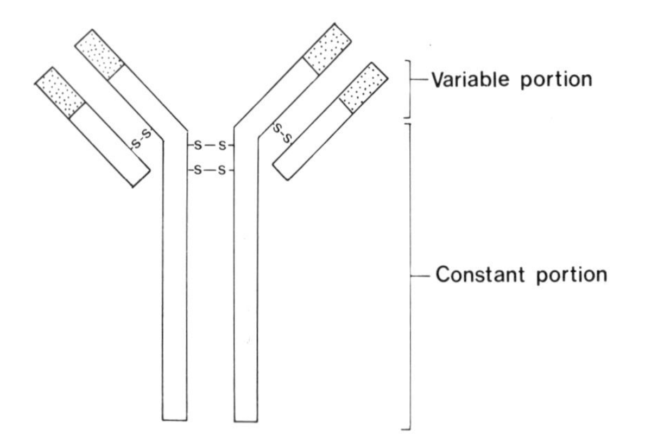

Fig. 2.15 Structure of the IgG antibody.

The antibodies (immunoglobulins) can be subdivided into five classes according to their molecular weight, IgG, IgA, IgM, IgD and IgE (Ig stands for immunoglobulin).

IgG are the most abundant antibodies and they are capable of diffusing across the placenta and so give immunity to the neonate for the first few months of life. They are present in colostrum (the first fluid the baby suckles) and they can be subsequently absorbed from the intestine. These antibodies are mainly concerned with combating bacterial infection and when they combine with bacteria they activate complement and are chemotactic to phagocytes.

IgA antibodies are present in many secretions, tears, nasal fluid, sweat, colostrum, saliva and in the lung and alimentary tract where they are the first line of defence against micro-organisms. The antibodies may act by preventing adherence of the micro-organisms to the mucosal cells, thereby preventing penetration.

IgM, together with IgG account for the major defence against bacteria. They cause agglutination and together with complement, induce lysis of foreign cells. Anti-A and anti-B blood group antibodies are included in the IgM class.

The IgD group of antibodies have been found on the surface of some lymphocytes in the blood and they may be involved in regulating lymphocyte activity.

IgE antibodies are present in small amounts in the blood and they appear to play a role in hypersensitivity and in combating parasitic infections.

In general antibodies can act by directly attacking the micro-organisms, although this effect is generally weaker than their action of activating the complement system whose actions are described in the section on non-specific immunity. Some antibodies, particularly IgE can attach to mast cells and cause them to release histamine, others (IgG), by their opsonizing action, enhance phagocytosis.

Antigens are often phagocytosed by macrophages in the tissues and then travel to the lymph nodes in the lymph. There the antigen is 'presented' to the B lymphocyte by the macrophage and causes it to divide (see Fig. 2.16). The daughter cells differentiate into lymphoblasts and then some may become 'memory' cells, which will then 'recognize' the antigen on a subsequent exposure. The other daughter cells become plasma cells which are rich in endoplasmic reticulum and can synthesize antibodies which then circulate to the infected site. Only a few of the B cells in the lymph node are capable of responding to a given antigen, and it seems that they represent a clone of cells, genetically programmed to synthesize a particular antibody and no other.

On a subsequent exposure to the same antigen, if 'memory' cells persist, then the response is faster and bigger than in the first exposure. This is the principle exploited in vaccination, where the body is exposed to the antigen in the vaccine so that should the micro-organism invade the body at a later

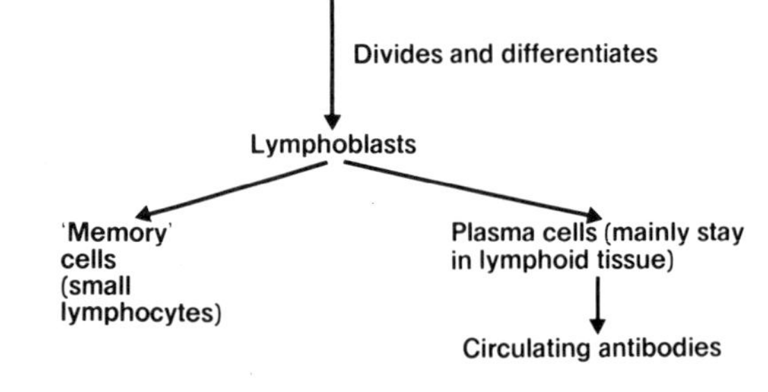

Fig. 2.16 Humoral immunity.

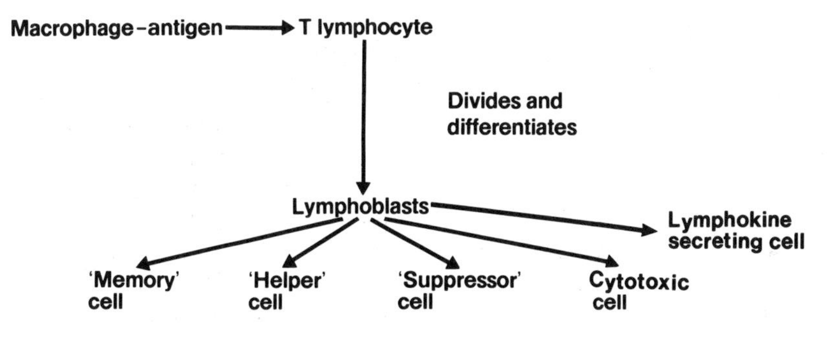

Fig. 2.17 Cell-mediated immunity.

date, 'memory' cells will respond to combat the infection more effectively. The first exposure results in the smaller, slower primary response, subsequent exposures result in the larger, faster secondary response.

Cell-mediated immunity is somewhat different in that the macrophages present the antigen to T cells in the lymphoid tissue; again only one clone can respond to a particular antigen. The effect of stimulation of the T cells by the antigen is that they are stimulated to divide and differentiate into lymphoblasts and then into a variety of 'sensitized' T lymphocytes (see Fig. 2.17). The cytotoxic cells circulate and pass to the site of infection or transplant, and the whole cell is involved in combining with the foreign antigen. Another group of T cells circulates, and after combining with the antigen, the cells release lymphokines. These are proteins which enhance the inflammatory response (analagous to the complement system). Lymphokines are chemotactic to neutrophils and monocytes; they cause aggregation of macrophages at the site of infection and they stimulate phagocytosis. Interferon is released from these T cells.

'Memory' cells are also produced as a result of antigenic stimulation so that a second exposure to the antigen has a much greater and faster effect. Two other types of T cells are also produced after antigenic stimulation, 'helper' cells and 'suppressor' cells, and these are involved in controlling the production of antibodies by the plasma cells.

Cell-mediated immunity is employed against viruses (in which case some of the host cells are also destroyed), in destroying cancer cells (which probably arise daily by the alteration of the genes of normal cells), in the rejection of transplants and in delayed hypersensitivity reactions such as is observed in the Mantoux test for tuberculosis.

3
The urinary system

The urinary system consists of the two kidneys, ureters, bladder and urethra (Fig 3.1).

The kidneys regulate the rates of elimination of water and electrolytes from the body and contribute to the maintenance of a constant blood pH. The kidneys are also responsible for the elimination of metabolic waste products and toxic substances whilst retaining those that are nutritionally useful.

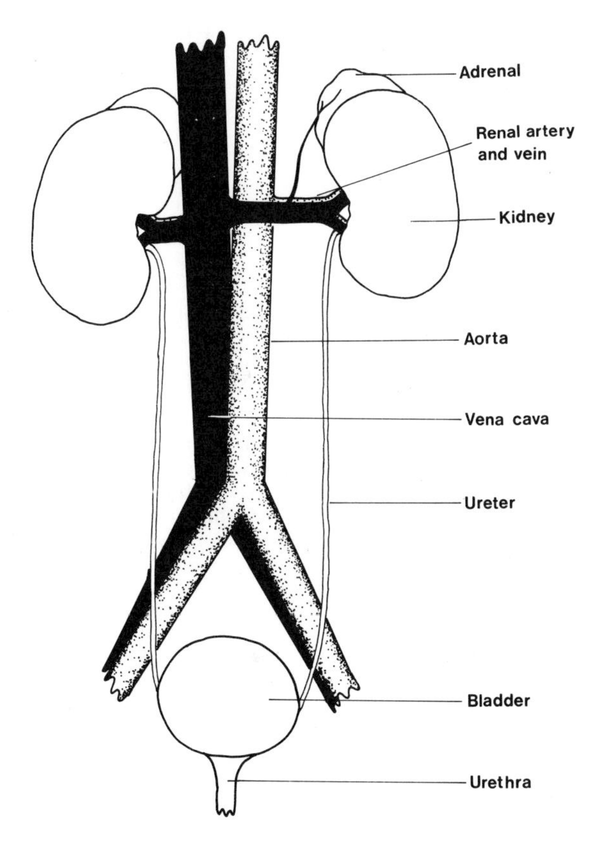

Fig. 3.1 General plan of the urinary system, viewed from the front.

STRUCTURE OF THE URINARY TRACT

Ureters

The ureters are two narrow, thick-walled, muscular tubes, 25–30 cm long, which originate in the pelves of the two kidneys. They pass down through the abdomen and open into the base of the bladder. Each tube is lined by transitional epithelium around which lie two layers (inner longitudinal, outer circular) of smooth muscle. The lower third of the ureter has an additional layer (external) of longitudinal muscle fibres. A fibrous coat or adventitia, continuous with the renal capsule, forms the outermost layer of the ureter.

Bladder

The ureters enter the bladder in an oblique manner, therefore as the bladder fills or contracts reflux of urine is prevented.

The bladder itself is a hollow muscular organ with a tetrahedral appearance when empty. A serosal (peritoneal) coat covers its superior surface. Below this lies a smooth muscle coat, with a layer of circular fibres sandwiched between two layers of longitudinal fibres. Internally, the bladder is lined with a mucosal coat of transitional epithelium.

Below the neck of the bladder lies the striated muscle which constitutes an external sphincter. The urethra proximal to the external sphincter is normally closed due to tension in the elastic fibres of the urethral wall (often an internal sphincter is described).

Urethra

The urethra extends from the neck of the bladder to the external meatus. In the male it is about 20 cm long, in the female, 4 cm. In both it is lined by transitional epithelium near to the bladder which gives way to stratified squamous epithelium. Smooth muscle is also present.

FUNCTION OF THE URINARY TRACT

Urine is carried down the ureters by peristaltic waves, which occur approximately every 10 seconds travelling down at 2 or 3 cm per second. Each wave sends a spurt of urine into the bladder. The bladder fills slowly over a long period of time, but is evacuated over a very small time period by the micturition reflex.

Micturition

Micturition is a complex act involving sensory and autonomic and somatic motor nerve pathways. It is regulated by centres in the cerebral cortex and spinal cord.

Two types of motor neurones are involved: parasympathetic fibres from the spinal cord travel to the bladder wall in the pelvic splanchnic nerves, while somatic fibres are carried in the pudendal nerves from the spinal cord to the external sphincter. In addition, sympathetic fibres also pass to the bladder wall from the hypogastric plexus but, while their exact role is uncertain, it is unlikely that they play a part in normal micturition.

The pelvic nerves also carry the most important sensory fibres to the spinal cord from stretch receptors in the bladder wall and posterior urethra. Much less important, afferent impulses from the neck of the bladder are carried in the pudendal nerves.

The process of micturition is regulated by a reflex which is initiated by impulses from stretch receptors in the bladder wall when the bladder contains about 300 ml of urine. The sensory neurones transmit the impulses to the spinal cord where they are relayed to the parasympathetic neurones which innervate the bladder wall and cause it to contract. Simultaneously, impulses travelling down the pudendal nerve to the external sphincter are inhibited, so that the sphincter relaxes and urine can be voided. The reflex pathways for this are shown in Figure 3.2.

Voluntary delay of micturition is achieved via descending pathways from the brain which inhibit the parasympathetic fibres to the bladder and stimulate the somatic fibres to the external sphincter.

Alternatively, micturition can be voluntarily initiated by descending pathways from the brain to the spinal centre whereby the parasympathetics are further stimulated and the somatics inhibited. Urine is thus forced into the neck of the bladder. This results in a volley of impulses in the pelvic nerves passing to the spinal cord and subsequently inhibiting impulses to the external sphincter.

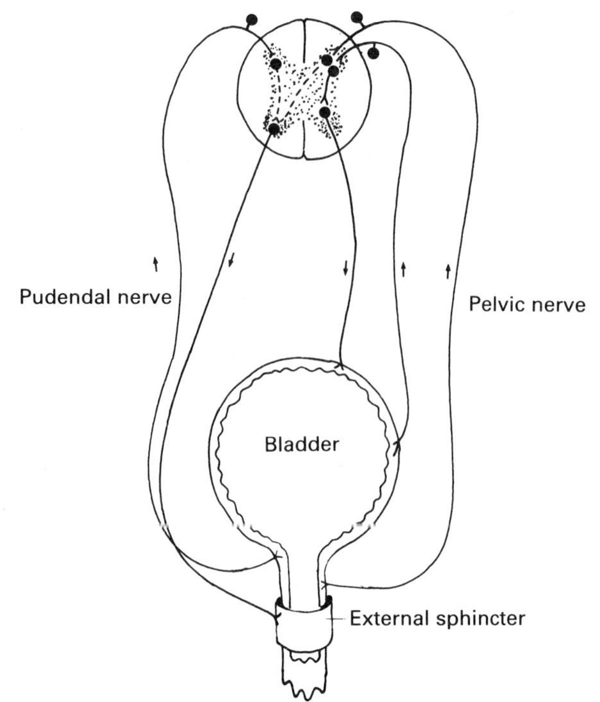

Fig. 3.2 Nerve pathways involved in the micturition reflex. (For simplicity, one pelvic and one pudendal nerve only are shown). The pelvic nerves carry sensory fibres from the bladder wall and posterior urethra to the spinal cord, as well as parasympathetic fibres from the spinal cord to the bladder wall. The pudendal nerves carry sensory fibres from the neck of the bladder to the spinal cord, and somatic motor fibres from the spinal cord to the external sphincter.

STRUCTURE OF THE KIDNEYS

Gross structure

The human kidneys are paired organs situated behind the peritoneum on the posterior wall of the abdominal cavity on either side of the vertebral column, extending from the 12th thoracic to the 3rd lumbar vertebrae. They each weigh between 120 and 170 g, being roughly bean-shaped, 10–13 cm in length, 5–6 cm wide and 3–4 cm in thickness. A deep indentation, the hilus, is found on the medial border, from which emerges the ureter, a large excretory duct to the bladder. The renal artery and renal vein also pass through the hilus.

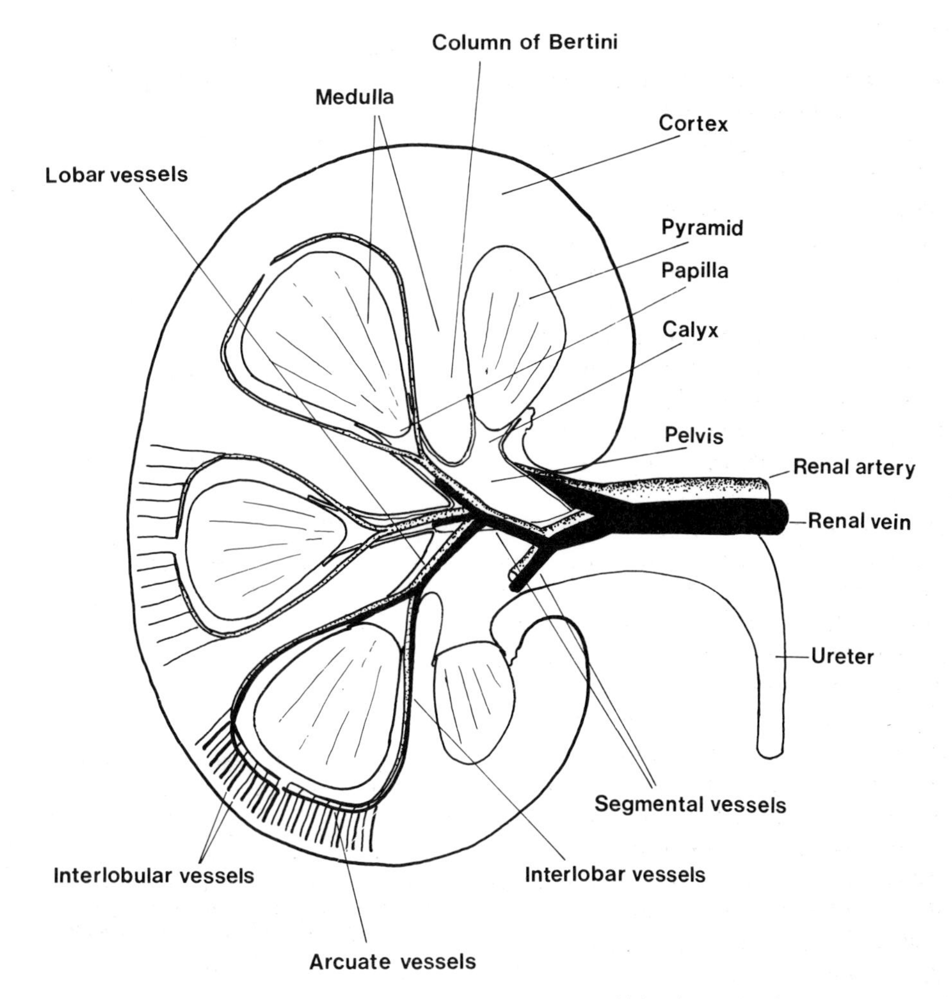

Fig. 3.3 Vertical section through the kidney to show the major areas and blood vessels. The upper part of the pelvis is cut open to show the insides of the calyces.

A section through the kidney (Fig. 3.3) shows a number of obvious anatomical regions. The outer cortex, which is covered on its outer edge by a dense connective tissue capsule, is easily distinguishable from the inner medulla. The latter is made up of between 8 and 18 renal pyramids whose apices (papillae) protrude into the innermost region, the pelvis. This is a funnel-like expansion of the upper end of the ureter, which possesses 2 or 3 outgrowths or major calyces. These, in turn have a number of smaller pockets called minor calyces. Each papilla protrudes into the lumen of a minor calyx.

The medullary pyramids are separated from one another by ingrowth of cortical tissue, the renal columns (or columns of Bertini.)

Blood vessels

Blood enters the kidney through the hilus, in the renal artery; the latter divides into two main branches which give rise to several segmental arteries (Fig. 3.3). Each lobe of the kidney is supplied by a lobar artery which divides into two or three interlobar arteries. These lie in the renal columns between the pyramids. At the junction of cortex and medulla, the interlobar vessels arch over to run parallel to the organ's surface as the arcuate arteries. Small interlobular arteries are given off radially from the arcuate arteries to the periphery. The interlobular arteries give off numerous afferent arterioles which lead into the glomeruli; blood leaves the glomeruli via efferent arterioles (see Fig. 3.4).

Efferent arterioles in the outermost part of the cortex break up to form the cortical peritubular capillary network. Blood is then drained towards the surface by radially situated superficial cortical veins which join the surface stellate veins. These are then drained by the interlobular veins.

In the deeper regions of the cortex, capillar-

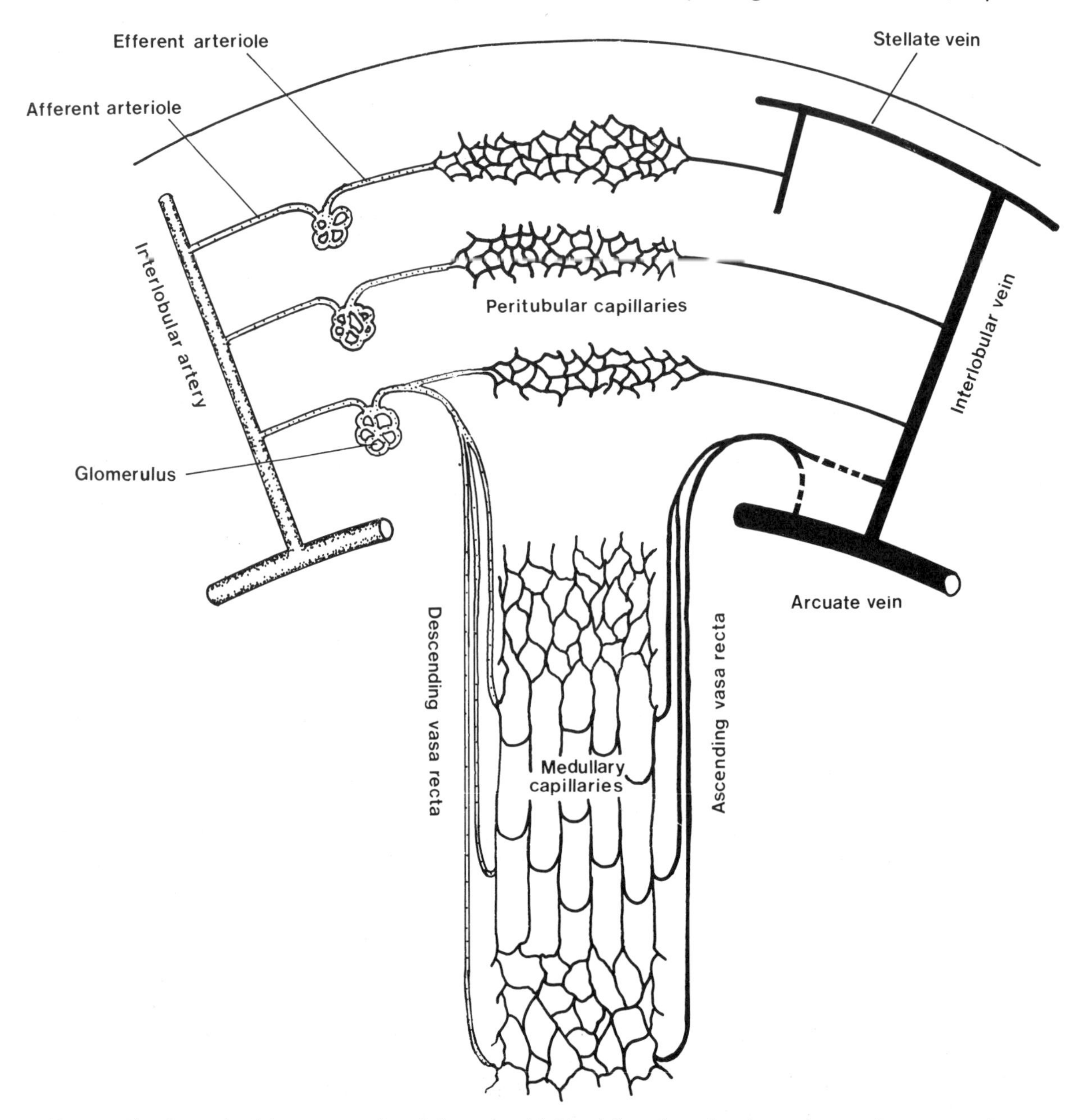

Fig. 3.4 Blood vessels of the cortex and medullary pyramid. Blood flows from the glomeruli into efferent arterioles which then lead into capillary networks before draining into the venous system. In the central medulla the descending vasa recta give rise to a number of parallel capillaries which lie close to the loops of the Henle of the nephrons.

ies empty into deep cortical veins (running parallel to interlobular arteries) which lead into the arcuate veins.

The efferent vessels of the juxtamedullary glomeruli are of large calibre and after supplying branches to the cortical peritubular network, they enter the medulla where they break into bundles of twelve to twenty-five large thin-walled vessels, the descending vasa recta. These form hairpin loops at various levels in the medulla and supply an elongated capillary meshwork which surrounds the loops of Henle and collecting ducts. The capillaries then drain into the ascending vasa recta, which return blood either to the interlobular or arcuate veins (Fig. 3.4).

Nerves

Sympathetic fibres enter the kidney from the coeliac plexus. They course alongside some of the major blood vessels and some fibres may reach the renal corpuscles. The exact distribution of the smallest branches is not known, although some are known to innervate the juxtaglomerular cells.

Fine structure

Each kidney contains at least one million nephrons (some estimates are as high as three million per kidney), together with their collecting tubules or ducts. The nephron is the functional unit of the kidney and represents the structure necessary for urine formation.

The nephron consists of a number of anatomically distinct regions; the Malpighian body, proximal tubule, loop of Henle and distal tubule (Fig. 3.5).

Malpighian body

The Malpighian bodies or renal corpuscles are located in the cortex and may be found at any level. Each corpuscle is composed of a tuft of capillaries, the glomerulus, which is inserted into the dilated, blind end of the renal tubule, the Bowman's capsule.

The outer wall (parietal layer) of the Bowman's capsule is composed of extremely flattened squamous epithelium. At the vascular pole this is reflected inwards to join the inner, visceral layer which is very closely applied to the glomerular capillaries.

The cells of the visceral layer (podocytes) are extensively modified, being composed of a small cell body (perikaryon) from which radiate a number of major processes. These in turn give rise to a large number of thin secondary processes (pedicels) which interdigitate with those of other podocytes. The perikaryon of the podocyte rarely lies directly on the basal lamina, but tends to stand away, so that in section pedicels (from another podocyte) may be seen to lie underneath (see Figs. 3.5 & 3.6).

The pedicels lie on a continuous basal lamina which they share with the glomerular endothelium on the inside. Adjacent pedicels are separated by gaps approximately 25 nm wide (filtration slits) bridged by a fine slit membrane.

The glomerular endothelium is very thin and is perforated by circular pores or fenestrae 50 nm to 100 nm in diameter.

Proximal tubule

The Malpighian body tapers into the proximal tubule, which is about 14 mm long. The first two-thirds constitutes the proximal convoluted tubule, whereas the last third is straight. The convoluted section follows a tortuous route through the cortex as a number of small loops, normally one larger loop is directed out towards the periphery. From here the straight part of the tubule courses down into the outer zone of the medulla where it tapers down into the loop of Henle.

The epithelial cells of the proximal tubule have a characteristic appearance with extremely well developed, closely packed microvilli on their apical surfaces. A large number of slender, lateral processes at the cell base extend under adjacent cells and occupy deep resesses in their base. The lateral borders of the cells are attached only at one point, just below the microvilli. Below this there is a gap

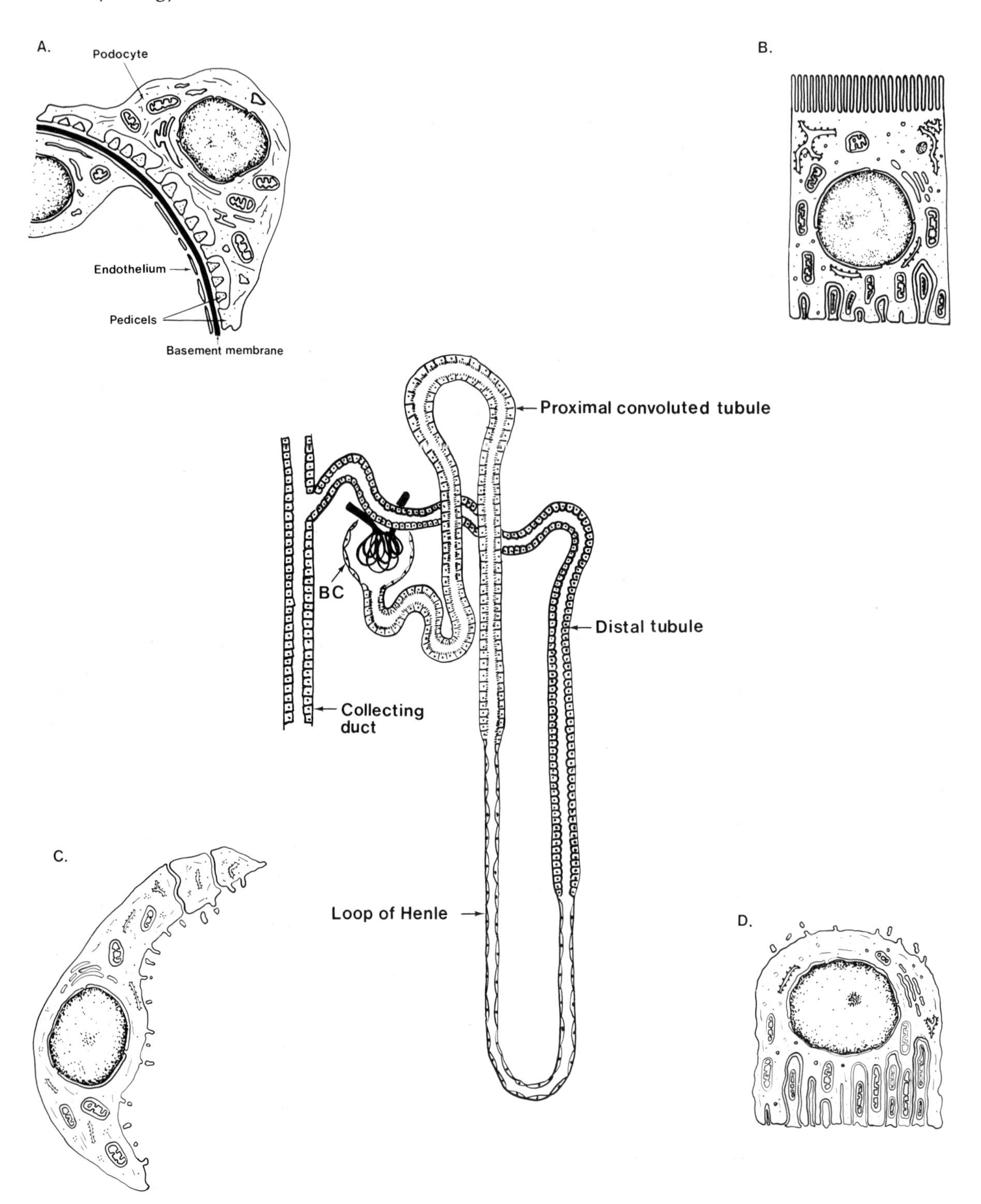

Fig. 3.5 General structure of the nephron with details of the cell types in the various sections of the tubule (central illustration; BC = Bowman's capsule). (A) Malpighian corpuscle, (B) Proximal convoluted tubule, (C) Loop of Henle, (D) Distal convoluted tubule.

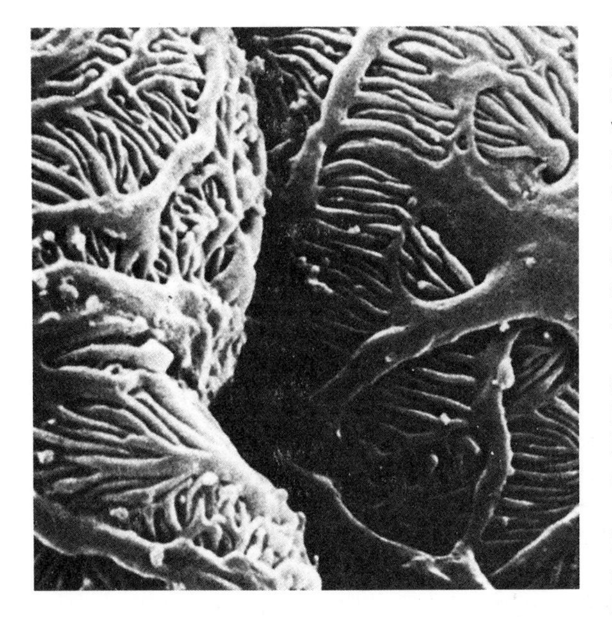

Fig. 3.6 Scanning micrograph of two adjacent capillary loops of rat glomerulus, covered with podocytes (reproduced with kind permission from Bloom & Fawcett, 1975).

between the cells forming an intercellular space.

Large numbers of mitochondria and the large surface area afforded by the microvilli reflect the intense reabsorptive activity taking place in the convoluted section. The cells in the straight part of the tubule have fewer microvilli and are less active.

Loop of Henle

The loop of Henle consists of the straight portion of the proximal tubule, the descending and ascending thin limbs and the straight portion of the distal tubule. The total length of the loop varies considerably although an average of 14–20 mm may be taken. In general the loops of Henle vary in length depending upon the position of their Malpighian body in the cortex.

The short loops (about seven times as numerous as the long ones) originate from corpuscles situated in the outermost parts of the cortex. In these tubules the bend is formed from the thick ascending limb and lies in the outer part of the medulla.

The deeper lying (juxtamedullary) corpuscles possess long loops which pass deep into the medullary tissue and here it is the thin limb which actually makes the loop. It is these nephrons which contribute to the formation of hypertonic urine. Functionally, these loops of Henle consist of a thin descending limb and an ascending limb which consists of a thin and a thick portion, all of which lie within the medulla.

The cells of the thin parts of the loop are simple squamous epithelium with elaborate interdigitations, thus in section small membrane bound areas of cytoplasm may be seen (see Fig. 3.5C).

The cells of the thick portion of the ascending limb which commences in the outer medulla resemble those of the distal convoluted tubule.

Distal tubule

According to some authorities the distal tubule commences with the thick ascending portion of the loop of Henle. Having reached the cortex, the tubule then follows a tortuous route back to the Malpighian corpuscle where it passes between the afferent and efferent arterioles on its way to its junction with the collecting duct. This section of the nephron is called the distal convoluted tubule.

Some sources consider that the distal convoluted tubule begins at the vascular pole, in which case it is about 5 mm long. Functionally however, it is convenient to regard the distal tubule as that section lying in the cortex between the loop of Henle and the collecting tubule. The cells are slightly flatter than those in the proximal tubule and lack well-developed microvilli. The basal portions of the cells are well compartmentalized due to interdigitating lateral processes (as in the proximal tubule).

Collecting tubules

The distal tubule leads into an arched collecting tubule which joins a straight collecting tubule in the cortex. These straight tubules fuse in the inner medulla and the new ones then fuse several times to (eventually) form the large papillary ducts or ducts of Bellini which open into the calyces.

The cells of the collecting tubules are cuboidal, becoming progressively taller in the larger ducts. The term 'collecting duct' is often used to refer to the straight tubules in the cortex and medulla. The average length of the collecting duct is about 20 mm.

Juxtaglomerular apparatus

In the wall of the afferent arteriole, immediately before its entry into the glomerulus, are a group of specially modified smooth muscle cells, the juxtaglomerular cells (see Fig. 3.7). These cells have a slightly basophilic cytoplasm containing a large number of distinct granules with a crystalline internal structure. The granules are known to contain renin (see below).

The distal tubule passes between the glomerular arterioles. At the point where the tubule touches the afferent arteriole the tubular cells are modified to form the macula densa.

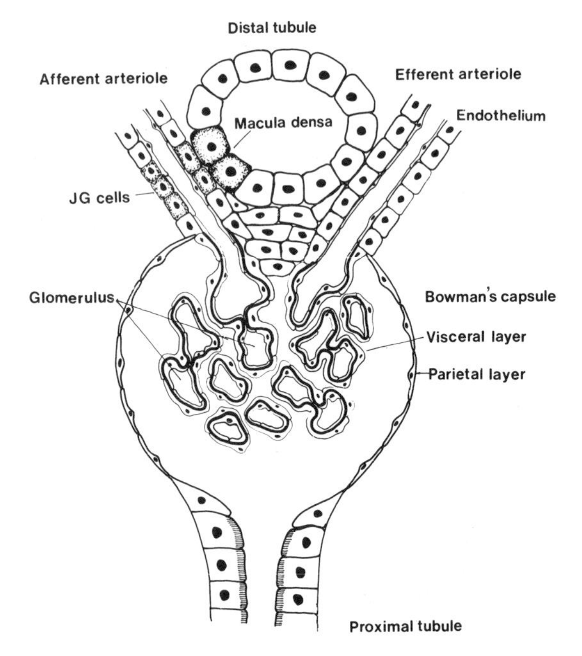

Fig. 3.7 Vertical section through a Malpighian corpuscle. Only a few capillaries are indicated.

The cells of the macula densa are rather taller than those typical of the distal tubule and the Golgi apparatus tends to lie close to the cell base in contrast to the typical distal tubule cells in which it lies close to the apex.

The juxtaglomerular cells, together with the macula densa comprise the juxtaglomerular apparatus.

FUNCTION OF THE KIDNEYS

The kidneys normally receive about one-fifth of the cardiac output per minute, thus renal blood flow is about 1200 ml per minute. About one-fifth of the plasma is filtered and the filtrate is then modified by the tubular cells until eventually urine is discharged into the ureter. Approximately 1 ml per minute of urine is produced by the two kidneys. Urine formation is a process of glomerular filtration followed by selective modification by the tubules.

Glomerular filtration

Filtration of the blood occurs in the Malpighian body. Blood enters the glomerular capillaries under relatively high pressure (approximately 50 mmHg) and is forced through the boundary membrane, into the lumen of Bowman's capsule. This high pressure is due to the fact that blood travels only a short distance from the high-pressure aorta and the glomeruli are drained by arterioles rather than venules.

The net filtration pressure is, however, less than the 50 mm blood pressure in the glomerular capillaries, because it is opposed by two other pressures acting across the capillary walls. The plasma proteins exert an osmotic pull tending to retain fluid in the blood, and the filtered fluid within Bowman's capsule exerts a back pressure against the glomerular capillaries.

It can be seen, therefore, that whenever glomerular capillary blood pressure exceeds 40 mmHg, fluid will be filtered into Bowman's capsule.

Experiments have shown that molecules of

net filtration pressure	=	glomerular capillary blood pressure 50 mmHg	−	plasma protein osmotic pressure 30 mmHg	−	capsular fluid pressure 10 mmHg	=	10 mmHg

molecular weight 10 000 will pass freely into the capsule. Larger molecules exhibit progressively less filtration until those of molecular weight 69 000 penetrate hardly at all. It is, therefore, hypothesized that pores of varying sizes exist in the filtering membrane and these will prevent free passage of proteins and cellular materials while allowing smaller molecules through. The glomerular filtrate is, therefore, deproteinized plasma.

The exact nature of the filtering membrane is not clear although it is probable that the fenestrations in the capillary endothelium allow free passage of molecules; the basement membrane acts as an intermediate gauge sieve; whilst the slit membranes between the pedicels are the finest filters.

The glomerular filtration rate averages 125 ml/min (180 litres/day) but can vary from a few ml to 200 ml/min.

Tubular function

Filtration is a relatively non-selective process and therefore the fluid entering the tubule contains all the constituents of plasma except protein.

Tubular activity involves the return of nutritionally useful substances to the blood and the elimination of the waste products of metabolism from the body. In addition, the tubule makes a major contribution to homeostasis by its ability to vary the elimination of physiologically important substances such as acid, salt and water according to their levels in the body.

Therefore, the volume and composition of glomerular filtrate and urine are substantially different. The changes are brought about by two processes, reabsorption and secretion.

Reabsorption

Reabsorption is the net transfer of solute or water from the tubular fluid back into the blood. Such substances travel through the tubular epithelium into the interstitial spaces and from there into the peritubular capillary network.

Any substances which are fitered, but which do not appear in the urine at all are reabsorbed completely in their passage through the tubules. They include substances which are of particular nutritional value such as amino acids, acetoacetate, vitamins and glucose.

Most of the substances in the glomerular filtrate are only partially reabsorbed and therefore appear in the urine. In some instances the amount reabsorbed is under hormonal control. Table 3.1 shows some constituents of glomerular filtrate and the amounts typically reabsorbed from the tubules.

Table 3.1 Renal handling of some plasma constituents (modified from Ganong, 1983).
Figures in brackets indicate the location of the activity. P, proximal convoluted tubule; L, loop of Henle; D, distal tubule; C, collecting duct

Substance	Percentage secreted	Percentage reabsorbed
Na^+	-	99.4 (P,L,D,C)
K^+	11 (D)	100 (P,L)
Cl^-	-	99.2 (P,L,D,C)
HCO_3^-	-	100 (P,D)
Urea	-	53 (P,L,D,C)
Urate	8 (P)	98 (P)
Glucose	-	100 (P)
Water	-	99.4 (P,L,D,C)

Uptake of material by the tubular cells may either be a passive or an active process.

Passive transport requires a favourable concentration or electrochemical gradient (or in the case of water, an osmotic gradient). However, it is a minority of substances which are carried by this method. Some negative ions are thought to 'follow' positive ions which are transported actively, attracted by the charge difference.

Water is reabsorbed down the osmotic gradients created by reabsorption of solutes.

Active transport involves the movement of solutes across a cell membrane by carrier molecules. A given membrane will contain a finite number of carrier molecules and therefore there will be a maximum number of molecules which could be transported at any one instant in time. This maximum rate of solute transfer is known as the tubular maximum or Tm for that particular substance, e.g. Tm value for glucose is 340 mg/min.

Secretion

The process of secretion involves the net transfer of solute from the peritubular capillaries, through the tubular cells and into the tubular fluid. It is therefore an additional, albeit minor route whereby substances can enter the urine.

Secretion may also be an active or a passive process. The major endogenous substances which are secreted by the tubular cells are potassium and hydrogen ions, both of which are secreted in exchange for sodium. Glucuronides, various ethereal sulphates and hydroxyindoleacetic acid (a serotonin derivative) are secreted by active transport in the proximal tubule. In addition, foreign substances such as penicillin, phenol red and para-aminohippuric acid (PAH) are also actively excreted. PAH is, in fact, excreted so completely by the tubular cells that it is completely removed from the blood.

PAH clearance (see below) is used clinically to estimate effective renal plasma flow.

Proximal tubule

About 65% of all reabsorptive and secretory activity takes place in this section, with water following the reabsorption of solutes, thereby maintaining isotonicity. The fluid leaving the proximal tubule and entering the loop of Henle is therefore still isotonic with plasma but is reduced to some 35% of the original volume.

The structure of the proximal tubular cells is adapted to their reabsorptive role. The exposed surface is increased twenty-fold by the apical microvilli and the channels between the cells enable the lateral surfaces to transport material with particular ease.

Nutritionally useful substances such as glucose, amino acids, acetoacetate and vitamins, which are normally completely reabsorbed from the filtate are all reabsorbed in the proximal convoluted tubule by active transport mechanisms. Much of the filtered potassium is also reabsorbed here. Any protein which has leaked through the glomerular membrane is

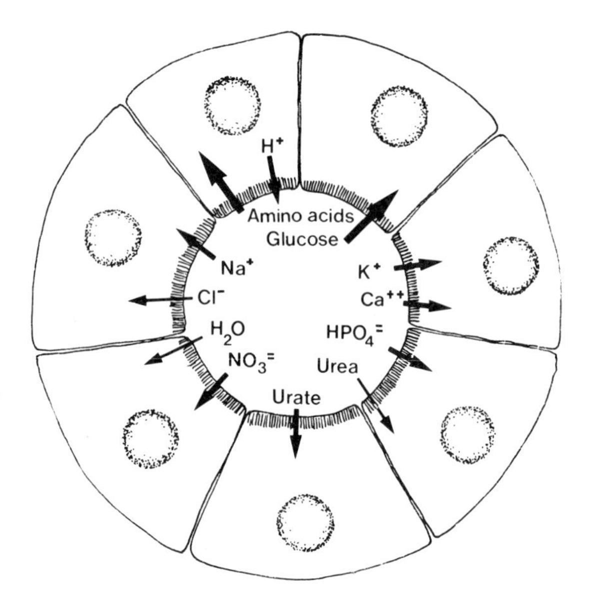

Fig. 3.8 Reabsorption and secretion of the major naturally occuring substances in the proximal convoluted tubule. *Extra thick arrows* — substances completely reabsorbed by active transport. *Thick arrows* — substances incompletely reabsorbed or secreted by active transport. *Thin arrows* — substances reabsorbed by diffusion.

completely reabsorbed by pinocytosis and is therefore normally absent from the urine.

Many substances are incompletely reabsorbed in the proximal tubule. These include sodium, chloride, phosphate, urate, nitrate and urea. The probable modes of transport are indicated in Figure 3.8. Bicarbonate is normally completely reabsorbed in the form of carbon dioxide (see *Bicarbonate reabsorption*).

The cells of the proximal tubule also actively secrete some metabolites, including hydroxyindoleactic acid, steroid glucuronides and some ethereal sulphates; and the foreign substances penicillin, PAH and phenol red. The proximal tubule is also the major site of hydrogen ion secretion (see *Renal regulation of acid-base balance*).

The loop of Henle and its role in urine concentration

The prime function of the juxtamedullary loops of Henle is to set up a favourable con-

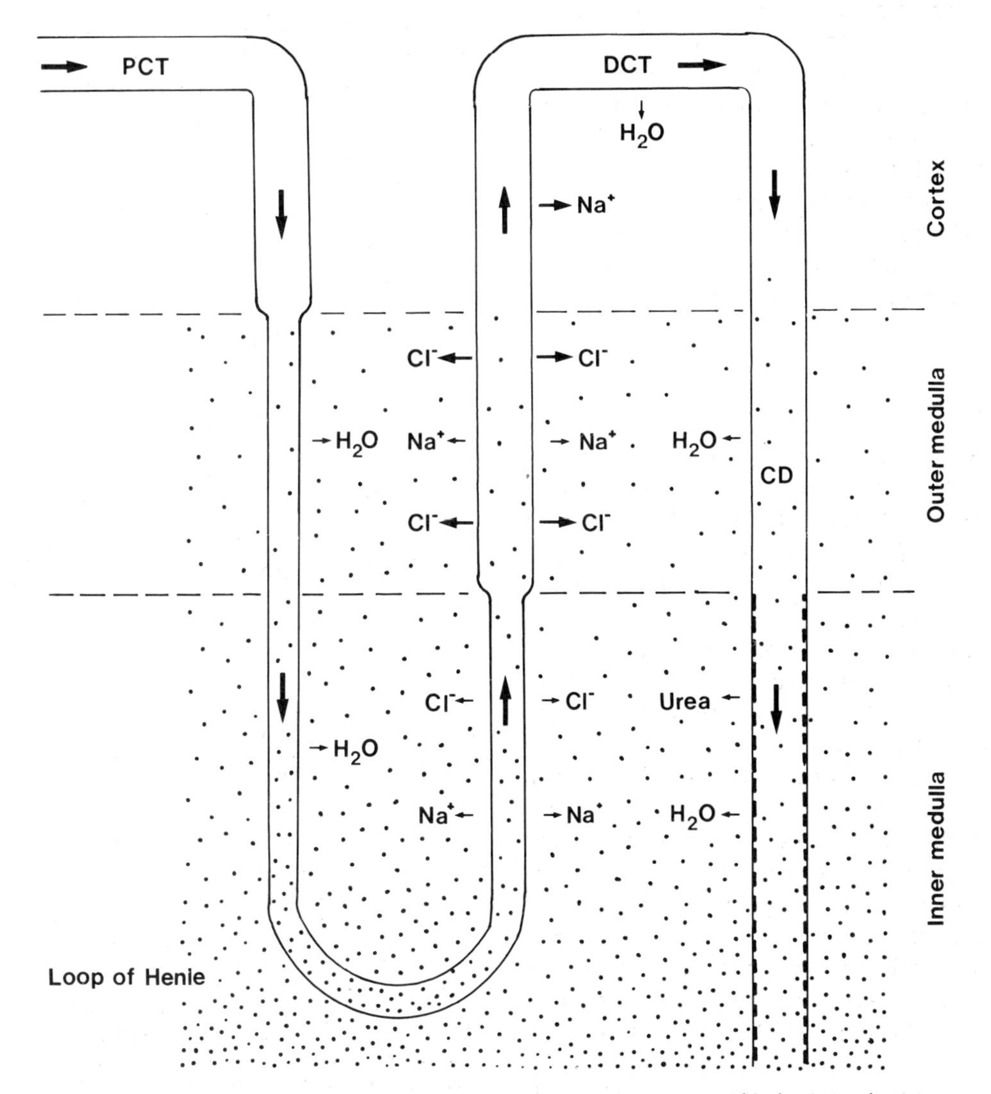

Fig. 3.9 Countercurrent mechanism. *Thick horizontal arrows* — active transport. *Thin horizontal arrows* — passive transport.

centration gradient which will result in the formation of a hypertonic urine in the collecting duct. This is achieved by the countercurrent mechanism (see Fig. 3.9).

Filtrate passes down through the descending limb of the medulla around the hairpin loop and then back to the cortex in the ascending limb.

As fluid passes up the thick ascending limb in the outer medulla, chloride ions are actively reabsorbed by the cells and passed into the interstitium. Sodium ions follow which maintains electrical balance, but since the entire ascending limb is impermeable to water, there is no reabsorption of water. The fluid which passes up the thick ascending limb becomes progressively less concentrated so that by the time it enters the distal convoluted tubule it is hypotonic to blood plasma.

The presence of increasing amounts of sodium and chloride in the interstitial fluid of the outer medulla causes the removal of water from (and possibly the addition of sodium and chloride to) the descending limb of the loop of Henle. The fluid draining towards the bottom of the loop therefore becomes increasingly hyperosmotic and as the fluid flows up towards the distal convoluted tubule, it loses solute and becomes hypotonic.

In the distal convoluted tubule further solute and water reabsorption occurs and the fluid entering the collecting duct is isotonic to plasma.

As the filtrate passes down the collecting duct, in the presence of antidiuretic hormone (ADH) water is drawn into the increasingly hypertonic interstitial fluid. Therefore as the fluid passes deeper into the medulla it becomes increasingly hypertonic. In the inner medulla the cells of the collecting duct become permeable to urea which has, by now become highly concentrated within the filtrate. Urea enters the interstitium of the inner medulla and draws water out of the thin descending portion of the loop of Henle. Therefore, immediately the loop of Henle turns back towards the cortex, sodium and chloride begin to diffuse passively into the surrounding environment. Thus the presence of urea, sodium and chloride serves to make the interstitium of the inner medulla extremely hypertonic.

It is evident that the entry of water into the interstitial fluid of the medulla from the loops of Henle and collecting duct would cause a decrease in the osmotic concentration. This decrease is prevented by the hairpin capillary loops, the vasa recta which abound in the medulla. The vasa recta act as 'countercurrent diffusion exchangers'. As blood passes down the descending limb of a vessel, water diffuses out and osmotically active particles (sodium ions, urea) diffuse in. As blood passes towards the cortex, water enters and osmotically active particles pass out. The net effect is therefore to trap solutes in the tip of the loop, in equilibrium with the medullary interstitium and the loop of Henle.

Thus the water which has been lost from the loops of Henle and collecting duct passes into the vasa recta and is returned to the cortex.

Distal tubule

The fluid entering the distal tubule is hypotonic to the surrounding interstitial fluid (and blood) so that water is reabsorbed thereby restoring isotonicity to the tubular fluid.

Sodium is reabsorbed from the tubule along with chloride and bicarbonate ions whilst hydrogen and potassium ions are secreted into the tubular fluid. Since more ions are removed from the tubular fluid than are added to it, an osmotic gradient is created which allows the further uptake of water.

The secretory and reabsorptive activities of the cells in this region are variable and depend largely upon the concentrations of the various substances in the blood. For example a rise in plasma potassium level results in an increased secretion of potassium (mediated by the hormone aldosterone); an increase in plasma sodium concentration results in decreased sodium reabsorption.

Acid-base balance is partially controlled in this section by the amount of hydrogen ion secretion and bicarbonate reabsorption that takes place. A more detailed account of sodium, potassium and acid-base balance is given in the section on the control of nephron function.

The quantity of water reabsorbed is determined by the degree of reabsorption of solute, since it follows by osmosis. It is now thought that ADH plays little if any role in the regulation of water reabsorption in the distal tubule in man.

Collecting duct

The main function of the collecting duct is the reabsorption of water and the formation of a hypertonic urine. The cells of the collecting duct are only freely permeable to water in the presence of ADH.

Sodium is reabsorbed from the collecting duct in the presence of aldosterone and chloride follows passively. Urea also is reabsorbed here, but only in the inner medulla. The walls of the ducts in the cortex and outer medulla are impermeable to urea, whilst those of the inner medulla are permeable in the presence of ADH.

As in the distal tubules the activities of the collecting ducts exhibit considerable variations which depend upon salt, water and acid-base balance within the body.

Control of nephron function

The elimination of water and salts by the kidney is largely regulated by endocrine organs. Water balance is under the influence of the posterior pituitary, and the adrenal cortices, which also regulate sodium and potassium levels. Calcium and phosphate levels are controlled by the parathyroid glands. The kidneys also contribute to the maintenance of the constant blood pH.

Regulation of water balance

Most water enters the body in liquids or solid foods, although a small amount (about 250 ml/day) is synthesized. In total, the daily intake of water is normally about 2.5 litres. In order that water balance is maintained an equal amount of water must be lost. Water is continually lost from the lungs on breathing out and through the skin by diffusion and then evaporation (insensible loss) and some is lost in the faeces. However, the two most important routes for water loss are the sweat glands, whose activities increase if body temperature rises, and the kidneys. It is through the kidneys that the body is able to regulate its water balance.

The volume of water reabsorbed by the kidneys is regulated primarily by antidiuretic hormone (ADH) released by the posterior pituitary. The hormone is actually synthesized within the cell bodies of the supra-optic and to a lesser extent the paraventricular nuclei both of which lie within the hypothalamus. Following synthesis, the neurosecretory material is transported along the axons to their terminals in the posterior pituitary (by a rapid transport system involving the microtubules). It is this stored hormone which is released directly into the blood stream.

Normally a small amount of ADH is constantly being released into the blood, thus the rate of its secretion may be increased or reduced.

A rise in the osmotic pressure of the plasma flowing to the brain initiates the release of ADH (see Fig. 3.10). The cells which are excited by this stimulus are termed 'osmoreceptors' and it has not yet definitely been established whether or not they are the actual ADH-synthesizing cells.

As a result of stimulation of the osmoreceptors, nerve impulses pass down the axons of the hypothalamo-hypophyseal tract at an increased rate and cause rapid release of ADH into the blood stream.

The hormone passes to the kidneys where it increases the permeablity of the collecting ducts to water. An increased amount of water is then drawn back from the tubules into the hypertonic environment of the medulla and then to the blood. As a result the osmotic pressure of the plasma is reduced and the stimulus removed.

A reduction in the osmotic pressure of the blood will reduce ADH release below the basal level and water will be lost from the body.

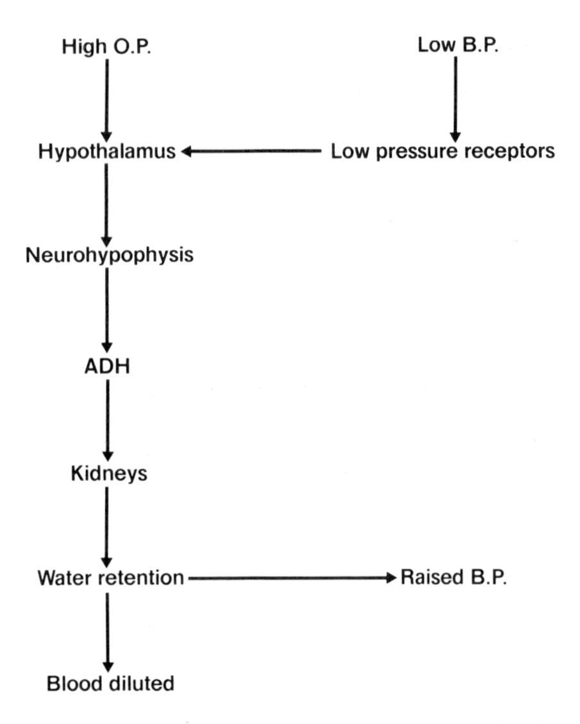

Fig. 3.10 Role of ADH in the regulation of water loss by the kidneys.

Thus in the first instance a small volume of hypertonic urine will be produced while in the second a large volume of dilute urine will be eliminated.

Even in the complete absence of ADH the collecting duct is not totally impermeable to water. In the presence of ADH up to 20% of the filtered fluid may be reabsorbed in the distal portion of the nephron, resulting in urine with a concentration of up to 1400 mosmol/l. When ADH is completely lacking 8% of the filtrate will still be reabsorbed and the concentration will fall to as low as 30 mosmol/l.

The hormone aldosterone also promotes water reabsorption from the distal nephron (see *Regulation of sodium and potassium balance*).

Although it is vitally important, the osmotic pressure of the blood plasma is not the only stimulus for ADH release. Changes in the extracellular fluid (e.c.f.) volume also exert a very significant influence.

When the e.c.f. volume rises, ADH output is reduced and when e.c.f. volume falls, ADH output is increased. The mechanism which regulates ADH output in these circumstances depends upon two types of receptors.

A small drop in the fluid volume within the circulation excites low pressure receptors in the great veins, atria and pulmonary system. Nerve impulses pass to the hypothalamus and as a result ADH is released. This, in turn, leads to a reabsorption of water by the kidneys and expansion of the e.c.f. volume (see Fig. 3.10).

A significant rise in blood pressure leads to excitation of baroreceptors within the carotid sinus and aortic arch. Impulses from these receptors inhibit hypothalamic activity and reduce ADH output.

Thus a change in the volume or composition of the extracellular fluid can be compensated by the kidneys through the activities of ADH.

The volume of water lost in the urine is also profoundly affected by the glomerular filtration rate (GFR) an effect which is independent of ADH.

A rise in the systemic blood pressure will increase the filtration pressure across the walls of the glomerular capillaries and thereby increase the GFR. The filtrate, because of its increased volume, will pass more quickly through the tubules so that less solute will be reabsorbed. As a result of the presence of extra solute in the tubular fluid, its osmotic concentration will be increased thereby causing water retention within the tubules. Thus the amount of water passing from the tubules will increase and urine output will rise. The diuretic effect of raised blood pressure, though very important, would be much greater were it not for the phenomenon of autoregulation whereby a rise in systemic blood pressure is automatically accompanied by a reduction in renal blood flow. The mechanism by which the kidney is able to regulate its own blood flow still remains controversial.

Regulation of sodium and potassium balance

Perhaps the most important, and certainly the best known regulator of sodium and potassium balance by the kidneys, is the hormone aldosterone which is synthesized and released by the outermost zone of the adrenal cortex (zona glomerulosa).

A drop in plasma sodium, or more especially a rise in plasma potassium, will directly stimulate the adrenal cortex to liberate aldosterone. The latter will then stimulate the cells of the distal tubules and collecting ducts of the kidneys to reabsorb sodium from the filtrate at an increased rate. The subsequent drop in the sodium concentration of the filtrate sets up an electrical imbalance, i.e. a surfeit of negative ions within the filtrate, which allows potassium ions to diffuse out of the cells. Thus aldosterone causes an increase in sodium reabsorption and a simultaneous increase in potassium secretion.

Since there are rather more sodium ions reabsorbed than potassium ions secreted, there is a net uptake of solute. Water is therefore osmotically withdrawn from the distal tubule.

It should be noted that hydrogen ions rather than potassium may be exchanged for sodium (see *Renal regulation of acid-base balance*).

Another stimulus which causes aldosterone release is a reduction in blood volume. It has been suggested that there may be a pressure sensitive area in the diencephalon which secretes an aldosterone-releasing factor, such that a drop in blood pressure will promote aldosterone secretion.

A drop in the blood flow in the afferent arterioles leading to the glomeruli will induce an indirect release of aldosterone. The juxtaglomerular cells which lie in the walls of the afferent arterioles at their entrance into the glomeruli are believed to be sensitive to stretch.

A decrease in the blood pressure or blood flow reduces the degree of stretch in these cells and, as a result, they release a peptide, renin, directly into the blood (see Fig. 3.11). Renin catalyses the conversion of a plasma protein angiotensinogen into angiotensin I which is, in turn, converted by a hydrolysing enzyme in the blood into the physiologically active angiotensin II. Since angiotensinogen and the converting enzyme are normally present in the blood, it is evident that the production of renin is the rate-limiting step in the formation of angiotensin II. The latter is a potent vasoconstrictor and produces a generalized vasoconstriction which helps to raise blood pressure. Angiotensin II also stimulates the release of aldosterone from the adrenal cortex, which promotes sodium and therefore water reabsorption. An increase in the plasma sodium concentration causes the withdrawal of water from the tissue spaces which leads to expansion of the plasma volume. A third effect of angiotensin II is to stimulate thirst, which may lead to the expansion of plasma volume by a different route.

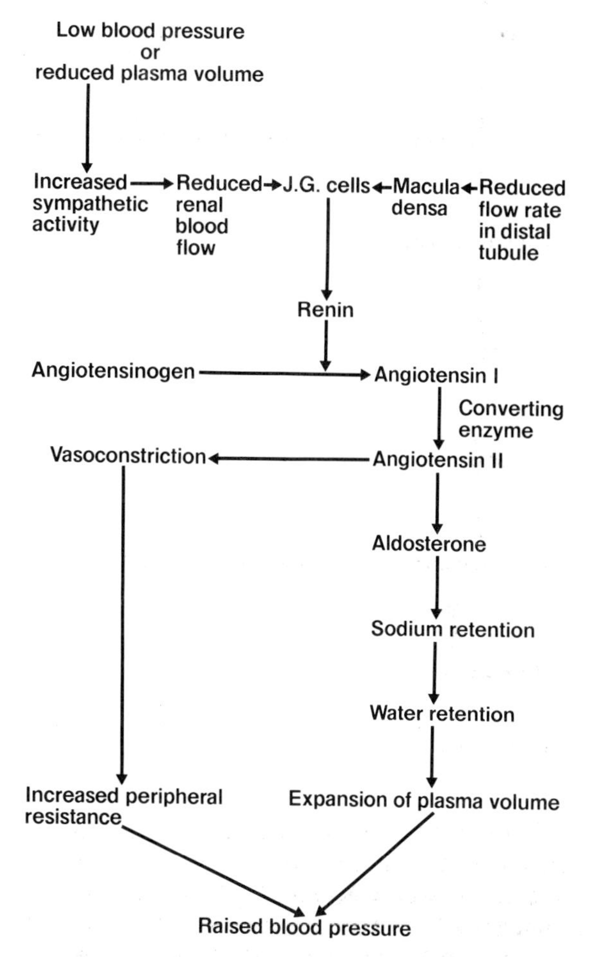

Fig. 3.11 Role of aldosterone in the regulation of blood pressure.

The juxtaglomerular cells also appear to be stimulated directly by sympathetic nerve fibres. Thus the reflex increase in sympathetic activity which occurs in response to low blood pressure may be accompanied by an increase in renin output.

Additionally, a change in the flow rate within the distal convoluted tubule may also induce the release of renin. The cells of the macula densa which are next to the juxtaglomerular cells are excited when the flow rates are low. The macula densa cells then appear to stimulate the juxtaglomerular cells directly in some way to release renin. This leads to the formation of angiotensin II, aldosterone release, sodium retention and expansion of plasma volume and an increased glomerular filtration rate. The exact mechanism by which the juxtaglomerular apparatus operates is not understood.

It is interesting to note that angiotensin II, as well as stimulating aldosterone release, also exerts a direct salt-retaining action upon the kidneys.

Regulation of calcium and phosphate balance

Calcium and phosphate ions are both filtered freely at the glomeruli and reabsorbed by active transport from the tubules. The amount filtered, as for all ions, depends upon the plasma concentration and the glomerular filtration rate. The amount reabsorbed depends upon the plasma level of parathyroid hormone.

Low levels of calcium in the blood stimulate the parathyroids to release their hormone. This then acts rapidly to reduce the reabsorption of phosphate by the tubules and more slowly to promote the reabsorption of calcium. It should be noted that in addition to the effects upon the kidney, parathyroid hormone exerts an important influence on bone metabolism and promotes the removal of calcium and phosphate from bone. Under the influence of parathyroid hormone there is a net rise in blood calcium, but a small drop in blood phosphate. Parathyroid hormone also influences the reabsorption of other ions, increasing the uptake of magnesium and hydrogen, but reducing the reabsorption of amino acids, sodium and potassium.

Renal regulation of acid-base balance

The blood pH (-log H^+ concentration) is maintained at a value of about 7.4 ($10^{-7.4}$ mol/l). Venous blood is about 0.03 pH units less than arterial. Constant pH is important because enzyme activity and therefore metabolism is particularly sensitive to hydrogen ion concentration. If the acidity of body fluids changes, then some chemical reactions are stimulated, others are depressed, so that a disordered state results.

To keep the acidity of body fluids at a constant level, acid must be eliminated and base must be conserved.

Acids (mainly sulphuric and phosphoric) are constantly being produced by metabolic processes. However, the major source of acid is carbon dioxide which forms carbonic acid in water.

$$H_2O + CO_2 \rightarrow H_2CO_3$$

As pH is a measure of the concentration of free hydrogen ions, then, provided that the hydrogen ions produced by metabolism are effectively removed by chemical combination with other substances, the pH will be unaffected. Such substances which can accept hydrogen ions and are reluctant to release them again are known as 'buffers'. For example, the phosphate buffer system involves the addition of a hydrogen ion to a biphosphate ion to produce a dihydrogen phosphate ion.

$$HPO_4^{=} + H^+ \rightarrow H_2PO_4^-$$

Hydrogen ions may be buffered by the blood (both in the plasma and in the red cells); by interstitial fluid and by intracellular fluid. When hydrogen ions enter cells, electrical neutrality is maintained by potassium or sodium ions leaving the cells in exchange.

Buffering is obviously only a temporary solution to the accumulation of acid and ultimately it must be excreted.

The major buffer system is the bicarbonate/carbonic acid one. This involves two consecutive reversible reactions.

1. When carbon dioxide is dissolved in water some of it forms carbonic acid.

2. Some carbonic acid dissociates into hydrogen ions and bicarbonate ions.

$$CO_2 + H_2O \overset{(1)}{\rightleftharpoons} H_2CO_3 \overset{(2)}{\rightleftharpoons} H^+ + HCO_3^-$$

Thus if acid is added to the blood it will 'push' the above reactions to the left using up bicarbonate and increasing the concentrations of carbonic acid and carbon dioxide.

The relationships between pH, bicarbonate and carbon dioxide concentrations at any one instant can be summarized from the Henderson-Hasselbalch equation.

$$\text{pH} \propto \frac{\text{bicarbonate concentration}}{\text{dissolved } CO_2}$$

It follows then that a rise in pH (less acid) could be achieved by an increase in bicarbonate ions, a fall in CO_2 levels and of course a fall in hydrogen ion concentration.

Bicarbonate reabsorption. Bicarbonate reabsorption has a threshold value around 28 mmol/litre. Below this plasma concentration, practically all filtered bicarbonate is reabsorbed. Above the threshold value the excess bicarbonate which is not absorbed appears in the urine.

Bicarbonate is reabsorbed indirectly as carbon dioxide and this is linked to sodium reabsorption and hydrogen ion secretion as shown in Figure 3.12A.

The sodium pump (1) actively extrudes sodium from the tubular cell into the interstitial fluid and sodium diffuses from the tubular fluid into the cell. Hydrogen ions are actively secreted into the tubular fluid (2) where they combine with the filtered bicarbonate to form carbonic acid.

This then splits to form carbon dioxide and water by the catalytic action of carbonic anhydrase (3) which is present on the cell membrane of the proximal tubule cell. The carbon dioxide then diffuses into the cell to join the 'pool' of dissolved gas which may diffuse out (4) or combine with water to form carbonic acid under the influence of carbonic anhydrase in the cytoplasm (5). The dissociation of carbonic acid provides the hydrogen ions for secretion (2) and bicarbonate for reabsorption by diffusion (6).

These events occur mainly in the proximal convoluted tubule where about 90% of the filtered bicarbonate is normally reabsorbed.

If blood potassium levels are raised then potassium ions will enter the body cells in exchange for hydrogen ions. In the kidney the loss of hydrogen ions into the interstitial fluid results in less hydrogen ions being available for secretion into the tubular lumen so that less bicarbonate reabsorption can occur.

Thus raised body potassium results in reduced hydrogen secretion and reduced bicarbonate reabsorption, whilst a fall in body potassium (hypokalaemia) results in increased hydrogen ion secretion and increased bicarbonate reabsorption.

Hydrogen ion secretion The amount of hydrogen ion secreted is much higher in the proximal convoluted tubule than in the distal nephron. These hydrogen ions are immediately reabsorbed as water (see Fig. 3.12) so that there is no loss of acid from this site. The importance of proximal tubular activity is not that acid is eliminated but that base is conserved.

Acidification of the urine is limited to a minimum pH of 4.5. Elimination of acid occurs by the secreted hydrogen ions being buffered in the tubular fluid and subsequently excreted in the urine. The two principal buffers are biphosphate (see Fig. 3.12B) and ammonia (Fig. 3.12C).

The phosphate buffer is filtered at the glomerulus and is therefore available in the tubular fluid to accept hydrogen ions. It may be however that the amount of biphosphate is small because it has already buffered hydrogen ions in the blood. Hydrogen ions will therefore accumulate in the tubular cells because of the build-up of unbuffered hydrogen ions in the tubular fluid.

The accumulation of acid within the tubular cells stimulates the production of ammonia from the amino acid glutamine (Fig. 3.12C).

The ammonia can then act as a hydrogen ion acceptor and enable the elimination of acid to take place. This process occurs in the

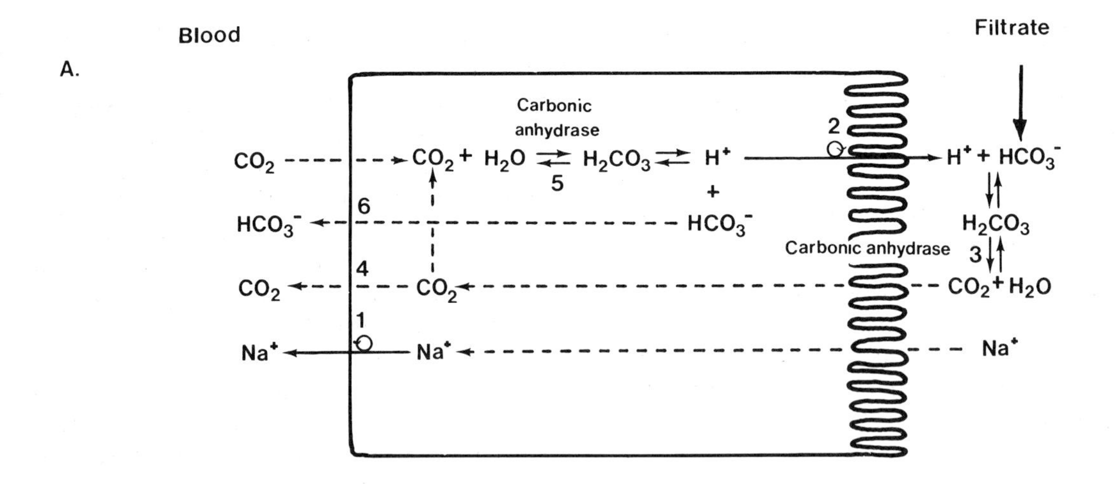

B.

Carbonic
anhydrase

$CO_2 \rightarrow CO_2 + H_2O \rightleftharpoons H_2CO_3 \rightleftharpoons H^+ \rightarrow H^+ + HPO_4^=$

+

$HCO_3^- \leftarrow HCO_3^-$

$H_2PO_4^-$

$Na^+ \leftarrow Na^+ \leftarrow Na^+$

C.

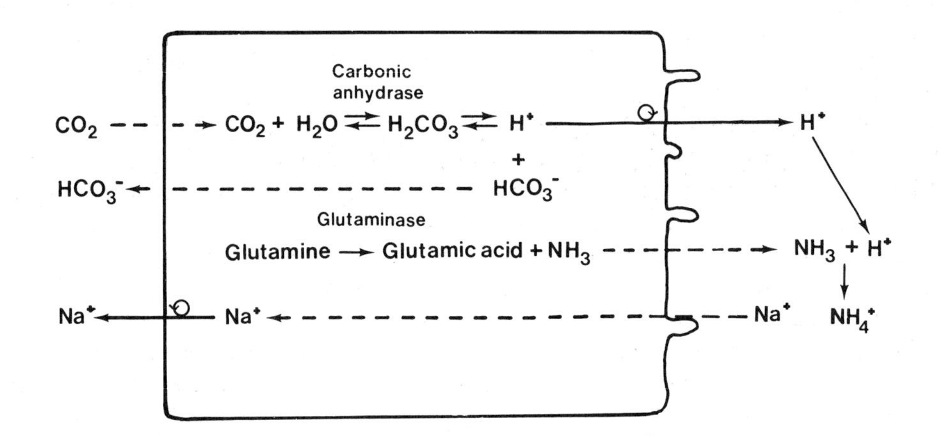

Fig. 3.12 Acid-base balance. (A) Role of bicarbonate buffer, (B) Role of biphosphate buffer, (C) Role of ammonia.

proximal convoluted tubule, the distal convoluted tubule and the collecting duct, but the major site is probably the distal convoluted tubule.

Respiratory and renal regulation of acid-base balance in altered states

It may be recalled from the Henderson-Hasselbalch equation that pH is directly proportional to bicarbonate ion concentration and inversely proportional to the amount of dissolved carbon dioxide.

$$pH \propto \frac{HCO_3^-}{CO_2}$$

The kidneys prevent acidosis by the constant reabsorption of bicarbonate and secretion of hydrogen ions, whilst the lungs eliminate CO_2 thereby effectively removing acid.

The interplay between the lungs and the kidneys in the maintenance of a constant pH may be illustrated by a brief consideration of altered acid-base states.

Metabolic acidosis may arise from an excessive production of hydrogen ions, e.g. in diabetes mellitus; or by a loss of alkaline fluids such as in diarrhoea. The fall in pH stimulates ventilation thereby eliminating CO_2 and tending therefore to reduce the increased acidity. Respiratory compensation however is incomplete, so that restoration of the normal pH requires the elimination of hydrogen ions and the conservation of bicarbonate ions by the kidneys.

Metabolic alkalosis can arise from persistent vomiting, whereby hydrochloric acid is lost from the stomach. The immediate effects are that blood pH and bicarbonate concentrations increase. The increased pH may reduce ventilation thereby raising blood carbon dioxide and acidity so that the pH will tend to fall.

The renal compensation is that the excess bicarbonate is lost in the urine.

Respiratory acidosis results from impaired elimination of carbon dioxide from the lungs, so that the primary effect is raised blood carbon dioxide level (hypercapnia). This leads to raised carbonic acid levels and a fall in pH of the blood.

The kidneys compensate for the change in acidity in the following manner. An increased amount of carbon dioxide diffuses into the tubular cells resulting in increased hydrogen ion secretion and bicarbonate reabsorption (see Fig. 3.12). Both these effects will attenuate the acidosis. Any excess hydrogen ions will combine with biphosphate and ammonia.

Respiratory alkalosis caused by chronic hyperventilation will reduce blood carbon dioxide and carbonic acid levels. Renal compensatory mechanisms are the reverse of those operating in response to respiratory acidosis. Less carbon dioxide enters the tubular cells from the blood in this case, and therefore less hydrogen ions are secreted and fewer bicarbonate ions are reabsorbed. The effect of the reduced tubular activity will be to conserve more acid and eliminate more base thereby counteracting the alkalosis.

Renal handling of individual substances

Sodium and chloride

These two ions contribute about 80% of the total osmotic activity of the glomerular filtrate. Generally sodium is reabsorbed by active transport and chloride follows passively, although recent work has indicated the existence of an active transport system for chloride in the thick ascending limb of the loop of Henle and possibly the distal convoluted tubule.

Generally about 99% of the filtered sodium is reabsorbed by the tubular cells; between 60 and 80% in the proximal convoluted tubule. The active transport systems for sodium are located on the basal and lateral surfaces of the epithelial cells (Fig. 3.13). Thus the removal of sodium from the tubular cells creates a low concentration for sodium within the cell cytoplasm and creates a concentration gradient between it and tubular fluid. As a result of this gradient, sodium ions diffuse into the cells and are then pumped into the interstitial fluid

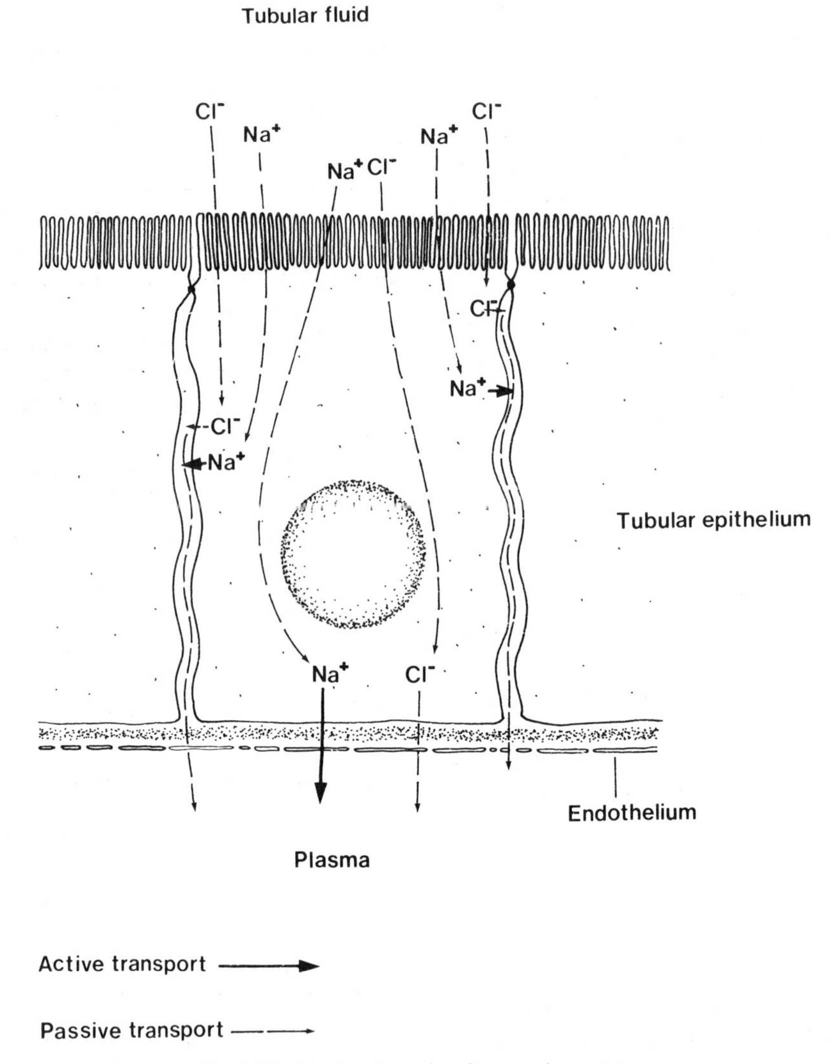

Fig. 3.13 Mechanism of sodium reabsorption.

which surrounds the cells. Sodium ions then diffuse down a concentration gradient from the interstitial fluid into the peritubular capillaries.

The reabsorption of sodium in the proximal convoluted tubule is, for the most part, accompanied by chloride ions, although some bicarbonate may be reabsorbed instead.

Water

Water is reabsorbed passively down an osmotic gradient from the proximal and distal convoluted tubules, the descending limb of the loop of Henle and the collecting duct.

In the published texts it is not always clear whether the values quoted for tubular reabsorption refer to animals or man. We have, therefore, found it necessary to quote a range of estimates in each case.

Between 65 and 75% of the filtered water is reabsorbed in the proximal tubule; this reabsorption is independent of hormonal control and for a given individual is relatively constant (obligatory reabsorption). 5–15% is reabsorbed from the loop of Henle and again this is not hormone dependent and is a fixed amount.

Reabsorption from the distal tubule and collecting duct together varies between 8 and 20%. This facultative reabsorption is controlled by aldosterone and ADH. Aldosterone stimulates sodium reabsorption which is followed by water in both the distal convoluted tubule and the collecting duct, whereas ADH promotes water reabsorption only from the collecting duct. Thus the rate of urine production can vary from less than one ml per minute to about 15 ml per minute according to salt and water balance in the body.

Potassium

Potassium is actively reabsorbed by the cells of the proximal tubule and the ascending limb of the loop of Henle. Thus most of the potassium has been removed from the filtrate when it enters the distal tubule and here it is secreted passively when sodium is actively reabsorbed. Potassium itself may also be reabsorbed in the distal tubule when, for example, there is a low dietary intake.

Urate

Urate is formed in the liver as a breakdown product of nucleic acids. Approximately 90% of the filtered urate is reabsorbed in the proximal tubule and loop of Henle. Urate is also secreted in the proximal tubule.

Creatinine

Creatinine is derived from the breakdown of muscle creatine and creatine phosphate and over a period of hours excretion is relatively constant.

Creatinine is filtered freely but is not reabsorbed at all by the tubules but is secreted to a small extent by the proximal tubule. The clearance value (see below) for creatinine is used to estimate the glomerular filtration rate.

Renal clearance

The concept of clearance provides a useful estimate of renal function, since it indicates how efficiently substances are removed from the blood by the kidney.

The clearance rate of a substance is defined as that volume of plasma completely cleared of a given substance in one minute (or that volume of plasma which contains the amount of substance which is excreted in the urine in one minute). By choosing a substance which filters freely but is neither reabsorbed nor secreted in the tubule the clearance value is found to correspond to the rate of filtration; such a substance is the polysaccharide inulin.

A large dose of inulin is given intravenously, and the following values are obtained:

Volume of urine/unit time = V(ml/min)
Inulin concentration in plasma = P(g/100 ml)
Inulin concentration in urine = U(g/100 ml)

$$\text{Amount of inulin excreted/min} = \frac{U \times V}{100}\ \text{g}$$

Each ml of plasma contains $\frac{P}{100}$ g of inulin

If z ml of filtrate is formed per minute,

then $\frac{zP}{100}$ g of inulin is filtered per minute.

Since inulin is neither secreted nor reabsorbed in the tubule then:

$$\frac{UV}{100} = \frac{zP}{100}$$

and $\frac{UV}{P}$ = z ml/min = filtrate formed/min = inulin clearance.

Example: V = 1 ml/min
P = 0.05 g/100 ml
U = 6.25 g/100 ml

$$z = \frac{UV}{P} = \frac{6.25 \times 1}{0.05} = 125\ \text{ml/min}$$

Thus the clearance/filtration value for inulin is 125 ml/min. A clearance value of less than 125 ml/min indicates that the substance is probably reabsorbed in the tubule, e.g. glucose = 0 ml/min (total reabsorption) and

Table 3.2 Concentrations of substances in glomerular filtrate (125 ml/min) and in urine (1 ml/min) (modified from Guyton, 1981)

Substance	Concentration in glomerular filtrate	Concentration in urine	Concentration in urine / Concentration in plasma (Plasma clearance per minute)
Na^+	142 mEq/l	128 mEq/l	0.9
K^+	5	60	12
Ca^{2+}	4	4.8	1.2
Mg^{2+}	3	15	5
Cl^-	103	134	1.3
HCO_3^-	28	0	0
$H_2PO_4^-$ / HPO_4^{2-}	2	50	25
SO_4^{2-}	0.7	33	47
Glucose	100 mg%	0 mg%	0
Urea	26	1820	70
Uric acid	3	42	14
Inulin			125
PAH			585

urea = 70 ml/min (partial reabsorption). Table 3.2 gives typical clearance values of the major substances in glomerular filtrate.

Creatinine clearance is often used to estimate glomerular filtration rate. Creatinine is filtered completely and is secreted (slightly) into the tubules. However, since the plasma creatinine level usually appears to have a higher value than it should (because plasma chromogens give a similar reaction) the clearance value is approximately 125 ml/min.

A value greater than 125 ml/min indicates tubular secretion, e.g. PAH = 585 ml/min, PAH filters freely at the glomerulus and any left in the blood is then removed by the tubular cells. Thus the blood leaving the kidneys contains no PAH at all; in other words, PAH is completely removed by the kidneys and its clearance represents the maximum possible. Since all of the plasma leaving the kidneys/min has been cleared of PAH then the clearance value must be equal to the plasma flow rate. Using this value, together with a knowledge of the proportion of blood cells in the renal circulation, allows the renal blood flow to be calculated. (Approx 1200 ml/min.)

4

The cardiovascular system

Every cell in the human body requires a constant supply of food materials and oxygen. In addition, each cell must rid itself of its waste products, including carbon dioxide. The circulatory system provides a means by which both of these functions can be fulfilled. Oxygen is transported in association with haemoglobin from the lungs to the tissues of the body; glucose, amino acids and lipids are carried in the plasma from the gut. Carbon dioxide and other metabolic wastes are transported to the lungs or kidneys, again carried by the blood.

Blood must be, therefore, a mobile fluid which circulates around the body. A pump (the heart) and a system of tubes (veins and arteries) are required for the efficient circulation of this fluid.

The circulatory system is effectively two systems. One, the pulmonary system carries blood from the right side of the heart, to and through the lungs where the blood is oxygenated, before being returned to the left side. The second, or systemic system carries oxygenated blood from the left side of the heart to the rest of the body, from which it returns to the right side (see Fig. 4.1).

THE HEART

The heart is a small muscular organ contracting in the order of 4200 times every hour. This represents an energy output of about 4.12 kJ, when each ventricle is pumping out 60 ml of blood per beat.

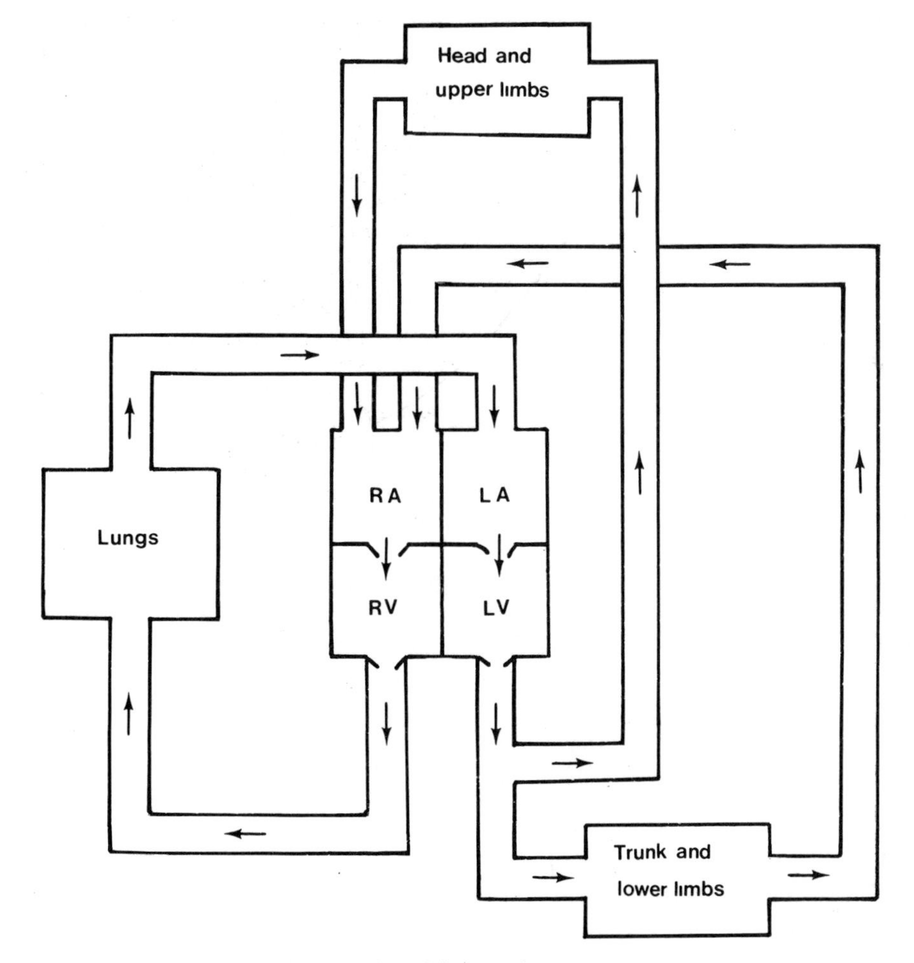

Fig. 4.1 Plan of the circulatory system.

The heart lies in the thoracic cavity immediately above the diaphragm and between the lungs.

The thoracic cavity is divided into two parts by a central cavity the mediastinum, which itself is subdivided into superior and inferior portions. The heart lies in the middle of the inferior mediastinum surrounded by a fibrous sac, the pericardium.

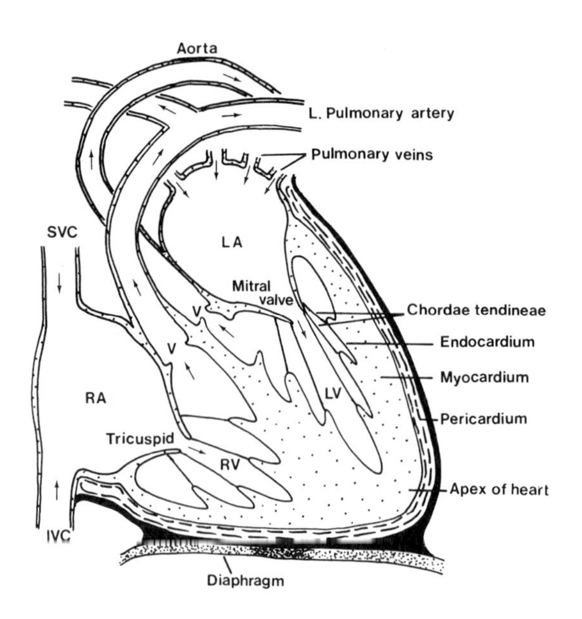

Fig. 4.2 Diagrammatic section through the human heart. The atria have been displaced slightly so that all four chambers can be seen in the same plane. SVC — superior vena cava, IVC — inferior vena cava, RA — right atrium, RV — right ventricle, LA — left atrium, LV — left ventricle, V — pulmonary and aortic valves.

STRUCTURE OF THE HEART

The human heart is a hollow, conical, muscular organ, placed obliquely in the chest behind the sternum, one third to the right, two-thirds to the left of the median plane. The apex is directed to the left.

In size it is approximately 12 cm × 8–9 cm × 6 cm, and weighs about 300 g.

It has four chambers whose divisions are indicated on its surface by grooves or sulci.

The pericardium

The heart is enclosed by a conical, fibroserous sac which also contains the roots of the great vessels. This sac comprises two layers, an outer fibrous layer and an inner serous layer.

The outer pericardium is a fibrous bag with a truncated apex, continuous with the external coat of the great vessels. Its base is attached to the central tendon of the diaphragm and a small part of the anterior face fuses with the chest wall. The posterior surface rests on the bronchi while the lateral faces are in contact with the pleurae of the lungs.

The inner or serous pericardium lines the fibrous layer and is invaginated by the heart. It consists of visceral and parietal layers. The visceral portion or epicardium covers the heart and great vessels and is reflected back to form the parietal layer which lines the fibrous pericardium (see Fig. 4.2).

Where it covers the major vessels, the pericardium is arranged so as to form two tubes. The aorta and pulmonary trunk are enclosed in one tube, the superior and inferior venae cavae and pulmonary veins in the other.

Atria

The right and left atria lie at the top of the heart, separated by an oblique interatrial septum, so that the right lies anteriorly to the left.

The right atrium receives blood from the superior and inferior venae cavae. Only the inferior vena cava exhibits a (poorly developed) valvular opening. The opening of the coronary sinus (which drains blood supplying the heart itself) lies between the opening of the inferior vena cava and the atrioventricular orifice. A small conical pouch, the auricle, projects towards the left from the upper anterior part of the atrium, overlapping the right side of the root of the ascending aorta.

The left atrium is rather smaller than the right, but has thicker walls. The cavity is largely derived from the proximal parts of the pulmonary veins. The auricle is directed forward on the left side of the pulmonary trunk and overlaps the root of this vessel.

Ventricles

The right ventricle extends from the atrium nearly to the apex of the heart. The wall is much thinner than that of the left, being thickest at the base and becoming thinner towards the apex. The interior cavity is divided into an inflowing and an outflowing region by a muscular ridge, the supraventricular crest, situated between the atrioventricular and pulmonary orifices.

The right atrioventricular orifice is a large oval aperture between the right atrium and ventricle. It is encircled by a fibrous ring to which are attached the three flaps of the tricuspid valve. The latter have smooth atrial surfaces but their lower surfaces are rough and irregular. The margins and lower surfaces of the cusps are attached to the ventricular wall by a number of delicate cords, the chordae tendineae. These are primarily anchored to irregular columns of muscle tissue, the trabeculae carneae (papillary muscles), projecting from the wall of the ventricle. These structures prevent eversion of the tricuspid valve, thus ensuring one-way passage of blood.

The pulmonary valve consists of three semilunar segments, attached by their convex margins to the wall of the pulmonary trunk at its junction with the ventricle.

The left ventricle is longer than the right and forms the apex of the heart. Its cavity presents a circular or oval transverse section, with walls approximately three times thicker than the right.

The left atrioventricular or mitral orifice is placed below and to the left of the aortic orifice and is guarded by two triangular cusps attached to a dense fibrous ring. The cusps are of unequal size and are larger and stronger than those of the tricuspid valve. Again chordae tendineae prevent eversion of the valve.

The aortic valve possesses three semilunar segments, similar in structure, but larger and thicker than those of the pulmonary valve.

Conducting system of the heart

The conducting system of the heart consists of specially differentiated muscle tissue — nodal fibres and Purkinje tissue. This system is responsible for the initiation and maintenance of normal cardiac activity (see Fig. 4.4).

The sinoatrial node (pacemaker) is a narrow structure situated in the upper part of the right atrium, in front of the opening of the superior vena cava. It is 10–20 mm long, 1 mm thick and 3 mm wide. The atrioventricular node is smaller and lies above the opening of the coronary sinus in the atrial septum. From the node, Purkinje fibres (see below) travel down in the slender atrioventricular (AV) bundle (of His), between the fibrous rings of the AV valves, to the ventricular septum. The AV bundle then divides into right and left branches which travel just underneath the endocardium down to the apex of the heart and then up towards the base again on the lateral walls. Purkinje fibres branch off from the main bundles and supply the myocardium.

Cellular organization

The heart proper is covered by the serous layer of the pericardium. This is firmly attached to the muscular wall of the heart and is termed epicardium. The contractile portion of the wall to which the epicardium is attached, the myocardium, forms the bulk of the heart tissue. It is lined by the endocardium which is composed of elastic tissue and a smooth endothelium.

The cardiac muscle cells of the myocardium are branched and striated and are arranged in two blocks, the atrial and ventricular syncytia (see Ch. 1). Within each syncytium action potentials can pass very rapidly from cell to cell, but transmission between the syncytia is only by the AV bundle (see Fig. 4.4).

Nodal fibres are cross-striated but are narrower than ordinary cardiac fibres. Purkinje fibres, on the other hand, are much broader and possess irregular striations. In addition, they are particularly rich in glycogen.

FUNCTION OF THE HEART

The cardiac cycle

The cardiac cycle is the series of pressure changes, valve actions and electrical potentials that bring about the movement of blood through the heart during one complete heart beat.

Oxygenated venous blood returns from the lungs to the left atrium via the pulmonary veins. Blood then flows from the atrium into the thick walled left ventricle through the AV valve. Ventricular contraction causes a rise in pressure which closes the AV valve and eventually opens the aortic valve, thus passing blood into the aorta and around the body.

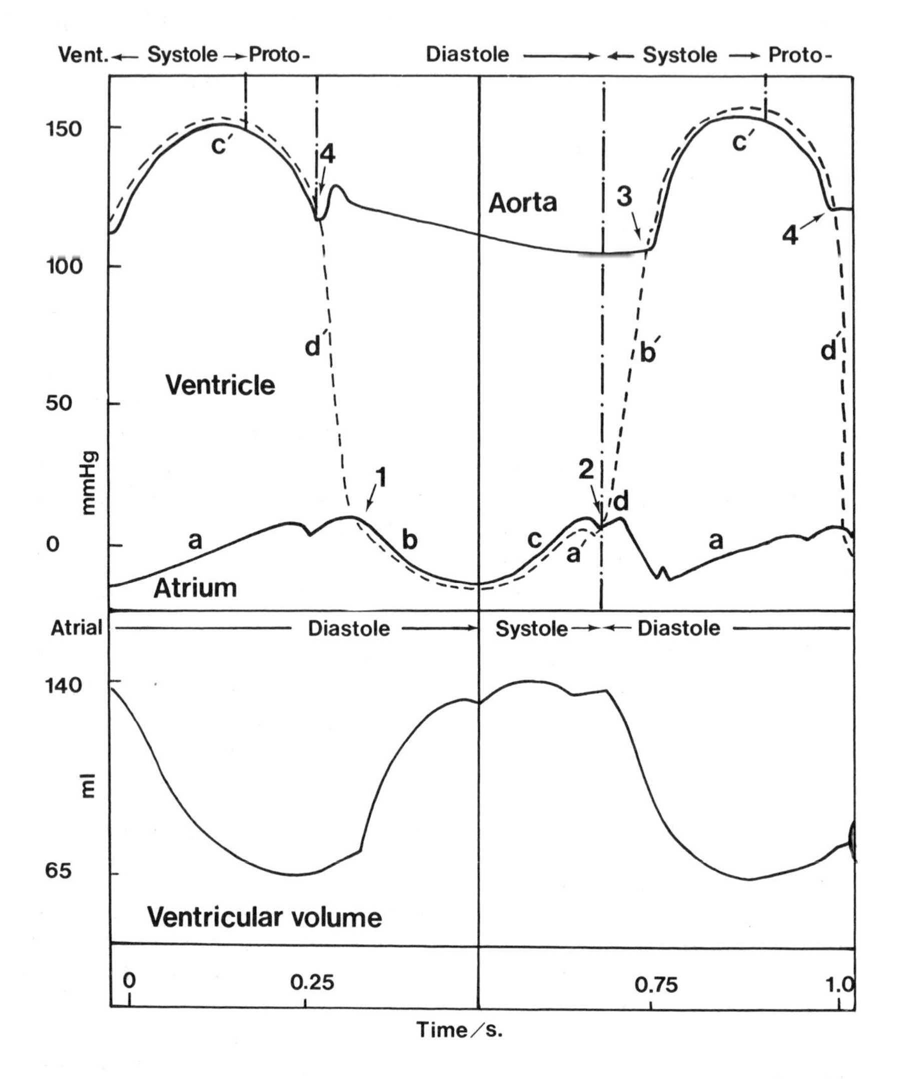

Fig. 4.3 Pressure curves of atrium, ventricle and aorta: left side only.
a–d : atrial pressure curve
a′–d′: ventricular pressure curve
1 : AV valve opens
2 : AV valve closes
3 : Aortic valve opens
4 : Aortic valve closes

Deoxygenated blood enters the right atrium from the superior and inferior venae cavae. Blood flows passively at first and then is pushed into the right ventricle when the atrium contracts. The ventricle then contracts and forces blood into the pulmonary artery and then to the lungs.

The atria contract almost simultaneously (atrial systole) during this cycle as do the ventricles (ventricular systole). In each case a diastolic or relaxation phase follows the contractions.

The sequence of events in the cardiac cycle is best studied from records of pressure changes within the chambers of the heart (see Fig. 4.3).

Atrial pressure

Pressure within the atrium rises as blood enters passively from the great veins into a chamber that is sealed shut by the AV valve (a). When atrial pressure just exceeds ventricular pressure, the valve opens (1), after which the pressure falls as blood flows into the ventricle (b). Pressure again begins to rise as atrial contraction forces blood into the ventricle (c). A third rise (d) is caused by ventricular systole and closure of the AV valve (2).

Ventricular pressure

Atrial systole forces blood into the relaxed ventricle causing a small rise in pressure (a′). Following closure of the AV valve (2), ventricular contraction produces a large pressure increase (b′) as the muscular wall attempts to compress the blood in the closed chamber. This is the isometric phase of contraction. Once the ventricular pressure rises above that in the aorta, the aortic valve opens (3). Blood is pumped into the aorta and the ventricle decreases in size so that pressure rises in both the aorta and the ventricle (c′).

During the last one-fifth of systole, the ventricular and aortic pressures fall as blood flows away from the aorta and very little is being added from the ventricle. This period is known as protodiastole. At the end of contraction, ventricular pressure falls below that in the aorta, and therefore the aortic valve closes (4). As the ventricle relaxes so the internal pressure drops, but as the AV valve remains closed then the volume remains constant (d′). Ventricular pressure continues to diminish to a value below that in the atrium, when the AV valve opens (1). Thus a new cycle is begun.

The curves of pressure changes in the two sides of the heart are similar, but those on the right are smaller (peak right ventricular pressure is about 25 mmHg). Both sides of the heart eject 60–70 ml of blood per beat at rest, leaving about the same volume behind (end-diastolic volume).

The muscular activity of the heart causes it to move within the chest cavity. In diastole the heart is soft, but in systole the ventricles harden, causing an increase in the antero-posterior and a decrease in the transverse diameter. The apex of the heart scarcely moves since it is anchored to the central tendon of the diaphragm. The base of the heart, how ever, moves downwards and forwards during ventricular systole.

The cardiac cycle also produces a number of sounds which can be heard with the aid of a stethoscope. The first heart sound (lub) is recorded when the cusps of the AV valve come together as ventricular pressure builds up. The second heart sound (dup) is due to closure of the aortic and pulmonary valves.

Initiation and conduction of the cardiac impulse

Individual cardiac muscle cells are capable of spontaneous rhythmic, self-excitation. If this auto-rhythmicity were not co-ordinated, however, the heart as a whole would not be an effective pump. Co-ordination is brought about by the specialized cardiac muscle cells of the sinoatrial (SA) node which have the highest rate of self-excitation of any in the heart.

A wave of excitation spreads out from the SA node through the muscle of the right atrium and into the left atrium at a rate of approximately 0.5 m/s (see Fig. 4.4). As a result

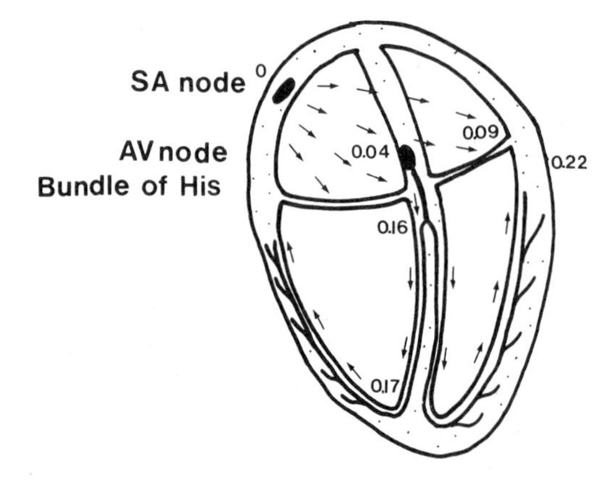

Fig. 4.4 Spread of excitation through the heart. Figures indicate elapsed time in seconds.

of this electrical activity the two atria contract more or less simultaneously.

At the bottom of the right atrium a second group of specialized cells, the atrioventricular (AV) node receives electrical excitation and after a short delay of about 0.1 sec passes the impulse to the bundle of His which passes through the fibrous rings separating the tissues of the atria and ventricles. The delay ensures that the atria have time to relax before the ventricles begin to contract. The impulse passes down the bundle of His which divides into right and left branches, eventually terminating in the Purkinje tissue deep within the ventricular myocardium. The wave of electrical excitation is thus carried rapidly to the myocardial cells (2.5 m/s) causing simultaneous contraction of the two ventricles.

Cardiac muscle cells have an extremely long absolute refractory period (0.3 s in ventricular cells) which allows them to relax completely between action potentials and prevents the development of sustained contractions. This is very important since effective heart function is dependent upon discrete contractions.

Electrocardiogram

Depolarization and repolarization of the cardiac muscle cells sets up electrical currents flowing through the body fluids which can be detected by attaching electrodes to the skin surface.

Conventionally three sets of electrodes are used which are attached to leads as follows: lead I (electrodes on right and left arms), lead II (electrodes on right arm and left leg) and lead III (electrodes on left arm and left leg). The leads are attached to a measuring device and the record produced by the heart's activity is known as the electrocardiogram (e.c.g.). In fact slightly different records are obtained depending upon which lead is chosen. Figure 4.5 illustrates a typical lead I e.c.g. from a healthy person.

Depolarization of the atria is indicated by a small P wave which is followed by a short interval during which time the atria are in a state of contraction. The wave and the short interval constitute the P-R interval (so named because the Q wave is frequently absent from the e.c.g.) and it lasts from 0.1 to 0.2 seconds. Atrial repolarization then occurs but is masked by the QRS complex which signals the commencement of ventricular contraction. Repolarization of the ventricular fibres commences 0.15 s, after initial depolarization but may not be completed for 0.3 s; as a result the T wave is usually more prolonged and flatter than the QRS complex.

Notice that the trace is flat during some of the P–R interval and the whole of the S–T in-

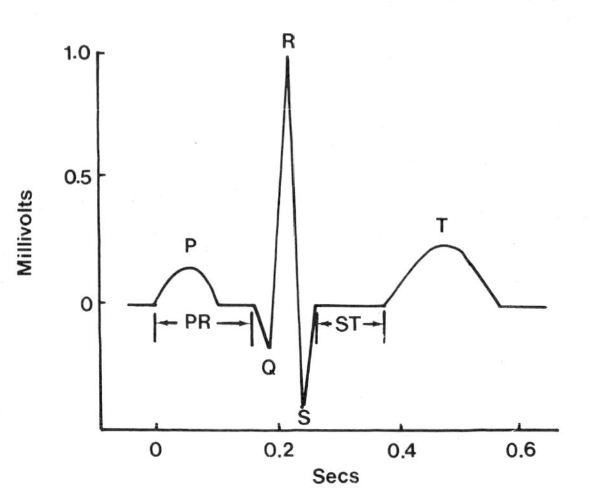

Fig. 4.5 Typical lead I electrocardiogram from a healthy person.

terval even though the atria, in the first case and the ventricles in the second, are in a state of contraction. A wave is only observed at the commencement of depolarization or repolarization.

Excitation is mainly passing from the base of the heart (SA node) towards the apex. The electrodes attached to the surface of the body record this wave as it radiates out from the heart through the body fluids. Using lead I, because the right arm electrode is closest to the base of the heart it receives the wave first and becomes negative with respect to the left arm electrode which is positive. The two electrodes are connected to a recording device so that an upward deflection is given when the right arm electrode is negative, therefore producing a positive R wave on the e.c.g. The negative Q wave is produced when the wave of excitation travels across the ventricular septum from left to right. This is because the left hand branch of the bundle of His is rather better developed than the right so that the left hand side of the septum depolarizes slightly ahead of the right. The negative S wave results from the wave travelling through the outer ventricular wall back towards the base.

It is evident from the above that it is the direction that the wave of excitation takes which determines the plane of deflection on the e.c.g. The ventricular repolarization wave also moves towards the apex and also produces a positive T wave.

Clinical determination of the e.c.g. is a much more complicated procedure than is suggested here. In addition to the three limb locations, as many as six locations on the chest may be used for the attachment of electrodes and as a result a number of different e.c.g. pictures may be obtained. Some or all of these pictures could be employed in the diagnosis of heart disease.

Cardiac output and its regulation

The volume of blood pumped out by each ventricle during one minute is known as the cardiac output. It is dependent upon the volume of blood pumped at each beat (stroke volume) and the number of beats during one minute (heart rate).

cardiac output = heart rate × stroke volume,

e.g. at rest
cardiac output = 70 beats/min × 70 ml/beat
= 4900 ml/min

An increase in cardiac output can therefore be brought about by an increase in heart rate and/or stroke volume. For example, a young athlete undergoing maximal exercise can attain a heart rate of 200 beats/min and a stroke volume of 150 ml/beat, resulting in a cardiac output of 30 litres/min.

Control of heart rate

The spontaneous rhythm of the heart is modified by the autonomic nervous system even under resting conditions. The heart receives fibres from the tenth cranial (vagus) nerves from the medulla and from sympathetic nerves arising in the upper thoracic regions of the spinal cord which pass to the heart via the cervical sympathetic ganglia. Fibres from the left vagus terminate mainly around the AV node while those of the right end close to the SA node. Sympathetic fibres from the right side primarily innervate the SA node while those of the left lead to the AV node and bundle of His.

Stimulation of the parasympathetic fibres reduces the heart rate; prolonged stimulation may even cause the heart to stop beating completely. Cutting the parasympathetics on the other hand will lead to an increase in the resting heart rate.

Stimulation of the sympathetic nerves induces an increase in heart rate, whereas cutting the fibres will result in a slowing down of the resting heart. Since the resting heart rate may be considered to be about 70 beats/min and the SA node alone elicits a rate of about 100 beats/min it is evident that, at rest, the parasympathetic system is dominant.

The autonomic pathways to the heart originate in a diffuse network of nerve fibres situated in the medulla oblongata. It has been established that some of these fibres initiate parasympathetic activity and this region is known as the cardioinhibitory centre. Another area, the vasomotor centre, is responsible for initiating vasoconstriction through the sympathetic nervous system. This area can also increase the heart rate.

The endocrine system also influences heart rate and the hormones adrenaline, and to a much lesser extent noradrenaline, will also increase the heart rate.

Control of stroke volume

The heart has an intrinsic mechanism of varying its output which is independent of nervous control. This mechanism enables stroke volume to be altered in response to changes in venous and arterial pressures.

A rise in the rate of filling of the heart (venous return) will result in a greater volume of blood being present in the ventricles at the end of diastole, i.e. there is an increase in end-diastolic volume. If cardiac muscle is stretched, it contracts more forcefully (rather like a piece of elastic) and therefore it ejects a greater stroke volume.

A rise in arterial pressure will result in similar, but not identical adjustments. If arterial pressure rises then resistance to outflow from the heart is increased and the heart will have to work harder to pump against this resistance. Initially the heart is not able to pump out as much blood as it receives and it therefore distends. Eventually output from the heart comes to equal input, but end-systolic and end-diastolic volumes are both increased, although stroke volume does not change.

In both of these situations it is clear that cardiac adjustments are mediated through volume increases. In both cases, the heart is contracting more strongly when cardiac volumes are raised and the muscle cells in the wall are stretched. The relationship between the degree of stretch of the muscle cells and the amount of work they produce is summarized by Starling's law of the heart; 'the force of contraction is proportional to the length of the muscle cells.' This means that within physiological limits, the heart adjusts its output of blood to keep pace with the inflow, thereby avoiding excessive amounts of blood either being retained in the venous system or shunted into the arterial system.

Changes in stroke volume may also be brought about under the influence of the sympathetic nervous system. Noradrenaline from the sympathetic nerve endings and also, incidentally, circulating adrenaline from the adrenal medullae increase the contractility of the cardiac muscle cells. Thus, for a given initial cell length noradrenaline and adrenaline will increase the strength of contraction and the final fibre length is reduced. In addition to this, since sympathetic activity and/or circulating adrenaline will also increase heart rate, there is less time available during each beat for cardiac filling and emptying. The rapid relaxation of the heart in diastole produces a very rapid drop in pressure which helps to suck blood into the ventricles. The increased strength of contractions leads to more complete ejection of blood and less blood is left in the heart at the end of each beat. The residual (end-systolic) volume, then, is reduced and stroke volume is raised.

Cardiac reflexes

Nerve impulses from many parts of the body converge on the cardiovascular centres in the medulla. The latter then initiate alterations in the functions of both heart and blood vessels.

The cardiovascular centres consist of two primary components, the vasomotor centre and the cardioinhibitory centre. The vasomotor centre is further subdivided into a depressor and a pressor area. The cardiovascular changes initiated by these centres may be illustrated by an analysis of their role in the baroreceptor reflexes.

Baroreceptors are spray-type nerve endings found in the walls of many of the major arteries of the thorax and neck; they respond to changes of blood pressure within those vessels.

They are most abundant in the walls of the aorta and the internal carotid arteries im-

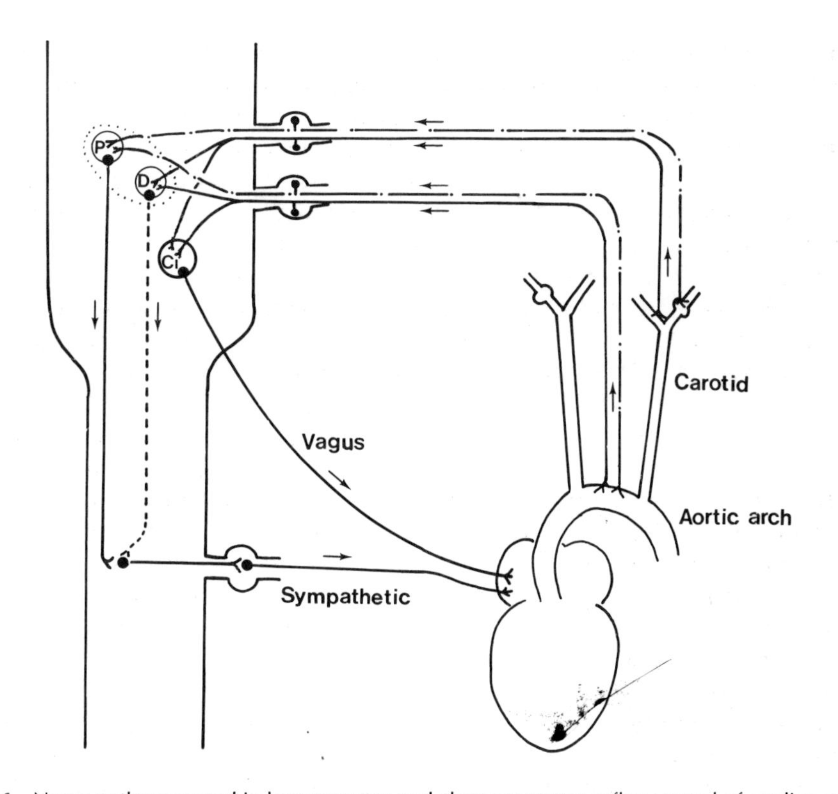

Fig. 4.6 Nerve pathways used in baroreceptor and chemoreceptor reflex control of cardiac activity.
Sensory pathways from chemoreceptors
Sensory pathways from baroreceptors
Motor pathways to the heart
inhibitory pathway
P — pressor area, D — depressor area of the vasomotor centre, Ci — cardioinhibitory centre.

mediately above their bifurcation, in the carotid sinuses (see Fig. 4.6). Nerves from the baroreceptors (vagi from the aortic receptors and Herings nerves which lead to the glossopharyngeal nerve in the case of the carotid sinus) exhibit tonic activity. Thus, at rest, there is a fairly low frequency of impulses passing to the medulla. If arterial blood pressure rises, then the baroreceptors are excited and the frequency of impulses in the nerves rises. This stimulates the depressor area of the vasomotor centre and stimulates the cardioinhibibitory centre (see Fig. 4.6).

The depressor area is connected by inhibitory neurones to the preganglionic sympathetic fibres in the spinal cord. When the centre is stimulated it reduces the number of impulses travelling in the sympathetic nerves to the heart. Stimulation of the cardioinhibitory centre increases vagal discharge to the heart. The combined effects of these two areas, therefore, slow down the heart and reduce stroke volume, thereby reducing cardiac output and lowering the blood pressure.

A reduction in blood pressure reduces the frequency of impulses and reverses the effect, increasing the cardiac output. These reflexes are very rapid and, for example, are responsible for reversing the pressure drop in the upper part of the body which occurs when standing up from the supine position.

In the long term the responsiveness of the reflex to the level of blood pressure may change. For example, in a person suffering from hypertension (high blood pressure), the initial increased baroreceptor response diminishes with time.

A second group of receptors which probably play only a minor role in cardiac regulation are the chemoreceptors. They respond to oxygen lack when the arterial blood pressure is in the range 40–80 mmHg. They are found in the aortic and carotid bodies; small bodies 1–2 mm diameter arising from the aortic arch and carotid bifurcation. They send impulses to the medulla via the vagi and glossopharyngeal nerves respectively. A decrease in the blood oxygen excites the receptors which, in turn, excites the pressor area of the vasomotor centre. The latter then increases sympathetic discharge to the heart causing a rise in the heart rate.

BLOOD VESSELS

Blood is pumped from the heart into the major arteries of the body. These branch to form increasingly smaller arteries and eventually arterioles. The latter are continuous with a close-meshed network of minute vessels, the capillaries. Blood then passes into larger venules which join with one another to form veins; these unite to form the largest veins which take blood back into the heart.

STRUCTURE OF BLOOD VESSELS

All blood vessels (except for capillaries) are composed of three major layers, the tunica intima (inner), tunica media (middle) and tunica adventitia (outer) (see Fig. 4.7).

The tunica intima is composed of two primary layers. The inner layer consists of endothelium which provides a smooth lining to the vessel. It serves to prevent blood clotting and to facilitate blood flow.

The tunica media is the muscular layer of blood vessels. It is composed of smooth muscle cells and connective tissue fibres wound spirally. Contraction of the smooth muscle cells consequently reduces the diameter of the blood vessel.

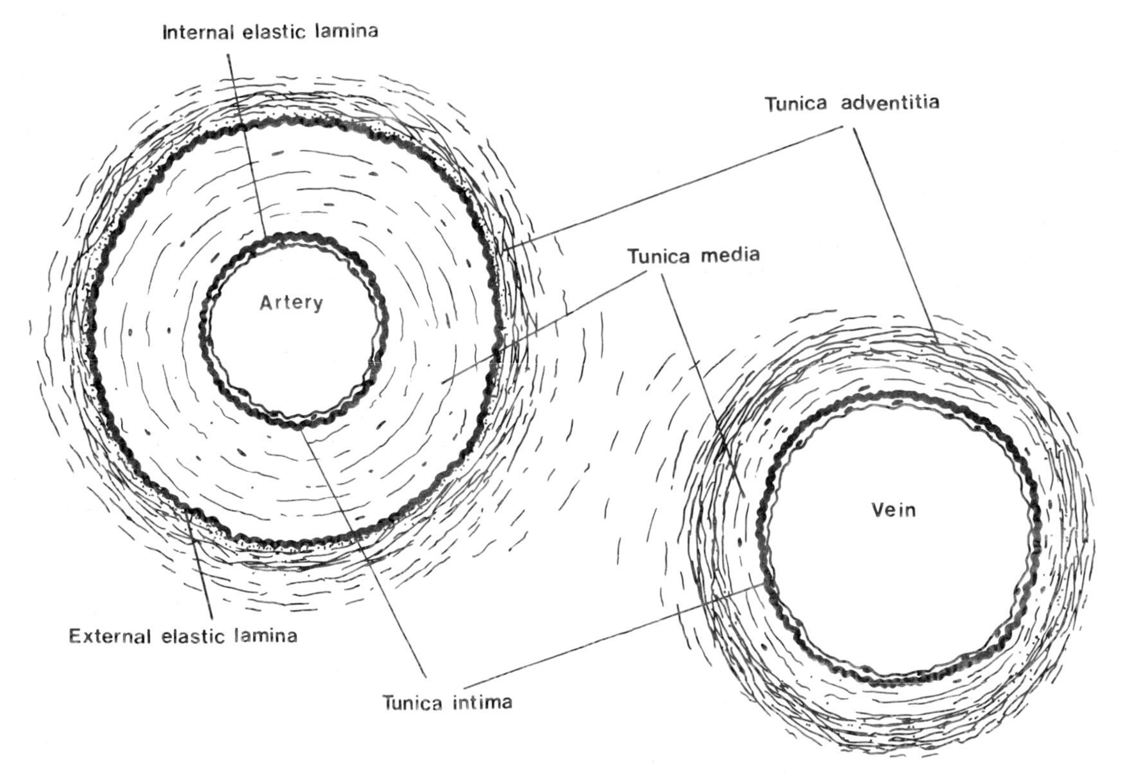

Fig. 4.7 Cross section through mesenteric (muscular) artery and vein.

The tunica adventitia is a connective tissue layer, containing elastic and collagen fibres, continuous with surrounding structures.

The composition and relative proportions of these three layers in the different blood vessels confers varying degrees of distensibility and contractility.

Elastic arteries (1–2.5 cm diameter)

These vessels (e.g. the aorta and pulmonary artery) receive blood in bursts from the heart. During systole, when additional blood is added to the arteries, the vessels distend, whilst during diastole the vessels recoil. This means that the flow of blood through the vessels is continuous despite the pulsatile nature of the heart beat.

The high degree of distensibility of these vessels is afforded by a large proportion of elastic fibres in the tunica media arranged in concentric sheets, separated by smooth muscle and collagen fibres. Additional elasticity is conferred on the vessel by the internal and external elastic laminae.

The internal elastic lamina consists of a perforated sheet of elastic tissue forming the subendothelial layer of the tunica intima. The outermost layer of the tunica media constitutes the external elastic lamina.

Muscular arteries (0.1–1 cm diameter)

The function of these vessels (e.g. brachial, femoral, radial arteries) is to distribute the blood under high pressure to the tissues. The walls are therefore strong and less elastic. The adventitia and media are thinner and the media contains a higher proportion of muscle cells.

Microcirculation

In arterioles (up to 100 μm diameter) the muscle layer is the predominant feature. Contraction or relaxation of the muscle fibres in the tunica media causes large changes in the diameter of these vessels and consequently they serve to regulate the amount of blood flowing through them to the capillaries and the pressure of blood behind them in the arterial system.

Capillaries (5–7 μm diameter) are composed of a single layer of endothelium with a little surrounding connective tissue. They are therefore structurally adapted for their role in the rapid exchange of materials between plasma and interstitial fluid. Blood from the capillary beds is drained into the venules, the smallest vessels of the venous system (20–30 μm diameter). These possess poorly developed walls with an almost indistinguishable structure.

Veins (up to 3 cm diameter)

The walls of the veins are composed of the same three layers as those of the arteries, but the thickness of the wall is a lower proportion of the vessel diameter compared with arterial vessels. They are less rigid and contain less muscle than arterial vessels. As a consequence, the veins offer little resistance to blood flow back to the heart, and they can accommodate relatively large changes in blood volume without a proportional increase in blood pressure.

The larger veins possess pocket-like valves on their inner surfaces. These valves aid the unidirectional flow of blood towards the heart when a vein is compressed by external pressure (see *Venous circulation*).

MAJOR ARTERIES OF THE BODY

Blood leaves the heart in the ascending aorta which passes over and behind the heart as the aortic arch and passes down as the descending aorta. The aortic arch gives off three major vessels. On the right side, the brachiocephalic artery arises which divides to form the right common carotid and right subclavian arteries. The left common carotid and subclavian arteries arise independently, directly from the aortic arch (see Fig. 4.8). Each subclavian artery arches laterally over the upper surface of the first rib, when it becomes known as the axillary artery. A little further down, below

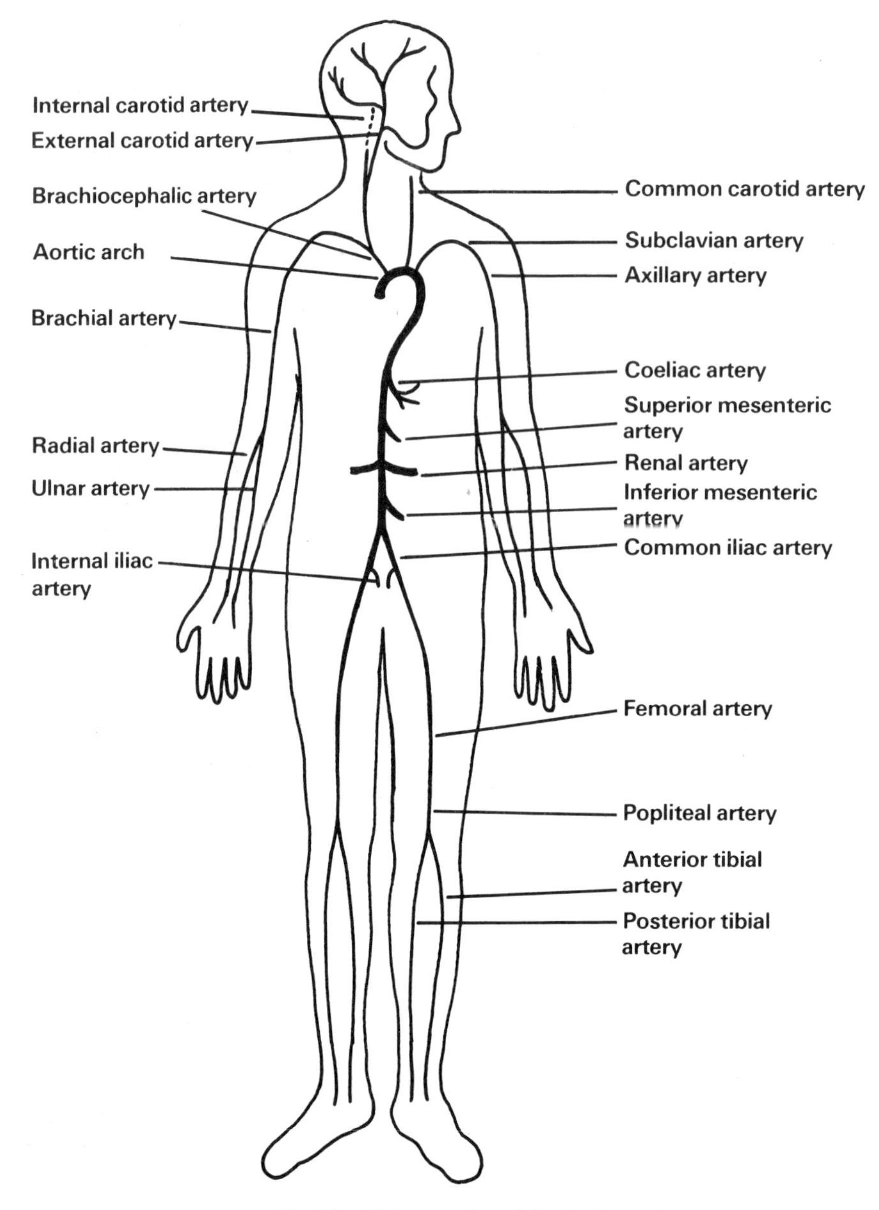

Fig. 4.8 Major arteries of the body.

the armpit, it becomes the brachial artery which divides at the elbow to form the radial and ulnar arteries carrying blood to the lower arm and hand.

Each common carotid bifurcates to form the internal and external carotid arteries. The internal branch supplies oxygenated blood to the cerebral hemispheres, communicating with the corresponding vessel on the other side through the circle of Willis (see *Cerebral circulation*). The external branch carries blood to the more superficial structures of the head, i.e. muscle, skin and bone.

Once the descending aorta has passed through the diaphragm, it becomes known as the abdominal aorta. Just below the diaphragm, the latter gives rise to three unpaired arteries, the coeliac trunk and the superior and inferior mesenteric arteries. The coeliac trunk has three branches, the hepatic, to the lower surface of the liver, the left gastric to the lesser curvature of the stomach and the lower oesophagus and the splenic artery which travels behind the stomach to the spleen (see *Circulation through the gut*). The superior mesenteric artery is the main distributing vessel to the small intestine and the first half of the large intestine, whilst the inferior vessel supplies blood to the distal portion of the large intestine.

A pair of renal arteries come off at right angles to supply the kidneys with blood. Immediately below are the testicular (in the male) or ovarian (in the female) arteries to the reproductive organs.

The abdominal aorta bifurcates above the pelvis to form the two common iliac arteries. Each of these then branches to form a small internal branch which supplies blood to the pelvis and a large external iliac artery which is the major trunk to the lower limb. In the region of the thigh this vessel becomes the femoral artery and at the knee becomes the popliteal artery before dividing into the anterior and posterior tibial arteries.

In addition to the systemic circulation described above, the heart also gives rise to a pulmonary circulation which carries blood from the right ventricle to the lungs. The pulmonary trunk passes upwards from the heart, parallel to the ascending aorta, and then divides into right and left pulmonary arteries, one to each lung.

MAJOR VEINS OF THE BODY

Blood from the upper regions of the body and the head drains through the superior vena cava into the right atrium, whilst that from the lower limbs and trunk passes through the inferior vena cava.

The internal jugular vein is the principal vessel which carries blood from the head and deeper tissues of the neck back towards the heart. The external jugular drains the superficial tissues of the neck. These two vessels drain into the brachiocephalic vein which then unites with the corresponding vessel on the other side to form the superior vena cava (see Fig. 4.9). Since the latter is situated to the right of the midline, the left brachiocephalic vein is significantly longer than the right.

In the upper limb, the main superficial vessels are the cephalic and basilic veins. The latter unites with the deep-lying brachial vein to form the axillary vein. This then joins with the cephalic vein to form the subclavian vein which carries blood towards the brachiocephalic vein and the heart.

Blood from the foot drains into the long saphenous vein and deep-lying anterior and posterior tibial veins. The last two unite at the knee to form the popliteal vein which becomes the femoral vein. This vessel and the long saphenous unite to form the external iliac, which like its counterpart in the arterial system combines with the internal iliac to form the common iliac vein; left and right common iliac veins then unite to form the inferior vena cava.

The major veins draining into the inferior vena cava include those from the kidney — the right and left renal veins, and the liver — the hepatic veins. The latter actually carry blood from most of the abdominal viscera, including the stomach, spleen, pancreas and intestine. Blood from these organs drains into the portal vein, which runs to the liver and then breaks up, eventually giving rise to a system of sinusoids within the body of the liver. The liver is then drained by the right and left hepatic veins which carry blood directly to the inferior vena cava.

The lungs are drained by a separate system of blood vessels, the pulmonary veins which carry oxygenated blood into the left atrium. There are two pulmonary veins draining each lung.

FUNCTION OF BLOOD VESSELS

Flow, pressure and resistance

Blood flows from areas of high pressure to areas of low pressure.

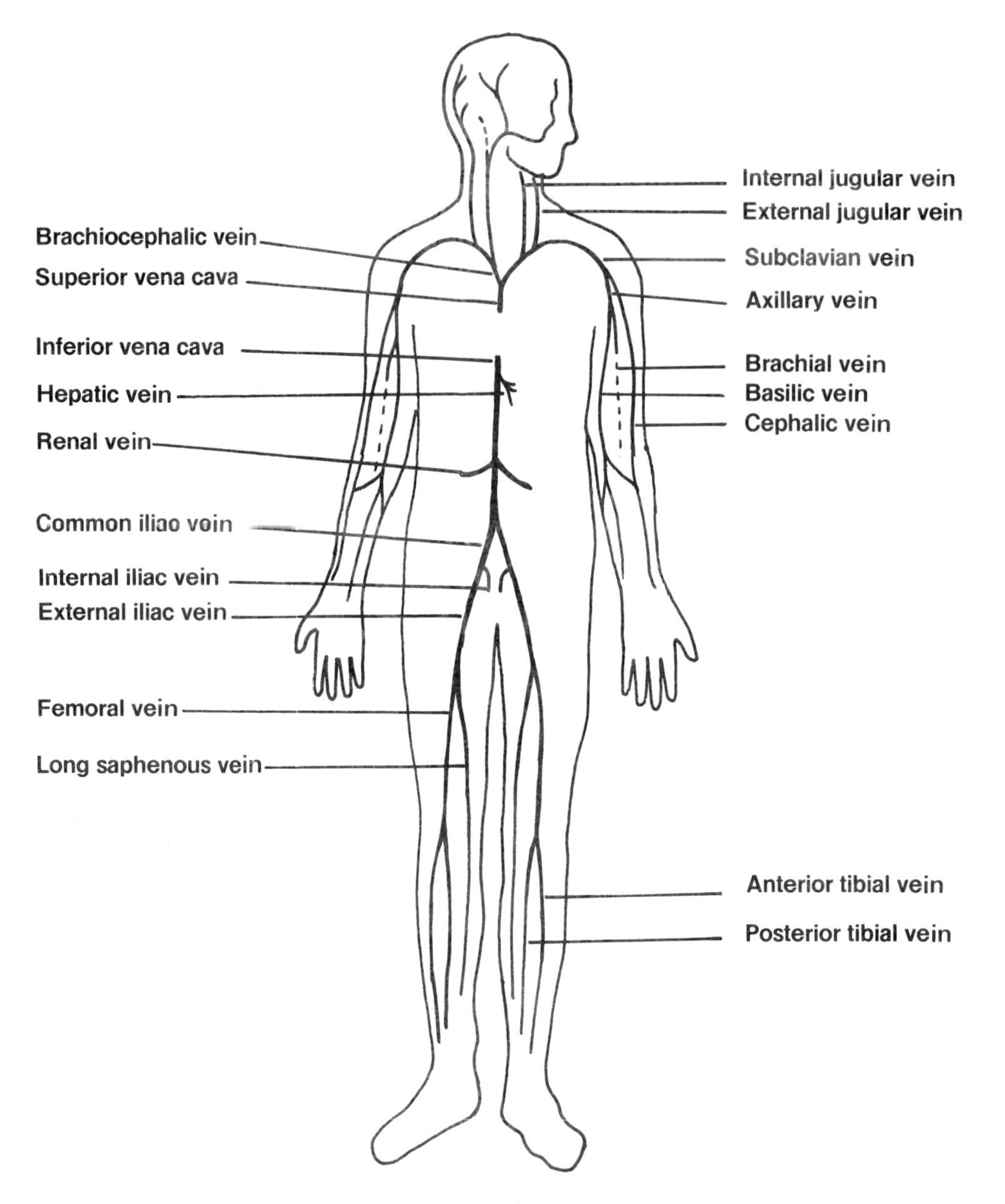

Fig. 4.9 Major veins of the body (--- deep veins).

Blood flow (F ml/min) is directly proportional to the difference in pressure between that at the start and that at the end of the journey ($\triangle$P).

$$F \propto \triangle P$$

The other factor which determines the rate of blood flow in a given blood vessel is the resistance (R) offered to the flow of blood by the vessels wall. If the resistance is high, it will reduce the volume of blood flowing through it, thus the rate of blood flow is inversely proportional to the resistance offered to it.

$$F \propto \frac{1}{R}$$

Putting these two equations together:

$$F = \frac{\Delta P}{R}$$

$$\text{or blood flow} = \frac{\text{pressure difference}}{\text{resistance}}$$

The resistance (R) to blood flow offered by the vessel wall is determined by three factors, the radius of the vessel (r), the length of the vessel (L) and the viscosity of the blood (η).

Poiseuille's formula relates the factors which affect the flow of fluids through tubes in general.

$$\text{Resistance} \quad \alpha \quad \frac{\text{viscosity} \times \text{length}}{\text{radius}^4}$$

$$\text{or R} \quad \alpha \quad \frac{\eta L}{r^4}$$

Thus a thick fluid (high viscosity) will have a higher resistance than a thin one, a long tube will offer more resistance than a short tube, and a narrow vessel will offer more resistance than a wide one. Physiologically, the importance of this formula is that resistance (and therefore blood flow) is markedly affected by small changes in vessel diameter. It can be seen from the formula that resistance is inversely proportional to the fourth power of the vessel radius. Thus doubling the radius will increase the blood flow by 16 (i.e. $2^4 = 2 \times 2 \times 2 \times 2 = 16$). Since the arterioles are structurally adapted for vasoconstriction or dilation, it is these vessels which primarily affect the flow of blood to the tissues beyond. They may be referred to as resistance vessels as they are the major determinants of the peripheral resistance.

The pressure of blood in the systemic circulation progressively falls from a mean value of about 100 mmHg in the aorta to approximately 0 mmHg in the right atrium. (These values are 'gauge' pressure so that a value of 0 means equal to atmospheric pressure.)

The decrease in arterial pressure in each part of the circulation is proportional to the resistance offered to the flow of blood. It can be seen from Table 4.1 that the most dramatic drop in blood pressure occurs as blood flows through the arterioles. They offer about half of the total peripheral resistance of blood vessels.

Table 4.1 Mean blood pressures (mmHg) in the systemic circulation.

Vessel	Start	End	Pressure drop
Aorta	100	100	0
Large arteries	100	96	4
Small arteries	96	85	11
Arterioles	85	30	55
Capillaries	30	10	20
Venules-veins	10	0	10

Arterial and arteriolar circulation

Arterial pulse

As the heart pumps blood into the arteries during systole, but not during diastole, blood flow through the arterial system is phasic. (The elastic recoil which occurs during diastole, however, ensures that blood still flows during this phase.)

The increase in pressure in the aorta during systole initiates a pulse wave which travels down the arterial wall at 4–5 m/s (i.e. ahead of the blood which typically travels at 40 cm/s). Pulse rate (or heart rate) can therefore be measured at sites where an arterial wall can be felt through the skin (e.g. wrist).

Measurement of arterial blood pressure

The most accurate measurements of blood pressure are made directly by inserting the measuring device into a vessel or by taking blood from the vessel to the device. These methods are unsuitable for routine use and so a non-invasive technique is used by employing the sphygmomanometer (developed by Riva-Rocci in 1896) (see Fig. 4.10).

The sphygmomanometer consists of a mercury-filled U-tube (manometer) one limb of which is extremely squat. The other, narrow limb is marked as a pressure scale in millimetres of mercury. The short limb is connected via a rubber tube to a rubber bag (18 cm × 12 cm) covered with cloth. This

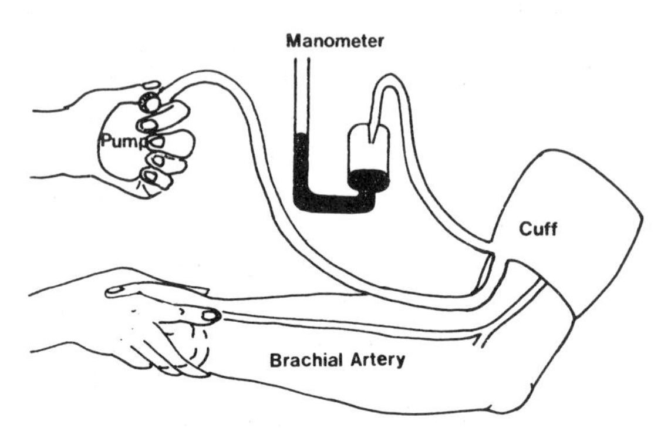

Fig. 4.10 Use of the sphygmomanometer.

is wrapped around the upper arm, above the elbow, and the bag is then inflated by means of a rubber bulb. As the pressure within the bag begins to rise, mercury begins to climb the narrow limb of the manometer. At a value above that of the systolic pressure (say 160 mmHg) pumping ceases and air is gradually let out of the bag via an air bleed valve. The bell of a stethoscope is placed lightly over the brachial artery at the bend of the elbow. When the pressure falls just below the systolic level, the artery opens momentarily (during systole) and a tapping sound is heard during diastole as the walls of the vessel come together. This sound becomes louder as the pressure is reduced, until suddenly it vanishes as the external pressure drops to diastolic pressure and normal blood flow is resumed.

Blood pressure is normally quoted as

$$\frac{\text{systolic pressure}}{\text{diastolic pressure}} \text{ or } \frac{SP}{DP}$$

and may be expected to be of the order of $\frac{120}{80}$ mmHg in a healthy young adult at rest. The difference between the two pressures, i.e. SP–DP is referred to as the pulse pressure. In this case it is 40 mmHg.

The exact magnitude of blood pressure is dependent upon a large number of factors, not least of which are the conditions under which it is taken. Thus basal blood pressure can only be taken 10 or 12 hours after a meal, after resting in a warm room for at least 30 minutes. Normally, casual blood pressure only is recorded.

Neural control of arteriole diameter

Arterioles are innervated by adrenergic sympathetic nerves which cause vasoconstriction. There is always some activity in these neurones causing tonic contraction or vasomotor tone. Increased activity in the sympathetic fibres causes vasconstriction, whereas decreased activity causes vasodilation. In general, neural control of arterioles is concerned with regulating peripheral resistance and hence blood pressure, rather than serving the metabolic needs of the tissues.

Sympathetic discharge to the vessels is controlled by the vasomotor centre in the medulla oblongata. It may be recalled that the centre is divided into a depressor and a pressor area. The depressor area is connected by inhibitory neurones (descending in the white matter) to the preganglionic sympathetic fibres in the spinal cord so that when the centre is active, the inhibitory influence on the preganglionic fibres reduces the rate of impulses travelling to the arterioles and thus results in vasodilation. The pressor area is connected by excitatory neurones to the preganglionic sympathetic fibres, so that an increased frequency of nerve impulses in these neurones will increase activity in the peripheral sympathetic neurones and cause vasoconstriction.

The resistance vessels in skeletal muscles have an additional sympathetic innervation by cholinergic neurones which cause vasodilation. These neurones are stimulated at the onset of physical exercise or even before it begins so that muscle blood flow increases in anticipation of the increased metabolic demand of active skeletal muscle. The control of activity of the sympathetic vasodilator neurones is directly from the hypothalamus and not via the vasomotor centre.

Local control of arteriole diameter

Local factors which increase or decrease arteriole diameter control the rate of blood flow in specific organs in accordance with the metabolic requirements. An increase in metabolic activity (e.g. in skeletal muscle contraction during physical exercise) is accompanied by increased blood flow through the active areas. This phenomenon is known as active hyperaemia. There are several possible candidates which have been shown experimentally to mediate this vasodilation directly; they include a fall in Po_2, a rise in Pco_2, H^+ conc., K^+ conc., adenosine, adenosine nucleotides and temperature. A similar affect is seen after a period of ischaemia induced by, for example, immersion of the arm in cold water, or the application of a tourniquet. During the ischaemia, metabolites produced by the tissues are not removed due to the lack of blood flow and they therefore accumulate. On removal of the tourniquet or the cold there follows a period of increased blood flow (reactive hyperaemia), producing a warm, red arm. The additional blood flow compensates for the period of deprivation and enables the excess metabolites to be washed away. Once this is achieved the stimuli for arteriolar dilation are removed and therefore blood flow returns to normal.

Circulating vasoactive substances

When the sympathetic nervous system is active, in addition to its effects on the heart and blood vessels, it also stimulates the release of adrenaline and noradrenaline from the adrenal medullae. Noradrenaline causes general vasoconstriction, whereas adrenaline causes vasoconstriction in most sites, but vasodilation in skeletal muscle.

Angiotensin II formed when renin is released from the kidney (when blood volume or pressure falls) causes generalized vasoconstriction.

Capillary circulation

The function of the capillaries is to allow the exchange of materials between the blood and the tissues and blood flow is varied according to the metabolic demands of the tissues. The microcirculation (see Fig. 4.11) starts at the level of the arterioles from which metarterioles branch almost at right angles. Initially, these vessels contain a discontinuous muscle layer but distally the muscle coat disappears and the vessel is known as a thoroughfare channel.

The true capillaries branch from the metarterioles and at the junction, one or two muscle cells encircle the capillary to form a contractile pre-capillary sphincter. Blood flow through a tissue can therefore either proceed through the thoroughfare channel or through the capillary bed. Which route is taken is controlled by the state of contraction of the pre-capillary sphincters. When these are closed, the blood will flow through the thoroughfare channels directly from arterioles to venules. As the metabolites from cells supplied by the capillary bed accumulate, they cause dilation of the pre-capillary sphincters (in the same way as they cause arteriolar dilation). Once the sphincters are dilated the blood will flow through the capillary bed until the metabolites are washed away thus removing the stimulus for dilation, so the sphincters close again.

In this way blood flow through the microcirculation alternates between the thoroughfare channels and the capillary bed. This constant change of route is known as vasomotion.

Thus the capillaries are passive vessels which simply receive whatever blood flow is delivered to them, regulated by the arterioles and the pre-capillary sphincters.

The velocity of blood flow through the capillaries is slow (0.07 cm/s), thereby enabling sufficient time for the exchange of nutrients and waste products between the blood and the tissues. The movement of substances across the capillary walls is always passive, i.e. by diffusion either through or between the endothelial cells. The tissues may actively transport material into or out of the cells thereby creating a concentration gradient between the blood and the interstitial fluid. Because of the passive nature of transcapillary transport, the composition of plasma and interstitial fluid is essentially the same (except for the amount of

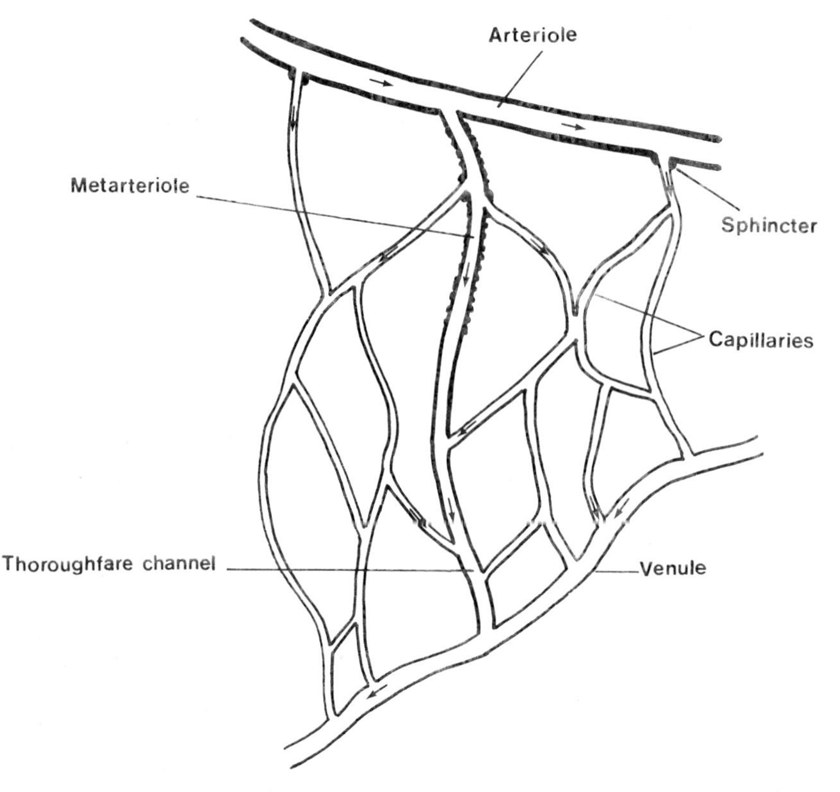

Fig. 4.11 The microcirculation.

protein, molecules of which are generally too large to pass out of the blood).

Formation and reabsorption of interstitial fluid

The interstitial fluid which bathes the cells of the body is their immediate source of nutrients and waste disposal route. Intersitial fluid is not, however, static but is continually being produced from the blood and is subsequently reabsorbed either directly into the blood vessels, or indirectly via the lymphatic system. The turnover of fluid is of the order of 20 litres per day (i.e. about seven times total plasma volume).

A mechanism for the formation and reabsorption of interstitial fluid was first suggested by E. H. Starling (1886–1927). Starling's hypothesis is that there is net filtration of fluid out of the blood into the interstitial compartment at the arterial end of capillary beds and a net reabsorption of fluid back into the blood at the venous end of capillary beds.

The movement of fluid is caused by a net pressure difference across the capillary wall. There are four pressures which are thought to operate across the wall — capillary hydrostatic pressure, plasma colloid osmotic pressure, interstitial fluid colloid osmotic pressure, and interstitial fluid hydrostatic pressure.

Capillary hydrostatic pressure (CHP) is simply the pressure of blood contained in the capillary and represents a force acting outwards across the vessel wall (see Fig. 4.12). The value of this pressure depends on the arterial blood pressure, the venous blood pressure and on the position of the capillary within a capillary bed. A capillary near to the arterial inflow will have a higher pressure than one at a site nearer the venous outflow. This is because the resistance to flow causes the pressure to fall.

Plasma colloid osmotic pressure (PCOP) is

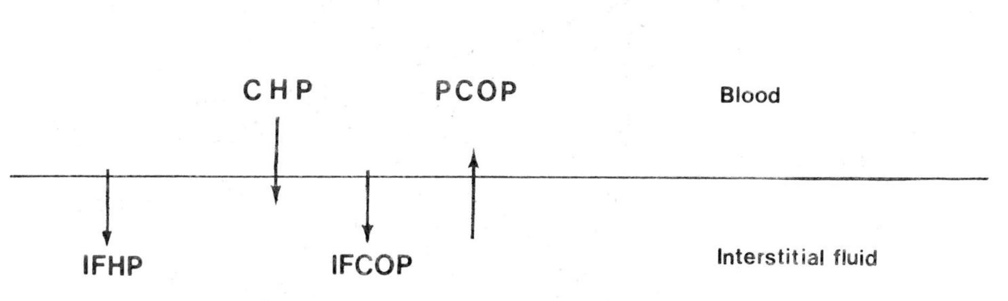

Fig. 4.12 Forces acting across the capillary wall.

the osmotic pressure exerted by plasma proteins. The effect of osmotic pressure is to draw solvent towards particles in solution. Thus the particles in plasma tend to draw fluid (strictly water, but solutes will follow by diffusion) into the blood, whereas particles in the interstitial fluid will tend to draw fluid out of the blood into the interstitial compartment.

As the composition of plasma and interstitial fluid differs essentially only in the amount of protein present, it is only the osmotic pressure exerted by the protein which is different on each side of the capillary wall and therefore acts as a force for fluid movement across the wall (see Fig. 4.12).

The albumin fraction of the plasma proteins exerts about 70% of the total colloid osmotic pressure.

Interstitial fluid colloid osmotic pressure (IFCOP) is the osmotic pressure exerted by protein in the interstitial fluid. The amount of protein present in the interstitial fluid varies according to the permeability of the capillary wall. Thus in some sites such as skeletal muscle, there is very little interstitial protein, whereas in the intestine and liver there is a much higher amount of protein present. These regional differences can be attributed to differences in structure of the capillary walls in the different sites. Capillaries in skeletal muscle have a thick basement membrane and no fenestrations or pores whereas in liver and intestine the basement membrane is much thinner, and there are fenestrations present. Passage of protein across such a structure is therefore easier.

IFCOP acts to draw fluid from the blood into the interstitial compartment (see Fig. 4.12).

Interstitial fluid hydrostatic pressure (IFHP) varies according to the volume of fluid present and how distensible the compartment is. It is now generally held that IFHP is subatmospheric i.e. at a negative pressure thereby acting as a force to draw fluid out of the blood (see Fig. 4.12).

Thus across the capillary wall there are three filtration forces (CHP, IFCOP and IFHP) and one absorptive force (PCOP). At the arterial end of capillary beds the sum of the filtration forces exceeds the absorptive force and therefore there is net fluid movement out of the blood into the interstitial compartment.

Pressure	*mmHg*
CHP	25
IFCOP	5
IFHP	6 (negative)
PCOP	28

Thus the filtration force = 25 + 5 + 6
= 36 mmHg

The net pressure is therefore 36 − 28
= 8 mmHg

At the venous end of the capillary beds the capillary hydrostatic pressure has fallen so that the filtration force is reduced.

Pressure	*mmHg*
CHP	10
IFCOP	5
IFHP	6 (negative)
PCOP	28

Thus the filtration force = 10 + 5 + 6
= 21 mmHg

The net pressure is therefore 21 − 28
= −7 mmHg

Fluid, therefore, moves from the interstitial compartment back into the blood.

It can be seen that the net pressure forcing fluid out of the blood at the arterial end of a capillary bed is greater than the net pressure causing reabsorption of fluid back into the blood at the venous end of a capillary bed (8 mmHg compared with -7 mmHg). The result of this is that more fluid is filtered into the interstitial compartment than is reabsorbed from it into the blood vessels. The excess fluid is reabsorbed into the lymphatic capillaries and subsequently returned to the venous system in the neck. The total volume of fluid transported in this way is of the order of 2–4 litres per day.

It will be recalled that lymphatic capillaries consist of overlapping endothelial cells. This structure confers an extremely important function on the lymphatic system, that is the transport of protein from the interstitial fluid back into the blood via the veins in the neck. Although the leakage of protein from blood capillaries into interstitial fluid is a slow process, if the protein were not constantly removed in the lymph, water would be osmotically retained in the interstitial fluid and oedema would result.

The flow of fluid within the lymphatic vessels is achieved in a similar way as the flow of blood in veins ascending to the heart. That is the muscle and respiratory pumps tend to draw fluid along the vessels which, like the veins, have valves ensuring unidirectional flow.

Venous circulation

Veins are known as capacitance vessels because they can accommodate a relatively large increase in blood volume without an equivalent rise in pressure. This is because their thin walls are more distensible than those of arterial vessels.

The sympathetic vasoconstrictor nerves supplying the veins can cause venoconstriction which reduces the distensibility of the vessel walls and thereby raises venous pressure. The slight reduction in vessel diameter causes a negligible effect on resistance to flow, thus overall the effect of venoconstriction is to increase venous return. Venous pressure is low compared with arterial pressure despite the fact that about 60% of the total blood volume is present in the venous system. Pressure in the venules is 10–15 mmHg so that the total pressure drop back to the right atrium is around 10 mmHg as right atrial pressure is 0–5 mmHg. These values apply when the subject is in the horizontal position.

If the subject stands upright then the pressures in the circulatory system are affected by the weight of blood in the vessels. The pressure in the veins of the feet is ≏ 90 mmHg because of the weight of the column of blood between them and the right atrium.

Conversely, the neck veins are collapsed by atmospheric pressure (so that the pressure is 0). The non-collapsible venous sinuses within the skull are at less than atmospheric pressure, e.g. − 10 mmHg in the sagittal sinus. The thoracic veins are not collapsed because intrathoracic pressure is lower than the pressure within these veins. Venous return from the lower parts of the body is aided in three principal ways — the presence of valves which prevent backflow of blood, and the thoracic and muscle pumps.

During inspiration, the diaphragm descends thereby raising abdominal pressure. This in turn will compress abdominal veins and thereby propel blood towards the heart. As the intrathoracic pressure falls (as the thoracic volume increases) this will aid the flow of blood into the thorax and hence to the heart.

The contraction of skeletal muscle will compress the veins running through it and thereby propel the blood along. The valves ensure that the blood flows towards the heart.

A rough estimate of central venous pressure can be made with the subject lying with the head and chest at an angle of 30° to the horizontal. The distended portion of the external jugular veins is clearly visible. The height of the upper limit of distention above the right atrium gives the venous pressure in mm of blood.

Alternatively peripheral venous pressure can be measured with the use of a catheter inserted into an arm vein at the level of the right atrium. The catheter is filled with sterile saline connected to a manometer. The mean pressure in the antecubital vein is about 7 mmHg.

INTEGRATED CARDIOVASCULAR FUNCTION

DETERMINANTS OF ARTERIAL BLOOD PRESSURE

The equation relating blood flow to pressure and resistance ($F = \frac{\Delta P}{R}$) can be applied to the vascular system as a whole (from the aorta to the venae cavae just entering the heart). In this case, the total blood flow is equal to the cardiac output; the pressure drop is equal to mean arterial pressure (100 mmHg) minus late vena cava pressure (0 mmHg). (Mean arterial pressure is the average pressure throughout the cycle. This is about 96 mmHg in a young adult, rounded up to 100 mmHg for simplicity.) Thus

$$\text{Cardiac output} = \frac{\text{Mean arterial pressure}}{\text{Total peripheral resistance}}$$

rearranged

$$\text{Mean arterial pressure} = \text{Cardiac output} \times \text{Peripheral resistance}$$

Therefore all the factors affecting cardiac output and peripheral resistance will influence arterial pressure. These factors are summarized in Figure 4.13.

Mean arterial blood pressure = cardiac output × peripheral resistance
Cardiac output = stroke volume × heart rate

Factors increasing stroke volume:

↑ venous return (Starling's law)
↑ sympathetic stimulation
↑ circulating adrenaline.

Factors increasing heart rate:

↑ sympathetic stimulation
↓ parasympathetic stimulation
↑ circulating adrenaline.

Peripheral resistance $\alpha \frac{1}{r^4}$ **(arterioles)**

Factors increasing arteriole radius:

↓ sympathetic vasoconstrictor nerves
↑ sympathetic vasodilator nerves (skeletal muscle)
↓ P_{O_2}
↑ P_{CO_2}
↑ H^+
↑ K^+
↑ adenosine
↑ adenosine nucleotides
↑ temperature
↑ circulating adrenaline (skeletal and cardiac muscle).

Factors decreasing arteriole radius:

↑ sympathetic vasoconstrictor nerves
↓ temperature
↑ circulating noradrenaline
↑ circulating adrenaline (in most sites)
↑ circulating angiotensin II.

Fig. 4.13 Principal determinants of arterial blood pressure.

CONTROL OF ARTERIAL BLOOD PRESSURE

The maintenance of arterial blood pressure within a range of values consistent with health is mediated by two types of response. There are rapid, short-term effects which moderate changes induced, for example, by standing up quickly after lying down. Such control is primarily brought about by changes in cardiac output and arteriole diameter. In the long-term, the level of arterial blood pressure is determined by the blood volume, which itself depends on a balance between fluid intake and fluid losses A 2% increase in blood volume can result in an increase in arterial pressure of as much as 50% and although the rapidly acting control systems will serve to reduce this change, the long-term adjustment will be by increased fluid loss by the kidneys.

Baroreceptor reflex

A fall in arterial blood pressure reduces the frequency of impulses emitted from the baroreceptors. This information is conveyed to the cardiovascular centres, which respond immediately by increasing sympathetic activity and reducing parasympathetic outflow to the heart, thereby increasing cardiac output and raising blood pressure.

Simultaneously there is increased sympathetic discharge to the veins resulting in venoconstriction, increased venous return and thus increased cardiac output. Sympathetic discharge to the arterioles also increases, thereby increasing peripheral resistance and raising blood pressure.

Pulmonary artery baroreceptors

There are baroreceptors present in the pulmonary arteries which operate over a lower pressure range than those in the systemic arteries. They elicit changes essentially the same as those resulting from the baroreceptor reflex, but over the range of pressures prevailing at this site (see *Pulmonary circulation*).

Atrial receptors

There are stretch receptors present in the atria which are variously described as baroreceptors or volume receptors.

Stretch of the atrial walls caused by, for example, increased blood volume, increased venous return or increased systemic arterial blood pressure, results in an immediate reflex dilation of peripheral arterioles. The consequences of this are (a) the peripheral resistance falls and therefore arterial blood pressure falls; (b) the increased blood flow to the capillary beds increases capillary hydrostatic pressure and therefore causes a slow transfer of fluid into the interstitial compartment which results in a reduction in blood volume and therefore blood pressure; (c) dilation of the renal afferent arterioles increases glomerular capillary hydrostatic pressure and therefore GFR which, within minutes, results in increased fluid loss in the urine.

A further consequence of atrial wall stretching is that the impulses from the receptors are conveyed to the hypothalamus which results in a reduction in ADH secretion rate from the posterior pituitary gland. The consequence of a reduced blood ADH concentration is to reduce renal water reabsorption, thereby increasing urinary fluid loss. Thus the increased GFR and the reduced blood ADH concentration both result in increased fluid loss from the body and will therefore reduce blood volume and blood pressure.

Chemoreceptor reflex

If the rate of oxygen supply to the chemoreceptors falls below a certain level, then reflex increases in blood pressure are mediated via the medullary cardiovascular centres in the same way as the baroreceptor reflex. This oxygen lack reflex acts as an emergency control system when blood pressure drops below 80 mmHg.

Sympathomimetic effect of adrenaline and noradrenaline

The reflexes which result in increased blood

pressure do so by stimulating increased sympathetic activity in nerves supplying the blood vessels and the heart. Simultaneously, sympathetic stimulation of the adrenal medullae is increased, resulting in a rise in blood adrenaline and noradrenaline concentrations for between one and three minutes. The effect of these hormones is to intensify the rapid cardiovascular responses induced by neural stimulation.

The role of the renin-angiotensin system in blood pressure regulation

A fall in renal blood flow (accompanying a fall in arterial BP) results in the release of renin from the juxtaglomerular cells of the afferent arterioles in the kidneys. Renin persists in the circulation for about one hour and initiates the sequence of changes resulting in the formation of angiotensin II from its precursor in plasma, angiotensinogen. Angiotensin II is a powerful vasoconstrictor and therefore raises blood pressure.

The maximal response takes about 20 minutes and is therefore slower to act than the nervous reflexes and the action of adrenaline and noradrenaline.

CARDIOVASCULAR RESPONSES TO PHYSICAL EXERCISE

At the onset of exercise when the motor cortex initiates muscle contraction, it also initiates increased activity in the sympathetic cholinergic vasodilator fibres supplying the arterioles in the active skeletal muscles. The neural pathway initiating vasodilation includes the motor cortex and the hypothalamus, but does not involve the cardiovascular centres. This neural vasodilation is of importance only at the beginning of exercise, and once exercise is under way, vasodilation is maintained primarily by local factors (e.g. the accumulation of metabolites and a fall in oxygen levels). The local factors also cause relaxation of the precapillary sphincters, so that perfusion of the active muscle is greatly increased. The rate of blood flow through active muscle can increase by as much as 18 times above the resting rate which is 4–7 ml/min/100 g muscle. This increased blood flow is partly due to the arteriolar vasodilation and increased number of 'open' capillaries, but it is also due to a rise in arterial blood pressure.

The increased arterial pressure is mediated primarily by an increased cardiac output. The motor cortex stimulates the cardiovascular centres to increase sympathetic activity and reduce parasympathetic activity to the heart, even in anticipation of the exercise. The resultant increase in stroke volume and heart rate raises arterial blood pressure.

Increased activity in the vasomotor centre also stimulates constriction in the veins and most arterioles. Venous return is increased by this venoconstriction and additionally by both the pumping action of the working muscles, and by the increased respiratory effort associated with physical exercise. An increased venous return will raise cardiac output by the Starling mechanism.

Arteriolar constriction reduces blood flow to areas such as the gastrointestinal tract and kidneys, thereby increasing the volume of blood available to flow into the active skeletal muscles. The coronary and cerebral circulations are little affected as they have relatively poor vasoconstrictor innervation.

If the vasodilation in skeletal muscle blood vessels is greater than the vasoconstriction in other areas then the total peripheral resistance falls. This means that there will be less increase in mean arterial blood pressure than might be expected, because the blood is flowing away more easily from the point of measurement. Diastolic pressure in particular will reflect any change in peripheral resistance as it is measured when no blood is being added to the artery. During strenuous exercise, therefore, diastolic pressure often falls.

The powerful sympathetic discharge to the blood vessels and the heart is also accompanied by increased stimulation of the adrenal medullae. Adrenaline both increases the vasodilation in skeletal muscle arterioles, and increases the force of contraction of the heart against the raised aortic pressure. Noradrenaline increases the general vasoconstriction induced by neural stimulation.

REGIONAL CIRCULATION

The distribution of blood to the different tissues and organs of the body is a reflection of their different levels of metabolic activity. Thus the kidneys, which have a combined weight of approximately 300 g, receive 22% of the cardiac output whereas the skeletal muscles (31 kg) at rest only receive about 15% (see Table 4.2). In this section a description of the regional blood supply to various parts of the body is given, together with an account of the mechanisms by which these blood flows are regulated.

CORONARY CIRCULATION

The myocardium of the heart is supplied principally by two coronary arteries which arise from sinuses situated above the cusps of the

Table 4.2 Approximate distribution of the resting cardiac output in an adult 65 kg male (at rest).

	Weight kg	ml/min	Blood flow ml/100 g/min	% Cardiac output
Heart	0.3	210	70	4
Brain	1.4	770	55	14
Splanchnic	4.0	1400	35	27
Skin	3.6	470	13	8
Kidney	0.3	1200	400	22
Skeletal muscle	31.0	840	3	15

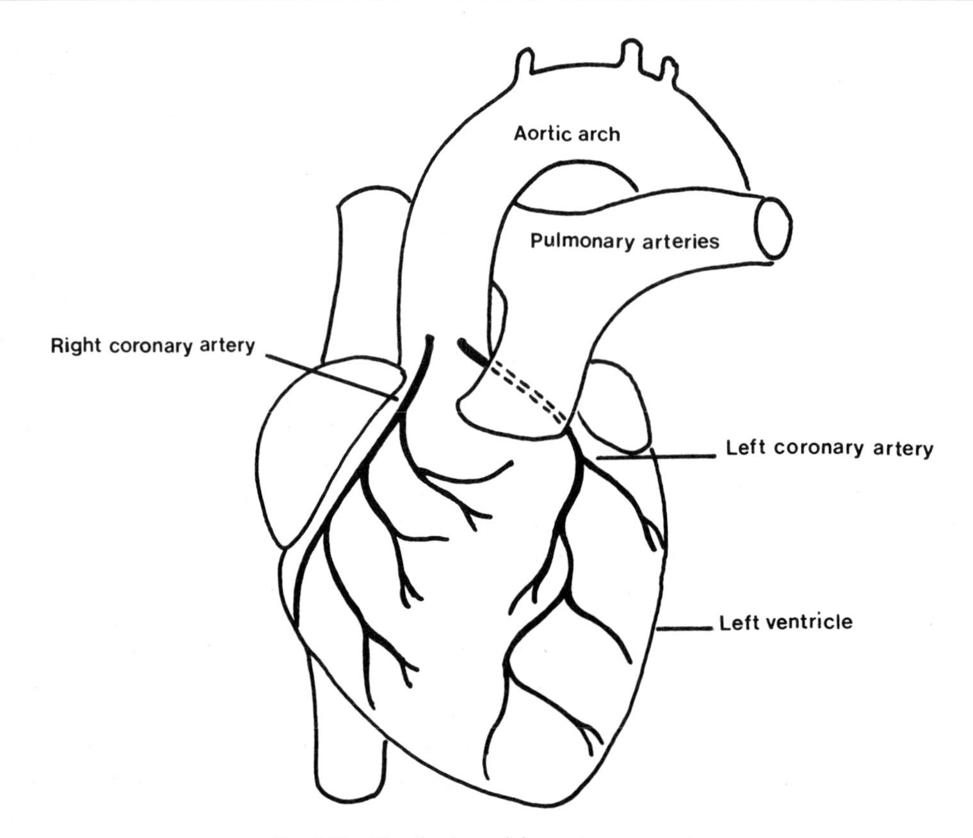

Fig. 4.14 Distribution of the coronary arteries.

aortic valve at the root of the aorta (see Fig. 4.14). The left coronary artery supplies the wall of the left ventricle with oxygenated blood, whereas the right coronary artery supplies the right ventricle and a portion of the left ventricle. Venous blood from the left ventricle drains via a coronary sinus into the right atrium; some blood from the right ventricle also drains into the right atrium through the anterior cardiac veins.

Blood flow through the coronary system represents about 4% of the output from the left ventricle (approx. 70 ml/100 g/min at rest). This flow tends to be intermittent, since during systole compression of the coronary vessels by the surrounding muscle inhibits the free passage of blood through the myocardium, especially in the left side.

Blood flow through the coronary system is regulated almost entirely by local factors, namely the oxygen levels of the blood and the presence of metabolites. When the body is at rest heart muscle extracts about 65% of the oxygen from the blood in the coronary vessels. This is near to the maximum amount which can possibly be removed. In exercise, when the oxygen requirement of the cardiac muscle is increased, there is an increased blood flow which provides the extra oxygen.

A four- or five-fold increase in blood flow may be observed and the stimulus for this increased blood flow is probably oxygen lack. The exact mechanism by which oxygen lack promotes vasodilation, thereby increasing blood flow, is not completely understood. It has been suggested that hypoxia promotes the released of vasodilator substances, e.g. adenosine, from the surrounding tissues. It will be recalled that sympathetic activity causes increased activity of the heart. Although sympathetic stimulation of the coronary blood vessels causes slight vasoconstriction, this effect is masked by vasodilation induced by the local metabolic factors.

PULMONARY CIRCULATION

The pulmonary artery extends for about 4 cm above the right ventricle before dividing into right and left branches. The latter are short and branch to supply the various parts of the lungs. All pulmonary vessels have larger diameters than their counterparts in the systemic circulation, additionally their walls are thinner and more distensible. Thus the compliance of the pulmonary circulations is as great as that of the systemic tree so that it can accommodate the entire output of the right ventricle (which is equal to the output of the left ventricle).

The pressure in the pulmonary artery is $\frac{25}{10}$ mmHg or less with a mean pressure of about 15 mmHg. The pressure in the left atrium is about 7 mmHg so that the total pressure gradient between the two ends of the system is approximately 8 mmHg. The low hydrostatic pressure throughout the system ensures that there is no loss of fluid into the alveolar spaces of the lungs.

The vessels of the lungs are not perfused evenly with blood. When the body is in a vertical position, blood pressure in the upper parts of the lung is about 20 mmHg less than in the lower parts. As a result of this there is a greater volume of blood flowing through the tissues in the lower parts of the lung than through those in the apex. Just as blood flow is uneven, so too is ventilation, since gas exchange is also greatest at the base of the lung.

Blood flow through the two lungs is approximately 5 litres/min and at any instant less than 1 litre of blood would be contained within the vessels, and of this only 100 ml would be within the capillary beds involved in gas exchange. The vessels within the lungs are normally in a state of vasodilation. However, further dilation can be brought about passively due to the extreme distensibility of the vessels or as a reflex due to increased baroreceptor activity in the carotid artery. In rapid exercise blood flow through the lungs can rise to 30 litres/min without significantly raising the pulmonary blood pressure. Local factors, primarily a fall in Po_2 will promote vasoconstriction. This means that poorly ventilated areas of the lung receive a correspondingly low blood flow. It has been suggested that this local control is the most important of all of the regulatory mechanisms.

CEREBRAL CIRCULATION

Blood enters the skull through the left and right internal carotid arteries and to a lesser extent through the two vertebral arteries. The latter unite to form the basilar artery which connects to the carotids which also link up to form the circle of Willis from which arises the arterial supply to the brain tissue (see Fig.4.15).

Cerebral vessels have less prominent elastic laminae and less well-defined muscle coats compared to systemic vessels. The grey matter of the brain is particularly rich in capillaries, while the white matter is less well provided.

Blood flows away from the brain via either deep or superficial vessels which drain into the venous sinuses of the cranial cavity.

Blood flow through the brain normally remains constant at a rate of 55 ml/100 g/min (750 ml/min) and only when the cardiac output drops below 3 litres/min does cerebral circulation become deficient. The constant nature of this circulation is very important since the tissues of the brain are very delicate but are housed within a rigid membrane, the dura, inside a rigid shell, the skull. The incompressibility of the brain tissue is one of the factors which actually helps to maintain a constant blood flow, i.e. the vessels can only dilate very slowly because of the tissues around them. Cerebral blood flow is regulated primarily by local factors and although the vessels are innervated by both sympathetic and parasympathetic fibres, they appear only to play a minor role.

There is some evidence to suggest that a drop in pressure within the vessels will initiate direct vasoconstriction. Better known however are the effects of blood gases and metabolites.

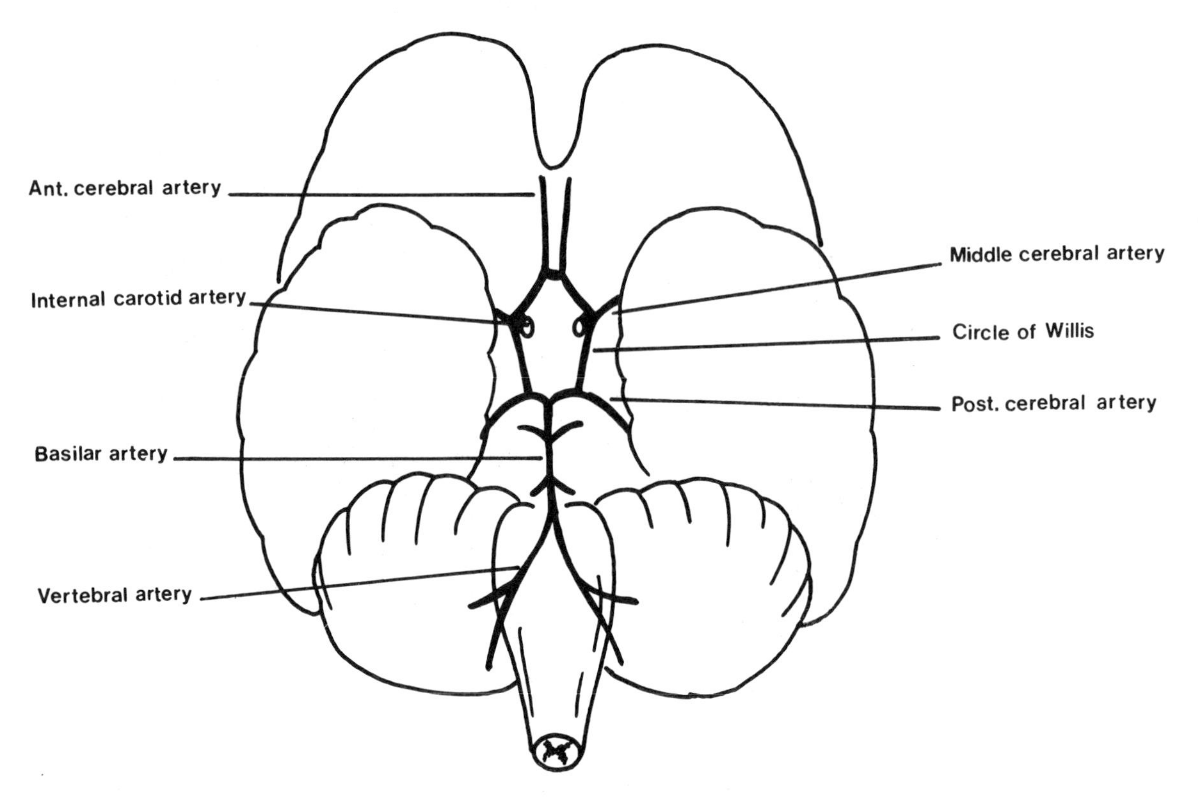

Fig. 4.15 Arteries at the base of the brain (viewed from beneath).

A rise in the CO_2 level of the blood, a drop in O_2 level and/or a rise in H^+ concentration will initiate vasodilation. A drop in CO_2 level and/or rise in O_2 level will initiate vasoconstriction. It has been demonstrated that a rise in the activity of a specific portion of the brain will increase the blood flow in that part of the brain alone so that overall blood flow will remain fairly constant. Equally if there is a change in blood gas concentration due to some external factor, e.g. breathing pure oxygen, then there will be an overall change in blood flow.

CIRCULATION THROUGH THE GUT

Blood flow to the gut is derived from several vessels, arising from the dorsal aorta (see Fig. 4.16).

Immediately below the diaphragm arises the coeliac artery which shortly divides into three branches. One of these branches forms the left gastric artery to the stomach whilst a second branch, the hepatic artery takes blood to the liver and part of the stomach wall. The third branch, the splenic artery carries blood

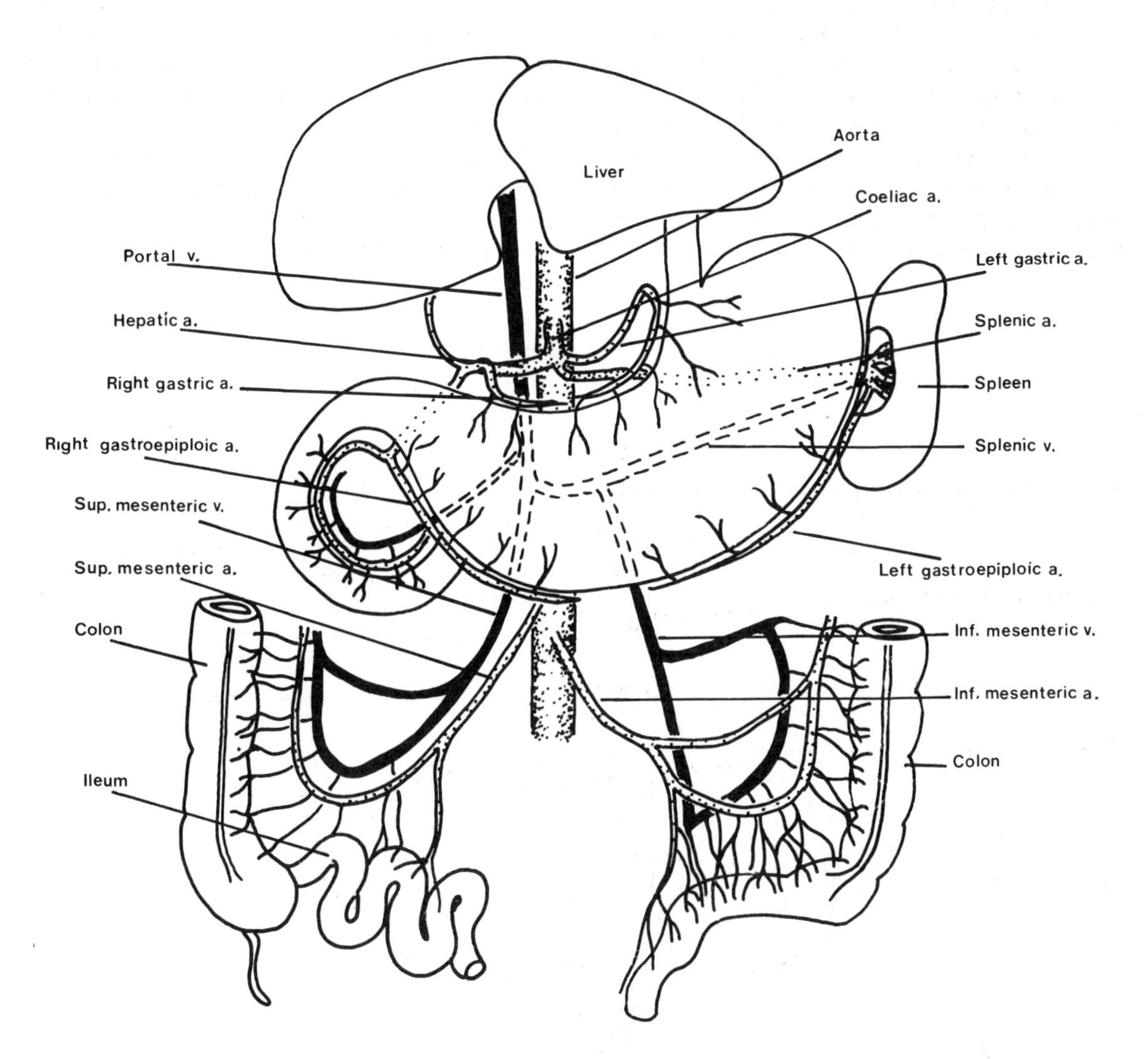

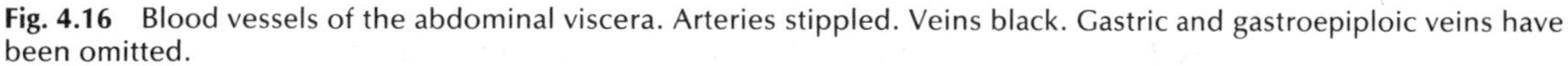
Fig. 4.16 Blood vessels of the abdominal viscera. Arteries stippled. Veins black. Gastric and gastroepiploic veins have been omitted.

to the spleen and pancreas. Also arising from the aorta is the superior mesenteric artery to the small intestine and proximal half of the large intestine and the inferior mesenteric artery to the lower half of the large intestine and rectum.

Venous blood from the alimentary tract drains largely through the portal vein and thence to the liver. The perfusion rate of the gut is normally in the range 50–80 ml/100 g/min which is about three times that which is required to supply adequate oxygen to the tissues. Blood flow is not evenly divided between the tissues, about two-thirds passing to the mucosa and submucosa and one-third to the muscle layers.

Blood flow to the two areas can be altered independently, for example when activity in the muscle layers is increased, blood flow in that area is increased. Likewise when glandular activity is increased, blood flow in the mucosa and submucosa also increases.

As in many areas of the body, blood flow appears to be mainly controlled by local needs, probably oxygen lack. However the increased blood flow which accompanies glandular secretion may be due to the production of chemical substances by these glands. This type of regulation is found particularly in the stomach and colon whilst vessels in the small intestine respond to the usual local factors. In addition, increased sympathetic activity, such as that which occurs in physical exercise, causes an overall vasoconstriction in the gut which may be so intense as to virtually inhibit blood flow completely.

SKIN CIRCULATION

Blood flow through the skin fulfils two functions. In common with blood passing through any area of the body it is responsible for nourishing the tissues. Secondly it serves as a heat exchanger. The rate of blood flow through the skin is extremely variable and depends largely upon the body's need to lose or retain heat. Thus when the body requires to retain heat, skin flow may be reduced to as low as 1 ml/100 g/min. Alternatively when heat is to be lost, blood flow may rise to as high as 150 ml/100 g/min.

The skin possesses capillary networks which are fed by small nutritive arteries, and in addition there are deeper lying venous plexuses, which, in certain parts of the body (hands, feet, lips, nose, ears) are connected to the nutritive system by arteriovenous anastomoses or shunts (see Fig. 4.17).

These anastomoses are regulated by sympathetic vasoconstrictor impulses which originate in the temperature regulating centre in the hypothalamus. These impulses constrict the anastomoses so that little or no blood flows into the venous plexus. Thus minimum blood flow to achieve nourishment occurs since blood is still flowing through the nutritive arteries but body heat is conserved. When the anastomoses are opened blood fills the venous plexus, warming the skin and allowing heat loss to the environment. Sympathetic activity also affects nutritive arterioles so that flow can be influenced to some extent irrespective of the state of the AV shunts. In the absence of anastomoses (i.e. over most of the body surface), blood flow will be varied exclusively through the nutritive vessels.

Although the nervous system is the most significant regulator of skin circulation, there are others. In conditions of extreme cold, for example, local dilation occurs which is probably due to cold inactivation of the sympathetic nerve endings possibly combined with the

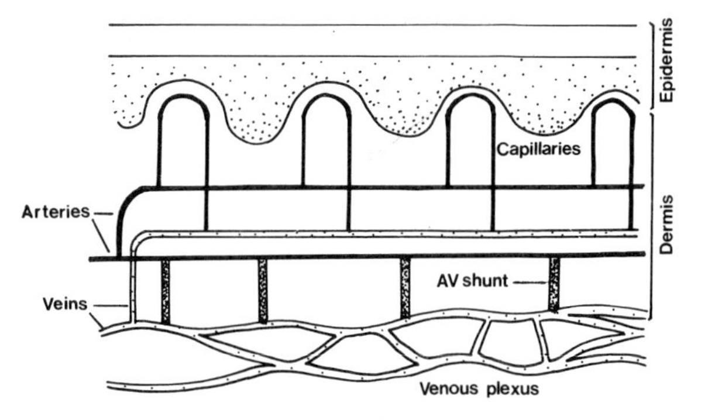

Fig. 4.17 Diagrammatic representation of the blood vessels in the skin of the hands, feet, lips, nose and ears. (Redrawn from Guyton, 1981.)

local accumulation of metabolites. The increased blood flow which results warms the tissues and prevents damage. In conditions where sweating occurs, the sweat glands release bradykinin, a powerful vasodilator which brings about a purely local vasodilation. When heat or cold is applied locally to the skin, receptors respond by sending impulses to the spinal cord, initiating a spinal reflex which does not include the vasomotor or temperature regulating centres.

5

The respiratory system

The primary function of the respiratory system is to bring about gaseous exchange, allowing the uptake of oxygen and elimination of carbon dioxide from the body.

The major respiratory structures are located within the thoracic cavity, i.e. part of the trachea, the bronchi and the lungs. However, the conducting passages of the nose, mouth, pharynx and larynx are also part of the respiratory system.

STRUCTURE OF THE RESPIRATORY SYSTEM

THORAX

The thorax is a closed cavity bounded by the vertebral column behind, the twelve pairs of ribs laterally, the sternum in front and the diaphragm below. The upper border is sealed by the tissues of the neck through which the trachea passes to allow communication between the lungs and the external atmosphere.

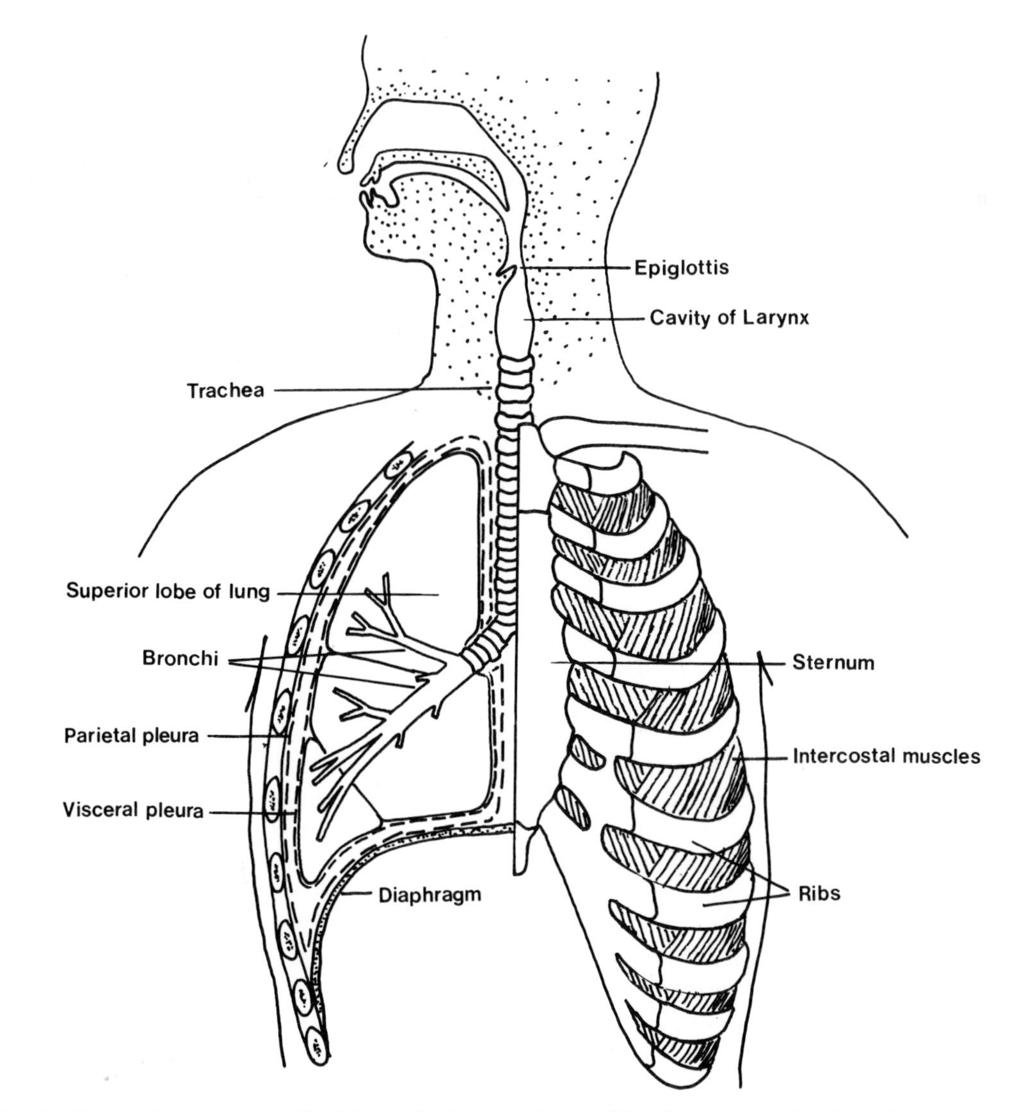

Fig. 5.1 The respiratory system. The left-hand side of the chest wall has been removed to show the contents of the thoracic cavity. The right-hand side illustrates the arrangement of the ribs and musculature.

Each of the two lungs is enclosed by two serous membranes (pleurae) in the form of a closed invaginated sac. Part of the sac most closely applied to the chest wall and diaphragm is termed the parietal pleura; that which is attached to the lung surface is the visceral pleura (see Fig. 5.1).

The free surfaces of the pleurae are smooth and moistened by serous fluid so that they will slide over one another during respiration. Under normal circumstances the two free surfaces of the pleurae are in contact and no pleural cavity exists.

CONDUCTING PASSAGES

Upper region

Air enters the respiratory tract via the external nose, whose inferior surface is pierced by two elliptical external nostrils or nares. Inside each opening is a vestibule lined by skin, with sebaceous and sweat glands; coarse hairs act as filters and help to prevent the entry of foreign bodies. The lateral wall of each nasal cavity has three elevations (superior, middle and inferior nasal conchae) which separate nasal passages (meatuses). As the air stream passes over the conchae it is disturbed so that turbulence develops; this allows small dust particles to precipitate out.

Apart from the vestibules, all other areas of the nasal cavities are lined by mucous membrane. In the respiratory region (i.e. most of the cavity) the membrane is covered by pseudostratified ciliated columnar epithelium with many goblet cells. Mucous glands are also present in the underlying connective tissue. The inside of the nasal cavity is therefore covered by a thin layer of mucus which traps dust from the air; this film is moved down and backwards towards the nasopharynx by the action of cilia.

Venous plexuses are present beneath the epithelium of the inferior nasal conchae and these warm the air before it passes down into the lungs.

In the uppermost part of the nasal cavity the epithelium is modified, containing olfactory cells; these are nerve elements which are sensitive to the presence of chemical substances. This area is known as the olfactory region.

Air passes from the nasal cavities into the nasal region of the pharynx through two small apertures, the internal nares or choanae. The pharynx is a musculomembranous tube shared by the digestive tract. It opens into the uppermost part of the true respiratory tract, the larynx.

Larynx and vocal apparatus

The larynx is situated in the upper, frontal region of the neck, between the root of the tongue and the top of the trachea, suspended from the hyoid bone by muscles, ligaments and membranes. It is made up of three pairs of cartilages, arytenoid, corniculate and cuneiform and three single cartilages, cricoid, epiglottis and thyroid. These nine cartilages are united by extrinstic and intrinsic ligaments and articulated by a number of muscles. The thyroid cartilage is the largest structure of the larynx and is visible externally as the 'Adam's apple', it wraps around the inner structures in a protective fashion (Fig. 5.2).

Internally, the larynx is lined by mucous membrane continuous with that of both pharynx and trachea. Its cavity is divided by two pairs of folds of mucous membrane, containing embedded fibrous and elastic ligaments, which run from front to back.

The lower or vocal folds are attached anteriorly to the thyroid cartilage and posteriorly to the two arytenoid cartilages. The latter are moved by muscle action to bring about abduction or adduction of the vocal folds, changing the shape of the gap or glottis which separates them.

In quiet breathing the glottis is V-shaped, becoming rounded in inspiration (see Fig. 5.3). When air is forcibly driven from the lungs the vocal folds vibrate producing sound in the column of air above.

The epiglottis, a leaf-shaped cartilage attached to the inside of the thyroid cartilage may be reflected back to close off the air passages during swallowing.

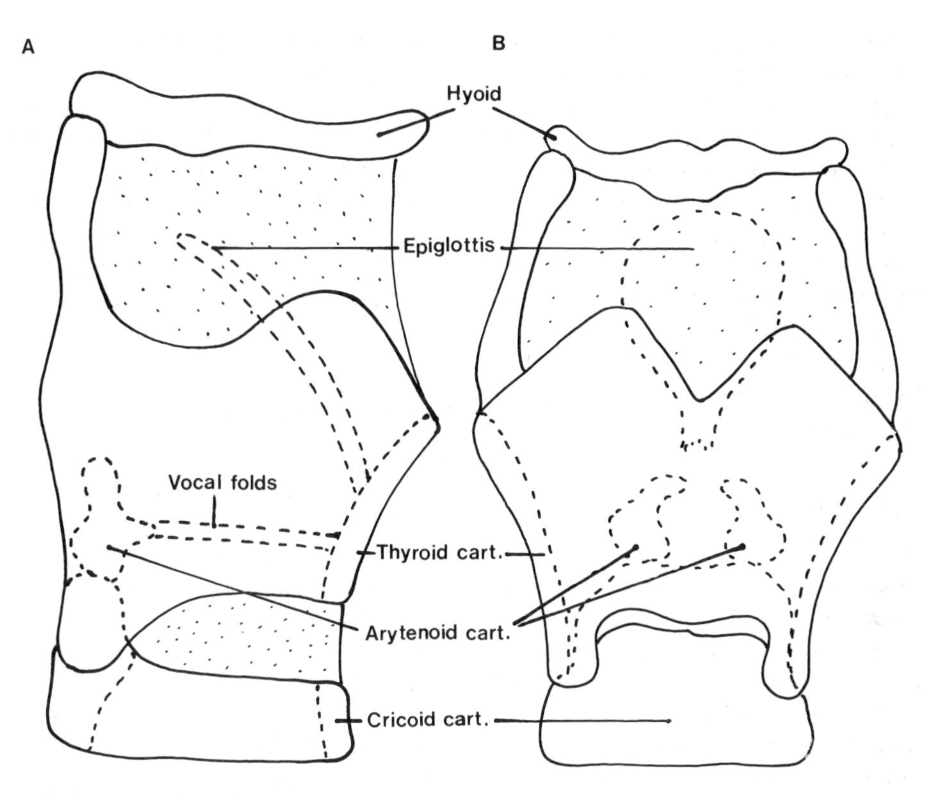

Fig. 5.2 The larynx, (A) lateral, (B) frontal. The principal internal structures are indicated by broken lines. The thyroid membrane and cricothyroid ligament (the latter shown only in A) are indicated by the dotted areas.

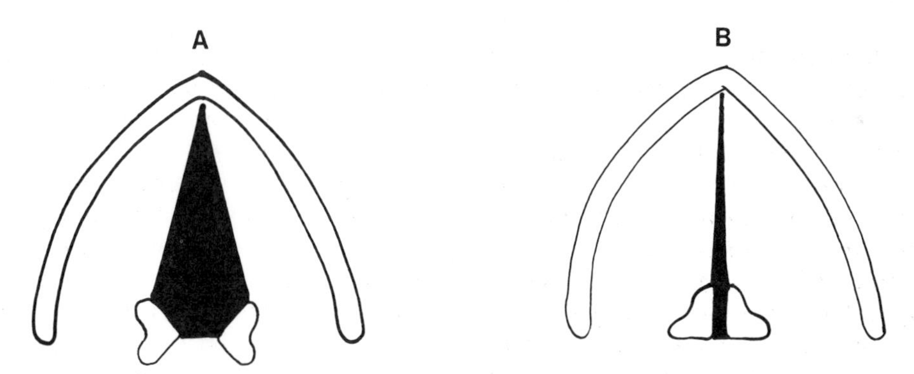

Fig. 5.3 The vocal folds and arytenoid cartilages. (A) widely abducted during forced respiration. (B) adducted during speech.

Trachea and bronchi

The trachea continues down from the lower part of the larynx for approximately 10 to 12 cm, where it divides to give rise to two bronchi. The major part of the wall is formed by sixteen to twenty horseshoe-shaped hoops of hyaline cartilage. These may branch obliquely so that adjacent hoops are joined; annular ligaments also span the intervals between them. These hoops function to maintain the patency of the airway. The posterior wall of each hoop is absent, being replaced by a thick layer of transverse smooth muscle bundles (trachealis muscle). The trachea is lined by pseudostratified columnar, ciliated epithelium containing goblet cells; in addition numerous mucous glands are present in the lamina propria.

The trachea divides at the carina to form two principal or primary bronchi. The right principal bronchus is wider, shorter and more vertically aligned than the left which predisposes the right lung to the entry of foreign bodies and the development of infections.

Once the principal bronchi enter the lung substance at the hilus, their cartilaginous hoops are replaced by cartilaginous plates which fulfil the same function. Both principal bronchi further divide into two smaller bronchi on the left side and three on the right. These give rise to secondary bronchi from which several orders of bronchioles (with a diameter of 0.5 mm or less) originate. These tubes do not have a cartilaginous component in their walls; mucous glands are also absent.

The smallest or terminal bronchioles form the final branches of the conducting portion of the lung. The air in this and all of the preceding regions of the respiratory tract constitutes the 'anatomical dead space' and does not exchange gases with the blood.

RESPIRATORY PORTION OF THE LUNG

The terminal bronchioles divide to form primary respiratory bronchioles, so named because they usually possess outpockets (alveoli) on their walls; they may further subdivide and give rise to secondary respiratory bronchioles. A respiratory bronchiole divides to give rise to a number (two to eleven) of alveolar ducts, each of which, in turn, gives rise to (usually) three alveolar sacs. Each sac has numerous terminal alveoli (see Fig. 5.4A).

The alveolar wall forms the major part of the respiratory surface of the lung. It is very thin and contains two types of cell which sit on a fine basement membrane (Fig. 5.4B). Type I cells are very flat simple squamous epithelium, and are mainly responsible for gas exchange, while type II (septal or alveolar) cells are cuboid and appear to have a secretory function. It is believed that they may be responsible for the synthesis and release of surfactant (see below).

Sometimes a direct communication exists between adjacent alveoli; these are termed Kohn's pores.

In addition to the alveolar epithelium, connective tissue which is an extension of the tunica propria of the bronchial wall is present. This fills the tissue interstices and helps to support the delicate alveoli.

Wandering polymorphonuclear cells and lymphocytes may also be observed and in addition free cells may be seen in the alveolar spaces, these are (phagocytic) dust cells which help to remove foreign particles that enter in the inspired air.

The respiratory bronchiole and its associated structures forms the functional unit of the lung called the primary or pulmonary lobule (0.5 to 2.0 cm diameter). These are aggregated together to form bronchopulmonary units, quite distinct units surrounded by connective tissue that is continuous with that of the visceral pleura.

Small groups (two to five) of bronchopulmonary units then associate to form a lobe. The right lung has three lobes, the left has two; each lobe is served by one of the large divisions of a principal bronchus.

The lungs as a whole have a double circulation. The pulmonary system is derived from the pulmonary arteries of the right side of the heart and drains via the pulmonary veins into the left atrium. Its function is to absorb oxygen and allow the elimination of carbon dioxide from the blood. The bronchial system is an extension of the systemic circulation and derives from the aorta, intercostal, subclavian and internal mammary arteries; it drains mainly into the azygous vein (on the right) and the superior intercostal vein (on the left), although some blood also drains into the pulmonary veins. The function of the bronchial system is to nourish the lung tissue.

The organization of blood vessels within the lung is complex, and a dense capillary network, mainly derived from the pulmonary system, surrounds each alveolus. The arteries and veins do not necessarily follow the bronchial

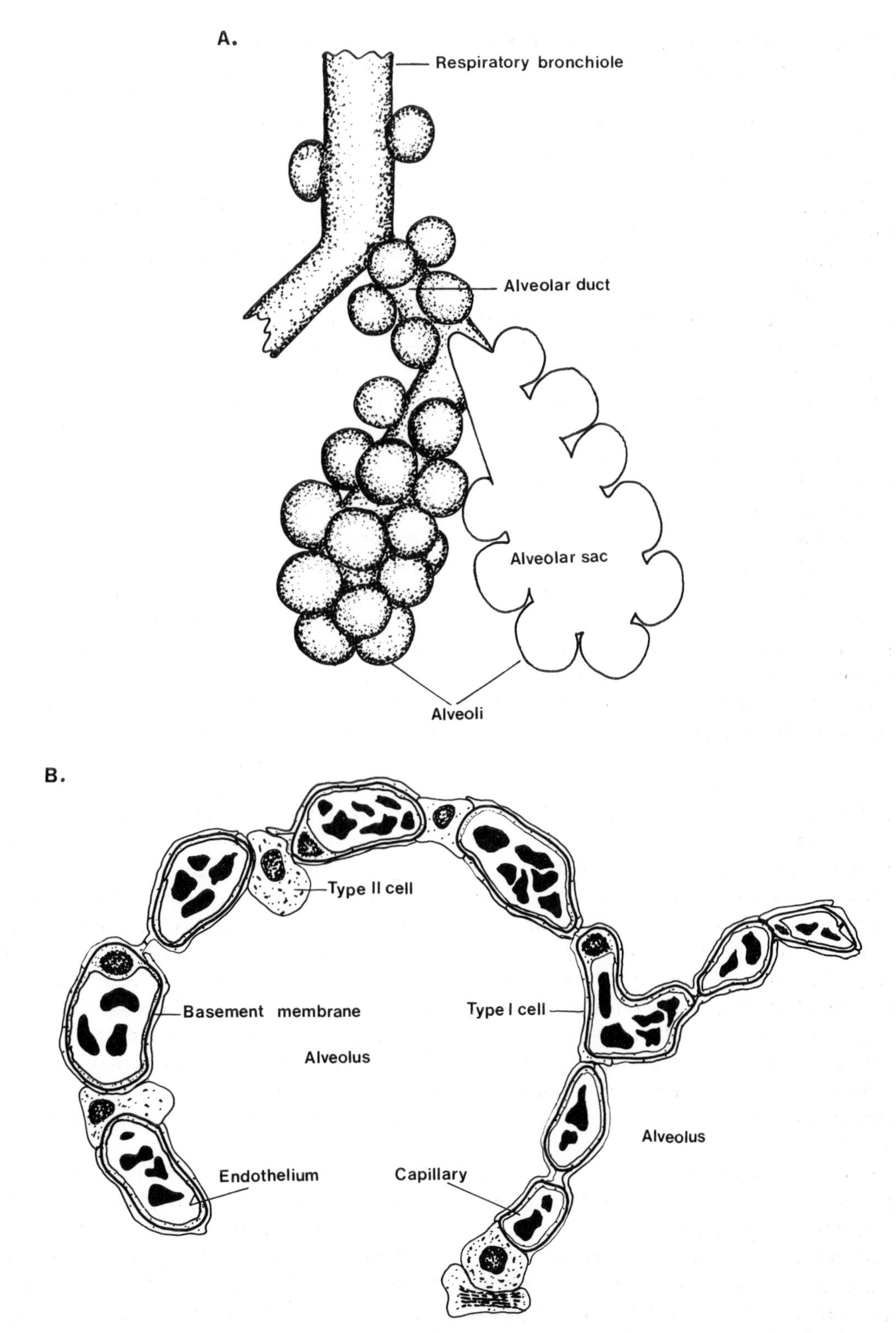

Fig. 5.4 Respiratory portion of the lung. (A) The respiratory bronchiole divides to give rise to alveolar ducts and alveolar sacs. (B) Reconstruction of an electron micrograph of a section through an alveolus.

system in a simple manner, so that the bronchopulmonary segment does not correspond to the bronchovascular unit.

The lung tissues and the pleural spaces are both drained by lymphatic tissue. Both drain independently into nodes at the hilus of the lung.

The nerve supply to the lung tissue is chiefly derived from the anterior and posterior pulmonary plexuses which are formed by branches from the sympathetic and vagus nerves. They supply efferent fibres to the bronchial muscle and afferent fibres to the bronchial mucous membrane and to the alveoli. Bronchoconstriction is probably brought about by the vagal fibres.

FUNCTION OF THE RESPIRATORY SYSTEM

PULMONARY VENTILATION

As a result of forces exerted by respiratory muscles upon the thorax, the intrathoracic pressure alters so that air is drawn into and expelled from the lungs. The lungs play an essentially passive role in these processes.

Inspiration

A large proportion of the thoracic enlargement in inspiration is brought about by contraction of the diaphragm. It consists of a dome-shaped sheet of striated muscle connected to a central tendon. Stimulation of the muscular component by the two phrenic nerves (which contain fibres from the 3rd, 4th and 5th cervical nerves) causes the diaphragm to flatten, thereby increasing the height of the thoracic cavity by as much as 10 cm. At the same time, the external intercostal muscles are stimulated by the intercostal nerves (ventral branches of the 1st to 11th thoracic nerves). This causes the bow-shaped lower ribs to move up and outwards increasing the anteroposterior diameter of the thorax.

When forced breathing occurs, the muscles of the neck may also be active; these include the scalene, sternomastoid and trapezius muscles.

Expiration

In quiet breathing, expiration is largely a passive process due to the elastic recoil of the lung tissues. However at very high rates of ventilation, contraction of the abdominal muscles (external oblique, rectus abdominis, internal oblique and transversus abdominis) increases intra-abdominal pressure and forces the diaphragm up, which actively expels air from the lungs.

The internal intercostal muscles may also contract, depressing the ribs and stiffening the intercostal spaces. This is particularly apparent in violent expiratory efforts, e.g. coughing.

Pressure relations within the chest

The visceral and parietal pleurae are separated only by a thin film of fluid, the volume of which is regulated by a balance between capillary filtration (which forms it) and lymphatic and capillary drainage (which remove it). The pressure within this pleural cavity is generally subatmospheric (and is sometimes said to be 'negative'), varying between – 7 cm of water on inspiration and – 2 cm of water on expiration.

The nature of this intrapleural pressure is due to elastic recoil, since the lungs and chest wall tend to pull away from one another. However, at the end of a quiet expiration (with the glottis open) the elastic forces of the thoracic cage and lungs are just balanced. If the chest wall was punctured and air allowed to enter to create a true pleural space, the lungs would collapse and the chest wall expand. This expansion would increase the thoracic volume by an amount which is approximately equal to 'normal' resting tidal volume, i.e. about half a litre.

During inspiration the activities of the respiratory muscles have to overcome the tendency of the lungs to collapse but they are assisted by the natural tendency of the chest wall to recoil outwards. Once inspiratory effort ceases, the elastic recoil of the lungs brings about collapse of the chest wall. Thus expiration occurs.

The extreme elasticity of the lungs is partly due to the presence of elastic connective tissue in the lung tissue framework and partly due to forces within the alveoli. The latter are lined by a thin layer of moisture and acting rather like soap bubbles, show a tendency to collapse. This tendency is reduced by the presence of an extremely thin film of lipoprotein, surfactant, which serves to reduce the surface tension at the alveolar walls.

During normal breathing the pressure inside the lung cavities (intrapulmonary pressure) changes as the chest volume alters. In inspiration it diminishes to about –3 mmHg (relative to atmospheric pressure) which causes air to enter. During expiration it rises to about +3 mmHg thereby causing air to be expelled.

Respiratory volumes

The volumes of gas breathed in and out of the lungs under various conditions can be measured using a spirometer (see Fig. 5.5). The subject breathes in and out of the air-filled bell (made weightless by a counter-weight) and a corresponding trace is made on a rotating drum (see Fig. 5.6).

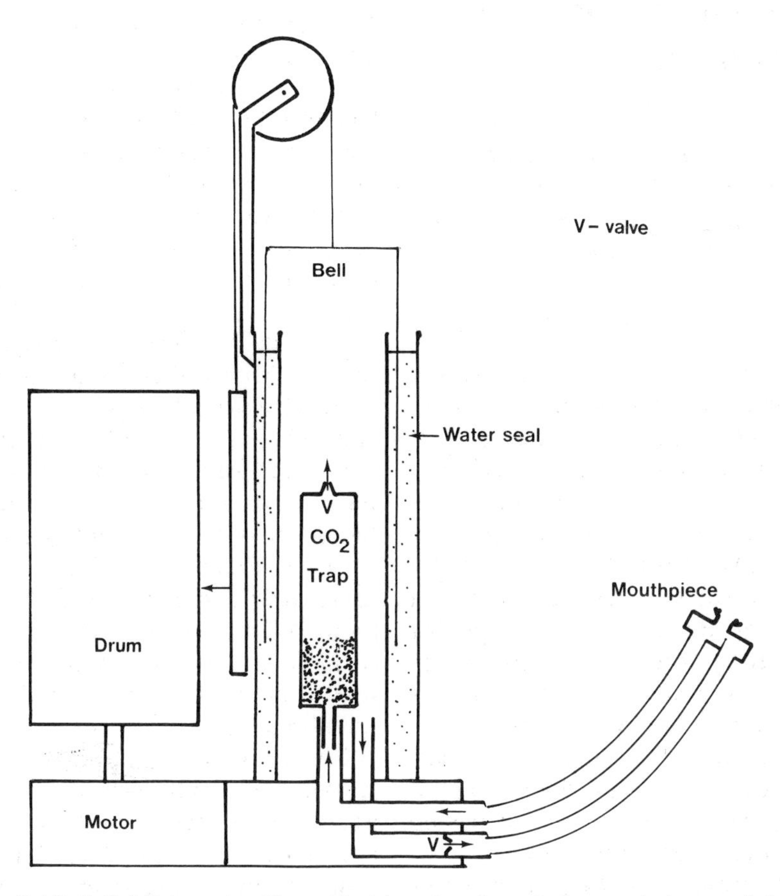

Fig. 5.5 The Benedict-Roth Bell Spirometer. The subject breathes through the mouthpiece, and air passes through a one-way valve system through the spirometer. As air enters the inner chamber the bell is displaced upwards, causing the pointer to move down. Air leaving the spirometer causes the reverse. One complete breath, therefore produces a rising and falling trace on the rotating drum (see Fig. 5.6).

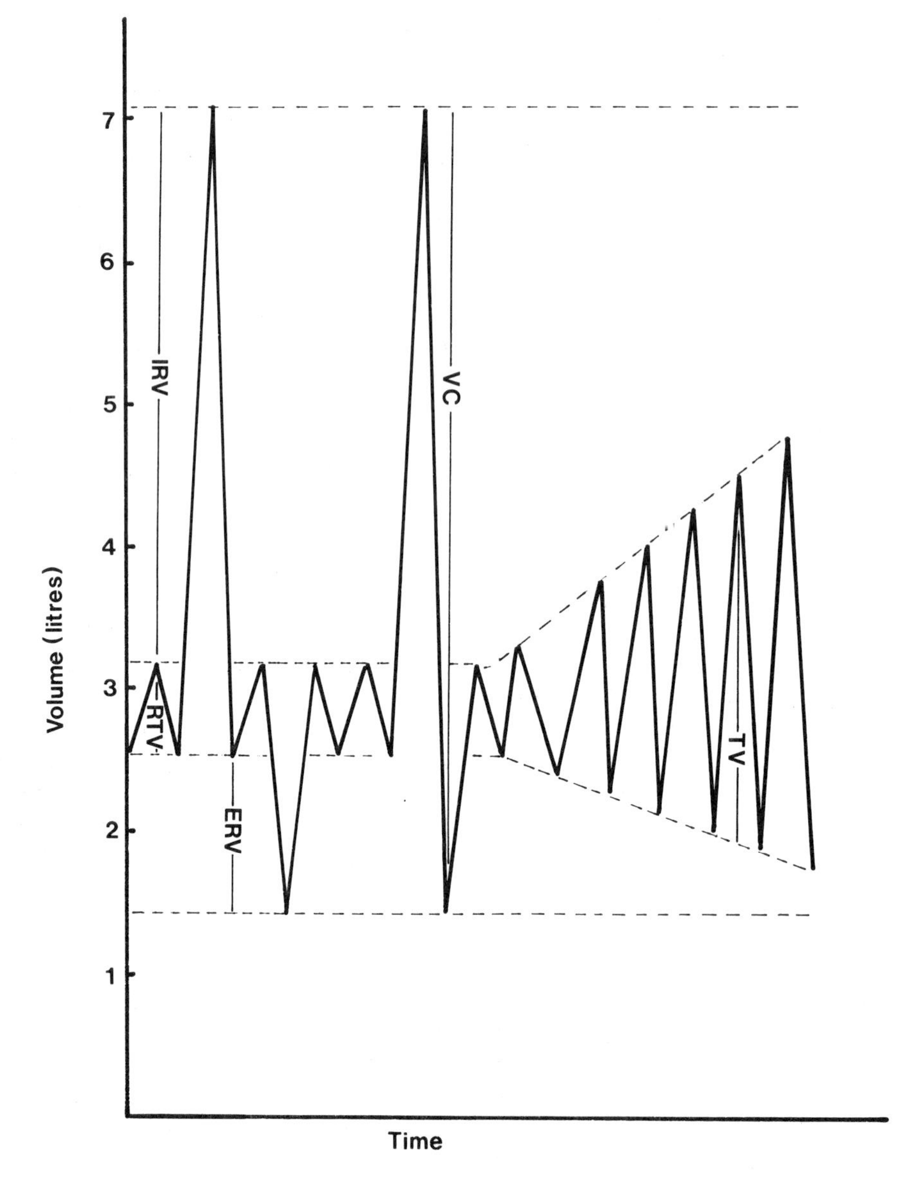

Fig. 5.6 Respiratory volumes: IRV — Inspiratory Reserve Volume, RTV — Resting Tidal Volume, ERV — Expiratory Reserve Volume, VC — Vital Capacity, TV — Tidal Volume.

Components of lung volumes:

1. *Tidal volume*. The volume of air that enters or leaves the lungs at each breath. Under resting conditions this becomes the resting tidal volume.

2. *Inspiratory reserve volume*. The extra volume of air that can be taken in by maximum respiratory effort, over and above the inspired tidal volume.

3. *Expiratory reserve volume*. The extra volume of air that can be expired by maximum respiratory effort, after the resting tidal volume has been expired.

4. *Forced vital capacity*. The volume of gas that can be expelled by the deepest possible expiration following the deepest possible inspiration.

5. *Residual volume*. The volume of air re-

maining in the lungs following the deepest possible expiration (not measured by spirometry).

6. *Total lung capacity.* The total volume of air that can be held by the lungs, i.e. the sum of the vital capacity and the residual volume.

7. *Resting minute volume.* The volume of air breathed per minute, i.e. the tidal volume multiplied by the respiration rate.

8. *Functional residual capacity.* The actual volume of air left in the lungs at the end of a quiet expiration (i.e. RV + ERV).

In addition a number of measurements are employed in a purely clinical situation. They include:

9. *Forced expiratory volume.* The volume of gas expelled from the lungs over a timed period (usually one second) when the subject makes a maximal expiratory effort from a position of full inspiration.

10. *Peak expiratory flow rate.* The maximal flow which can be sustained for a period of 10 msec during a forced expiration starting from total lung capacity. It is usually measured with a Wright peak flow meter.

The population at large exhibits tremendously wide variations in respiratory volumes, which are dependent upon height, weight, age, sex etc. Even within a particular group of individuals considerable variations in, say, vital capacity will be observed. The examples given in Table 5.1 therefore only serve as very approximate guides to quantitative measurements.

ALVEOLAR VENTILATION

Tidal volume is the total volume of gas inhaled or exhaled per breath. The volumes of air inspired and expired are not usually exactly equal. This is because the volume of carbon dioxide added to the lungs from the blood is less than the volume of oxygen taken up by the blood from the alveoli. An adult man at rest consumes about 250 ml oxygen and produces about 200 ml carbon dioxide per minute. The ratio of the volume of carbon dioxide produced to the volume of oxygen absorbed is known as the respiratory exchange ratio (R) and is usually less than 1.

Because inspired and expired volumes are different, it is the convention to express ventilation in terms of *expired* gas volume (V_E). Volume per minute is abbreviated with the use of a dot over the V, i.e. ($\dot{V}_E$).

The tidal volume can be regarded as consisting of two components, dead space (V_D) and alveolar space (V_A). Only the alveolar gas participates in gas exchange. Whilst pulmonary ventilation is measured by the volume of gas expired per minute ($\dot{V}_E$), alveolar ventilation equals pulmonary ventilation minus dead space ventilation

$$\dot{V}_A = \dot{V}_E - \dot{V}_D$$

After a normal quiet expiration, a residual amount of alveolar gas remains in both the alveolar and dead spaces of the lungs (see Fig. 5.7). This is the functional residual capacity and is of the order of 2400 ml in a

Table 5.1 Respiratory values

	Male (aged 20–30) (ml)	Female (aged 20–30) (ml)
Total lung capacity	6000	4200
Vital capacity	4800	3200
Expiratory reserve volume	1200	800
Inspiratory reserve volume	3100	2000
Resting tidal volume	500	400
Functional residual capacity	2400	1800
Residual volume	1200	1000
	(breaths/min)	(breaths/min)
Respiratory frequency at rest	11.7	11.7

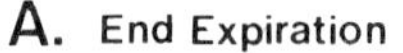

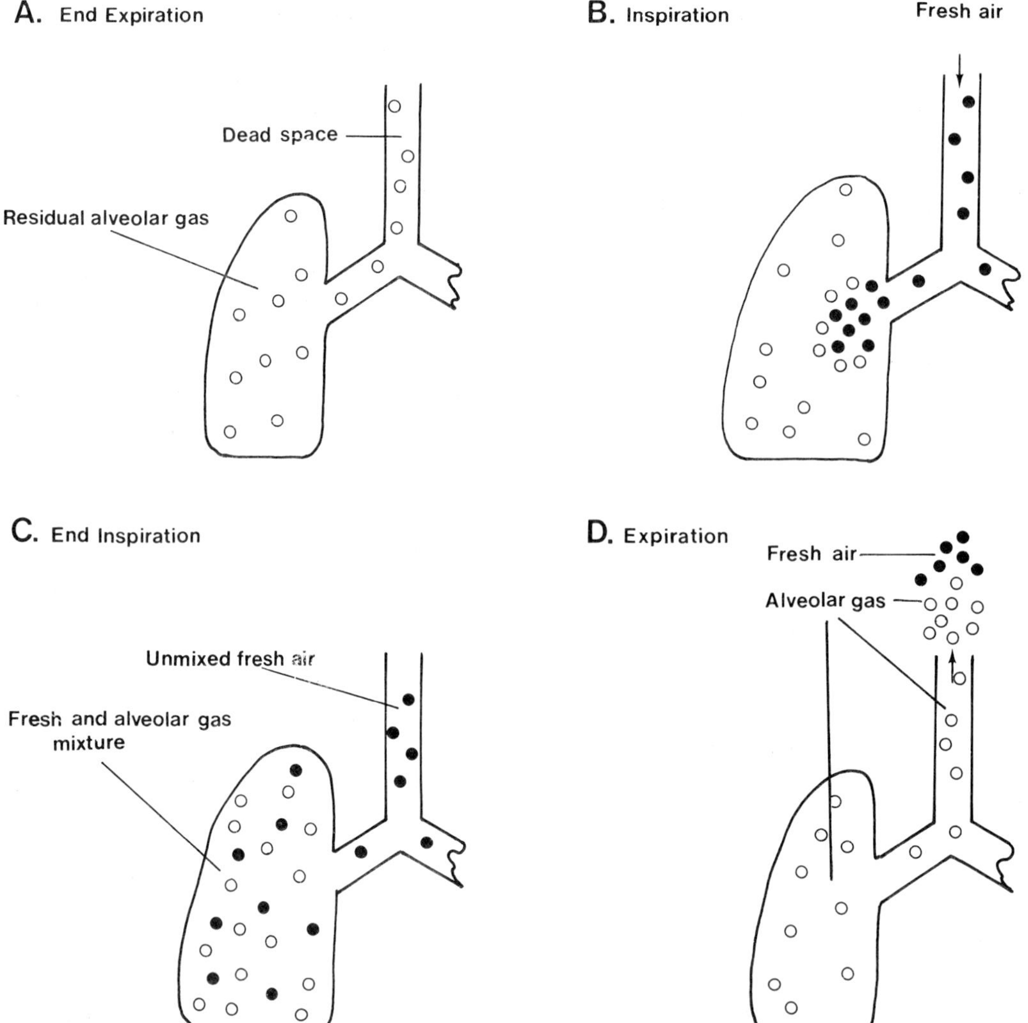

Fig. 5.7 Gas mixing. (A) At the end of expiration residual alveolar gas is distributed throughout the lungs and dead spaces. (B) On inspiration dead space gas is displaced by fresh air, a proportion of which also enters the lungs. (C) Fresh air mixes with the alveolar gas already in the lungs, but remains unchanged in the dead spaces. (D) On expiration fresh air is expelled from the dead spaces followed by alveolar gas. Thus expired gas is a mixture of alveolar gas and fresh air. The residual alveolar gas will now have the same composition as that in A.

young adult male. Approximately 150 ml of this volume is dead space. If the next inspiration were to admit 450 ml of fresh air, then 150 ml of stale air would be displaced into the alveolar region from the dead space, followed by 300 ml of fresh air, leaving the remaining 150 ml of inspired air in the dead space. Thus the percentage of air inspired which actually reaches the alveoli is

$$\frac{300}{450} \times 100 = 66\%$$

This example represents a typical resting value. However, if the inspired volume is increased by physical effort (thereby using some of the inspiratory reserve volume), the percentage of fresh air reaching the alveoli will be greater. If the inspired volume was 1500 ml, then the percentage of it reaching the alveoli would be

$$\frac{1500 - 150}{1500} \times 100 = 90\%$$

The amount of alveolar gas refreshed per breath in the resting example would be

$$\frac{300}{2400} = \frac{1}{8} = 12.5\%$$

With moderate exercise the amount of alveolar gas refreshed per breath would be

$$\frac{1350}{2400} = \frac{1}{1.8} = 55.5\%$$

Thus, because of the anatomical dead space and the residual volume of gas remaining in the lungs, alveolar gas is only partially refreshed by each breath of inspired air. The rate of turnover of alveolar gas increases if the tidal volume is increased.

COMPOSITION OF RESPIRATORY GASES

Inspired air

Inspired or atmospheric air contains nitrogen, oxygen, carbon dioxide, water vapour and traces of other gases. In respiratory physiology, the trace gases are usually designated as nitrogen because they are regarded as inert. The amount of water vapour present varies with the temperature of the air and is therefore variable in amount. It is usual to express the composition of atmospheric air in terms of dry gas. Table 5.2 shows the percentage composition of the constituent gases in air.

As air travels down the respiratory tract, moisture is added to it from the mucous membrane and it becomes saturated with water vapour which thereby changes its composition. A comparison of the values in the first two rows of Table 5.2 shows the difference in composition between dry air and saturated air.

Alveolar gas

The gas in the respiratory portion of the lung receives carbon dioxide from venous blood in the lungs and gives up oxygen to the blood. Its composition will therefore differ from inspired air in that it contains more carbon dioxide and less oxygen. Comparison of the composition of inspired air (saturated) and alveolar gas can be made in Table 5.2.

Expired gas

The first portion of gas to be expired is that part which occupied the anatomical dead space and will therefore have the same composition as atmospheric air because no gas exchange has taken place since the last inspiration.

As expiration proceeds, gas from the alveolar portion of the lung will be expelled, and this of course has a different composition from atmospheric air. If the expired gas is collected its percentage composition is found to lie between inspired air and alveolar gas because it is a mixture of the two (see Table 5.2 and Fig. 5.7).

GAS EXCHANGE

The term gas exchange is used to describe the movement of oxygen from the alveoli to the pulmonary blood and the simultaneous movement of carbon dioxide in the opposite di-

Table 5.2 Composition of respiratory gases (vol. %) (Adapted from Guyton, 1981)

	N_2	O_2	CO_2	H_2O
Atmospheric air (dry)	79.01	20.95	0.04	0.00
Inspired air (saturated)	74.09	19.67	0.04	6.20
Alveolar gas	74.90	13.60	5.30	6.20
Expired gas	74.50	15.70	3.60	6.20

rection. In the tissues, the gases exchange between intracellular fluid and plasma via interstitial fluid. Gas exchange takes place by diffusion of molecules of dissolved gas.

Diffusion

It may be recalled that diffusion is a passive process, that is, one which does not require energy. For molecules to diffuse from one solution to another, there must be more molecules in the first solution than the second, and the movement will tend to equalize the concentration of the solutions.

Water diffuses from a dilute solution (lower osmotic pressure) to a stronger solution (higher osmotic pressure) thereby reducing the difference in osmotic pressure between the two solutions. Water is said to diffuse along an osmotic pressure gradient.

Passive diffusion of a solute (something dissolved in a liquid) is from a region of high concentration to low concentration (i.e. down a concentration gradient). If the solute carries an electrical charge, then it will be attracted to a region where there is an excess of the opposite electrical charge. In the case of ions, then, there are two factors which can influence diffusion, concentration and electrical charge. The two may act in the same or in opposite directions. The net effect of the two is described as an electrochemical gradient.

Gases diffuse down a partial pressure gradient. The following sections describe the nature of partial pressures of gases in a mixture (as in the lungs) and in solution (as in body fluids).

Partial pressures of gases in a mixture

Dalton's law states that if several gases exist together in an enclosure, the total pressure that the mixture will exert is the sum of the partial pressures that each component gas would exert if it were alone in that enclosure.

$$P_T = P_1 + P_2 + P_3 + \ldots \text{etc.}$$

where P_T is the total pressure of the gas mixture and P_1, P_2, and P_3 are the partial pressures of constituent gases 1, 2, and 3.

The partial pressure of a gas is in proportion to its composition in the mixture and is therefore determined by multiplying its percentage composition by the total pressure of the mixture. Standard atmospheric pressure is 760 mmHg so the total pressure of air is 760 mmHg. In SI units (pascals), standard atmospheric pressure is 101.3 kPa. The partial pressure of nitrogen with a percentage composition of 74.09 will therefore be:

$$P_{N_2} = \frac{74.09}{100} \times 760 = 563.1 \text{ mmHg } (75.1 \text{ kPa})$$

The percentage composition and the partial pressures of the gases in inspired air are given in Table 5.3.

Table 5.3 Percentage composition and partial pressures of gases in inspired air (saturated)

Gas	% Composition	Partial pressure (mmHg)	(kPa)
N_2	74.09	563.1	75.1
O_2	19.67	149.5	19.9
CO_2	0.04	0.3	0.0
H_2O	6.20	47.1	6.3
Total	100.00	760.0	101.3

Partial pressures of gases in solution

Gas exchange involves the diffusion of gases between the gas in the alveoli and that in solution in the blood. The partial pressure of a gas dissolved in a liquid is the same as that of the gas above the liquid's surface.

Gases dissolve until the number of molecules escaping from the surface of the liquid equals the number of molecules dissolving in the liquid. The partial pressure of a dissolved gas may therefore be regarded as the pressure of molecules of gas attempting to escape from solution.

The *volume* of gas dissolving in a solution depends both on the partial pressure of the gas above the surface of the liquid (the higher the partial pressure of the gas the greater the force driving molecules into solution), but it also depends on the solubility of the gas in the particular solvent. A highly soluble gas will have less tendency to escape from solution than a less soluble one and therefore much more will dissolve before the partial pressure of the gas in solution equals the partial pressure of the gas above it.

These factors are summarized by Henry's law which states that, for a given temperature, the volume of gas (V) going into solution is equal to the partial pressure (P) of the gas above it multiplied by the solubility of the gas in the solvent (S).

$$V = P \times S$$

This can be rearranged to give an expression for partial pressure

$$P = \frac{V}{S}$$

Thus the partial pressure of a gas in solution is proportional to the volume dissolved and inversely proportional to its solubility. For a particular gas therefore, the greater the amount of gas dissolved in the plasma the greater the partial pressure it exerts. This can be seen in Table 5.4. Arterial blood is oxygen-rich and has a Po_2 of 95 mmHg, whereas mixed venous blood which is oxygen-poor has a Po_2 of 40 mmHg.

Arterial blood has a lower Po_2 than alveolar gas because a small amount of venous blood ($\simeq$ 1–2% of the cardiac output) is added to blood which has equilibrated with alveolar gas in the pulmonary capillaries. This venous admixture consists of blood draining poorly ventilated alveoli as well as tissues supplied by the bronchial circulation. (The bronchial veins drain into the pulmonary veins.)

Table 5.4 Partial pressures of oxygen and carbon dioxide mmHg (kPa)

	Po_2	Pco_2
Alveolar gas	100 (13.3)	40 (5.3)
Arterial blood	95 (12.6)	40 (5.3)
Mixed venous blood	40 (5.3)	46 (6.1)
Tissues	<30 (4.0)	>50 (6.7)

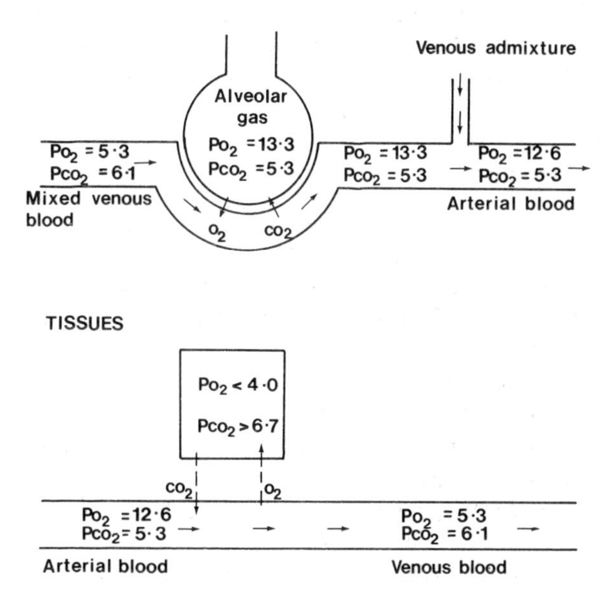

Fig. 5.8 Gas exchange by diffusion down partial pressure gradients (kPa)

Gas exchange takes place by diffusion down partial pressure gradients (see Fig. 5.8). Oxygen diffuses from the alveoli into venous blood down a gradient of 100–40 = 60 mmHg. In the tissues, the gradient for oxygen diffusion between arterial blood and intracellular fluid is at least 65 mmHg. Elimination of carbon dioxide from the tissues into the blood is achieved by diffusion of carbon dioxide down a partial pressure gradient of 10 mmHg or more, and elimination of carbon dioxide from mixed venous blood into the alveoli takes place down a gradient of 6 mmHg.

TRANSPORT OF RESPIRATORY GASES BY THE BLOOD

Oxygen

98% of the blood's oxygen content is chemically combined with the respiratory pigment haemoglobin in the red cells, the remaining 2% being carried in solution.

In arterial blood each 100 ml blood contains only 0.3 ml of oxygen in solution, and in mixed venous blood there is about 0.1 ml O_2 per 100 ml blood.

The oxygen carried in solution by the blood is important however, because it is this alone which exterts the partial pressure or tension.

Each molecule of haemoglobin has the capacity to bind four molecules of oxygen. If all the oxygen-binding capacity of the haemoglobin is utilized, the haemoglobin is said to be saturated. Since 100 ml of blood normally contains 15 g of haemoglobin, and 1 g of haemoglobin can bind 1.34 ml oxygen, the carrying capacity of haemoglobin in 100 ml blood is 20 ml. This is usually expressed as 20 vol %. Normally, arterial blood contains about 19 ml of oxygen and the haemoglobin is 97% saturated. Mixed venous blood contains about 14 ml of oxygen per 100 ml blood, and the haemoglobin is 70% saturated.

Haemoglobin acts as a store of oxygen, combining rapidly with oxygen in the pulmonary capillaries and releasing oxygen in the tissue capillary beds for consumption by the cells. The amount of oxygen carried by haemoglobin depends on the Po_2 in solution in the plasma. A graph plotting the % saturation of haemoglobin against Po_2 (see Fig. 5.9) shows that the relationship between the two is not linear but sigmoid. As the Po_2 falls (as is the case in the capillary beds when the tissues extract oxygen from the blood) the % saturation of haemoglobin falls as oxygen is released from oxyhaemoglobin into the plasma and thence to the cells. Oxyhaemoglobin *dissociates* into haemoglobin and oxygen.

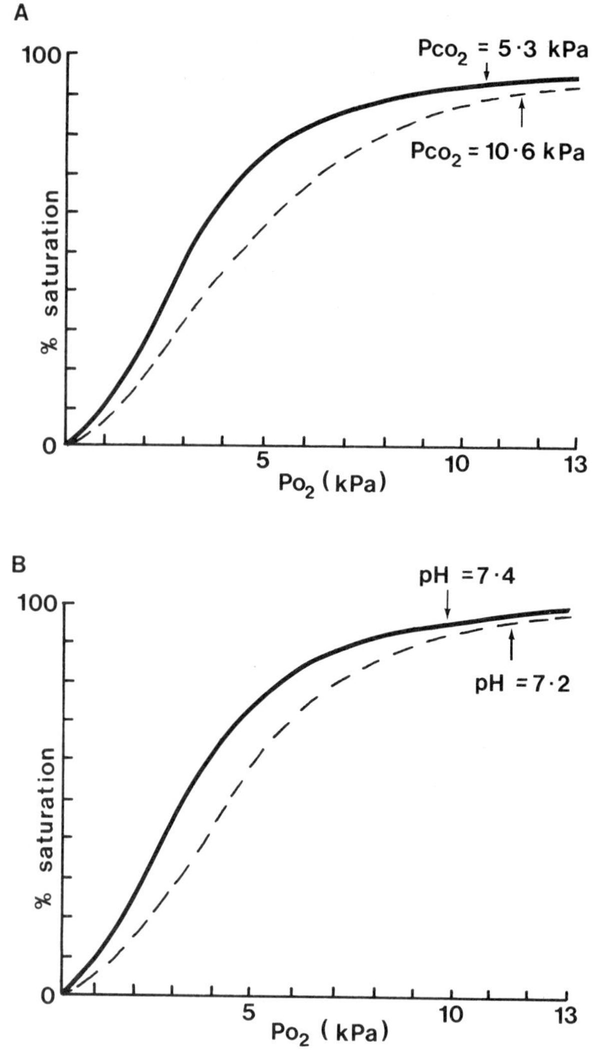

Fig. 5.9 Oxyhaemoglobin dissociation curves. (A) Effect of raised Pco_2 (B) Effect of lowered pH (increased acidity).

The upper, flat part of the dissociation curve covers the range of partial pressures found in oxygenated blood leaving the lungs (Po_2 12.5–14.0 kPa). Arterial Po_2 in turn depends on alveolar Po_2 and the shape of the curve means that a decrease in alveolar Po_2 will not result in much reduction of the % saturation of haemoglobin and therefore very little reduction in the amount of O_2 carried by arterial blood.

The steeper part of the dissociation curve covers the range of partial pressures found in reduced or venous blood. At this point on the curve it can be seen that large amounts of oxygen can be dissociated from oxyhaemoglobin with only a small change in blood Po_2. This means that in the capillary beds when oxygen is extracted, the blood Po_2 remains high enough to provide a driving force for oxygen diffusion from the blood to the cells even though the % saturation of haemoglobin has dropped.

The position of the oxyhaemoglobin dissociation curve depends on the pH, temperature and Pco_2 of the blood. If the blood becomes more acid, and/or has a higher Pco_2 the curve shifts to the right (the Bohr effect — see Fig. 5.9). A rise in blood temperature also shifts the curve to the right. Therefore, for a given Po_2 warmer, more acid blood and blood with a raised Pco_2 level will contain less oxyhaemoglobin. All these effects accompany metabolic activity. In exercising muscle cells, for example, heat, lactic acid and carbon dioxide are produced and will cause oxygen to dissociate from haemoglobin into the plasma and thence to the active cells. The Bohr effect then means that more oxygen is delivered to cells which have a high metabolic rate.

Red muscles cells have an additional store of oxygen when it is combined with the pigment myoglobin. The dissociation curve for myoglobin is steeper and to the left of that for haemoglobin. It can therefore store and release oxygen at relatively low Po_2 values, and thereby provide an extra source of oxygen to the cells when the blood Po_2 has fallen to very low levels in, for example, very severe exercise.

Carbon dioxide

Mixed venous blood contains a total of about 52 vol % of carbon dioxide. 4 ml carbon dioxide per 100 ml blood is given off at the lungs leaving an average of 48 vol % of carbon dioxide in arterial blood.

Carbon dioxide is carried in the blood in three forms — simple solution; in combination with protein as a carbamino compound ($-NHCOO^-$); and as bicarbonate (HCO_3^-). The percentage distribution is given in Table 5.5.

Simple solution

Carbon dioxide is more soluble than oxygen in water (2.4 vol % of carbon dioxide in arterial blood compared with 0.3 vol % of oxygen). Dissolved carbon dioxide, however, only accounts for 5–6% of the total carbon dioxide carried. The importance of this small amount of carbon dioxide in solution is that it exerts a partial pressure which in turn determines the amount of diffusion of the gas between fluid compartments. Approximately one molecule in a thousand combines with water in plasma to form carbonic acid. In the erythrocytes, however, the enzyme carbonic anhydrase catalyses the conversion of carbon dioxide and water to carbonic acid.

$$CO_2 + H_2O \overset{\text{carbonic anhydrase}}{\rightleftharpoons} H_2CO_3$$

Table 5.5 Distribution of carbon dioxide between plasma and erythrocytes in venous and arterial blood (% of total carbon dioxide carried)

	Arterial blood	Mixed venous blood
Plasma		
Dissolved	3.2	3.4
Bicarbonate	63.2	61.7
Total % CO_2 in plasma	66.4	65.1
Red cells		
Dissolved	2.2	2.3
Bicarbonate	26.4	25.4
Carbamino	5.0	7.2
Total % CO_2 in red cells	33.6	34.9
Plasma and red cells		
Dissolved	5.4	5.7
Bicarbonate	89.6	87.1
Carbamino	5.0	7.2

Bicarbonate

Approximately 90% of the total carbon dioxide carried in blood is in the form of bicarbonate ions. These are produced in the red cells from carbonic acid (formed from carbon dioxide and water under the influence of carbonic anhydrase).

$$CO_2 + H_2O \underset{}{\overset{\text{carbonic anhydrase}}{\rightleftharpoons}} H_2CO_3 \rightleftharpoons HCO_3^- + H^+$$

The hydrogen ions produced in this reaction are buffered by oxyhaemoglobin (see Fig. 5.11). The removal of H^+ by haemoglobin promotes increased dissociation of carbonic acid. The consequent increase in concentration of bicarbonate ions creates a concentration gradient between the erythrocytes and the plasma and therefore bicarbonate ions diffuse out. The movement of bicarbonate ions out of the cell transfers negative charge out, but this is balanced by an equivalent number of chloride ions diffusing into the erythrocyte down a favourable electrochemical gradient (chloride shift).

Carbamino compounds

The amount of carbon dioxide forming carbamino compounds with plasma protein is very small (less than 1% of total CO_2 carried). A higher proportion of carbon dioxide is carried in combination with amino groups on haemoglobin (see Table 5.5).

The carbon dioxide combines with different chemical sites on haemoglobin from oxygen; however, the two gases do not readily coexist on the same molecule so that a high carbon dioxide content in blood favours dissociation of oxyhaemoglobin, an appropriate effect for gas exchange in the tissues. Conversely, a raised Po_2 level will promote unloading of carbon dioxide from haemoglobin in the lungs (Haldane effect, see Fig. 5.10).

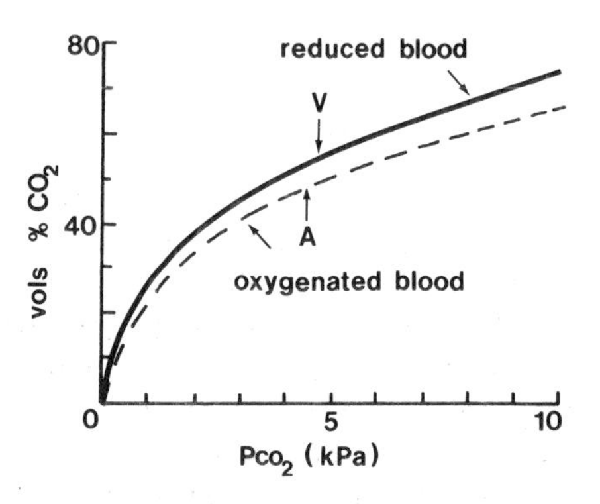

Fig. 5.10 The carbon dioxide-blood dissociation curves of oxygenated and reduced blood. The positions of typical venous and arterial blood values are shown as V and A respectively. The shift of the curve to the right caused by a raised Po_2 is known as the Haldane effect.

Summary of carbon dioxide transport

The fate of carbon dioxide added to blood from the tissues is summarized in Figure 5.11.

Dissolved carbon dioxide diffuses from the tissues into the plasma down the partial pressure gradient. As the Pco_2 in plasma rises, the gas is driven into the red cell where some molecules are converted into carbamino haemoglobin while others are converted into carbonic acid and thence to bicarbonate ions. Most of these then return to the plasma by diffusion down a concentration gradient in exchange for chloride ions.

The conversion of carbon dioxide to bicarbonate and carbamino haemoglobin produces free hydrogen ions in each case. These are buffered by oxyhaemoglobin which then releases its oxygen (Bohr effect). The dissolved oxygen then diffuses out of the red cell down the partial pressure gradient between red cell and plasma and then into the tissues down the partial pressure gradient between plasma and intracellular fluid (via interstitial fluid).

When venous blood arrives in the pulmonary capillaries the Pco_2 gradient between the plasma and alveolar gas results in diffusion of dissolved carbon dioxide from the plasma into the alveoli. Plasma Pco_2 therefore falls and causes carbon dioxide to diffuse out of the red cell into the plasma.

The removal of dissolved carbon dioxide from the red cell causes the various chemical reactions shown in Figure 5.11 to be reversed (e.g. reduced haemoglobin combines with

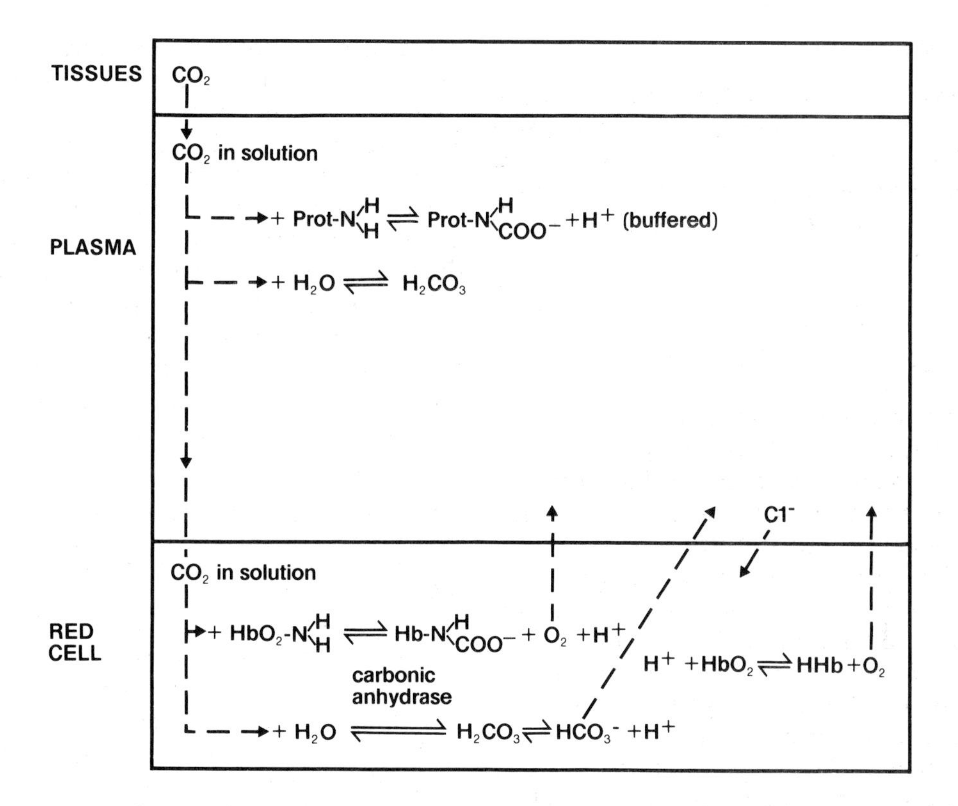

Fig. 5.11 Carbon dioxide transport from the tissues into the blood.

oxygen to give oxyhaemoglobin and free hydrogen ions which then combine with carbamino haemoglobin to give oxyhaemoglobin and dissolved carbon dioxide).

It must be remembered, however, that arterial blood contains a considerable amount of carbon dioxide and only about 11% of the total gas carried in venous blood is eliminated during its passage through the pulmonary capillaries.

CONTROL OF RESPIRATION

The respiratory muscles do not possess inherent rhythmicity like that of the heart. These muscles have striated fibres and are activated by impulses from the brain and spinal cord. The frequency of these impulses may vary, so that in quiet breathing muscular activity is regular while in talking and many other activities it becomes extremely irregular.

Since the prime function of the lungs is to allow exchange of gases between alveolar gas and the pulmonary circulation, it is understandable that blood gas composition will have a feedback effect on the central nervous system. Many other types of feedback are also known.

The respiratory centres

Two of the lower portions of the brain, the medulla and pons are known to contain groups of neurones which regulate respiratory activity although the exact arrangement and function of these neurones is still controversial. A model is presented here which derives from some of the experimental evidence; it must be remembered that alternative models do exist. The medulla contains a respiratory 'centre' which may be divided into two regions one of which is active during inspiration (see Fig. 5.12), the other during expiration. The

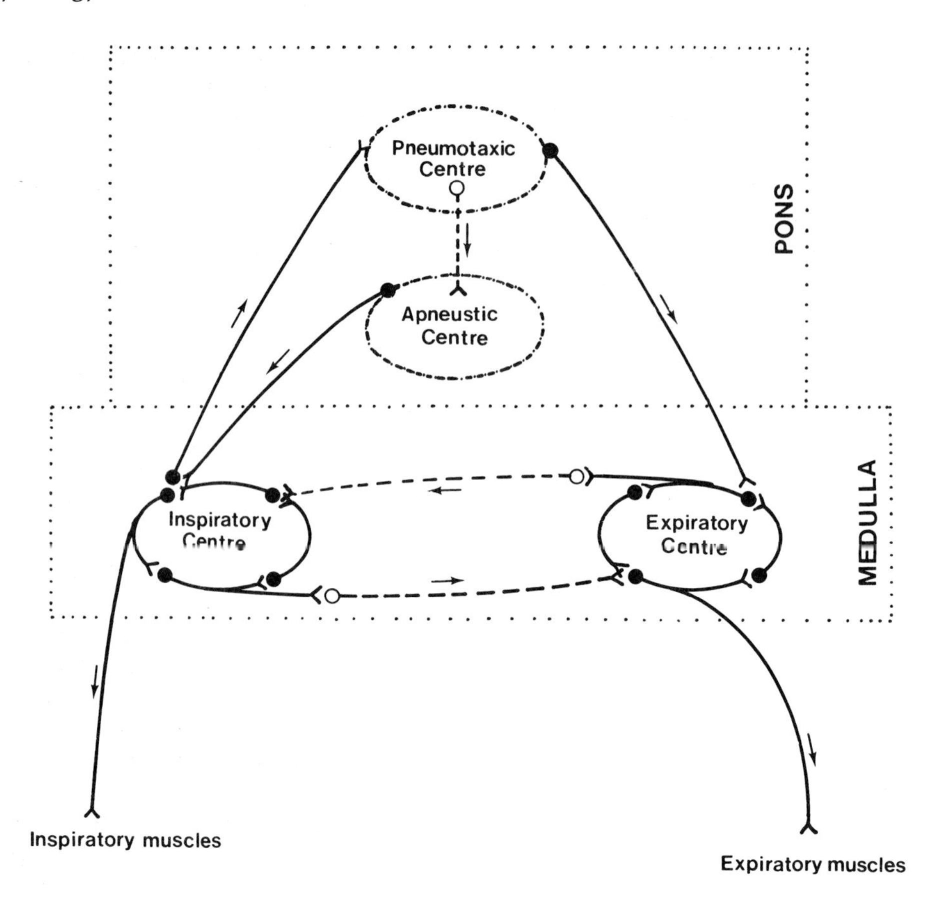

Fig. 5.12 This diagram summarises some current theories of the regulation of ventilation. Alternate oscillatory activity of the inspiratory and expiratory 'centres' is modified by the pneumotaxic and apneustic 'centres' of the pons. Nerve pathways drawn with continuous lines are excitatory, those with broken lines are inhibitory.

two centres exhibit alternate excitability which produces the rhythmic drive for respiration (medullary rhythmicity).

Inspiration occurs as a result of nerve impulses which arise spontaneously and pass around the neuronal circuit of the inspiratory centre, and from there to the inspiratory muscles. Nerve pathways which connect the inspiratory and expiratory centres include inhibitory synapses, and so when the inspiratory centre is active, the expiratory centre is inhibited.

The inspiratory centre is only active for a short while and when its activity ceases, the inhibition is lifted from the expiratory centre which then becomes active. When the expiratory centre is active, the inspiratory centre is inhibited, the inspiratory muscles relax and quiet expiration occurs. In forced expiration, impulses pass in addition from the expiratory centre to the expiratory muscles.

If the medullary centres are isolated from other areas in the brain, gasping respirations are observed. This is because areas in the pons, immediately above the medulla, act to produce a smooth respiratory cycle.

In the middle and lower pons lies the apneustic centre, while the upper pons contains the pneumotaxic centre. The apneustic centre stimulates the inspiratory portion of the medullary centre and is itself periodically inhibited by the pneumotaxic centre and by the impulses of the inflation reflex (see below). The pneumotaxic centre is stimulated by the inspiratory centre and is itself able to stimulate the expiratory centre (see Fig. 5.12). The simple interplay between the inspiratory and expiratory centres is therefore modified by activities higher up the brain stem to produce a smooth pattern of respiration.

Although the brain stem provides the basic respiratory drive, the modifications of breathing which occur in normal activity depend upon information from the higher centres of the brain and from various receptors located in the thorax and other parts of the body.

Sensory information

Receptors which influence the rate and depth of respiration fall into two main groups, chemoreceptors which respond to changes in the chemical environment and mechanoreceptors which respond to pressure and tissue deformation.

Chemoreceptors

Chemoreceptors are located centrally in the medulla, just below the ventral surface, and peripherally in the carotid and aortic bodies. Ventilation is stimulated by an increase in arterial P_{CO_2}, a fall in arterial pH (increase in hydrogen ion concentration), and to a lesser extent a fall in arterial P_{O_2}. The increased ventilation will tend to reduce the changes back towards normal. The central and peripheral chemoreceptors act in different ways.

An increase in the partial pressure of carbon dioxide in the blood results in hyperventilation, e.g. a rise of 1 mmHg in the arterial P_{CO_2} may cause a 2.5 litre increase in the minute volume. Breathing carbon dioxide-enriched air will increase the rate and depth of breathing, but only up to a point; above about 10% the carbon dioxide-sensitive areas of the brain become depressed.

Both peripheral and central receptors respond to changes in arterial P_{CO_2}. However, those situated in the carotid and aortic bodies, while exhibiting a rapid response, are considerably less sensitive than those of the medulla. The central receptors are thought to be sensitive to the hydrogen ion concentration in the cerebrospinal fluid (c.s.f.). When arterial P_{CO_2} rises, the CO_2 crosses the blood-brain barrier. Some of the CO_2 in the c.s.f. combines with water to form carbonic acid, which then dissociates to form hydrogen ions and bicarbonate ions. Since there is a relatively small amount of protein in the c.s.f. to act as a buffer for the hydrogen ions, then free ions remain in solution and stimulate the medullary receptors and increase ventilation.

In experimental animals it has been shown that a certain amount of carbon dioxide must be present in the arterial blood (and presumably, therefore, the c.s.f.) in order that breathing continues, i.e. if carbon dioxide levels are artificially lowered, breathing may cease completely. Such a clear-cut picture is not true of man where this effect is not usually observed. It is also difficult to understand the significance of the response to raised levels of carbon dioxide in man, since such rises do not actually often occur. Even in muscular exercise, which is accompanied by varying degrees of hyperventilation, there is not a measurable increase in arterial P_{CO_2}.

An increase in hydrogen ion concentration in arterial blood will stimulate the peripheral chemoreceptors and increase ventilation. The hydrogen ions do not cross the blood-brain barrier and therefore do not stimulate the central chemoreceptors.

A decrease in the partial pressure of oxygen in the blood results in hyperventilation. However, the mechanisms which detect changes in blood oxygen are relatively insensitive; only when the oxygen content of inspired air is decreased below about 8% is there a significant increase in the rate and depth of respiration.

The receptors which are primarily sensitive to oxygen lack are those of the carotid and aortic bodies, and there is little evidence to suggest that the medulla itself has similar receptors.

There is very rapid flow of blood through the carotid and aortic bodies (20 ml/g/minute) and so the receptor cells normally have a large supply of oxygen. However, a decrease in the amount of oxygen presented to these cells in a given time period will allow them to become excited. Such excitations, therefore, can arise from a lowering of the arterial Po_2 or a decrease in blood flow. Haemorrhage, therefore, may induce hyperventilation.

Mechanoreceptors

Several types of receptors are present in lung tissue itself and within the thoracic wall. Within the lung three discrete types of receptor have been identified: (1) Pulmonary stretch receptors in the smooth muscle lining of the air passages. (2) Lung irritant receptors in the intrapulmonary epithelium. (3) J-receptors in the alveolar walls.

1. During inflation of the lungs the pulmonary stretch receptors become excited and a stream of impulses pass to the respiratory centres in the medulla, where they inhibit the inspiratory drive (the Hering-Breuer reflex). These receptors may therefore help to regulate the duration and depth of respirations. In man the effect is rather weak so that the receptors only become excited when the tidal volume is above one litre.

In addition to the above, inspiration is also accompanied by a reflex dilation of the trachea and bronchi, which tends to lower airway resistance. This is also brought about by the pulmonary stretch receptors.

2. The presence of abnormal chemical substances, e.g. cigarette smoke, ammonia, brings about excitation of the lung irritant receptors. However, any abnormal mechanical disturbance, e.g. extremes of inflation or deflation will produce the same effect. Similar receptors in the trachea and larger air passages respond in a similar way and bring about coughing; they are known, therefore, as 'cough receptors'.

Excitation of the true lung irritant receptors does not bring about coughing, instead it induces hyperventilation. For example when the lungs are vigorously inflated, a contraction of the diaphragm may be seen, this is the inspiratory augmentation reflex which is supposed to help counteract lung collapse. In addition, excitation of these receptors causes a reflex bronchoconstriction. The physiological role of this reflex is not understood.

3. J-receptors apparently play little part in the normal regulation of lung function, but may become active when very large inspirations are made, for example, when there is a pathological increase in the fluid content of the air spaces. They appear to induce a decrease in the rate and depth of breathing.

The activities of the three types of lung receptor described above can be separated for experimental purposes. However, in vivo all three types of receptor will be active, feeding information into the brain.

If lung inflation is taken as an example it will be seen how they may function in sequence.

Inspirations of one litre or more may inhibit the inspiratory drive via the pulmonary stretch receptors which mediate the Hering-Breuer reflex. Larger volumes may excite the lung irritant receptors which reverse the effects of the stretch receptors via the inspiratory augmentation reflex. Very large inspirations can stimulate the J-receptors which can again inhibit the inspiratory drive.

The net response to lung inflation therefore depends upon the change in lung volume and the thresholds of the receptors involved.

Higher centres of the brain

Although breathing has been considered primarily as an involuntary activity, it is evident that some conscious control can be exerted over the process. For example in talking, irregular breaths are taken which are then let out slowly as the sound is made. Neural information can also be passed to the respiratory centres from the higher centres of the brain. The immediate increase in ventilation at the onset of exercise is thought to be partly due to stimulation of respiratory centres by the motor cortex.

Other stimuli

Information from many other parts of the body may also be received by the respiratory centres. For example, it is believed that the increase in ventilation observed at the onset of exercise may be brought about by excitation of the proprioceptors in and around the limb joints.

A rise in body temperature initiates an increase in the rate of respiration. This is due partly to a rise in the temperature of the blood and partly to warming the skin. Sudden cooling of the skin, on the other hand, induces a sudden inspiration followed by hyperventilation.

A rise in blood pressure stimulates the baroreceptors in the carotid sinuses and aortic arch, and this probably directly depresses respiration.

6

The digestive system

The digestive system comprises the alimentary tract, which extends from the mouth to the anus, together with several associated glands (see Fig. 6.1). Thus starting at the mouth, the system consists of the oral cavity, which contains the teeth and the tongue and receives saliva from the salivary glands; the pharynx; oesophagus; stomach; small intestine, which receives pancreatic juice from the pancreas, and bile from the liver; and the large intestine, which conveys the residues of ingested food to the anus.

The digestive system breaks down food, both physically and chemically, into molecules which can then be absorbed into the blood and lymph, mainly from the small intestine. The residue, together with some added waste material, is then excreted in the faeces.

The particular roles played by each section of the alimentary tract and glands, together with their structural features, are considered in this chapter, and a preliminary overview of the chemical nature of food and its digestion is given here.

THE CHEMICAL NATURE OF FOOD AND ITS DIGESTION

Carbohydrates

Carbohydrates are composed of one or more units of monosaccharide, each of which

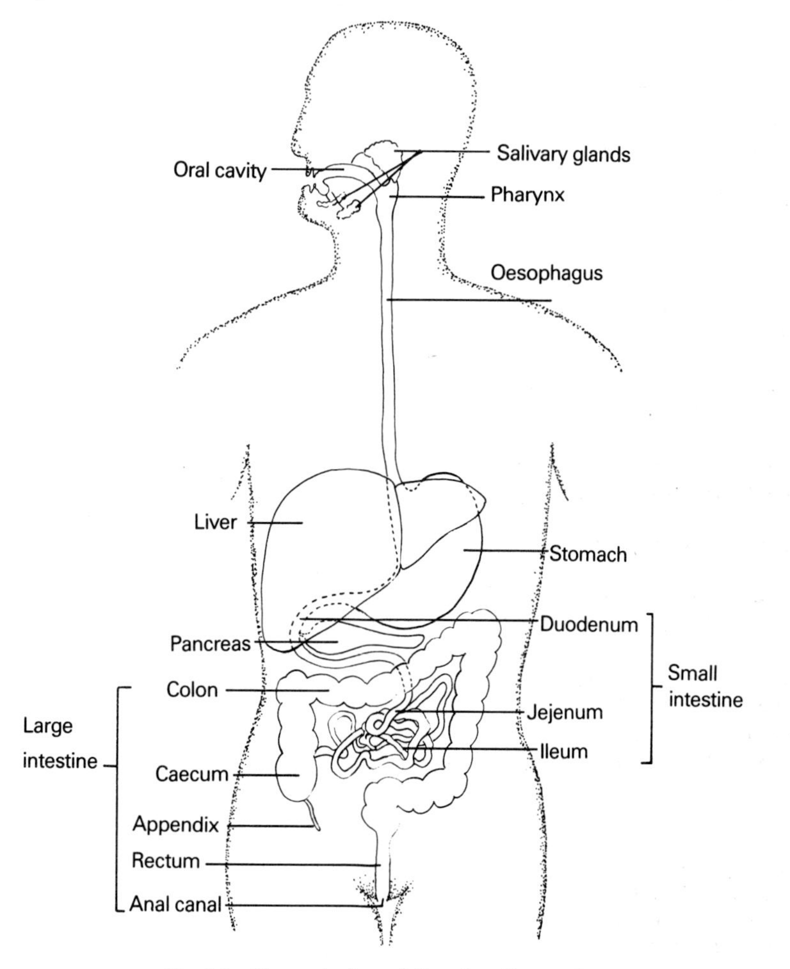

Fig. 6.1 General plan of the digestive system.

contains between three and six carbon atoms. All monosaccharides contain the atoms carbon, hydrogen and oxygen in the proportions $C_nH_{2n}O_n$. The most common monosaccharides in the diet are the hexoses, which contain six carbon atoms, giving a general formula of $C_6H_{12}O_6$. The commonest hexoses are glucose, fructose and galactose and they are usually present as part of larger molecules.

In Figure 6.2 it can be seen that four or five of the carbon atoms form a ring with one of the oxygen atoms, and the remaining atoms form side groups, projecting upwards or downwards from the ring. Glucose and galactose differ only in the orientation of the hydrogen (H) and hydroxyl (OH) groups on carbon 4 (C4).

Monosaccharides are denoted dextro- (D) or laevo- (L) forms according to the orientation of H and CH_2OH groups on C5. Alpha and beta forms are determined by the orientation of H and OH on C1.

Pentoses have five carbon atoms per molecule. Two such molecules, ribose and deoxyribose, are found as constituents of nucleic acids (see Figs. 1.13 and 1.14).

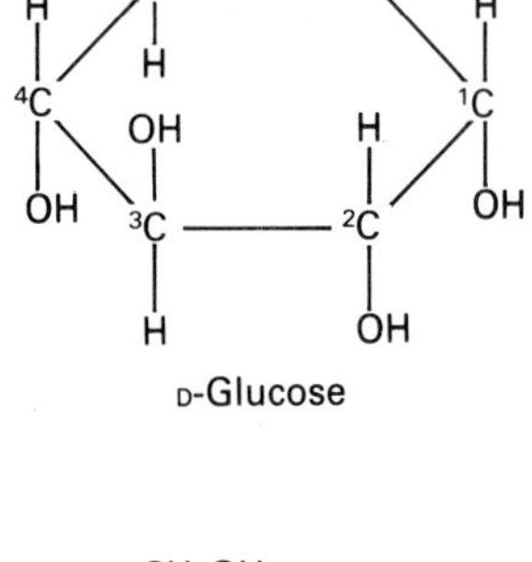

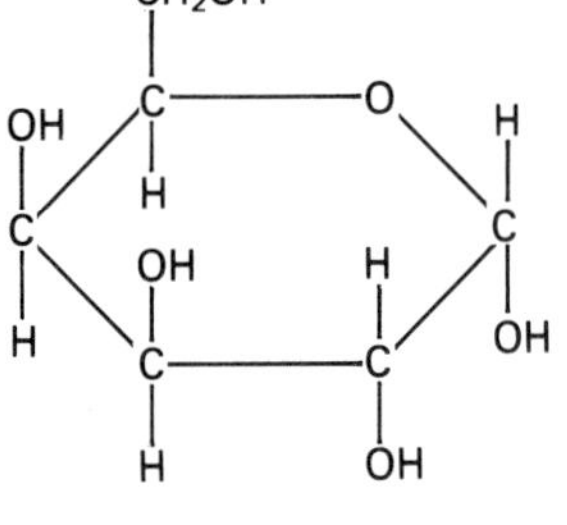

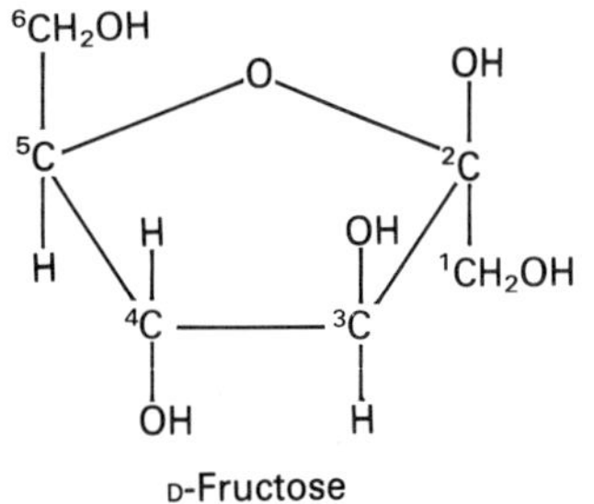

Fig. 6.2 Structure of glucose, galactose and fructose.

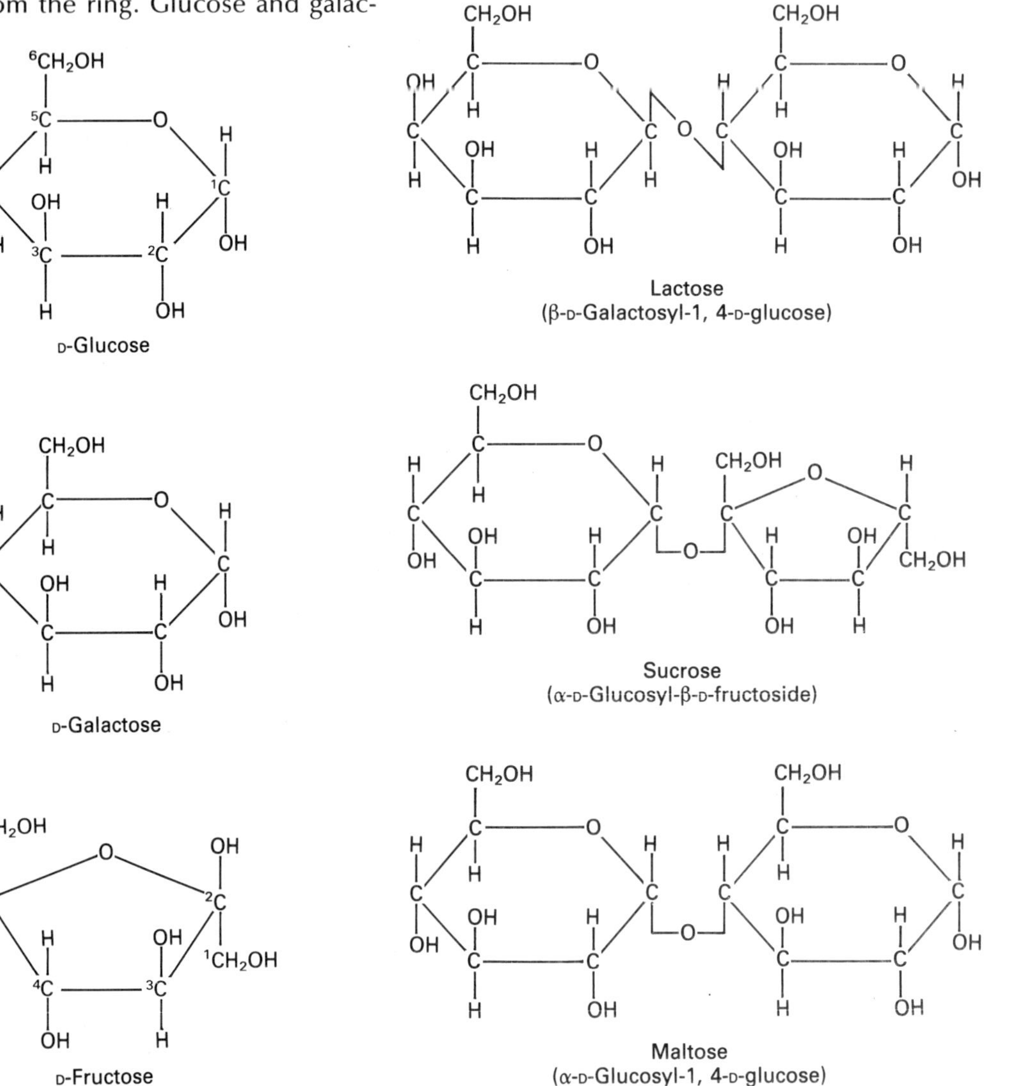

Fig. 6.3 Structure of sucrose, maltose and lactose.

Disaccharides (formed from two monosaccharides) in the diet include sucrose (the principal one), lactose and maltose. Their structures are shown in Figure 6.3.

Polysaccharides in the diet include starch, glycogen and cellulose. All of these molecules are polymers of glucose. Monosaccharides are joined together by taking H and OH from two different molecules to make water (a condensation reaction). Figure 6.4 shows how this results in three different types of link found in starch, cellulose and glycogen.

Starch is the main dietary source of carbohydrate and is stored by plants in granules. These are insoluble in water, unless they are cooked. Starch is composed of glucose chains formed by two different linkages. The α-1,4 links form straight chains of 25 to 2000 glucoses, called amylose, and this makes up about 20% of the starch molecule. Eighty per cent of starch is made up of amylopectin, which is a branched chain. Twenty-five to thirty glucose molecules are joined by α-1,4 links and then an α-1,6 link occurs, which gives a branch point.

Glycogen resembles amylopectin, but the branch points occur more frequently. This is the animal storage form of carbohydrate, found in liver and muscle. Cellulose, a component of plant cell walls, is a straight chain polymer of some 3000 glucose molecules joined by β-1,4 links.

The splitting of disaccharides and polysaccharides is achieved by the addition of water in an enzyme-catalysed reaction known as hydrolysis. Figure 6.5 gives a summary of the

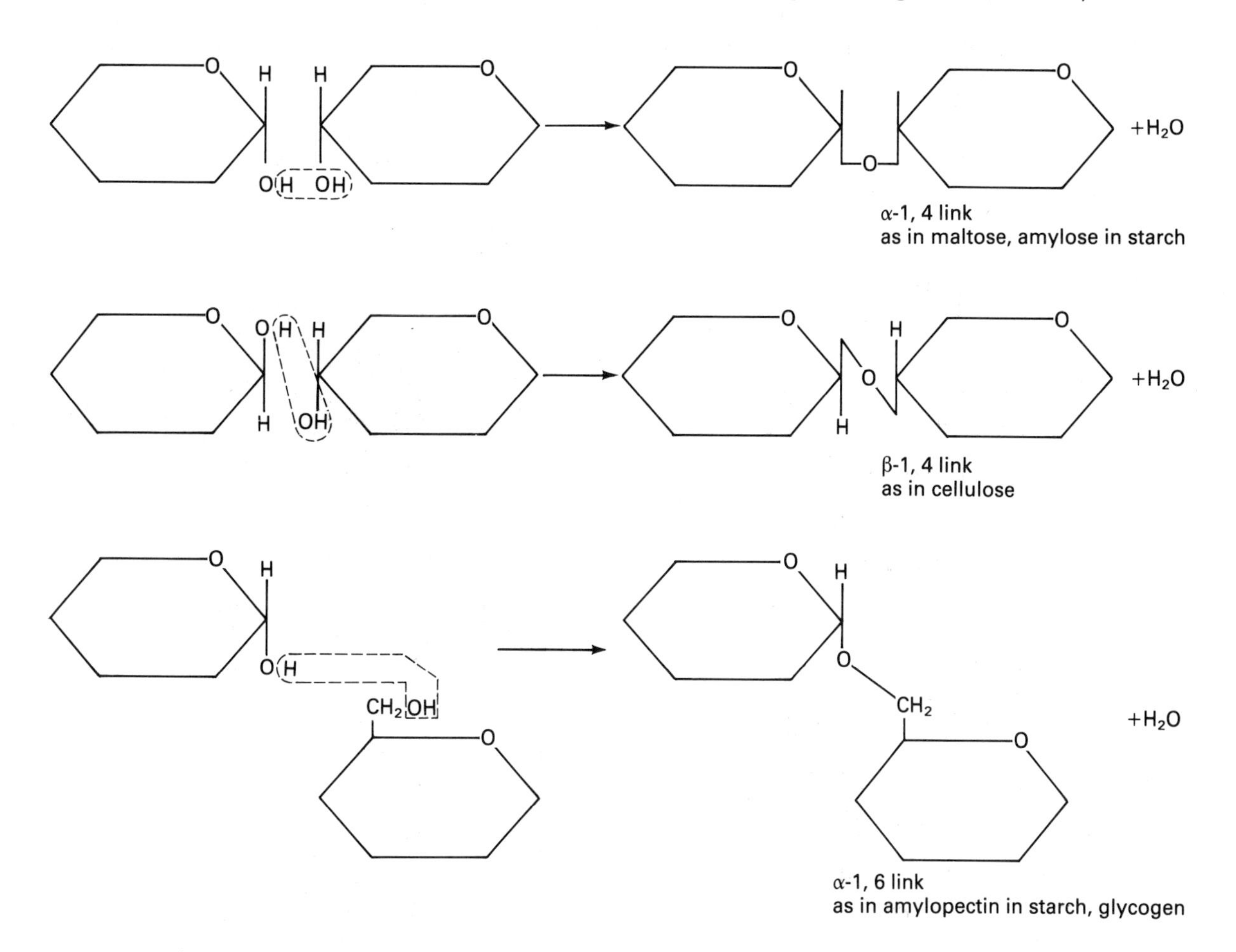

Fig. 6.4 Types of linkages employed in the formation of disaccharides and polysaccharides.

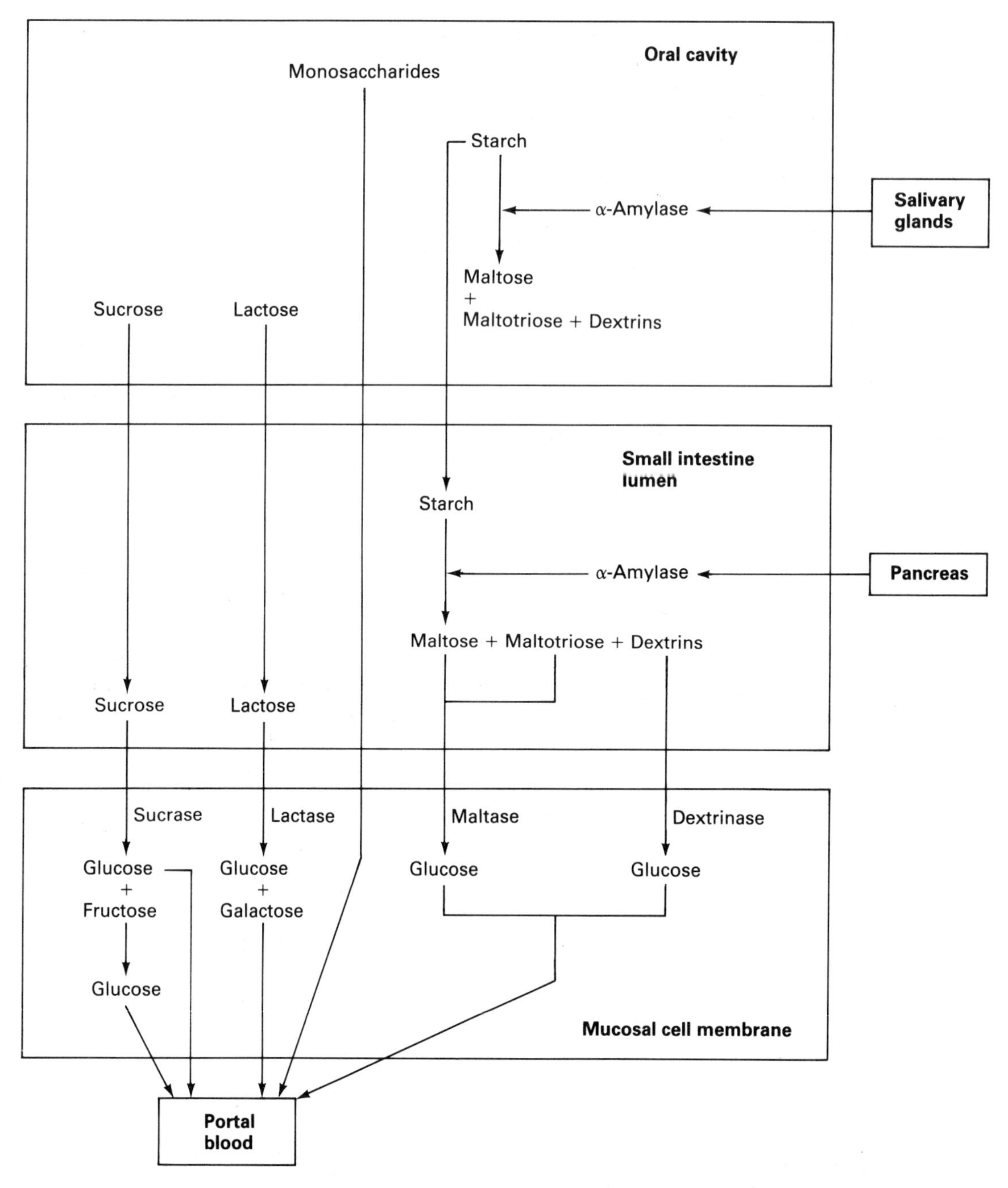

Fig. 6.5 Summary of carbohydrate digestion.

events in carbohydrate digestion. Salivary and pancreatic α-amylases aid the spliting of α-1,4 links in starch to give maltose, maltotriose and dextrins. These consist of a variable number of glucose molecules (five or more, usually eight) around a branch point. It may be noted that α-amylase will neither catalyse the splitting of α-1,6 links nor the final splitting of maltose to free glucose molecules. Hydrolysis of maltose, maltotriose and dextrins, together with any disaccharides in the diét is mediated by oligosaccharidases present on the cell membranes of cells covering the villi in the small intestine (see *Digestion and absorption by the small intestinal epithelium*).

Man has no enzyme capable of catalysing the hydrolysis of the β-1,4 links found in cellulose and so this component of the diet contributes bulk or 'roughage'. The importance of this material in promoting bowel movements and apparently reducing the incidence of several disorders, is currently receiving considerable attention.

The monosaccharides glucose and galactose are absorbed by an active transport mechanism linked to sodium absorption. Fructose is absorbed by facilitated diffusion. All monosaccharides are absorbed into the portal blood draining the small intestine.

Lipids

Triglycerides

The majority of dietary fats are triglycerides (triacylglycerols). A triglyceride consists of an alcohol, glycerol, linked to three (usually different) fatty acids (see Fig. 6.6). Butter, for example, contains butyric acid, oleic acid and stearic acid linked to glycerol.

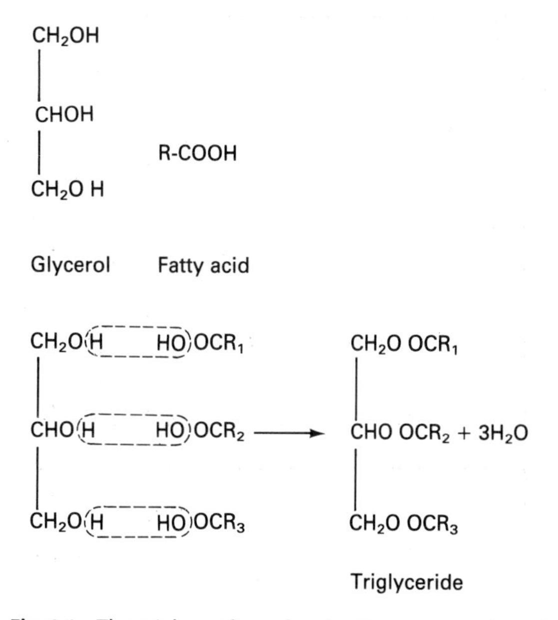

Fig. 6.6 The triglyceride molecule, its components and formation.

Fatty acids comprise a hydrocarbon chain and an acidic carboxyl group (COOH). Fatty acids can be subdivided into saturated or unsaturated types. The former type, present in animal fats, are molecules within which all the available sites on the hydrocarbon chain are occupied by hydrogen atoms (see Fig. 6.7).

Saturated chain

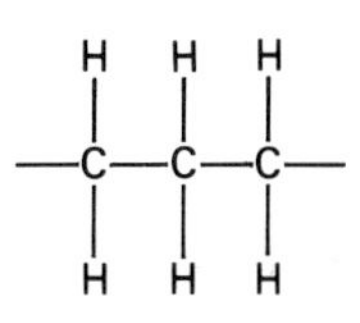

Saturated fatty acids $C_nH_{2n+1}COOH$

Butyric	C_3H_7COOH
Stearic	$C_{17}H_{35}COOH$
Palmitic	$C_{15}H_{31}COOH$

Unsaturated chain

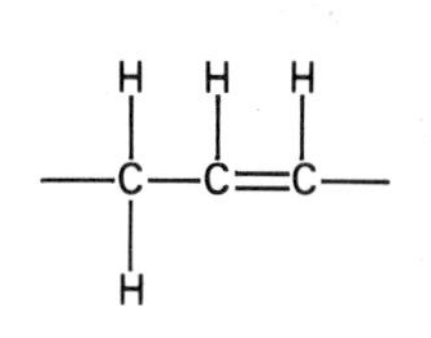

Unsaturated fatty acids

One double bond C_nH_{2n-1}
e.g. oleic acid $C_{17}H_{33}COOH$

Two double bonds C_nH_{2n-3}
e.g. linoleic acid $C_{17}H_{31}COOH$

Three double bonds C_nH_{2n-5}
e.g. linolenic acid $C_{17}H_{29}COOH$

Four double bonds C_nH_{2n-7}
e.g. arachidonic acid $C_{19}H_{31}COOH$

Fig. 6.7 The classes of fatty acid molecules.

These fats tend to be solid at room temperature. Unsaturated fatty acids, which are the type found in plants and fish (and present in other animals too, as well as the saturated fats), have one or more double bonds present in the hydrocarbon chain, and therefore contain less hydrogen (see Fig. 6.7). These fats are liquid at room temperature.

Some unsaturated fatty acids are essential in the diet (e.g. arachidonic acid) and their importance in lowering blood cholesterol level and thereby reducing the incidence of

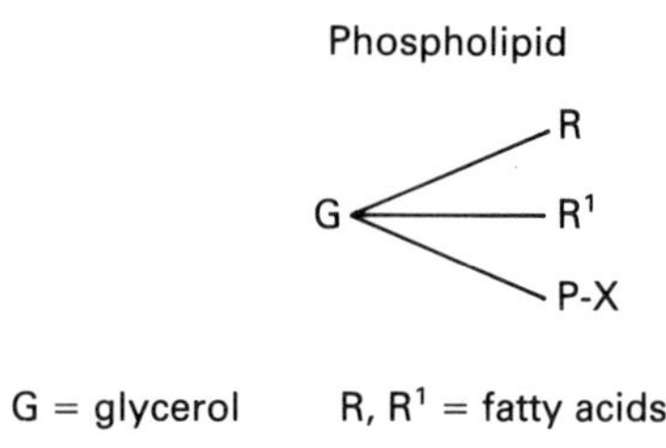

Fig. 6.8 General structure of a phospholipid molecule.

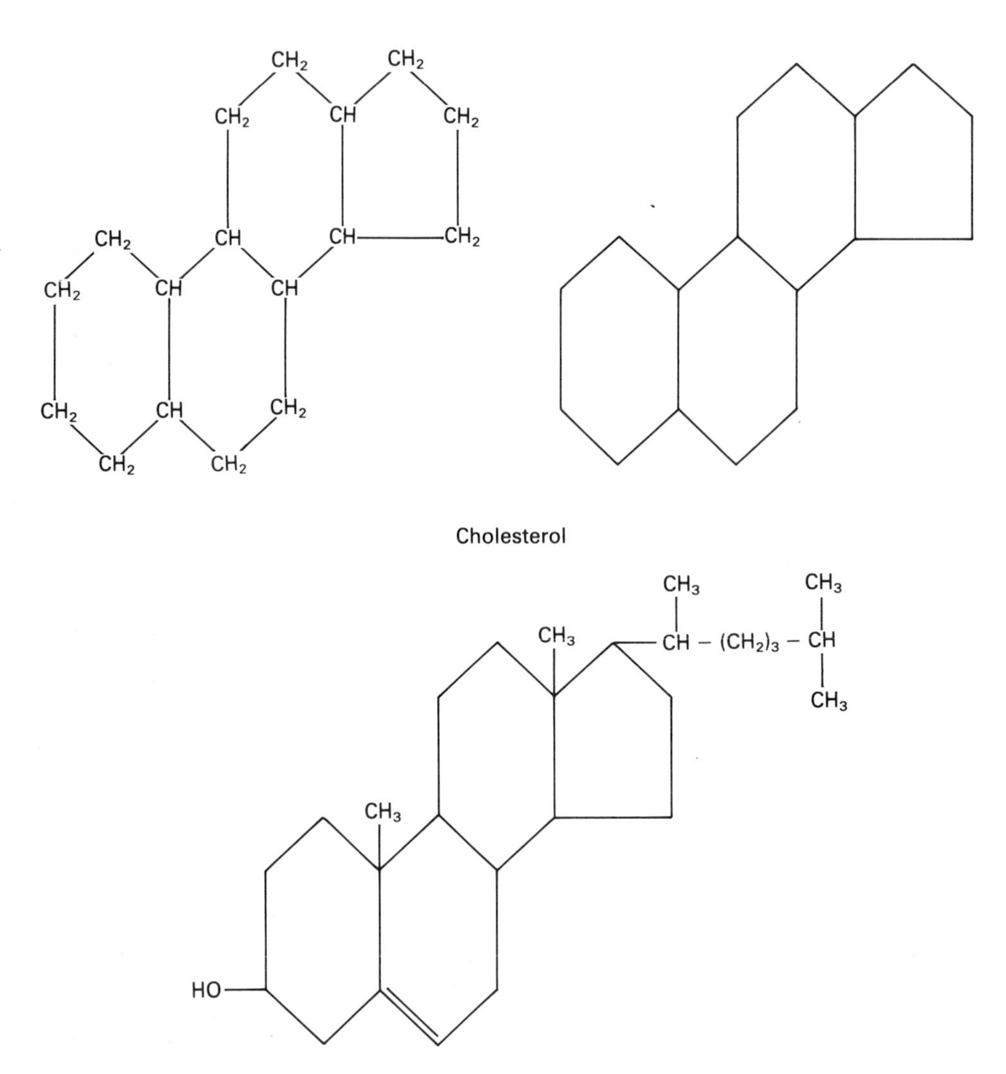

Fig. 6.9 The steroid nucleus (top left), its abbreviated form (top right), and the structure of cholesterol.

atherosclerosis (cholesterol deposits in arteries) has long been recognized.

Phospholipids

This is a very large group of fats, the components of which are glycerol, two fatty acids and a nitrogenous group (see Fig. 6.8). Lecithins, which are a type of phospholipid found in cell membranes, incorporate choline in their structure.

Steroids

Steroids are complex lipids, with a hydrocarbon ring structure as shown in Figure 6.9. Cholesterol, an alcohol, is found in all animal foods, particularly eggs. Although extremes of variation in dietary intake of cholesterol can influence blood plasma levels, normally a fall in dietary intake is compensated for by increased production of cholesterol from the liver and vice versa. A diet high in saturated fatty acids raises plasma cholesterol concentration, and conversely a diet high in unsaturated fatty acids lowers plasma cholesterol level. Several hormones can influence plasma cholesterol levels, e.g. lack of thyroid hormones raises it as does lack of insulin. Male sex hormones raise blood cholesterol, whereas female ones lower it.

Lipid digestion and absorption

Lipid digestion is summarized in Figure 6.10. No digestion occurs in the mouth or stomach, although the churning action of the stomach assists in the formation of a coarse emulsion. Further emulsification occurs in the small intestine owing to the mixing movements of the intestine itself, and the combined action of bile salts, lecithin and cholesterol as well as the fat digestion products themselves.

The problem in fat digestion and absorption is not the number of enzymes involved (only one) or difficulty in absorption (fat is relatively easily transported through cell membranes); but the fact that fat and water do not mix so that fat tends to form a separate layer, or large droplets in water. Pancreatic lipase is a protein, which is water soluble, and therefore can only act on the surface of a fat droplet. It is therefore vital that the fat is broken up into a fine emulsion consisting of large numbers of very small droplets so that the lipase has a sufficiently large surface area on which to act.

The triglycerides are hydrolysed to free fatty acids and monoglycerides (the middle fatty acid stays attached to the glycerol), which then form minute structures called micelles with bile salts and other fats. Further details are given in *Secretion of bile* and *Digestion and absorption by the small intestinal epithelium*.

Monoglycerides and free fatty acids are absorbed by diffusion, whereas the bile salts, after being used many times, are finally absorbed actively.

The longer chain fatty acids are reassembled into triglycerides inside the intestinal epithelial cells, and there they are packaged into spherical structures called chylomicra (sing. chylomicron). These enter the lymph and are thereby conveyed to the blood via the thoracic duct. Chylomicra are mainly broken down in adipose tissue and the liver. Shorter chain free fatty acids are not assembled into triglycerides and are absorbed into the portal blood rather than the lymph.

Proteins

Like polysaccharides, proteins are polymers, but in this case the repeating units are amino acids. These are named because of the presence of an amino group (NH_2) and a carboxyl group (COOH) in the molecules (see Fig. 6.11). Amino acids may be further classified: as acidic, if there are two carboxyl groups; basic, if there are two basic groups; or neutral, if there is one amino group and one carboxyl group.

Figure 6.11 shows the 20 amino acids present in the diet, some of which are classified as essential, because they cannot be synthesized in the body.

Although protein is used less than carbohydrate and fat to produce energy, about 30 g

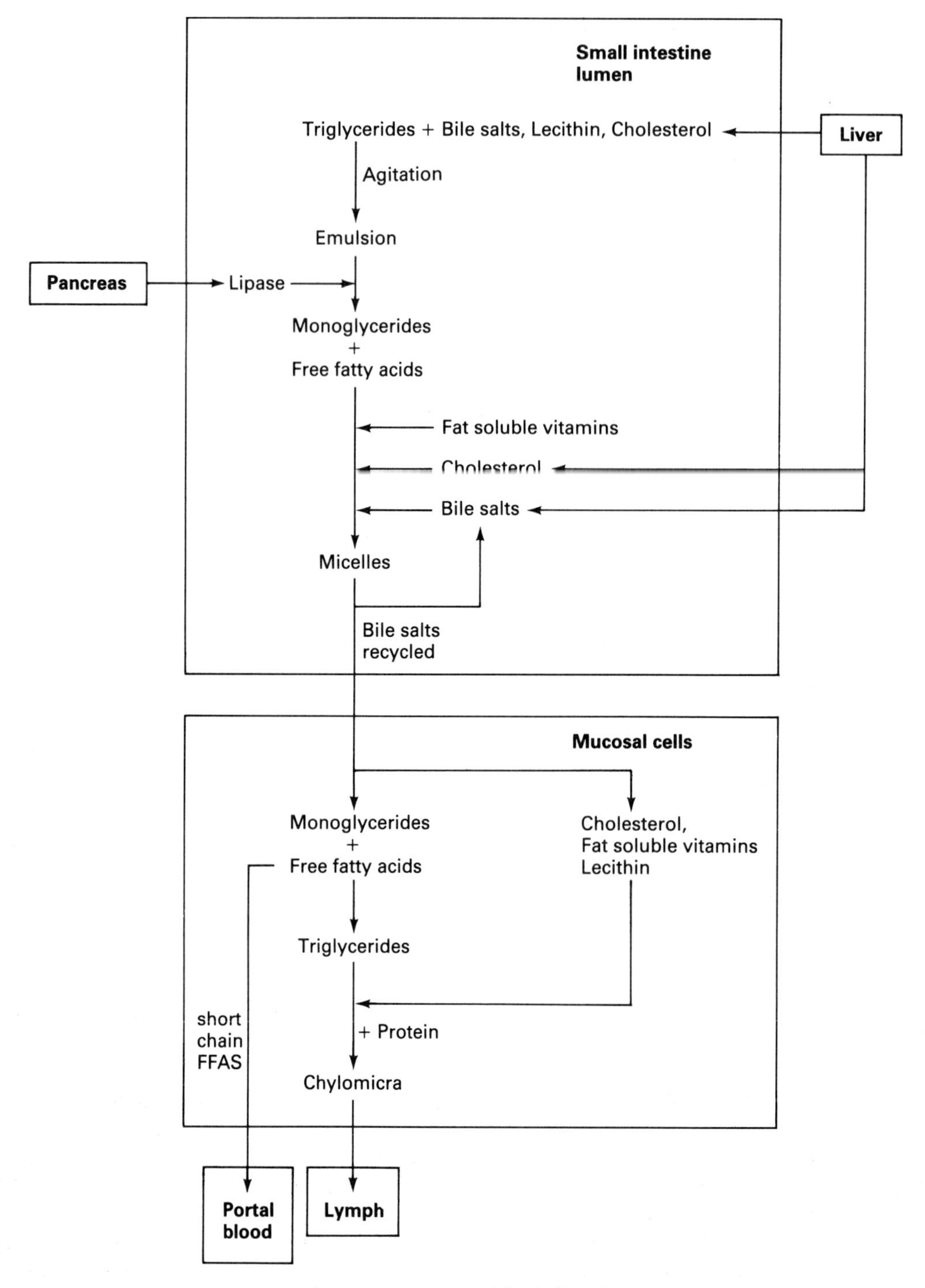

Fig. 6.10 Summary of lipid digestion.

is broken down each day and so needs to be replaced in the diet in order to avoid negative nitrogen balance.

Peptide chains are built up by peptide bonds (see Fig. 6.11), and the final protein may consist of only one polypeptide chain or

General structure of an amino acid

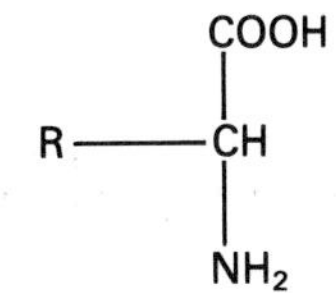

Formation of a peptide bond

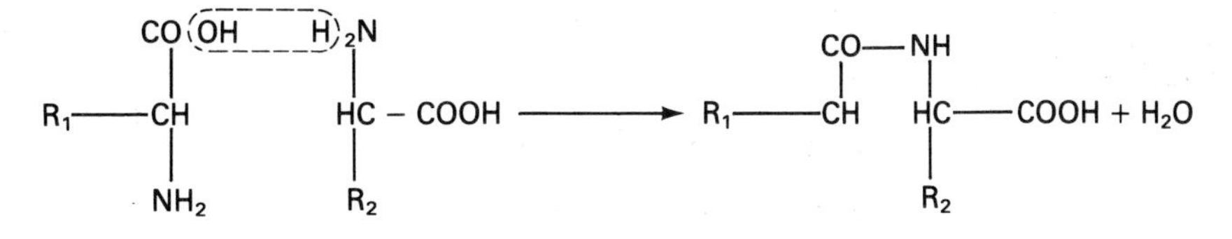

Eight essential amino acids required by adults

Valine
Leucine
Isoleucine
Threonine
Lysine
Phenylalanine
Tryptophan
Methionine

Two essential amino acids additionally required by children

Arginine
Histidine

Ten non-essential amino acids

Glycine
Alanine
Serine
Aspartic acid
Glutamic acid
Ornithine
Tyrosine
Cysteine
Proline
Hydroxyproline

Fig. 6.11 The general structure of an amino acid and the formation of a peptide bond. Essential and non-essential amino acids.

as many as four of them, linked together at various points. The chains may be folded in a complex three-dimensional arrangement, the shape of which is important for the function of the particular protein. Globular proteins are so named because the peptide chains are coiled and folded into a globular shape. These proteins are water-soluble and include albumin, globulins, haemoglobin and most cellular enzymes.

Other proteins are classified as fibrous, because their polypeptide chains are elongated and the separate chains are linked together to form bundles. Such proteins include collagen and elastin in connective tissue, actin and myosin in muscle, and keratins in skin, nails and hair.

Conjugated proteins are those which are combined with non-protein substances, e.g. nucleoproteins (protein and nucleic acid), proteoglycans or mucoproteins (protein and polysaccharide), lipoproteins (protein and lipid) and phosphoproteins (protein and phosphorus).

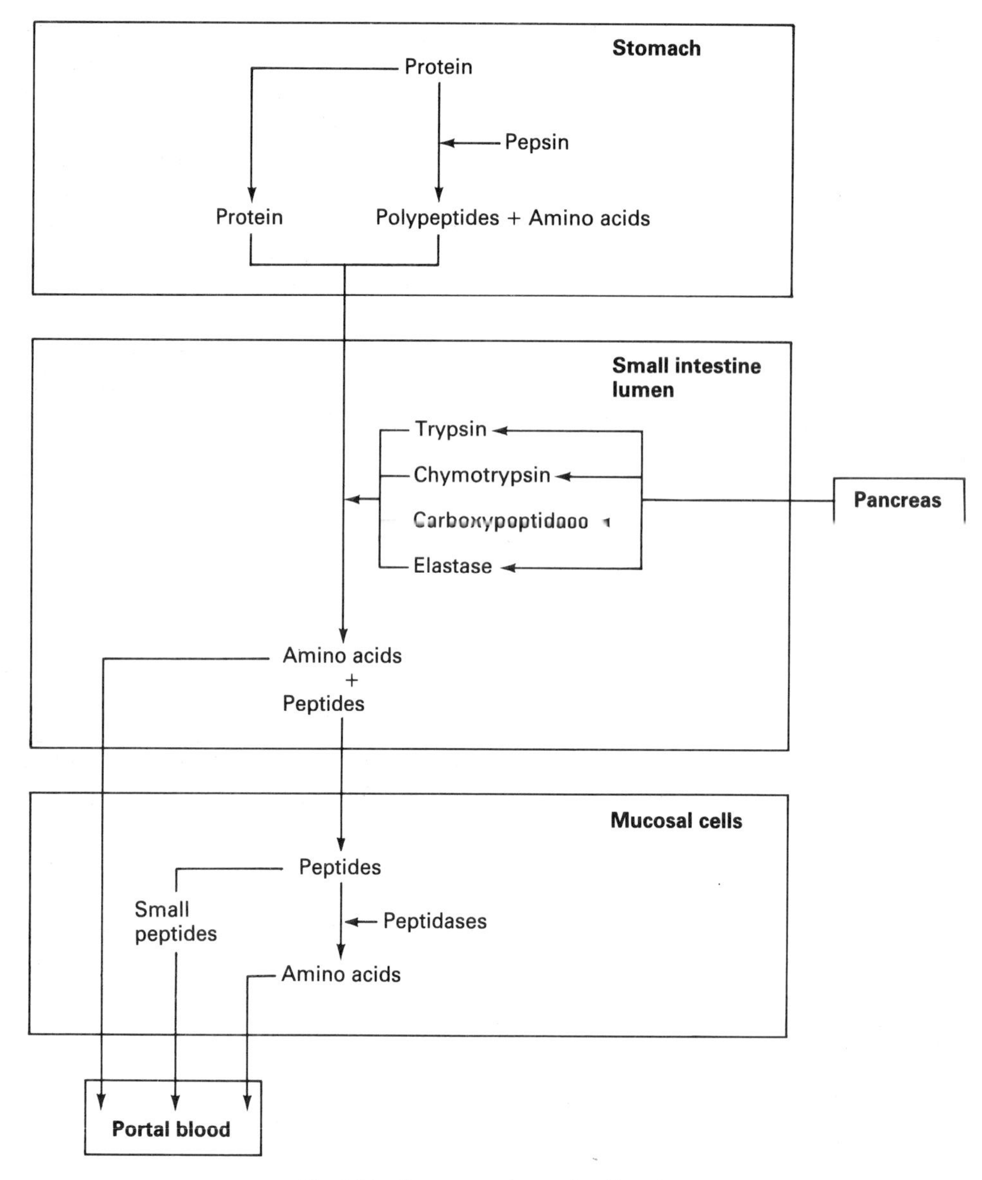

Fig. 6.12 Summary of protein digestion.

Protein digestion is summarized in Figure 6.12. It begins in the stomach with the hydrolysis of protein by pepsins to produce polypeptides and a few free amino acids. Once in the duodenum, chyme is subjected to a variety of proteolytic enzymes produced by the pancreas, which releases free amino acids and small peptides (see *Digestive functions of the pancreas*).

Final hydrolysis to free amino acids is catalysed by peptidases on the intestinal cell membranes or in the cytoplasm, depending on the peptides concerned (see *Digestion and absorption by the small intestinal epithelium*).

Amino acids (and some small peptides) are absorbed into the portal blood. D- Amino acids are absorbed passively, whereas L-amino acids are absorbed actively by a sodium-linked carrier mechanism.

GENERAL STRUCTURE OF THE ALIMENTARY TRACT

The various parts of the alimentary tract exhibit certain common structural features. The wall consists of four principal layers: the mucosa, the submucosa, the muscularis externa and the outer serosa (see Fig. 6.13). The detailed structure of each layer varies somewhat in the different parts of the gut according to the specialized function of each section.

The mucosa consists of an epithelial lining, enclosing a lamina propria of loose connective tissue which is especially rich in blood and lymphatic capillaries and sometimes contains glands. Smooth muscle cells and lymphoid tissues are also present. A thin layer of smooth muscle, the muscularis mucosae, constitutes the deepest layer of the mucosa.

The submucosa is also composed of connective tissue with many large blood and lymph vessels and a nerve (Meissner's) plexus. It also contains glands and lymphoid tissues.

The muscularis externa comprises an inner circular layer and an outer longitudinal layer, both of smooth muscle. Between the two layers lies a network of nerve fibres, the myenteric (Auerbach's) plexus. Blood vessels and lymphatics also found between the muscle layers. The two intramural plexuses (Meissner's and Auerbach's) extend the whole length of the alimentary tract and are interconnected with each other.

The outer serosa consists of loose connective tissue, rich in blood and lymph vessels and adipose tissue, covered by a layer of simple squamous epithelium (mesothelium). The serosa covering the viscera is continuous with the peritoneum lining the abdominal cavity. The moist, smooth mesothelium covering these two layers enables the viscera to glide over each other as well as against the abdominal wall. The serosa gives rise to and is continuous with the mesenteries which support the intestine and to the lesser and greater omenta which are attached to the stomach.

The gut is innervated by the autonomic nervous system. The parasympathetic innervation consists of the two vagus nerves as far as the proximal half of the colon, and the pelvic splanchnic nerves which innervate the more distal sections. The preganglionic fibres synapse in the intramural plexuses and the postganglionic neurones supply the muscle and glands in the gut wall. The parasympathetic system in general promotes glandular secretion and increases motility of the gut wall. The sympathetic neurones supplying the

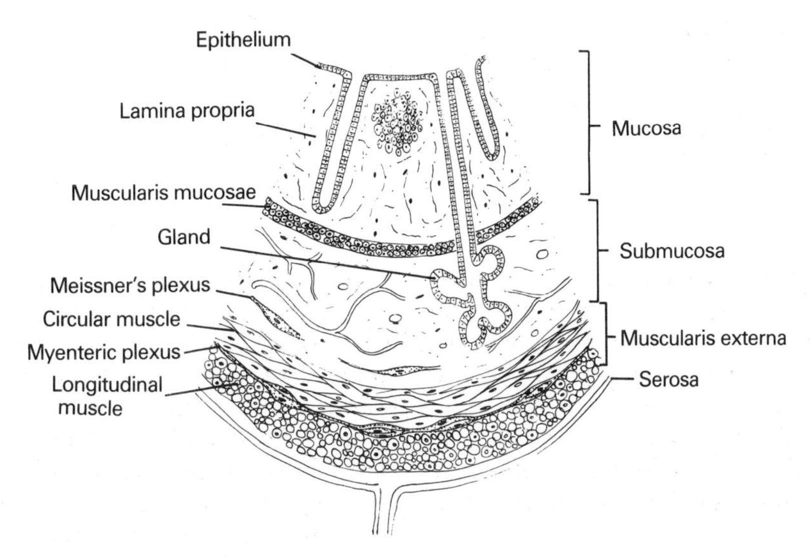

Fig. 6.13 General structure of the gut wall.

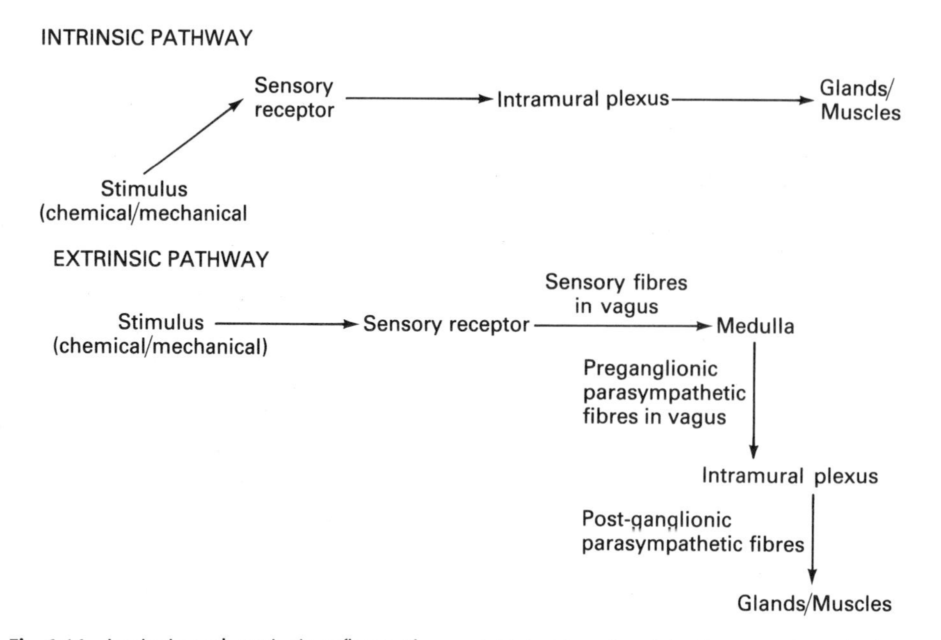

Fig. 6.14 Intrinsic and extrinsic reflex pathways initiated by stimuli in the gastrointestinal tract.

gut synapse in the coeliac and superior mesenteric ganglia outside the gut wall and then travel in the splanchnic nerves to supply the blood vessels and the muscle of the gut wall. Some sympathetic fibres terminate in the myenteric plexus. Sympathetic stimulation causes reduced muscle activity and vasoconstriction.

Some of the sensory neurones in the gut have their fibres within the plexuses, whereas others run in the autonomic nerves to the brain. This gives rise to two types of reflex pathway, intrinsic and extrinsic. The intrinsic reflex pathway is confined to the gut, whereas the extrinsic reflex pathway includes the brain (see Fig. 6.14). Pain fibres travel in the sympathetic nerves to the brain.

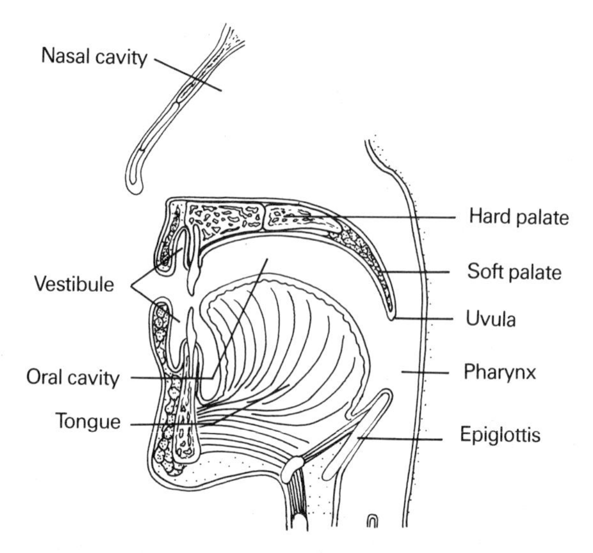

Fig. 6.15 Section through the lower head to show the oral and nasal cavities and neighbouring structures.

STRUCTURES ASSOCIATED WITH CHEWING, SALIVATION AND SWALLOWING

The oral cavity and its boundaries

The oral or buccal cavity consists of two parts, an outer smaller vestibule being separated by the teeth and gums from the inner oral cavity proper (Fig. 6.15).

The outer wall of the vestibule is formed by the inner surfaces of the lips and cheeks. Superiorly and inferiorly it is bounded by reflexion of the mucous membrane lining the cheeks and covering the jaw bones. Posteriorly the vestibule communicates with the oral cavity proper through the space between the

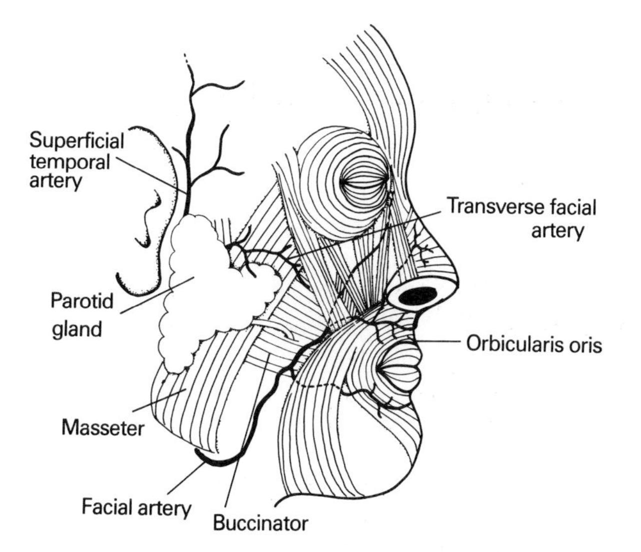

Fig. 6.16 Principal muscles and arteries of the face.

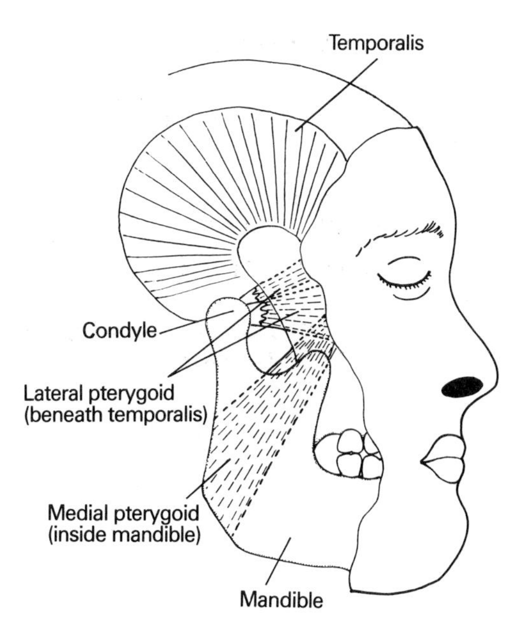

Fig. 6.17 The temporalis and pterygoid muscles used in chewing.

last molar teeth and the mandible (when the teeth are clamped together). Numerous small, mucus-secreting buccal glands open into the vestibule.

The lips are formed primarily by the orbicularis oris muscle (Fig. 6.16) and are covered externally by skin and internally by mucous membrane. They contain many minute blood vessels and nerves.

The cheeks are formed principally from the buccinator muscles with variable amounts of adipose tissue.

The jaw movements involved in chewing are brought about primarily by the actions of the masseter (Fig. 6.16) and the temporalis and pterygoid muscle (Fig. 6.17).

Below the teeth lie the gums or gingivae which are made of dense fibrous tissue. They are highly vascular and covered by thinly keratinized squamous epithelium. In young people the gums are attached to the enamel covering the teeth, but with age, they recede and become attached to the cement covering the root.

The surface of the floor of the mouth is formed by the mucous membrane connecting the tongue to the mandible. The underside of the tongue is connected to the floor of the mouth by a median fold, the frenulum. The sublingual glands lying beneath the mucous membrane form a sublingual fold on each side running backwards and laterally from the frenulum.

The roof of the mouth is formed by the palate. The anterior two-thirds is bony and is known as the hard palate whilst the posterior one-third forms the soft palate. The latter is attached to the rear of the hard palate and has a free posterior margin from which the uvula hangs down.

Tooth structure

In adults there are up to 16 permanent teeth in each jaw. Each half-jaw has two incisors, one canine, two premolars and up to three molars (Fig. 6.18). In children the deciduous (milk) dentition consists of 10 teeth in each jaw, so that each half-jaw exhibits two incisors, one canine and two molar teeth.

All teeth have the same basic structural organization (Fig. 6.19) but differ in shape and size. Each tooth has one or more roots and a crown which is covered by hard translucent enamel which gives the tooth its cutting or grinding surface.

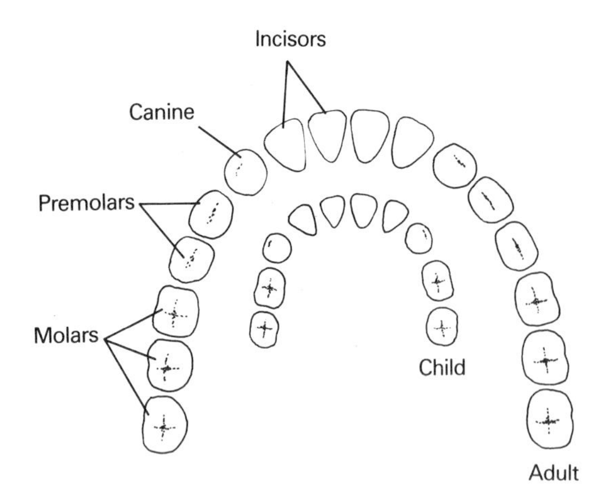

Fig. 6.18 Diagram to show the dentition of an adult and a child.

The bulk of the tooth is constructed of a mineralized yellow-white substance called dentine which surrounds a central pulp cavity. The latter contains nerve fibres and blood and lymph vessels which enter through an apical foramen at the tip of each root. The root is completely surrounded by a periodontal ligament which anchors the tooth to the bone of the jaw. It is attached to the cement which covers the dentine of the root.

The shape of the crown varies depending upon the position of the tooth in the mouth and its function. At the front of the mouth are the incisors, chisel-like teeth whose function is to cut food. Behind them lie the canines, each of which has a single sharp cusp. Their function is to tear food, but in humans they are so reduced in size as to have little specialized function. Behind the canines lie the premolars, each with two cusps and behind them the molars which usually have four cusps. The premolars and molars are so arranged that when the jaws are clamped there is maximum contact between the cusps and fissures of the upper and lower molars. When food is chewed it is ground between the molars and reduced to a pulpy consistency.

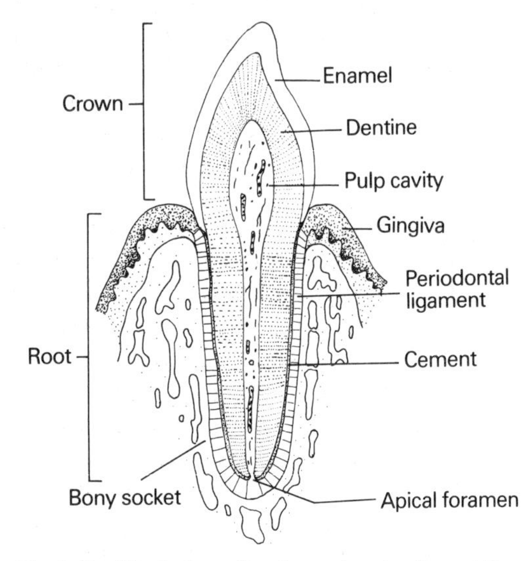

Fig. 6.19 Vertical section through a tooth and its socket.

Tongue

The tongue is a very mobile organ which bulges up from the floor of the mouth. It is covered by stratified squamous epithelium and consists of a mass of striated muscle with some fat and many glands.

The dorsum (upper surface) of the tongue is separated into palatine (front) and pharyngeal (rear) parts by the V-shaped sulcus terminalis (Fig. 6.20). The thick mucous membrane of the palatine part is rough and covered by papillae. The pharyngeal part on the other hand is smooth and thinner, being finely nodular owing to the presence of small lymph follicles in the submucosa.

The palatine surface has three types of papillae. Largest are the vallate papillae which lie immediately anterior to the sulcus terminalis, each being a short cylinder sunk into the surface of the tongue. The opposing walls of the surrounding trench are studded with taste buds.

Smaller, numerous fungiform papillae are seen on the tip and margins of the tongue and are scattered over the entire dorsum. They have knob-like extremities and carry taste buds.

In addition there are numerous minute pointed filiform papillae covering all of the palatine surface and the margins of the tongue. The apices are cornified and may be broken up into thread-like processes. Filiform papillae do not carry taste buds.

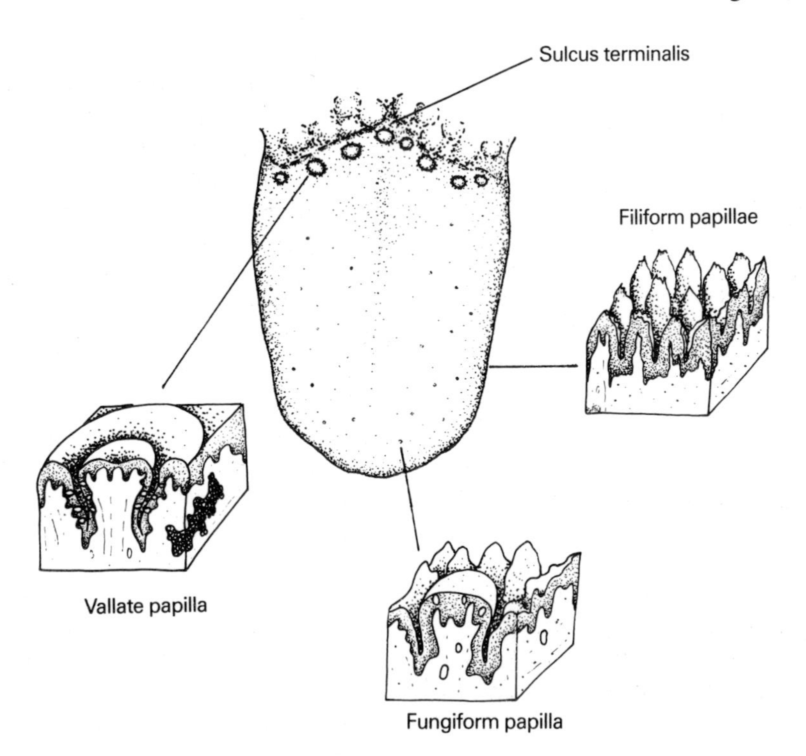

Fig. 6.20 The dorsal surface of the tongue and enlarged views of the types of papillae found there.

Pharynx

The pharynx forms the passage between the oral cavity and the major parts of the respiratory and digestive tracts. Lined by mucosa, it is a tube about 13 cm long which extends from the base of the skull down behind the larynx to the level of the cricoid cartilage. It is about 3.5 cm wide at the top and 1.5 cm at its junction with the oesophagus.

The submucosa contains numerous pharyngeal mucous glands and nodules of lymph tissue; aggregations of lymph follicles form pharyngeal, tubal and palatine tonsils.

The pharynx contains three constrictor muscles, the superior, middle and inferior constrictors (Fig. 6.21). Three other muscles run longitudinally into the pharyngeal wall, the stylopharyngeus, the salpingopharyngeus and the palatopharyngeus. The last-named forms an arch and extends from the soft palate to form the inner layer of the wall of the pharynx.

Salivary glands

There are two types of salivary glands associated with the buccal cavity; they can be differentiated into major and minor glands. There are three pairs of 'major' glands, the parotid, submandibular (submaxillary) and sublingual glands (Fig. 6.22). The 'minor' glands are found in the lips, cheeks and roof of the mouth as well as in the substance of the tongue.

The major glands are composed of secretory units known as adenomeres which are supported by connective tissue and are richly supplied with blood and lymph vessels and nerve fibres. Each adenomere has a secretory and a conducting portion (Fig. 6.23). The secretory portion consists of groups of glandular cells of various types. There are serous cells with small, discrete eosinophilic granules and mucous cells containing large, densely packed, ill-defined granules. In addition there are seromucous cells which have an inter-

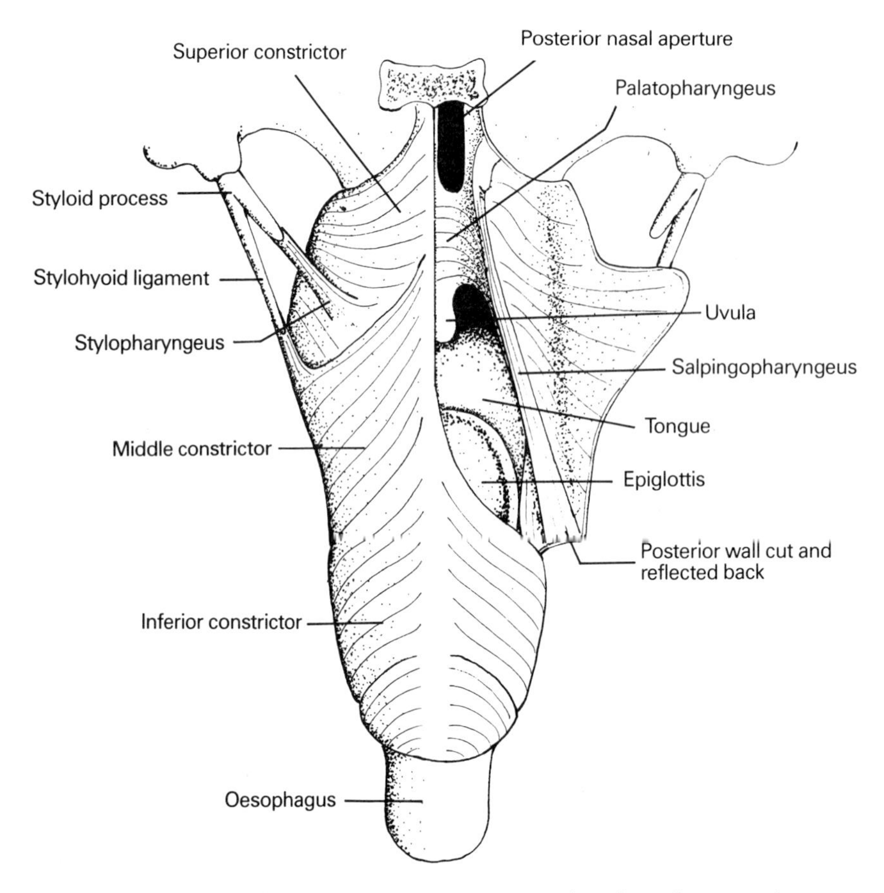

Fig. 6.21 Muscles of the pharynx (posterior view), opened to show the internal structures.

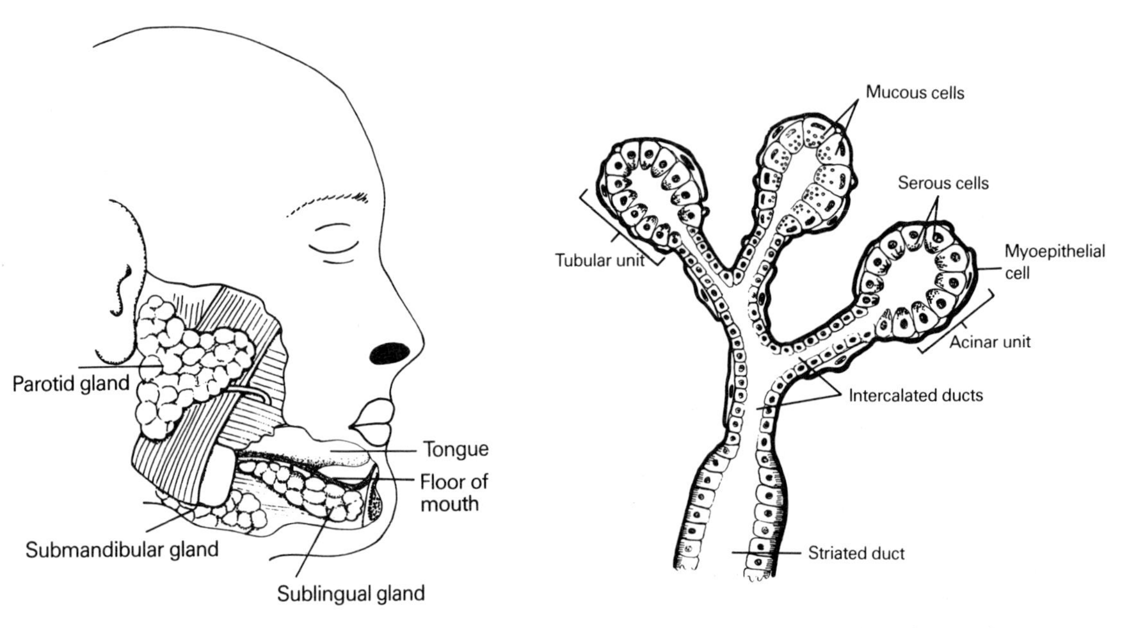

Fig. 6.22 The salivary glands.

Fig. 6.23 General structure of a salivary adenomere.

mediate appearance. The secretory units are acinar or tubular in shape, and intermediate tubulo-acinar types also occur.

The conducting portions of the glands consist of three types of ducts; intercalated ducts carry secretions from the glandular cells to striated (intralobular) ducts which drain into the excretory ducts. The cells of the various ducts have both a secretory and an absorptive function.

The parotid glands are the largest of all, with combined weights of about 50 g. Each gland lies below the ear and between the mandible and the sternocleidomastoid muscle. The secretory portion consists almost entirely of seromucous cells containing granules which have a high amylase activity.

Each parotid duct opens through a papilla on the cheek opposite the crown of the second upper molar.

Each submandibular gland is about the size of a walnut and lies inside and beneath the mandible, one on each side of the face. The ducts open at the summits of the sublingual papillae which are situated on either side of the frenulum of the tongue. The secretory portions of the glands contain both mucous and seromucous cells and show weak amylase activity.

The sublingual glands are located beneath the mucous membrane of the floor of the mouth. They are small, with combined weights of 6–8 g. Each gland has 8–20 ducts, most of which open on the surface of the sublingual folds on the floor of the mouth. About two-thirds of the cells in these glands are mucus-secreting and the remaining one-third are serous cells.

Oesophagus

The oesophagus is a muscular tube about 25 cm long connecting the pharynx with the stomach. The upper part lies behind the trachea and in front of the vertebral column. Lower down in the thorax, below the level of the trachea it passes behind the right pulmonary artery, the left principal bronchus and the pericardium. It then passes through the diaphragm and joins the stomach.

The mucosa of the oesophagus is arranged in folds which disappear when the tube is distended. The inner lining is composed of non-keratinized stratified squamous epithelium. This layer is separated from the submucosa by a typical muscularis mucosae, although the latter is absent at the top of the oesophagus. The submucosa contains the oesophageal glands, mucus-secreting glands which open into the lumen by long ducts.

The muscularis externa is relatively thick and consists of the typical two layers — outer longitudinal and inner circular. The upper two-thirds contains striated muscle while the lower third is smooth muscle only. In the middle section the striated muscle is replaced gradually by smooth muscle. Surrounding the outside of the oesophagus is a layer of fibrous tissue.

There is no anatomically distinct sphincter separating the oesophagus from the stomach. However, the combination of circular muscle fibres adjoining the junction, the shape and disposition of surrounding tissues and organs, and the proximity of the diaphragm, all appear to contribute to the presence of a 'physiological sphincter'.

CHEWING, SALIVATION AND SWALLOWING

In the mouth, food is broken down physically by the process of chewing (mastication) and mixed with saliva to bring about the formation of a moist ball or bolus. Chewing softens the food so that it passes more easily through the alimentary tract. By breaking the food up, chewing also increases the surface area available to digestive enzymes.

The presence of food in the mouth and its taste and smell, apart from producing pleasant sensations, also stimulates the secretion of saliva, gastric and pancreatic juices and bile by means of parasympathetic pathways.

Mastication

Pieces of food are bitten off from the main mass by the incisors, the sharp chisel-like teeth at the front of the mouth. Once inside the mouth the food is then moved around by the tongue and brought into contact with the occlusal surfaces of the premolars and molars, which crush and grind it into small pieces. This function is particularly important in the breakdown of meat and any other foods with a fibrous consistency. It is also an important prerequisite for the digestion of plant material because the cellulose wall is indigestible and must be ruptured to release nutrients from the cell.

Mastication is primarily a grinding process which is dependent upon the ability of the lower jaw (the mandible) to rotate with respect to the upper jaw (the maxilla) on the temporomandibular joint.

Chewing is a reflex activity initiated by the presence of food in the mouth which stimulates touch receptors in the gums and the front of the hard palate. The sensory information is relayed to the chewing centre in the medulla oblongata of the hindbrain, which brings about reflex inhibition of the masticatory muscles (masseter and temporalis) and the jaw drops. The subsequent stretching of the muscle fibres then initiates a stretch reflex on the same side, causing contraction of the muscles which raise the lower jaw and the food is thereby compressed between the teeth. The contraction of the muscle in the cheeks and around the lips helps to force the food between the teeth as does the movement of the tongue. The tongue also rolls the softening food into a bolus, mixing it with saliva preparatory to swallowing.

Salivation

Saliva is secreted by the paired parotid, sublingual and submandibular glands. It is a watery fluid containing ions, mucin and the digestive enzyme salivary α-amylase (previously known as ptyalin). In addition to its role in digestion, saliva also has several other functions. It keeps the mouth moist which aids in the movements of the lips and tongue in speech and also enables molecules to dissolve on the surface of the tongue and stimulate the taste buds. It keeps the teeth clean and the enzyme lysozyme has an antibacterial action. In addition, at its normal pH (7.0) it is saturated with Ca^{++} which prevents calcium from dissolving out of the teeth. Usually, a total of between 1000 and 1500 ml of saliva is secreted per day. The rate of secretion varies with the body's activities, with a basal rate of about 0.5 ml per minute rising to as high as 4.0 ml per minute during mastication.

The osmolality of saliva varies with its rate of secretion. At low rates it is very hypotonic (about 50 mOsm/l) compared with plasma (about 300 mOsm/l), whereas at higher rates of secretion it approaches isotonicity.

The concentration of sodium and chloride ions in saliva is lower than in plasma, whereas the concentrations of potassium and bicarbonate ions are higher. The formation of saliva is summarized in Figure 6.24. The acinar cells produce the primary fluid by a process involving active transport. In the intercalated ducts leading from the acini, ionic exchange

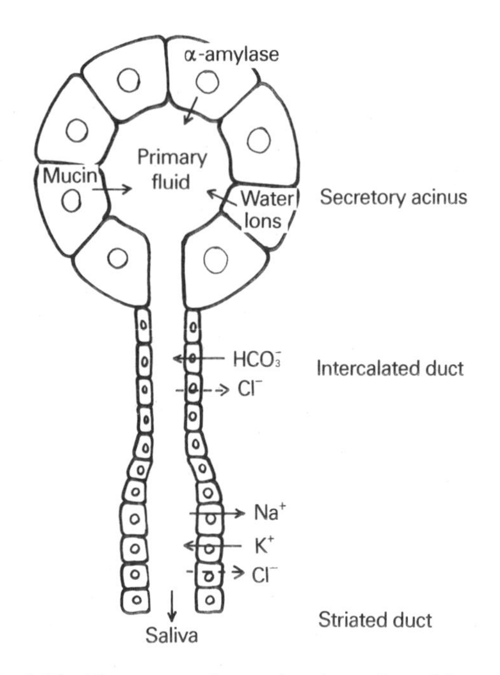

Fig. 6.24 Diagram to show saliva formation. Primary fluid is produced in the secretory acinus and the secretion is then modified in both the intercalated and the striated ducts.

takes place so that the composition of the fluid then resembles plasma. In addition, bicarbonate ions are secreted into the fluid from the cells lining the intercalated ducts. Carbon dioxide and water combine in the cells to form carbonic acid, a reaction that is catalysed by the enzyme carbonic anhydrase and this produces bicarbonate ions which are then secreted in exchange for chloride.

In the striated ducts, sodium ions are absorbed and potassium ions added in an exchange that favours sodium ions. Chloride ions are then passively absorbed following the gain of the positively charged ions. The sodium/potassium transport is stimulated by aldosterone in the same way as in the distal nephron of the kidney.

The salivary glands are stimulated by both the sympathetic and parasympathetic divisions of the autonomic nervous system, although the most copious response results from parasympathetic stimulation. Activation of the sympathetic system causes vasocontriction and a scant, viscous secretion. The precise role of the sympathetic system in the control of salivation is uncertain.

Sensory stimuli which initiate salivation include the presence of food in the mouth, taste and smell, although perhaps the must potent stimulus is acid as in, for example, fruit. Impulses are conveyed to the salivary centre in the medulla oblongata which initiates reflex secretion of saliva via the parasympathetic pathway to the glands. In dogs, the sight and smell of food and conditioned reflexes can evoke secretion, but these effects are relatively slight in man. Salivation may also be induced by the presence of irritating materials in the stomach or small intestine.

Stimulation of the salivary glands releases the enzyme kallikrein into interstitial fluid, where it activates bradykinin. The latter then causes vasodilation which may enhance the secretory activity of the glands. There is some evidence that the blood vessels are innervated by parasympathetic vasodilator fibres, which may be an additional cause of the vasodilation accompanying parasympathetic stimulation of the glands.

During mastication, as the food is ground between the teeth and is mixed with saliva, the water content of the saliva helps to moisten and soften the food, while the mucin acts as a lubricant. Any cooked starch within the food is liable to attack by salivary α-amylase. This is activated by chloride ions and catalyses the hydrolysis of starch to maltose, maltotriose and dextrins. The amount of cooked starch which is hydrolysed by salivary α-amylase depends upon how much the food is mixed with the saliva, how long it stays in the mouth and how long it takes before the acidity of the gastric juice inhibits the enzyme after the food has been swallowed. The enzyme is active down to a pH of 4, and up to 50% of the starch may have been hydrolysed before inactivation occurs.

The food is rolled by the tongue into a bolus, which is ready for swallowing.

Deglutition (swallowing)

Once the food has been formed into a bolus it is passed through the pharynx and down the oesophagus to the stomach. This is achieved by the act of swallowing (deglutition).

Deglutition is a complex reflex which is regulated by a 'swallowing centre' in the medulla. It is initiated when the tongue muscles contract and push the bolus upwards and backwards into the pharynx. This is a voluntary activity. The soft palate is elevated and comes into contact with the posterior wall of the pharynx, thereby closing off the rear openings of the nasal cavities.

The pharyngeal stage of deglutition is the beginning of involuntary reflex activity and starts when the vocal cords approximate to close the glottis. Respiration is inhibited and the larynx is pulled upwards and forwards by the myohyoid muscles (which run from the tongue to the larynx). Food is thus prevented from entering the respiratory tract by the closure of the glottis. The bolus pushes the epiglottis back over the glottis which confers extra protection to the respiratory tract. The movement of the larynx pulls the oesophagus up and opens the hypopharyngeal sphincter.

The oesophageal phase of deglutition involves the constrictor muscles of the pharynx contracting and initiating a peristaltic wave which pushes the bolus into the oesophagus. At rest, the opening to the oesophagus is closed by the hypopharyngeal sphincter. This is a 3 cm long segment at the top of the oesophagus in which the fibres are arranged so that when the muscles are relaxed the sphincter is closed. During swallowing the muscles contract and cause the sphincter to open, thereby allowing the bolus to pass down into the oesophagus. A wave of contraction in the circular muscle layer then propels the bolus down to the stomach. This peristaltic wave travels at a rate of about 4 cm per second.

Fluid and semi-solid food usually pass down the oesophagus ahead of the peristaltic wave in upright subjects simply due to the effect of gravity.

At the bottom end of the oesophagus, the last 4 cm or so acts as a functional sphincter (i.e. it appears structurally identical to adjacent areas) which is normally in a state of tonic contraction. As the peristaltic wave approaches the sphincter, the muscle relaxes (receptive relaxation) and allows the food to enter the stomach. The sphincter then closes again and prevents regurgitation of gastric material back into the oesophagus.

STRUCTURE OF THE STOMACH

The stomach is a muscular sac which is divided into four areas, the fundus, cardia, body and the pyloric antrum (Fig. 6.25). Most commonly the empty stomach is found to be J-shaped and when it fills it tends to expand downwards and forwards, thereby displacing the intestine. The opening of the oesophagus into the stomach is called the cardiac orifice and is surrounded by a functional sphincter, the cardiac sphincter. The pyloric orifice connects the pyloric antrum with the duodenum and at the junction is found the pyloric sphincter or pylorus which consists of a thickened layer of circular muscle.

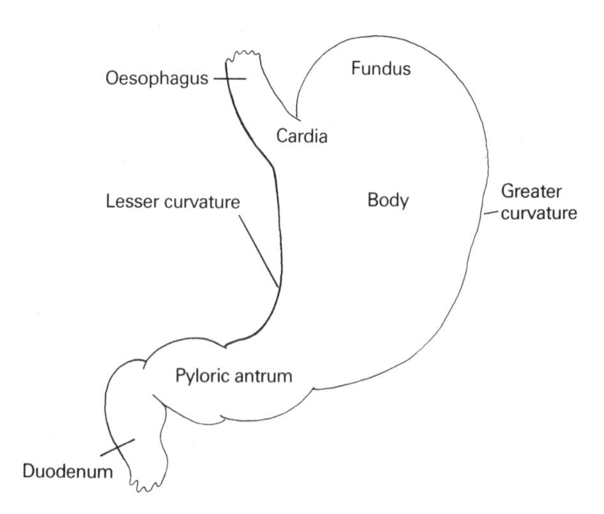

Fig. 6.25 The regions of the stomach.

The stomach wall exhibits the usual four layers (see Fig. 6.26). The outermost layer, the serosa, covers almost the entire surface of the organ, being absent only where the two layers of peritoneum which form it come together along the greater and lesser curvatures. The lesser omentum is a double layer of perito-

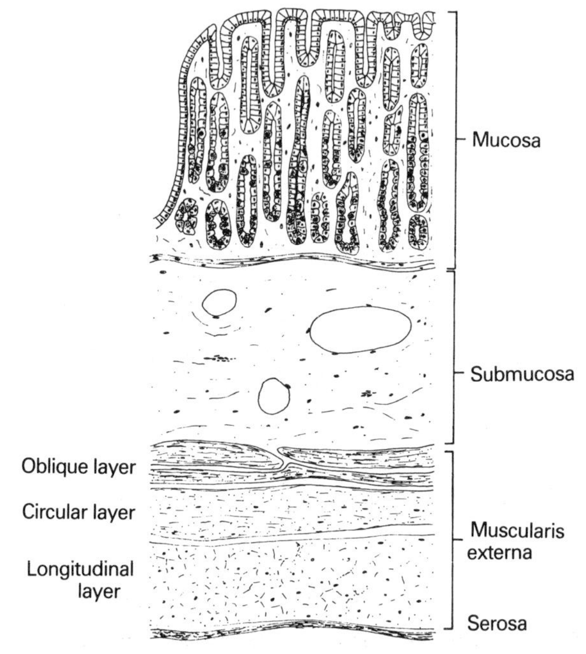

Fig. 6.26 The layers of the stomach wall.

neum connecting the lesser curvature with the liver, and the greater omentum extends from the greater curvature to form an apron-like fold in front of the small intestine before being reflected backwards and upwards to wrap around the organs of the upper abdomen. As the greater omentum is a double layer folded back on itself it contains four layers of serous membrane.

The muscularis externa has the usual longitudinal and circular muscles layers as well as an additional oblique layer on the inside. These oblique fibres are found predominantly in the body of the stomach and are particularly well-developed near to the cardiac orifice.

The lining of the empty stomach is thrown into folds or rugae which flatten out when the stomach distends. The surface epithelium consists of a single layer of columnar epithelial cells which continually secrete neutral mucus onto the surface. The mucosa is covered by small (0.2 mm diameter) depressions or gastric pits which contain the openings of the gastric glands (Fig. 6.27). The latter exhibit structural differences in the various areas of the stomach and it is therefore conventional to describe cardiac glands, glands of the fundus and body and pyloric glands. In all regions the glands are located within the lamina propria of the mucosa and do not penetrate the muscularis mucosae.

The cardiac glands consist almost entirely of mucus-secreting cells. The glands may be simple or branched and are present in relatively small numbers.

The fundus and body of the stomach contain the most highly developed gastric glands. They are present in large numbers and there are from three to seven opening into each pit. Each gland may be simple or branched and contains three histologically recognizable areas, the isthmus, neck and base.

The isthmus is the upper region of the gland and is lined by mucous cells which are similar to those found in the surface mucous epithelium of the stomach.

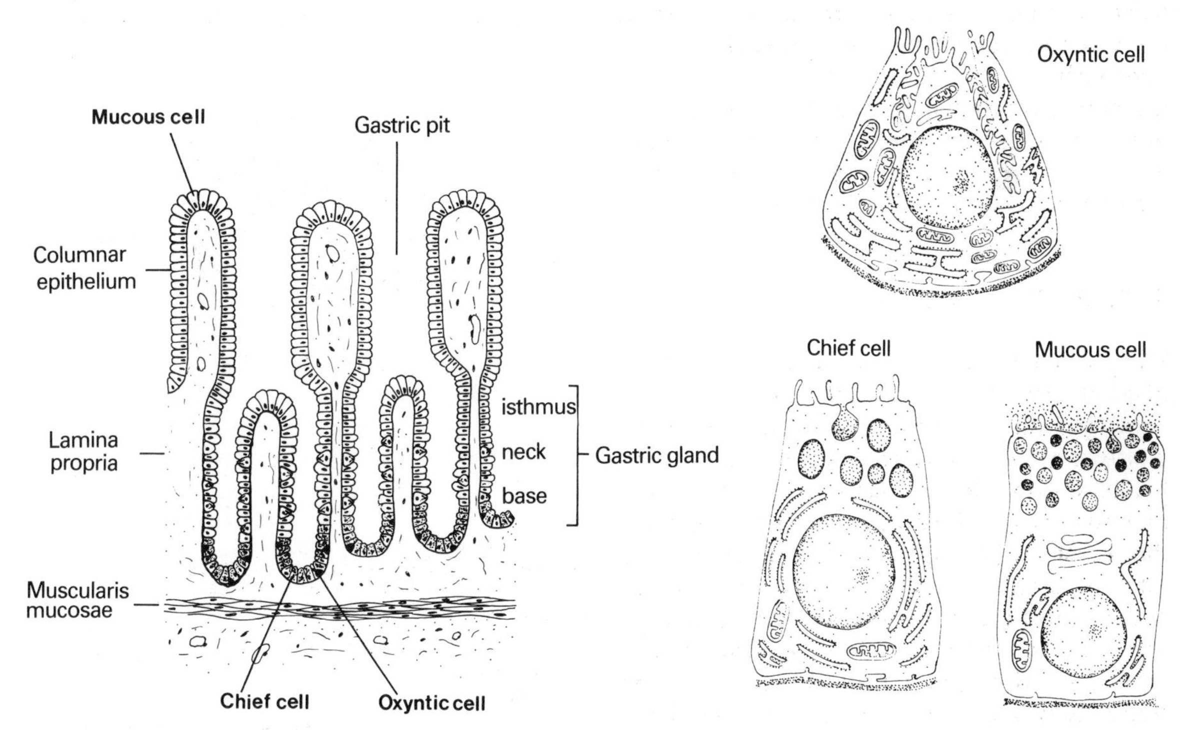

Fig. 6.27 The gastric mucosa showing the gastric pits and glands together with the secretory cells. (The gastric pits have been widened for clarity.)

The neck region, which occupies a position between the isthmus and the base also contains mucous cells. However, they are structurally rather different from those of the surface, being more irregular in shape. This region also exhibits acid-secreting oxyntic (parietal) cells. These are easily distinguished from the mucous neck cells by their intensely eosinophilic cytoplasm. The electron microscope shows that they have a characteristic deep circular invagination (canaliculus) in the apical cytoplasm, that in the active cell is lined by well-developed microvilli. Additionally, the cell contains large numbers of mitochondria.

Oxyntic cells may also be found in small numbers in the base region, although enzyme-secreting zymogenic (chief) cells are normally predominant. These cells are basophilic, contain a well-developed endoplasmic reticulum, large numbers of ribosomes and numerous secretory granules.

Argentaffin (enterochromaffin) cells are also found in the base region lying between the zymogenic cells and the basal lamina. They secrete a variety of hormones, the physiological roles of which are still ill-understood.

The wall of the pyloric region of the stomach is studded with deep gastric pits containing the openings of glands that are similar to those found in the cardiac region. The cells are predominantly mucus secreting, although oxyntic cells also occur in small numbers. In addition, the pyloric glands contain G-cells which secrete the hormone gastrin into the surrounding blood vessels, rather than into the gastric juice.

The stomach is a richly vascular organ receiving blood through the right gastric and gastro-epiploic arteries from the common hepatic artery, the left gastric artery from the coeliac artery and the left gastro-epiploic and short gastric arteries from the splenic artery. These vessels form anastomoses at all levels but especially in the submucosa where a well-developed submucosal plexus is found.

Blood drains away from the stomach into the splenic, superior mesenteric veins and hepatic portal veins (see Fig. 4.15).

The stomach is innervated by both sympathetic and parasympathetic nerve fibres. The sympathetic supply is derived from the coeliac plexus, while the parasympathetic fibres are carried in the vagus nerves. In both cases the nerve supplies are broken up into numerous small branches which enter the stomach wall at a number of points.

FUNCTIONS OF THE STOMACH

The stomach stores the food eaten during a meal and later releases it at a rate which is optimal for digestion. Food is mixed with gastric juice, thereby changing its consistency so that it will be more easily transported to subsequent sections of the gut. In addition, food is exposed to enzymes which begin the digestion of proteins. The acid in gastric juice is bacteriocidal and also converts ferric iron in the diet to the ferrous form in which it is subsequently absorbed in the intestine. Furthermore, gastric glands produce intrinsic factor which is necessary for the absorption of vitamin B_{12} from the diet.

Storage, mixing and emptying

The volume of the empty stomach is of the order of 50 ml. As food is swallowed, it passes into the stomach and is laid down in concentric circles in the fundus and body. Filling continues to a maximum volume of about one litre and is accompanied by relaxation and stretching of the wall. The latter ensures that the intragastric pressure does not rise, since this would bring about premature expulsion of the stomach contents into the duodenum.

A patch of longitudinal muscle along the greater curvature, midway between the fundus and the pylorus, exhibits a spontaneous electrical activity known as basic electrical rhythm (BER) with a fairly steady frequency of three bursts per minute. Depolarizations may not reach threshold in an empty stomach, although, if they do, feeble contractions in the form of a concentric wave travelling down in the circular muscle layer towards the pyloric region are initiated every

20 seconds or so. In hunger, the contractions become more vigorous since there is a greater number of impulses in each burst of electrical activity.

When the muscle is stretched by swallowed food, peristaltic contractions are stimulated via both intrinsic and extrinsic pathways to parasympathetic fibres supplying the muscularis externa. About three of these contractions arise every minute at the level of the cardia and travel down towards the pyloric antrum, increasing in force as they go.

The pyloric sphincter is relatively weak and is normally partly open. The peristaltic wave pushes some of the gastric contents through the orifice and into the duodenum. The orifice then closes and the remainder squirts back into the body of the stomach. By this means the swallowed food is mixed with gastric juice, develops a pasty consistency and becomes known as chyme.

The rate of emptying of the stomach depends on the fluidity of the chyme, the degree of opening of the pyloric orifice, the force of the peristaltic contractions in the wall and inhibitory influences initiated by food arriving in the duodenum. All of these factors ensure that the stomach empties at a rate that is optimal for the digestion of its contents both in the stomach and subsequently in the duodenum.

The degree of contraction of the pyloric sphincter is reduced by both parasympathetic activity and gastrin which therefore increase the aperture of the orifice and allow more chyme to pass through. The force of peristaltic contractions is increased by distension of the stomach wall as well as by gastrin which can also increase the BER.

A variety of stimuli initiated by food entering the duodenum result in reduced peristaltic activity and therefore a reduced rate of gastric emptying. These stimuli include distension of the duodenal wall, acid, digestion products and hypertonic fluid. The precise mechanisms of inhibition remain uncertain, although it is known that both neural and hormonal components are involved. The neural component is known as the enterogastric reflex and involves both extrinsic and intrisic pathways. The hormone has been called enterogastrone, but it may be that it is, in reality, a group of several duodenal hormones including secretin, cholecystokinin-pancreozymin and gastric inhibitory peptide.

Secretion of gastric juice

Gastric juice contains hydrochloric acid, intrinsic factor, mucus and a number of proteolytic enzymes or pepsins. The major ions apart from H^+ and Cl^- are Na^+, K^+ and HCO_3^-. There are small quantities of other enzymes including gastric lipase and gelatinase, but these have only a very minor role in digestion. The total volume of gastric juice secreted per day is of the order of two to three litres, it is slightly hypotonic to plasma, and contains an amount of H^+ that is proportional to the rate of secretion. At high rates of secretion the H^+ concentration is around 150 mmol/l.

Hydrochloric acid

The quantity of H^+ secreted by the oxyntic (parietal) cells is over one million times greater than that found in plasma. The presence of these large quantities of acid in gastric juice give it a bacteriocidal function. In addition, the acid activates pepsins (secreted as inactive pepsinogens by the chief cells), and provides an optimal pH for pepsin activity. Hydrochloric acid also converts ferric iron (Fe^{+++}) to ferrous iron (Fe^{++}), which is more soluble in the prevailing pH in the intestine where it is absorbed.

The precise mechanism of acid secretion is controversial, but it is thought that both chloride and hydrogen ions are actively transported into the canaliculi of the oxyntic cells, and then pass to the gastric lumen (see Fig. 6.28). The hydrogen ions are derived from water molecules as a product of the metabolic processes of the oxyntic cell. The splitting of water molecules also gives rise to hydroxyl ions which combine with further hydrogen

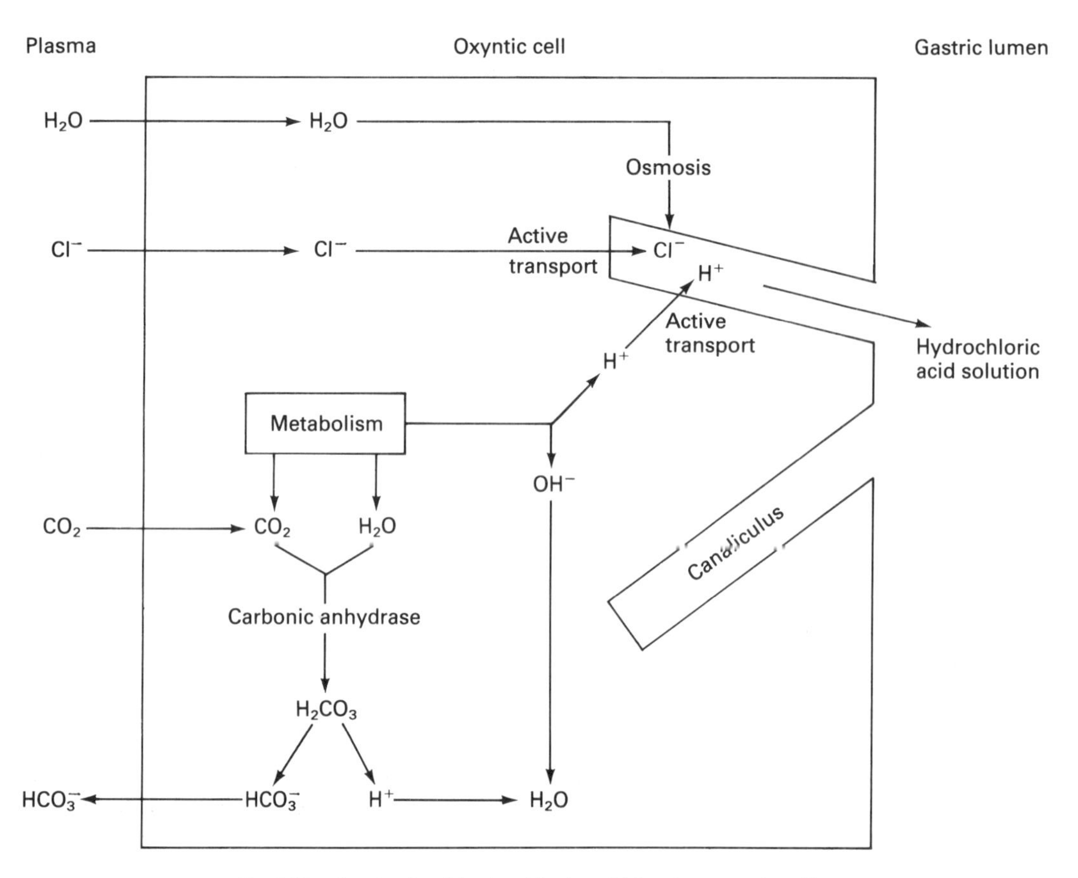

Fig. 6.28 Synthesis of hydrochloric acid by the oxyntic cell.

ions formed by the breakdown of carbonic acid. The latter is derived from the combination of carbon dioxide (either from plasma or cellular metabolism) and water, with the aid of the enzyme carbonic anhydrase. Bicarbonate ions, which are also formed when carbonic acid breaks down, diffuse out of the cell into the plasma in exchange for chloride. This chloride is then actively pumped into the canaliculus as the second component of the hydrochloric acid. The presence of the hydrochloric acid in the canaliculus creates an osmotic gradient which leads to the passage of water through the cell. Therefore a solution of hydrochloric acid is formed inside the canaliculus which then passes into the lumen of the stomach.

As blood flows through the stomach, it loses carbon dioxide and chloride ions and gains bicarbonate ions. Venous blood draining the stomach therefore has a higher pH than that which enters.

The oxyntic cells are stimulated by parasympathetic fibres in the vagus nerves, gastrin and histamine. The role of histamine in acid secretion is not known, but if histamine H_2 receptors in the stomach are blocked by the drug cimetidine, then both histamine and gastrin fail to stimulate acid secretion. This does suggest that histamine is involved in the mechanism by which gastrin stimulates the cells.

Intrinsic factor

The oxyntic cells also produce a mucoprotein, intrinsic factor, which is vital for the absorption of vitamin B_{12} from the diet. Vitamin B_{12} is bound to protein in food and is released from it either by cooking or by proteolytic enzymes. Intrinsic factor combines with the free vitamin B_{12} and protects it from digestion. Absorption of vitamin B_{12} occurs in the ileum and intrinsic factor is required for this process. Absence of intrinsic factor leads to pernicious anaemia (see *Erythropoiesis* in Ch. 2).

Mucus

The surface cells of the stomach mucosa secrete a neutral, viscous mucus which forms a lining layer about 1 mm thick. This affords protection to the mucosal cells from the gastric contents. Mucus is also produced by the neck cells of the gastric glands in the fundus and body, and from most of the cells of the cardiac and pyloric glands.

The surface cells have a low permeability to H^+ and tight cell junctions. They are replaced every 1–3 days.

Mucous cells are stimulated by parasympathetic fibres in the vagus nerves, and directly by mechanical or chemical irritation.

Pepsins

Seven slightly different pepsins are secreted in inactive form as pepsinogens by the chief cells, which are thereby protected from digestion by their own products. Once exposed to the acid in the gastric lumen, the pepsinogen molecules are cleaved and the active pepsins are released. Pepsins act as autocatalysts and thereby promote activation of more pepsins.

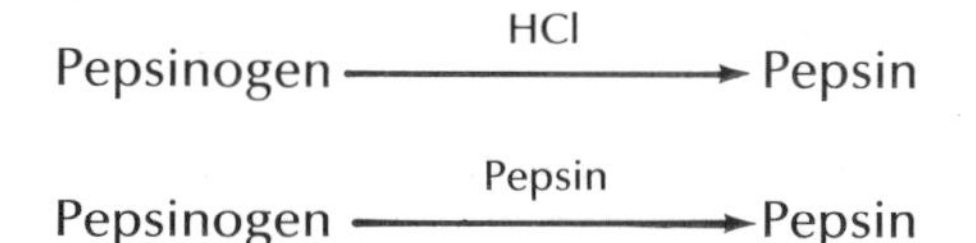

The pepsinogens are stored within the cells in granules, and these are secreted if the cells are stimulated by parasympathetic fibres, histamine or gastrin (weak response). If the granules are depleted and stimulation continues, then the cells discharge their secretion without forming storage granules. Following stimulation, the chief cells continue to synthesize pepsinogen and replenish the storage granules.

Pepsins catalyse the hydrolysis of dietary proteins forming polypeptides and a few free amino acids.

$$\text{Protein} \xrightarrow{\text{Pepsin}} \text{Polypeptides} + \text{Amino acids}$$

Control of secretion of gastric juice

Gastric juice secretion is stimulated by both intrinsic and extrinsic nerve reflexes, and by the hormone gastrin released from the G-cells in the pyloric glands. A small amount of gastrin is also produced by glands in the duodenum.

Gastrin is, in reality, two different peptides known as G-17 and G-34 (the numbers indicating how many amino acids are present), of which G-17 is the more abundant. Gastrin does not stimulate the production of all of the components of gastric juice equally, exhibiting a more powerful influence upon acid than upon pepsinogen release.

Parasympathetic stimulation increases the output of acid, mucus and especially pepsinogen by direct effects upon each of the cell types that produce them.

Even in a fasting individual, there is a basal rate of secretion (about 15% of maximum) which is probably caused by histamine stimulation.

The blood flow to a resting stomach is about 0.5% of the total cardiac output, whereas in an active stomach the blood flow can amount to some 20% of the cardiac output.

Gastric secretion is subdivided into three phases according to where the stimuli originate: the cephalic, gastric and intestinal phases.

Cephalic phase

The cephalic phase depends upon a range of stimuli which cause secretion before the food arrives in the stomach. It accounts for about 10% of the total secretion associated with a

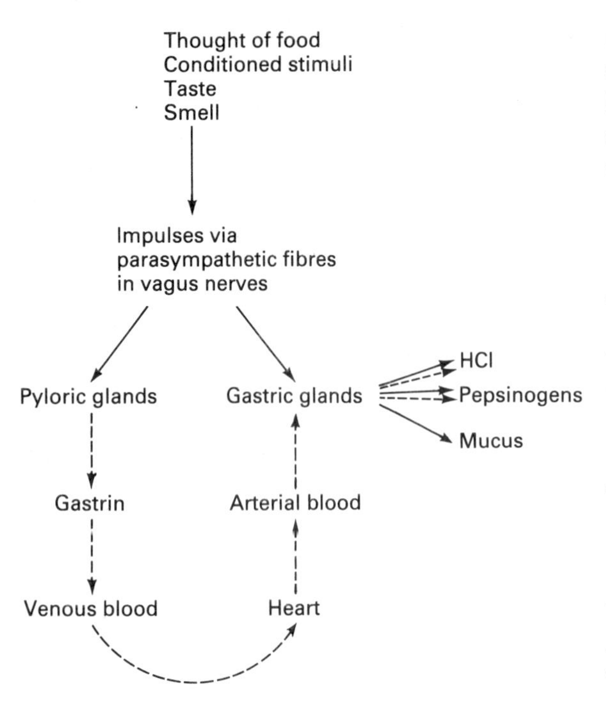

Fig. 6.29 The cephalic phase of gastric secretion.

meal. Thinking about food, or conditioned stimuli such as the clink of cutlery, as well as the taste and smell of food can bring about the secretion of what is called 'appetite juice'. There are two mechanisms involved in stimulating secretion, one is neural, the other hormonal. Figure 6.29 shows the neural pathway which involves the vagus nerves from the brain to the gastric glands, which when stimulated secrete at an increased rate. Vagal stimulation also extends to the pyloric glands and they respond by releasing gastrin into the blood. When gastrin returns to the stomach in the blood it stimulates the gastric glands to release acid and some pepsinogens.

Gastric phase

The presence of food in the stomach elicits about 80% of the total secretion of gastric juice. The food stretches the stomach wall and this distension stimulates the glands by both intrinsic and extrinsic reflex pathways. Distension of the pyloric antrum results in the release of gastrin into the blood by an intrinsic reflex. Some substances in the food, known as secretagogues, elicit the release of gastrin by an intrinsic reflex. Such substances include alcohol, protein digestion products and caffeine.

Intestinal phase

As long as food is present in the duodenum, it causes the secretion of gastric juice above the basal rate. This effect is hormone-mediated, although the nature of the hormone is uncertain. Intestinal gastrin is one possibility, a second hormone named entero-oxyntin has also been suggested.

Inhibition of gastric secretion

There is a feedback mechanism between acid and gastrin release. If the pH of gastric juice falls, then gastrin release is inhibited. Conversely, if the pH rises, gastrin release is stimulated. A fall in pH is associated with an excess of acid compared with the quantity of food which is present (the latter tends to neutralize acid) and so a feedback system reducing further acid secretion is appropriate and avoids ulcer formation.

The presence of food in the duodenum initiates an enterogastric reflex which is mediated via both intrinsic and extrinsic nerve pathways. The stimuli from the duodenum include distension, protein breakdown products, mechanical irritation and acid.

It is also known that a hormonal mechanism is involved in the inhibition of gastric secretion initiated by a variety of stimuli including acid, fat, protein breakdown products, hypertonic or hypotonic fluids or irritation. The possible hormones involved include secretin, cholecystokinin-pancreozymin and gastric inhibitory peptide.

Once food arrives in the duodenum then, it reduces both the rate of emptying of the stomach (see *Storage, mixing and emptying*) and the rate of secretion of gastric juice. These mechanisms ensure that food does not enter the small intestine too fast to ensure its digestion.

Absorption in the stomach

The amount of food absorbed through the stomach mucosa is inevitably small, partly because most of it is only partially digested at this stage, and partly because of the lack of special transport systems in mucosal cells.

Substances which can be absorbed are small and lipid-soluble and therefore able to diffuse through cell membranes. One such substance is alcohol, another one is aspirin (salicylic acid). The latter is a potential hazard to the stomach wall, because once aspirin leaves the very acid environment in the gastric juice and enters the mucosal cells it dissociates to a greater extent and thereby becomes a stronger acid. The acid can damage the mucosal cells and lead to bleeding from the lower layers of the stomach wall.

STRUCTURE OF THE SMALL INTESTINE

The small intestine is a muscular tube which, in the living adult, is approximately 5 m long (6–7 m in its relaxed state after death). It is divided into a number of anatomically recognizable areas: the short, curved duodenum (approximately 20 cm long); the jejunum, which constitutes nearly 40% of the total length; and the ileum, which represents the distal 60%.

The duodenum, which is the widest section of the small intestine, encloses the head of the pancreas and contains the common opening of the bile and pancreatic ducts.

The jejunum has a diameter of about 4 cm and is very vascular, which gives it a reddish appearance. It has a particularly well-folded mucosa, which is covered by large villi.

The ileum has a diameter of about 3.5 cm and a thinner wall than the jejunum.

The jejunum and ileum are connected to the posterior abdominal wall by the mesentery. This is a double layer of peritoneum between which lie the branches of the superior mesenteric artery and vein, together with lymph vessels (lacteals) and nerves. Lymph nodes and fat are also present in the mesentery.

The parasympathetic nerve supply to the small intestine is via the vagus nerves, whereas the sympathetic fibres travel in the splanchnic nerves and synapse in the coeliac and superior mesenteric ganglia.

The superior mesenteric vein drains the splenic vein, which in turn takes blood to the hepatic portal vein to the liver (see Fig. 4.15). The right and left hepatic veins drain into the inferior vena cava.

A few centimetres from the pylorus as far as the midpoint of the ileum, the inner surface of the small intestine is thrown into circular folds (plicae circulares). These generally extend to between one- half and two-thirds of

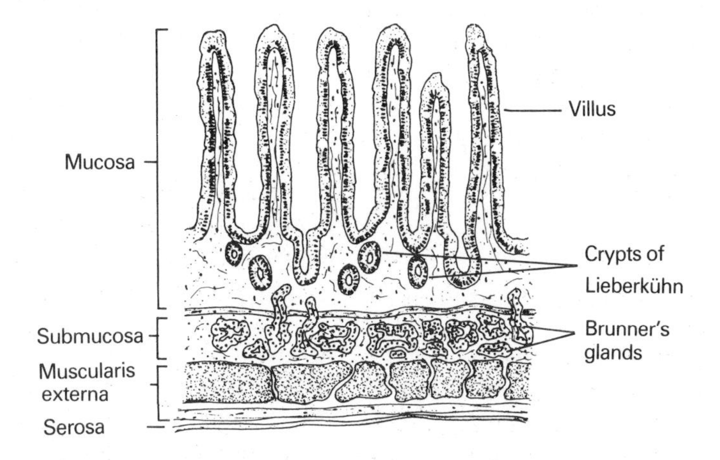

Fig. 6.30 Section through the wall of the duodenum.

the circumference of the intestine, and reach a maximum depth of 8 mm. They serve to increase the surface area available for absorption, as well as retarding the speed with which chyme passes through this region.

The four layers of the wall of the small intestine differ somewhat from those of the stomach (see Fig. 6.30). The muscularis externa contains a thick, inner circular layer with a much thinner external longitudinal layer. The mucosa is much thicker in the upper than in the lower small intestine and, unlike the stomach, it is studded throughout with minute projections, the villi. These are leaf-like in the upper intestine, becoming finger-like lower down.

Villi are between 0.5 and 1.5 mm long and there are between 10 and 40 per square millimetre. Scattered between the villi are small openings of simple tubular intestinal glands, the crypts of Lieberkühn.

Each villus consists of a single layer of columnar epithelial cells covering a strip of lamina propria (see Fig. 6.31). This epithelial covering is constantly renewed by cells arising by cell division in the lower parts of the crypts of Lieberkühn. The new cells migrate slowly over the villus surface and are lost from the apical region. It has been estimated that as much as 30 g of cells are lost each day from the wall of the alimetary tract and that it takes from two to four days to renew the epithelial lining of the gut.

The surface epithelial layer consists mainly of absorptive cells. These are columnar with apical surfaces that are covered by microvilli, which are about 1 μm long and there are approximately 600 of these microvilli per cell. Recent evidence suggests that absorption occurs only at the tips of the villi. Scattered between the absorptive cells are the goblet cells, which are particularly numerous in the epithelium of the ileum.

The lamina propria forms the core of the villus and it contains an extensive capillary network and a central lymph vessel, the lacteal. Smooth muscle fibres allow the villus to change its length, and also to move laterally.

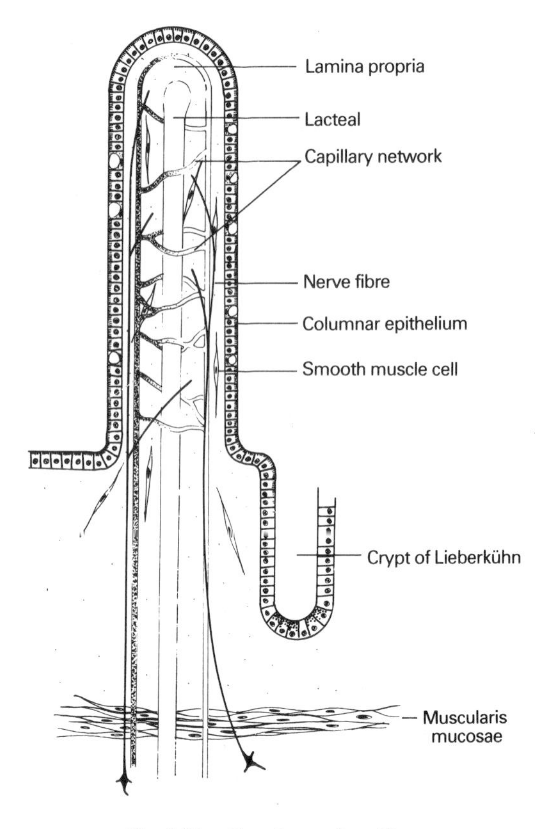

Fig. 6.31 Structure of a villus.

Intestinal glands comprise mainly undifferentiated cells, a few Paneth cells which produce enzymes, and argentaffin cells, like those in the stomach, which secrete hormones.

In the duodenum, the openings of additional glands are found lying between the villi. These are the Brunner's glands (compound, acinotubular), which lie in the submucosa below the muscularis mucosae. They are mucous glands containing some argentaffin cells.

Solitary and aggregated lymph nodes (Peyer's patches) are also found in the mucosa of the intestinal wall, especially in the lower ileum.

MIXING AND SECRETORY FUNCTIONS OF THE SMALL INTESTINE

After it leaves the stomach, chyme is mixed with intestinal secretions as well as with bile and pancreatic juice. Digestion is completed and the products are absorbed through the villi of the intestinal wall. Fat is mainly transferred to the lacteals, whereas protein and carbohydrate digestion products are transferred to the blood capillaries in the villi.

The lymphatic system conveys the fat to the venous circulation via the thoracic duct, and the blood draining the small intestine flows first to the liver and thence back to the heart in the inferior vena cava.

Waste materials are left in the intestinal lumen and are then passed into the large intestine.

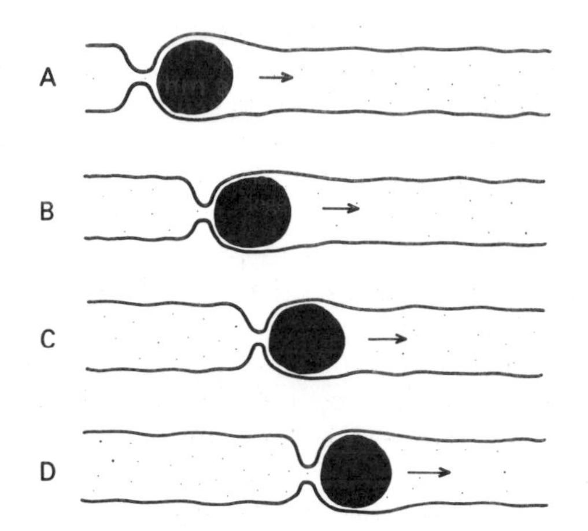

Fig. 6.32 (A–D) A peristaltic wave of contraction moves along the intestine.

Movements of the small intestine

The wall of the small intestine is capable of several different types of movement, the two principal ones being peristalsis and segmental contractions. Both of these movements are rhythmic and they depend upon the basic electrical rhythm (BER) generated by cells in the longitudinal muscle layer. In the duodenum, BER is around 12 per minute and this reduces along the length of the small intestine to about 8 per minute in the terminal ileum. The higher frequency of BER in the duodenum is thought to contribute to the analward propulsion of chyme.

Both peristalsis and segmentation contractions depend upon the myenteric plexus for effective function, although weak activity has been demonstrated after experimental pharmacological blocking of the myenteric plexus. The autonomic nervous system serves only to modify these muscular contractions. The main propulsive movement of the gut is peristalsis which is initiated by distension of the gut wall. This results in the contraction of several centimetres of the longitudinal muscle layer, followed by contraction of a ring of circular muscle above the point of stimulation, and 'receptive relaxation' of the circular layer below the point of stimulation. This response is known as the 'law of the gut' and causes chyme to be propelled analward. The contractile ring appears to move forward for a few centimetres before dying out (see Fig. 6.32). This is achieved by adjacent sections of circular muscle contracting in sequence.

The speed of a peristaltic wave is quite slow (0.5 to 2 cm/s) and so it can take chyme up to 10 hours to travel through the small intestine. Intense irritation of the small intestine, for example by pathogens, can cause an unusual form of peristaltic wave (a peristaltic rush) which travels rapidly and for a greater distance, with the result that the contents of the small intestine are rapidly shunted into the colon.

Mixing of the intestinal contents is brought about by segmentation contractions. These consist of contractions of short lengths of the circular muscle, about 1 cm long, regularly spaced over a short section of gut which cause the chyme to be broken up into segments. After a few seconds, the muscle fibres relax and contraction occurs at points intermediate between the previous ones (see Fig. 6.33). Thus the segments are alternately split and recombined, and this serves to mix the chyme with the various secretions in the small intestine. The contractions travel analward for a short distance, and so they also contribute to the propulsion of food.

The longitudinal muscle fibres contract over relatively long segments of the gut, causing a reduction in the overall length of the gut ('living' and 'dead' gut lengths are very

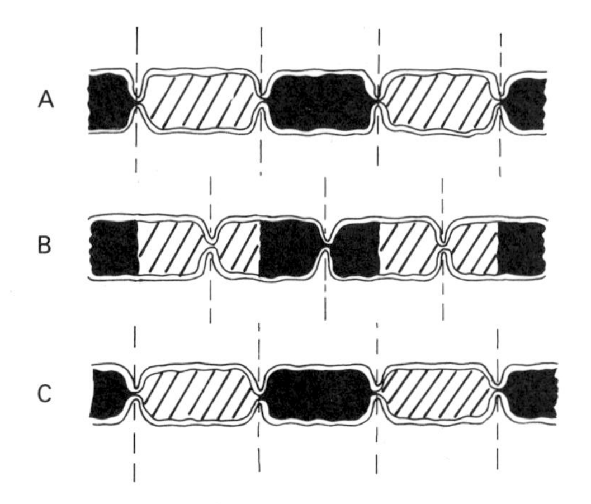

Fig. 6.33 Segmentation contractions in the intestine. (A) The intestine is divided into segments by rings of contracting circular muscle. (B) The original rings relax and others contract to form new segments. (C) Further relaxation and contraction at the original points.

different). Each contraction, though, is of short duration.

Intestinal motility may be increased by stimulation of the extrinsic parasympathetic fibres, although the intrinsic nerve reflexes are generally responsible for most of the activity. In the gastroenteric reflex, distension in the stomach results in increased peristalsis in the small intestine. This is probably mediated by the myenteric plexus, although gastrin may be involved.

Stimulation of the sympathetic nerve supply inhibits gut motility both directly, by the effect of noradrenaline on the muscle cells, and indirectly, by inhibiting the parasympathetic neurones in the myenteric plexus. The vasoconstriction caused by sympathetic stimulation also tends to reduce activity. The sympathetic fibres are also involved in the intestino-intestinal reflex, in which distension of one part of the small intestine results in relaxation of the rest.

The mucosa is thrown into folds by contractions of the muscularis mucosae. These folds serve to increase the surface area of the intestinal lining for the absorption of digested material.

The intestinal villi are also capable of movement since they contain smooth muscle fibres. They are able to shorten and lengthen and move from side to side. These movements are brought about by local reflexes initiated by the presence of chyme. It is also believed that a hormone, known as villikinin, is secreted by the intestinal mucosa into the gut lumen and then reabsorbed into the blood and subsequently it stimulates the villi. Movements of the villi help to bring fresh material to the mucosal surface and also to promote movements of lymph within the lacteals and blood within the capillaries.

Intestinal secretions

The crypts of Lieberkuhn secrete an isotonic fluid, at a rate of about 2000 ml per day. Intestinal juice is essentially extracellular fluid, which does not contain digestive enzymes secreted by the intestinal glands.

Digestive enzymes do, however, find their way into the intestinal lumen by a more indirect route. The epithelial cells which line the mucosa contain a number of digestive enzymes on their plasma membranes and these cells are constantly sloughed off from the apices of the villi into the lumen of the intestine. It is the presence of these enzymes that has given rise, in the past, to the belief that the small intestine secretes an enzyme-rich fluid sometimes called the succus entericus.

The release of fluid by the crypts of Lieberkühn is primarily under the control of local factors. Distension, for example, will result in copious secretion by the crypts. Parasympathetic stimulation will also increase the rate of secretion but only two- or threefold.

The Brunner's glands of the duodenum secrete a thick alkaline mucus which helps to protect the duodenal lining from the acid chyme which passes down from the stomach. These glands are also responsive to local factors and are stimulated by the vagus nerves. Sympathetic stimulation, on the other hand, has an inhibitory effect.

Additionally, mucus is also secreted by the

goblet cells which are found in large numbers in the mucosal epithelium. These cells are responsive to direct stimulation by the contents of the gut.

The mechanisms involved in the final stages of digestion, and the absorption of the products, are described in *Digestion and absorption by the small intestinal epithelium*. Before, this, however, the roles of the pancreas, liver and gall bladder are considered.

STRUCTURE OF THE PANCREAS

The pancreas is a soft gland, lying transversely across the abdominal cavity and usually measuring between 12 and 15 cm in length. It has a bulbous 'head' region which fits into the curve of the duodenum, and the main section or 'body' connects with the tapered 'tail' which extends out to the spleen (see Fig. 6.34). The bulk of the tissue in the pancreas is composed of exocrine cells. These produce pancreatic juice, which is conveyed into the duodenum through the pancreatic duct. Scattered among the exocrine tissue are groups of hormone-secreting cells, the islets of Langerhans.

The organ is covered by a thin connective tissue capsule from which septa arise to penetrate the tissue, dividing it into lobules. Within the lobules, the cells are arranged to form acini around a central intercalated duct (see Fig. 6.35). The acinar cells are pyramidal, secretory cells, with a well-developed endoplasmic reticulum and plentiful zymogen gran-

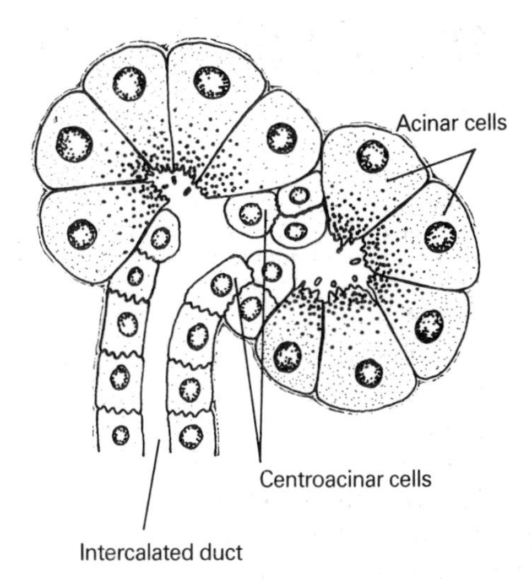

Fig. 6.35 A secretory unit of the pancreas. (The lumeni are enlarged for clarity.)

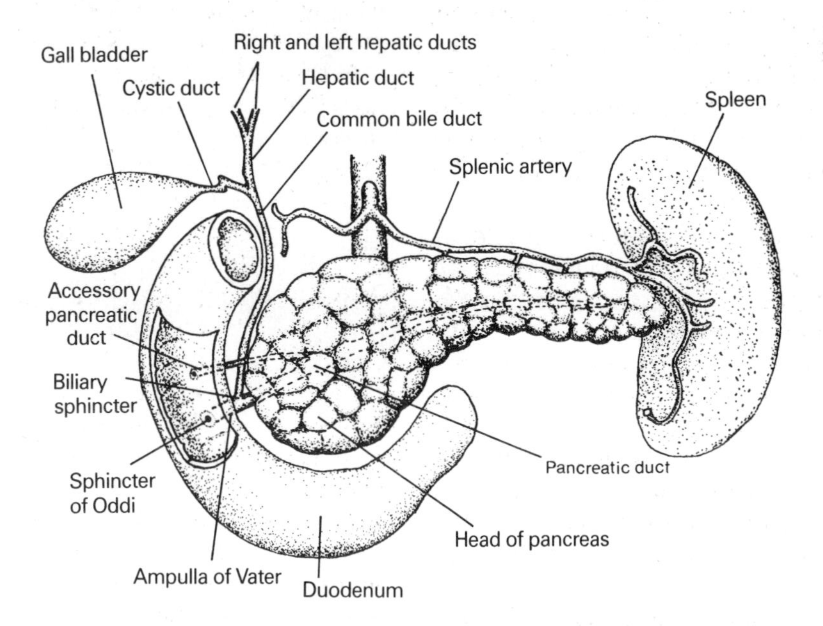

Fig. 6.34 The pancreas and neighbouring structures.

ules. It is these cells which produce the enzyme-rich component of pancreatic juice.

The cells of the first part of the ducts within the acini are known as centroacinar cells and they are cuboidal in shape. Outside the acini the cells are columnar. The intercalated ducts drain into interlobular ducts and eventually to the centrally located main pancreatic duct which drains into the duodenum. The other component of pancreatic juice is an alkaline-rich fluid and this is producéd by the cells of the smaller ducts.

The pancreatic duct usually fuses with the bile duct just before it enters the duodenum to form a hepato-pancreatic ampulla (Ampulla of Vater). This opens at the top of the major duodenal papilla, which itself protrudes into the duodenal lumen (see Fig. 6.34). In many individuals there is an accessory pancreatic duct which collects secretions from the lower part of the head of the pancreas. This duct opens into the duodenum on a small, minor, duodenal papilla.

The blood supply of the pancreas is derived primarily from the splenic artery and the pancreatic-duodenal arteries. Blood drains from the the pancreas into the portal splenic and superior mesenteric veins.

The pancreas is innervated by parasympathetic fibres from the right vagus nerve and sympathetic fibres from the coeliac ganglion.

DIGESTIVE FUNCTIONS OF THE PANCREAS

The exocrine part of the pancreas is responsible for the production of several of the major digestive enzymes. About 1200 ml of pancreatic juice are secreted daily.

Composition of pancreatic juice

Pancreatic juice is a watery alkaline fluid, isotonic with plasma and rich in digestive enzymes. Na^+ and K^+ are present in similar concentrations to plasma. The alkalinity of the fluid is due to the presence of a great deal of HCO_3^- and this acts to neutralize the acid chyme passing into the duodenum from the stomach. The amount of Cl^- present varies inversely with the amount of HCO_3^-, so that the total amount of the two ions is constant.

HCO_3^- derives from plasma CO_2 which, under the influence of carbonic anhydrase in the duct cells, combines with water to form carbonic air and then HCO_3^- and H^+. The hydrogen ions are actively transported out into the plasma, and bicarbonate ions diffuse out into the duct lumen, along with sodium ions which are expelled by active transport. Water passes down an osmotic gradient.

The digestive enzymes present in pancreatic juice include ones which will break down proteins, carbohydrates and fats.

The proteolytic enzymes from the pancreas are secreted in an inactive form. This prevents the enzymes from digesting the cells that produce them. The activation of each proteolytic enzyme is catalysed by a specific enzyme. The enzyme trypsinogen is converted into active trypsin by enterokinase present on the intestinal villi. Chymotrypsin is then converted into the active form by trypsin. The latter can also activate proelastase and procarboxypeptidase and is also able to act as an autocatalyst and activate trypsinogen. Thus once some trypsin has been formed it will lead to the formation of more trypsin. These events are summarized in Table 6.1.

Table 6.1 Proteolytic enzymes produced by the pancreas

Inactive form	Catalyst	Active form
Trypsinogen	Enterokinase Trypsin	Trypsin
Chymotrypsinogen	Trypsin	Chymotrypsin
Proelastase	Trypsin	Elastase
Procarboxypeptidase	Trypsin	Carboxypeptidase

In addition to the powerful proteolytic enzymes described above, the acinar cells produce a trypsin inhibitor. This is an extra precaution against activation within the pancreas and its subsequent digestion. The actions of proteolytic enzymes are summarized in Table 6.2

Table 6.2 Pancreatic enzymes and their actions

Enzyme	Substrate	Products
Trypsin	Proteins, large peptides	Smaller peptides Amino acids
Chymotrypsin	Proteins, large peptides	Smaller peptides Amino acids
Elastase	Elastin	Peptides Amino acids
Carboxypeptidase	Proteins and peptides containing acidic amino acids	Acidic amino acids
α-Amylase	Starch	Maltose Maltotriose Dextrins
Lipase	Triglycerides	Mono-glycerides Free fatty acids
Ribonuclease	RNA	Nucleotides
Deoxyribonuclease	DNA	Nucleotides

There is also α-amylase present which attacks raw as well as cooked starch to release maltose, maltotriose and dextrins.

At least three lipolytic enzymes are secreted by the pancreas. The most important of these is lipase, which hydrolyses triglycerides (glycerol esters), giving rise principally to free fatty acids and monoglycerides.

Lastly, the pancreas secretes two enzymes which attack nucleic acids, breaking them down to release free mononucleotides. These are ribonuclease, which splits RNA, and deoxyribonuclease, which attacks DNA.

The actions of all of the major pancreatic enzymes are summarized in Table 6.2.

Control of secretion of pancreatic juice

The release of pancreatic juice is regulated by both neural and hormonal mechanisms. Activity in the parasympathetic fibres in the vagus nerve to the pancreas increases the production of the enzyme-rich component of the juice. Gastrin has a similar effect. A small quantity of enzyme-rich fluid is thereby released during the cephalic and gastric phases of the control of gastric secretion.

Once the chyme arrives in the duodenum, it stimulates the release of a much larger volume of pancreatic juice. This is brought about by two hormones, secretin and CCK-PZ.

The release of secretin into the blood from the intestine is stimulated by acid. Once secretin reaches the pancreas, it stimulates the duct cells to produce copious amounts of the alkaline component of pancreatic juice. This serves to neutralize the acid chyme and provide a suitable pH for the pancreatic enzymes (pH 7.6–8.2).

As soon as fat and protein digestion products appear in the small intestine, they stimulate the release of the hormone CCK-PZ. This hormone reaches the pancreas through the blood stream and stimulates the acinar cells to release the digestive enzymes. Thus secretin and CCK-PZ regulate the output of different components of the pancreatic juice. The process of pancreatic secretion continues until the digested chyme has left the small intestine. Control of secretion of pancreatic juice is summarized in Figure 6.36.

STRUCTURE OF THE LIVER

The liver is the largest organ in the body, weighing up to 1.8 kg in the adult male and 1.4 kg in the female. It is situated in the upper part of the abdominal cavity beneath the diaphragm and is divided into two parts, a large right lobe (about five-sixths of total) and a much smaller left lobe (see Fig. 6.37).

Like the alimentary tract, the liver is almost entirely covered by a layer of peritoneum. Beneath this is a fibrous capsule which is continuous with areolar connective tissue situated within the substance of the liver. This areolar tissue forms a tree-like structure which carries branches of the hepatic portal vein, the hepatic artery, bile ducts and lymphatics. These vessels enter and leave the liver through the porta hepatis, a short, transverse fissure on the inferior surface of the liver.

Liver cells are arranged in a large number of minute, polygonal hepatic lobules, approximately 1 mm in diameter. Each lobule consists

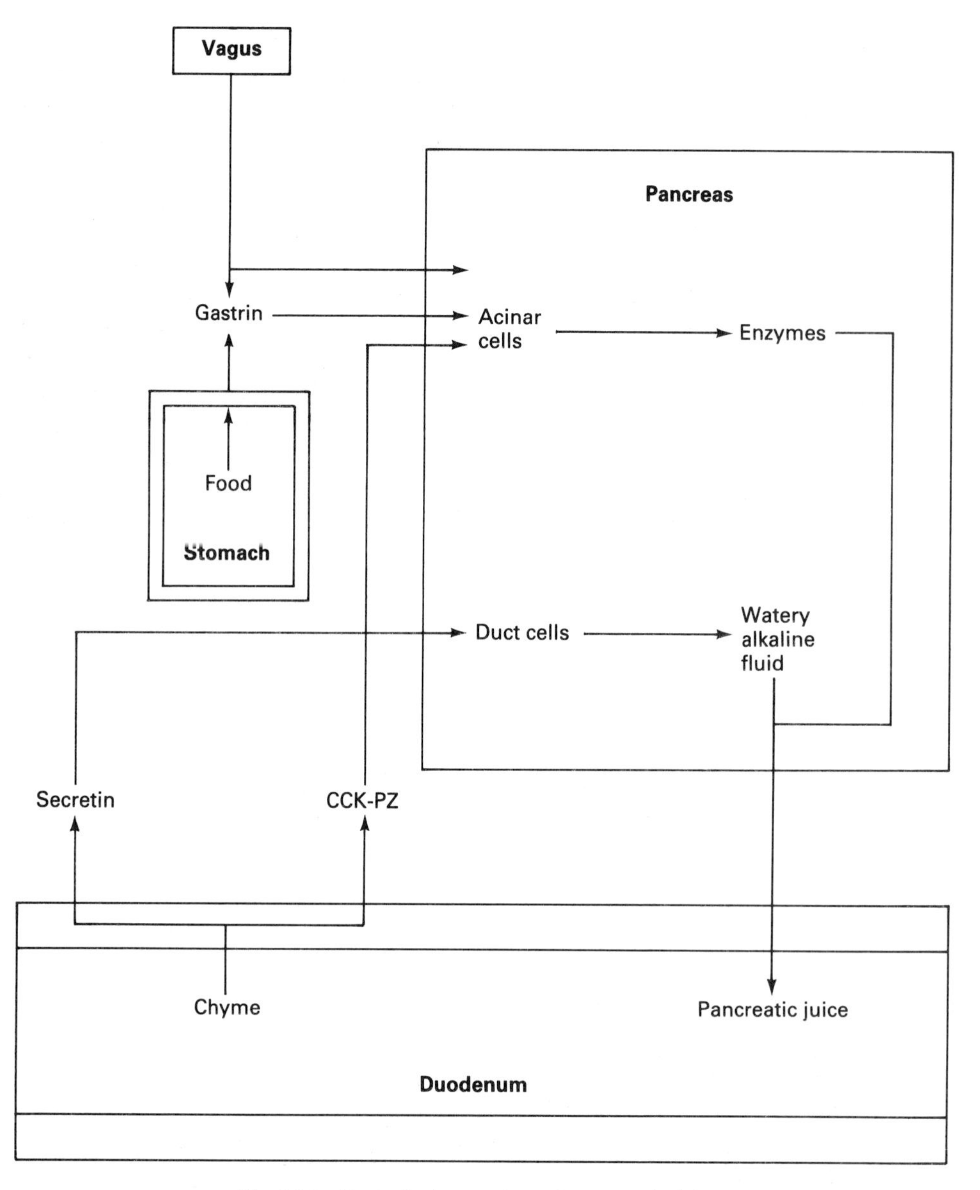

Fig. 6.36 Control of secretion of pancreatic juice.

of a central vein from which radiate sheets of liver cells (hepatocytes). At the corners of the lobules are arranged groups of three vessels; a branch of the portal vein, a branch of the hepatic artery and a bile duct. These three vessels constitute a portal triad (see Fig. 6.38). The connective tissue containing these vessels is called a portal canal and is separated from the main lobule structure by a single layer of hepatocytes, the limiting plate. The portal canal also encloses a variable number of lymph vessels.

Hepatocytes are polygonal cells arranged in sheets (laminae), one-cell thick, which interconnect with each other (see Fig. 6.38). They are rich in glycogen, with a well-developed endoplasmic reticulum and numerous mitochondria. More than one nucleus is often present. The laminae are separated by spaces or lacunae which contain venous sinosoids

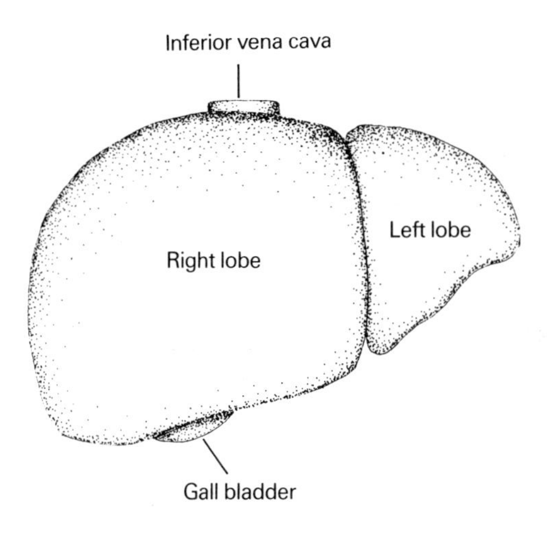

Fig. 6.37 Anterior view of the liver.

carrying blood from vessels at the periphery of the lobule towards the central vein. The sinusoids are lined by fenestrated endothelial cells which are separated from the hepatocytes by a small space, the space of Disse. Irregular microvilli projecting from the hepatocyte surface occupy this space, and there is little collagen or other supporting material present. Scattered among the endothelial cells are numerous macrophages, the Kupffer cells.

At the abutment of two adjacent hepatocytes, there is a small space and since spaces between cells are coincidental a system of narrow tubules is formed which is completely enclosed within the hepatocyte sheet (see Fig. 6.39). These tubules are known as bile canaliculi and they carry bile towards the portal triads.

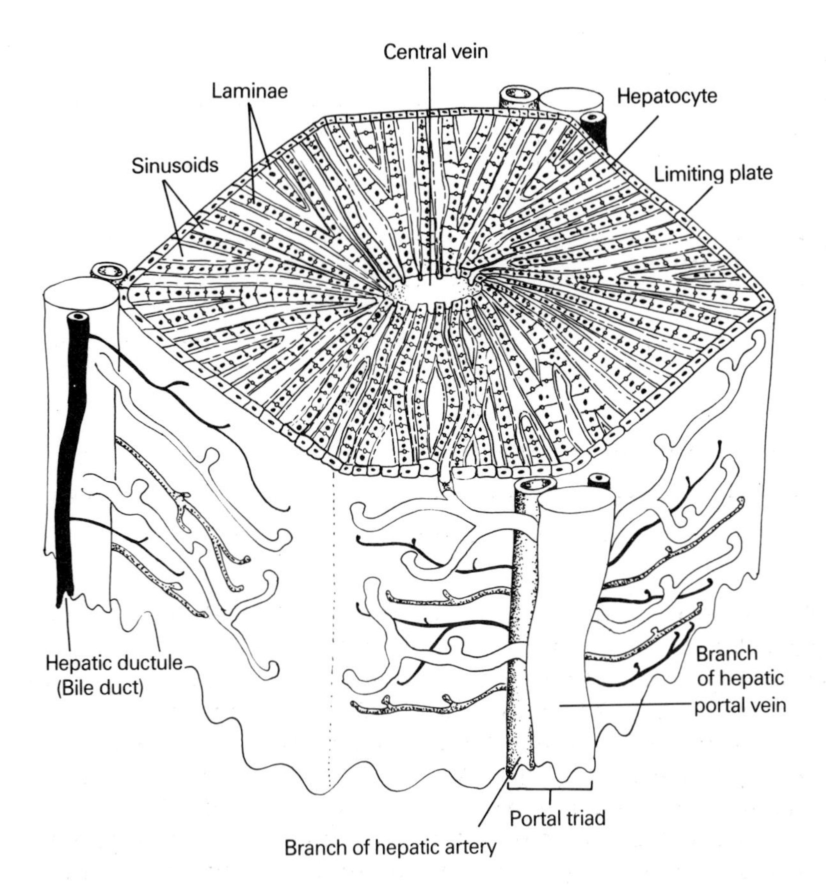

Fig. 6.38 An hepatic lobule.

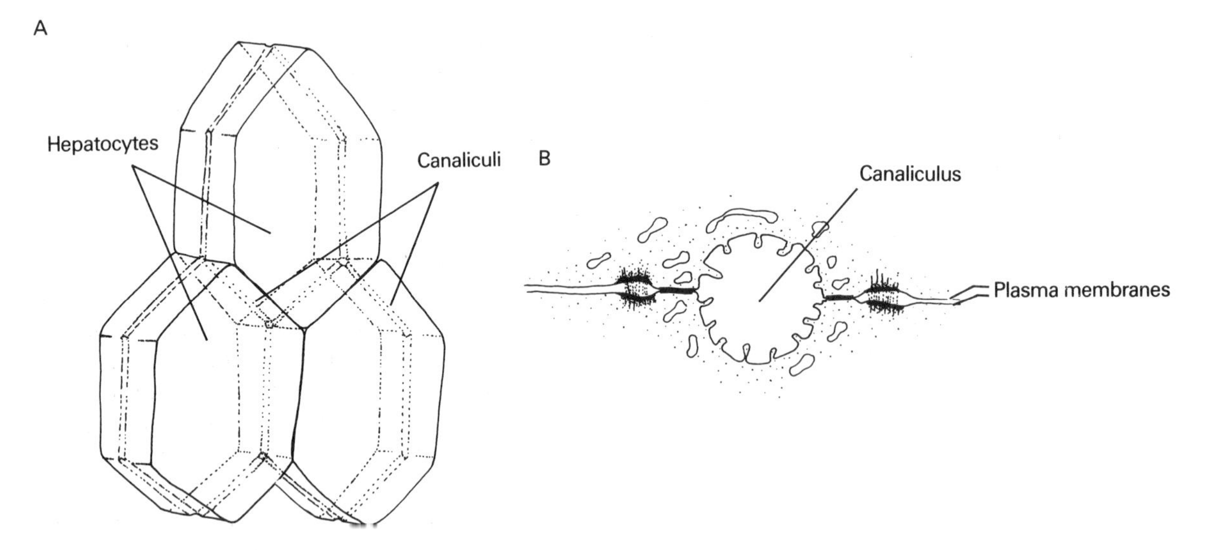

Fig. 6.39 (A) Diagram to show the path of bile canaliculi around and between hepatocytes. (B) Ultrastructural appearance of a bile canaliculus in section.

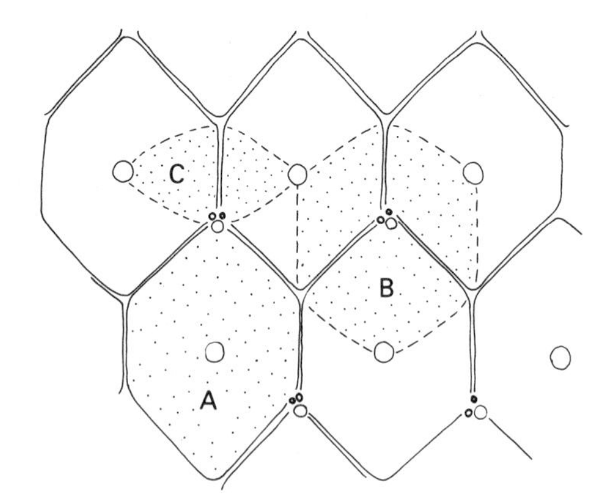

Fig. 6.40 Diagram to illustrate three different functional units in the liver: (A) hepatic lobule, (B) portal lobule, (C) portal acinus.

While it is usual to describe the hepatic lobule as the unit of liver structure, other interpretations also exist (see Fig. 6.40). The portal lobule takes the portal triad as its centre and the central veins of three adjacent lobules form the corners. Thus a triangular structure is created with the bile canaliculi draining into the centre.

A third structure, the portal acinus, has also been described. This unit is defined as the volume of liver tissue served by a distributing branch of an hepatic arteriole.

About 70% of the blood entering the liver derives from the portal vein which carries blood from the gut, spleen, pancreas and gall bladder. Within the liver the portal vein branches repeatedly before sending portal venules into the portal triads. These venules give rise to branches, the distributing veins, which run around the periphery of the lobules. Small inlet venules then penetrate the lobule limiting plate and carry blood to the sinusoids. The latter drain into the central vein of the lobule which then carries blood down into a larger sublobular vein. The sublobular veins gradually converge and fuse to form a variable number of hepatic veins which lead into the inferior vena cava.

The hepatic artery carries about 30% of the blood to the liver. Within the liver substance it branches repeatedly, eventually giving rise to the interlobular arteries of the portal triads. These, like the venules, give rise to distributing branches which carry oxygenated blood around the outside of the lobule. From here arterial blood is passed to the sinusoids where it mixes with blood from the portal system.

At the periphery of the hepatic lobules, the

bile canaliculi fuse to form slender intralobular bile ductules (canals of Hering). These penetrate the limiting plates to enter the interlobular ductules of the portal triads.

The bile that is carried by the bile ductules leaves the liver in the right or left hepatic duct which shortly fuse to form the common hepatic duct. The latter combines with the cystic duct from the gall bladder to form the common bile duct which drains into the duodenum (see Fig. 6.34).

The liver is supplied with sympathetic and parasympathetic fibres. These also travel in the portal canals.

STRUCTURE OF THE GALL BLADDER

The gall bladder is a flask-shaped sac, slate-blue in colour, which lies on the inferior surface beneath the right lobe of the liver. It is attached to the liver by connective tissue, and its sides and bottom are covered by a layer of peritoneum which is continuous with that covering the liver. The gall bladder measures 7–10 cm long, 3 cm broad and has a capacity between 30 and 50 ml.

The wall is composed of three layers; the peritoneum, or serous layer; the fibromuscular layer; and the mucous layer.

The fibromuscular layer contains fibrous tissue and smooth muscle. The mucous layer is thrown into rugae and is lined by columnar epithelium with microvilli. Many capillaries are present beneath this layer. Some of the epithelial cells secrete mucus into the lumen of the gall bladder.

FUNCTIONS OF THE LIVER AND GALL BLADDER

The liver exhibits a wide variety of complex activities. It has many important functions related to the metabolism of carbohydrates, fats and proteins. It stores iron and several vitamins. It is the site of detoxification of many drugs and poisons and the breakdown of some hormones. The manufacture of bile by the liver is important both for the digestion of fats within the small intestine and for the excretion of bilirubin, a breakdown product of haemoglobin.

Secretion of bile

Bile is a watery fluid containing a variety of organic and inorganic substances, but no enzymes. Bile is not, therefore, a digestive juice like that secreted by the pancreas. As much as one litre of bile may be secreted per day by the liver and this is stored in the gall bladder prior to its release into the duodenum. While the bile is stored in the gall bladder, selective reabsorption of some of its constituents occurs. Ions are absorbed, followed by water, but the organic components are not, so that their concentration increases. Overall, the volume of bile may be reduced to as little as one-tenth of its original value. Table 6.3 shows the composition of bile released from the gall bladder into the duodenum.

Table 6.3 Composition of bile (as released by the gall bladder). *Values from Guyton 1981*

Water	92.0 g%
Bile salts	6.0 g%
Bilirubin	0.3 g%
Cholesterol	0.3–0.9 g%
Fatty acids	0.3–1.2 g%
Lecithin	0.3 g%
Na^+	130 mEq/l
K^+	12 mEq/l
Ca^{++}	23 mEq/l
$C1^-$	25 mEq/l
HCO_3^-	10 mEq/l

Bile salts

Bile salts are derived from cholesterol and the first stage of their synthesis is the production of two primary bile acids, chenic acid and cholic acid (see Fig. 6.41). These acids are conjugated with glycine or taurine in the liver. The four acids so produced (glycochenic, taurochenic, glycocholic and taurocholic acid) are secreted in the bile and 75% of them are unchanged as they travel along the small intestine. They are actively reabsorbed in the

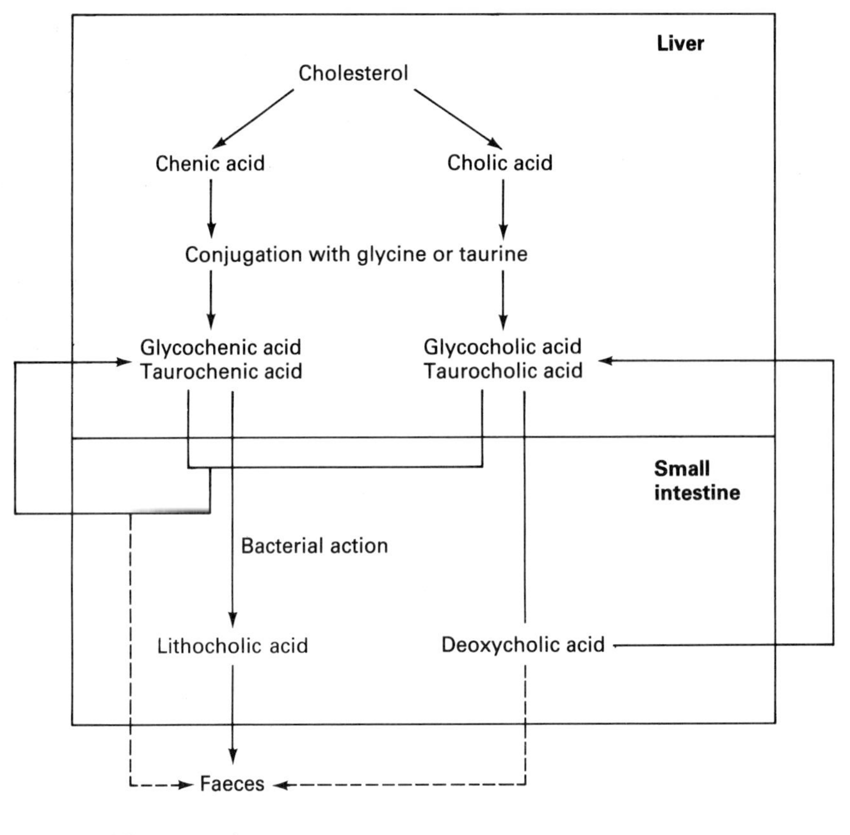

Fig. 6.41 Synthesis and fate of bile salts.

terminal ileum where they enter the portal circulation and return to the liver for recycling. This is called the enterohepatic circulation of bile salts and means that each bile salt is used over and over again (about 18 times on average).

Two secondary bile acids, lithocholic acid and deoxycholic acid, are produced in the small intestine by the action of bacteria on the conjugated primary bile acids. Deoxycholic acid is mostly absorbed and recycled in the liver, but lithocholic acid is poorly absorbed and is therefore excreted.

Bile salts are the negatively charged dissociation products of the acids. They are effectively hybrid ions with a fatty part (derived from cholesterol) and a polar end which is hydrophilic. As a result of this structure, the ions are able to distribute themselves between fat and water. Thus the bile salts are able to emulsify fats, that is they break down the fat droplets entering the small intestine from the stomach into smaller droplets (0.5–1.0 μm diameter). Other substances aid the emulsification process including cholesterol, lecithin and the fat digestion products themselves, monoglycerides and fatty acids.

Emulsification is important because it greatly increases the surface area of fat available for digestion by pancreatic lipase. The latter, which is the enzyme which catalyses the hydrolysis of triglycerides, is a water-soluble protein and can therefore only act on the surface of fat droplets.

A second role of bile salts in digestion is that they form micelles, minute structures (4–5 nm diameter) which contain fat digestion products in sufficiently small numbers that

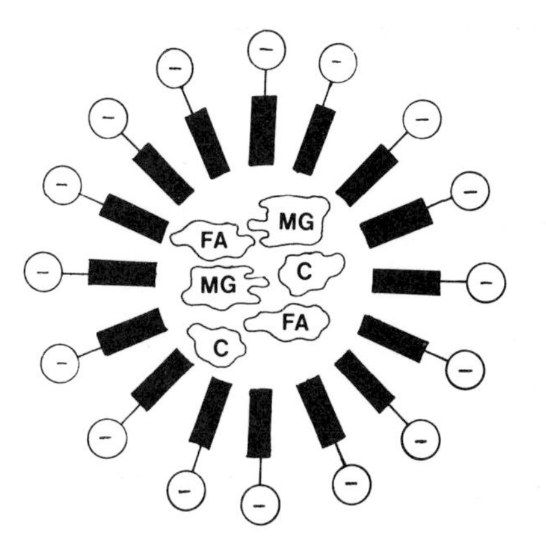

Fig. 6.42 A micelle, showing fatty acids (FA), monoglycerides (MG) and cholesterol (C) within a coat of bile salts.

they are effectively in solution. Micelles are cylindrical structures, the bile salts forming the shell, with their polar ends projecting into the water and the core containing the fat digestion products (free fatty acids and monoglycerides) as well as cholesterol and fat soluble vitamins (see Fig. 6.42).

As bile salts are water-soluble derivatives of cholesterol, they serve to excrete it from the body in this form. Another function of bile salts is that they stimulate bile secretion by the hepatocytes.

Bilirubin

Most of the pigment bilirubin present in the bile derives from haemoglobin, although up

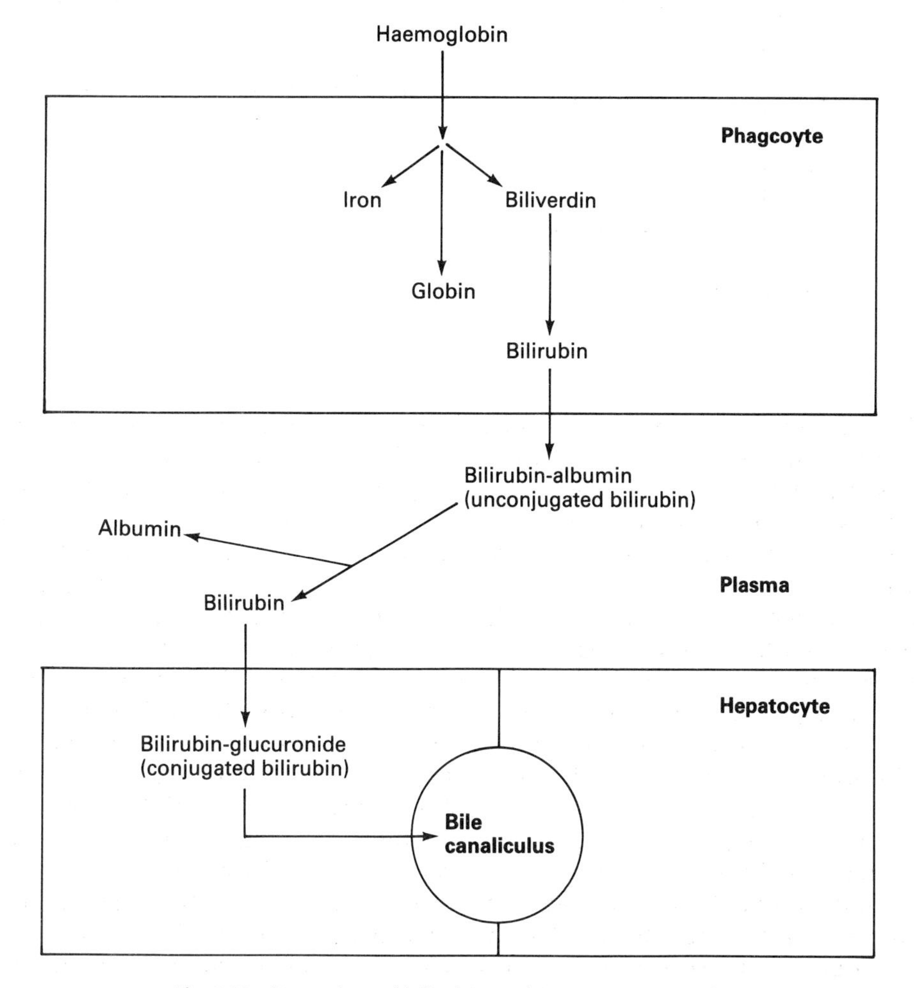

Fig. 6.43 Formation of bilirubin and its excretion into bile.

to 20% comes from other sources such as myoglobin.

Haemoglobin from worn-out red cells is broken down by phagocytic cells mainly in the spleen. The bilirubin released from the phagocytes becomes attached to plasma albumin in which form it is relatively insoluble and cannot cross the glomerular membrane. Bilirubin is split from albumin at the hepatocyte surface, and enters the cell (see Fig. 6.43). Inside the hepatocyte, bilirubin is conjugated with glucuronide which renders it soluble. Most of this conjugated bilirubin is actively transported into the bile canaliculi and thence to the bile; some, however, enters the blood via the lymph in the space of Disse. In obstructive jaundice, this latter route is used much more and so raised levels of conjugated bilirubin are found in the blood and the urine.

In the large intestine, bacterial action on conjugated bilirubin produces the urobilinogens, which are colourless. Some of these are absorbed and either returned to the liver or excreted by the kidneys (see Fig. 6.44). The colour of stools is due to the presence of conjugated bilirubin which has not been altered by gut bacteria.

Control of bile secretion

The active secretion of bile acids into the bile canaliculi causes water to follow by osmosis and this in turn creates a diffusion gradient for ions. Some ions are thought to be added as the fluid flows along the bile ducts within the liver.

The concentration of bile salts in the blood affects the rate of bile secretion by the liver. After a meal, when bile has been added to the duodenum and subsequently most of the bile salts have been reabsorbed in the ileum, the bile salt concentration in hepatic portal

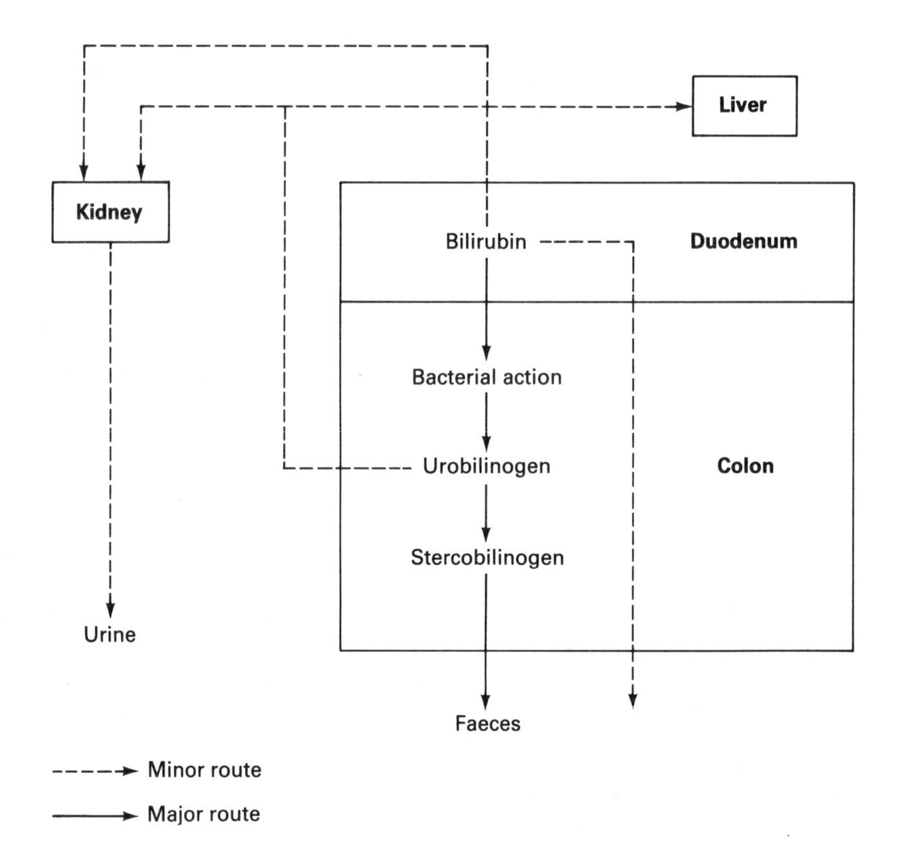

Fig. 6.44 Fate of bilirubin.

venous blood rises. This increase stimulates bile flow from the liver. Vagal stimulation increases bile secretion rate as early as the cephalic phase of digestion.

The hormones gastrin, secretin and CCK-PZ are all thought to stimulate bile secretion. Secretin promotes a bicarbonate-rich secretion, as it does in the pancreas.

Emptying of the gall bladder

Bile is delivered to the duodenum by contraction of the gall bladder, which thereby raises the pressure of bile in the bile ducts and causes the sphincter of Oddi to open.

The principal stimulus for gall bladder contraction appears to be the hormone CCK-PZ. This is released from the duodenum in response to the presence of fat and protein digestion products. As well as causing gall bladder contractions, CCK-PZ causes relaxation of the sphincter of Oddi.

Vagal stimulation increases gall bladder contraction at the same time as promoting secretion of bile from the liver in the cephalic phase, but is physiological role in this connection is probably minor.

Metabolic functions of the liver

The hepatic portal vein delivers blood from the gut and other viscera directly to the liver sinusoids. These are in close proximity to the hepatocytes, enabling an easy exchange of substances between the hepatocytes and blood to occur. The phagocytic Kupffer cells remove bacteria and other debris from the blood draining the gut. The arterial blood supplying the liver appears to mix with the venous portal blood in the sinusoids.

The hepatocytes extract nutrients and other substances from arterial and portal bloods and carry out a great number of different chemical reactions. Some of the products are used by the hepatocytes themselves, but a great number are used elsewhere in the body, and are added to the hepatic venous blood draining from the organ.

Carbohydrate metabolism

The liver plays a crucial role in keeping the blood glucose level within quite narrow limits (normally 3–5 mmol/l). This is important because all tissues rely on adequate provision of glucose, principally for oxidation which produces energy. Brain tissue is particularly sensitive to fluctuations in blood glucose level, as glucose is its preferred substrate for energy provision (rather than fat or amino acids).

After a meal, in the absorptive state, there is likely to be a considerable quantity of glucose entering the portal vein from the small intestine. The hepatocytes prevent the blood glucose level from rising too high by absorbing some glucose and converting it to glycogen, a process catalysed by the hormone insulin. The liver also converts glucose to triglycerides, which can be stored in adipose tissue or oxidized in the liver itself.

In the postabsorptive state, when there is no glucose being absorbed from the gut, the blood sugar level tends to fall. The liver helps to limit this fall by releasing glucose into the blood following the breakdown of glycogen (glycogenolysis). This process is catalysed by the hormone adrenaline as well as by stimulation of the hepatocytes by the sympathetic nervous system.

Glucose is synthesized from amino acids during the postabsorptive state. The production of glucose from non-carbohydrate sources is called gluconeogenesis, and the effect of this is to increase the total amount of carbohydrate in the body.

The monosaccharide galactose, which is delivered to the liver in the portal vein, is converted into glucose in the hepatocytes. Fructose, is largely converted into glucose by the intestinal cells themselves. The term blood sugar, therefore, normally means blood glucose (except in the hepatic portal vein, where fructose is also present).

Lipid metabolism

Just as hepatocytes takes up glucose and fructose from the hepatic portal blood, so also they take up lipids in the form of chylomica. The triglycerides which are released from the latter are hydrolysed in the liver and the fatty acids produced are either oxidized or used to synthesize other lipids. An excessive amount of fatty acid oxidation in the liver results in an accumulation of the breakdown products, ketoacids (acetoacetic and β-hydroxybutyric acids), thereby producing the condition known as ketosis. Several types of fat are synthesized in the liver, including cholesterol and its derivatives, phospholipids and lipoproteins.

Carbohydrates and amino acids can be converted into lipid in the liver.

Protein metabolism

Some (non-essential) amino acids can be synthesized from others in the liver, by a process called transamination.

Many plasma proteins are synthesized in the liver, including albumin (about 3 g/day), globulins, complement and clotting factors I, II, V, VII, IX and X.

Amino acids are broken down in the liver, and the reactions include deamination; this is the removal of the amino group (NH_2), which is converted to ammonia (NH_3) and then to urea (NH_2CONH_2). Urea is relatively harmless compared with the toxic ammonia, and the latter's conversion to urea is also employed elsewhere in the body, e.g. to detoxify ammonia produced by gut bacteria in the large intestine.

Inactivation of drugs, poisons and hormones

Hepatocytes are able to inactivate drugs and poisons by rendering them water soluble for excretion in the bile or urine. Various hormones are inactivated in the liver including the steroids cortisol, aldosterone and sex hormones; thyroid hormones and antidiuretic hormone.

Storage functions

The liver stores iron and the fat-soluble vitamins A, D, E and K as well as the water-soluble vitamins B_{12} and folate. It may be noted that vitamin K is not only stored in the hepatocytes, it is also used in the synthesis of clotting factors II, VII, IX and X.

DIGESTION AND ABSORPTION BY THE SMALL INTESTINAL EPITHELIUM

Carbohydrates

The digestion of starch starts in the mouth, brought about by the action of salivary α-amylase, to produce maltose, maltotriose and dextrins. The process is continued in the small intestine by the action of pancreatic amylase (Fig. 6.5).

Disaccharides in the diet (sucrose, lactose, maltose) are undigested at this stage. These, together with the digestion products of the amylases, are finally hydrolysed to monosaccharides by enzymes present on the external

Table 6.4 Enzymes produced by the intestinal mucosa

Enzyme	Substrate	Products
Membrane peptidases	Peptides	Amino acids
Cytoplasmic peptidases	Peptides	Amino acids
Maltase	Maltose, maltotriose	Glucose
Lactase	Lactose	Glucose, galactose
Sucrase	Sucrose	Glucose, fructose
Dextrinase	Dextrins	Glucose
Nuclease	Nucleic acids	Pentoses, nitrogenous bases

surface of the intestinal cell membranes (on the microvilli). Table 6.4 lists these enzymes and their actions.

The absorption of monosaccharides by the intestinal epithelium has been the subject of a number of different theories. The sodium-gradient hypothesis was proposed by R.K. Crane (1968), and seemed to apply to the absorption of amino acids and several other substances in addition to the monosaccharides glucose and galactose.

Glucose and galactose are both able to attach themselves to a particular carrier molecule, present in the luminal surface of the intestinal cell membrane. The carrier attaches to one monosaccharide molecule and also attaches to a sodium ion (see Fig. 6.45). The latter is present in a high concentration in the gut lumen, but a low concentration within the cell. Sodium, together with the attached carrier and glucose or galactose molecule, diffuses rapidly into the cell down its concentration gradient. Once across the membrane, the glucose and sodium split from the carrier, which can then be reused. The sodium is pumped out of the basolateral faces

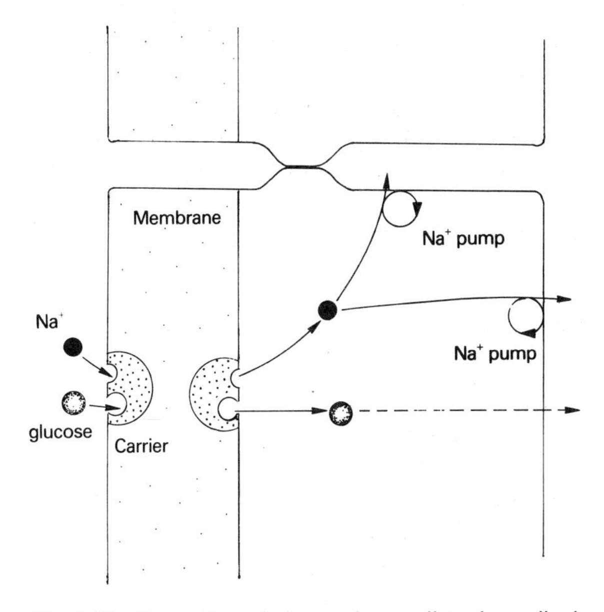

Fig. 6.45 Absorption of glucose by a cell in the wall of the small intestine (see text for details).

of the cell, and glucose and galactose diffuse into the portal blood. Therefore, although glucose and galactose are not themselves actively transported through the cell, their transport is linked to that of sodium ions which are actively pumped out of the cells.

The monosaccharide fructose is thought to be transported through the intestinal cell membrane by a carrier molecule which does not carry sodium ions. Fructose is transported by facilitated diffusion rather than by active transport. It is then converted into glucose within the intestinal cell before passing into the portal blood.

Lipids

It may be recalled that pancreatic lipase breaks down triglycerides principally into monoglycerides and free fatty acids (Fig. 6.10). Bile salts serve both to emulsify the triglyceride droplets and to form micelles with the digestion products, cholesterol and fat-soluble vitamins.

The fat digestion products are conveyed to the surface of intestinal cells as micelles and there the various constituents split up. The bile salts are used many times, providing a 'shuttle service' to the absorptive cells before being finally absorbed in the terminal ileum.

Free fatty acids and monoglycerides are absorbed mainly in the duodenum and proximal jejunum by diffusion. Once inside the cells, long-chain fatty acids (i.e. containing more than 10–12 carbon atoms) are reassembled into triglyceride molecules on the endoplasmic reticulum. They are then packaged by the Golgi apparatus into spherical structures called chylomicra, which consist of the triglycerides surrounded by a coat of protein, phospholipid and cholesterol. These may be as large as 100 nm diameter. They are released from the base or sides of the intestinal cells by exocytosis and from there they are able to enter the lacteals, rather than the blood capillaries.

Short-chain free fatty acids pass into the portal blood and are not assembled into triglycerides.

Proteins

Pepsins in gastric juice catalyse the hydrolysis of protein, liberating polypeptides and a few free amino acids. Several proteolytic enzymes in pancreatic juice continue this process, leaving a mixture of small peptides and free amino acids.

Peptidases are found both on the intestinal cell membranes and within the intestinal cells. It appears that peptides containing arginine, lysine, methionine and leucine are hydrolysed by the enzymes on the microvilli, whereas peptides containing glycine, aspartic acid and glutamic acid are hydrolysed by cytoplasmic peptidases.

The D and L forms of amino acids are absorbed differently. The D isomers are absorbed by diffusion, whereas the L isomers are absorbed in a similar manner to glucose, i.e. by carrier-mediated, sodium-linked transport. At least four carriers are involved. Some research workers believe that transport out of the epithelial cells, on the way to the portal circulation, may involve more than simple diffusion.

Some small peptides (less than four amino acids long) can be absorbed intact by an active transport mechanism which is also sodium dependent.

Nucleic acids

Nucleic acids are digested in the duodenum by ribonuclease and deoxyribonuclease in the pancreatic juice. The resulting nucleotides are split at the mucosal surface into phosphoric acid and nucleosides which are then, themselves, further split into their constituent parts. The purines and pyrimidines released at this final stage of digestion are then absorbed by active transport.

Absorption of ions, water and vitamins

Sodium and chloride

Sodium ions are actively transported out of the lumen of the small intestine in association with glucose and amino acids (see above). The quantities involved are massive, some 25 to 35 g per day, most of which derives from the secretions added to the alimentary tract. Chloride ions diffuse passively along the electrochemical gradient created by sodium absorption along most of the small intestine, but in the lower ileum (and colon) chloride ions are able to be actively absorbed in exchange for bicarbonate ions.

Potassium

Potassium ions diffuse out of the intestine and into the blood. The movement is proportional to the potential difference between the blood and the intestinal lumen. This potential difference is lowest in the upper part of the small intestine and highest in the colon.

Calcium

Calcium ions are absorbed by active transport in the upper part of the small intestine. The process is dependent upon the presence of calcium-binding protein associated with the microvilli. The synthesis of the protein is induced by active vitamin D (1,25-dihydroxycholecalciferol) which, in turn, requires parathyroid hormone for its activation in the kidney. Calcium absorption is aided by the presence of bile acids in the intestinal lumen, but inhibited by fatty acids with which calcium forms insoluble soaps.

Iron

Most dietary iron is in the ferric (Fe^{+++}) form and needs to be converted into the ferrous (Fe^{++}) state in order that it can be absorbed. This reduction occurs in the stomach in the presence of hydrochloric acid.

Having passed through the stomach, most of the iron is absorbed in the upper part of the small intestine. Some of the iron passes through the mucosal cells directly into the blood stream. Most, however, becomes bound to the protein apoferritin, forming ferritin within the mucosal cells.

Iron will pass into the blood stream when

plasma levels are low; thus the cellular ferritin is in equilibrium with plasma iron, which is bound to the protein transferrin. The mucosal cells regulate the level of plasma iron, in that when the blood is saturated with iron there is no movement out of the cells. Under these circumstances iron will accumulate within the cells and then be lost to the body when the cells are shed into the intestinal lumen.

Water

Water is able to diffuse across the wall of the intestine in both directions. It diffuses both into the cells and between them through the tight junctions. The amount of water absorbed depends upon the relative osmolarities of the luminal contents and the blood plasma. Normally nearly all of the water added to the alimentary tract in secretions (about seven litres) and in the diet (about two litres) is absorbed.

Vitamins

Most vitamins are absorbed in the upper part of the small intestine. An exception to this is vitamin B_{12} which is absorbed in the ileum. Vitamin B_{12} absorption depends upon the presence of intrinsic factor, secreted by the gastric mucosa.

Many water-soluble vitamins are thought to be absorbed in the same way as glucose and L amino acids, that is by an active, sodium-linked transport system.

The absorption of fat-soluble vitamins (A, D, E and K) is passive (see *Lipids*).

STRUCTURE OF THE LARGE INTESTINE

The large intestine has a length of about 1.5 m and extends from the end of the ileum to the anus. It may be divided into a number of regions which are distinguished by both their anatomical structures and positions within the abdominal cavity (see Fig. 6.46).

The first section is the caecum, a blind-ended bulbous sac, approximately 6 cm long and 7.5 cm in diameter which leads into the ascending colon. The ileum opens into the caecum at right angles; the entrance being guarded by the ileocaecal valve which regulates the passage of digested food material from the small to the large intestine. A short (average length 9 cm) blind-ended tube, the vermiform appendix, protrudes from the outer wall of the caecum. The function of the appendix is unknown, although, since it contains a high proportion of lymphoid tissue, it may be involved with the body's immune mechanisms.

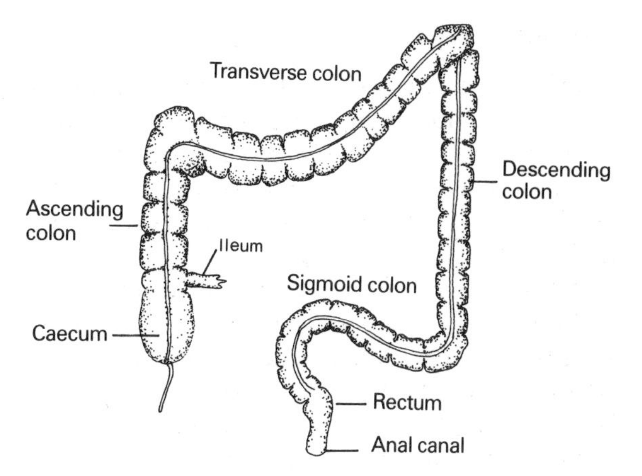

Fig. 6.46 The sections of the large intestine viewed from the front.

The caecum leads directly into the ascending colon, which is about 15 cm long and passes up the right side of the abdominal cavity to the lower surface of the liver. Here it bends, at the right colic flexure, before passing to the left and forwards for about 50 cm as the transverse colon. The latter terminates on the left-hand side of the abdominal cavity, in the vicinity of the spleen, at the left colic flexure. The descending colon passes down for about 25 cm, curving medially and leading into the sigmoid colon. This, the last part of the colon, is extremely variable in both length and distribution. The sigmoid colon leads into the rectum which commences near the bottom of the abdominal cavity at the level of the third sacral vertebra. The rectum is about 12 cm long with a smooth outer wall and a well-folded mucous membrane (when the cavity is empty). The rectum leads into the narrower anal canal which is normally less than 4 cm in length and is characterized by the presence of a mass of muscular sphincters. The internal anal sphincter comprises the upper part of the (thickened) circular muscle layer of the anal canal. The external anal sphincter is striated muscle and surrounds the whole length of the anal canal. The sphincters seal off the end of the alimentary tract and are therefore normally constricted. Their activity must be inhibited in order to allow the passage of faeces.

The wall of the large intestine has the same basic structure as that of the small intestine, but there are considerable differences of detail (Fig. 6.47). The outermost layer, the serosa, is not present in all parts. Both muscle layers are always present, however, and are represented by a thin continuous layer of circular muscle with a longitudinal layer of uneven distribution. In the walls of the caecum and colon the longitudinal muscle layer is thickened into three bands (6–12 mm wide), the taeniae coli; between these bands the longitudinal layer is much thinner. Towards the lower end of the sigmoid colon the muscle bands coalesce to form a continuous layer which then extends into the rectum.

The taeniae coli appear to have a shorter length than the rest of the large intestine, so that the wall appears to be puckered, forming a series of pouches or haustrations. The circular muscle fibres are found to be especially well developed in the intervals between haustrations.

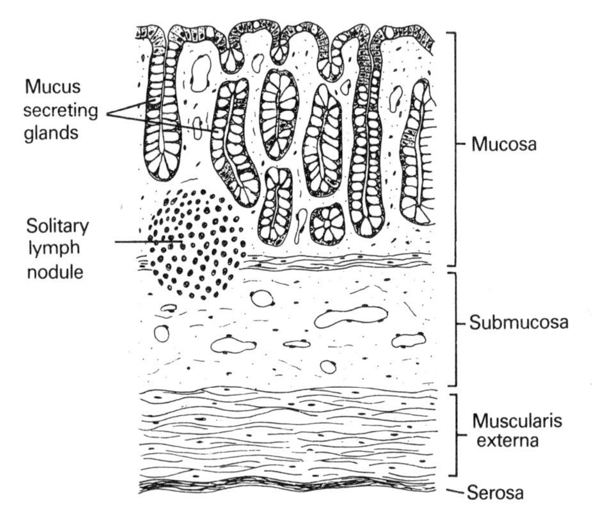

Fig. 6.47 The layers of the large intestine wall.

The submucosa has a structure similar to that found in the small intestine, but the mucosa is somewhat different. There are no villi present although the mucosa is folded. The epithelium consists primarily of columnar absorptive cells with scattered goblet cells. There are numerous simple tubular glands whose prime function appears to be the secretion of mucus. Solitary lymph nodules are scattered throghout but are especially numerous in the caecum and vermiform appendix.

The blood supply for the bulk of the large intestine is derived from the colic branches of the superior mesenteric artery. The left-hand (distal) end of the transverse colon, the descending and sigmoid colons and the rectal area all receive blood from the inferior mesenteric artery and the middle rectal artery. Blood drains away from the two areas into the corresponding veins, i.e. the superior and inferior mesenteric veins.

The large intestine receives its nerve supply from both the sympathetic and parasympathetic systems. The caecum, ascending colon and most of the transverse colon are innervated by sympathetic fibres from the coeliac and superior mesenteric ganglia, while the parasympathetic supply derives from the vagus. The left-hand end of the transverse colon and distal parts as far as the upper half of the anal canal, derive their sympathetic supply from the lumbar region of the spinal cord and from the superior hypogastric plexus. The parasympathetic supply of this part of the gut is derived from the pelvic splanchnic nerves. (The nerve supply to the anal sphincters is described in the section on defaecation).

FUNCTIONS OF THE LARGE INTESTINE

The large intestine transforms chyme into semi-solid faeces, by the absorption of ions and water, and stores them until they are removed by defaecation. It is also able to secrete large quantities of water and electrolytes in response to irritation, e.g. bacterial infection. The colon harbours an extensive population of micro-organisms (mainly bacteria) whose presence is necessary for the normal development of lymphoid tissue in the large intestine and thus resistance to infection. The bacteria synthesize vitamin K, which is necessary for the synthesis of several clotting factors by the liver.

Movements of the large intestine

Chyme enters the caecum through the ileocaecal valve. This valve is normally in a state of tonic contraction, but it opens briefly to allow a small amount of chyme to pass through each time a peristaltic wave arrives.

Various types of movement occur in the colon, including peristalsis, segmentation, haustral shuttling and mass movements. The BER in the colon has a low frequency compared with the small intestine, and the pacemaker is located half-way along the colon. This may account for the fact that peristalsis in the proximal colon moves towards the caecum, rather than the rectum. In the transverse and descending colon, the commonest movements are segmentation contractions, which may be stationary or propulsive. Haustral shuttling involves apparently random contractions, which move the contents backwards and forwards and bring about mixing.

Mass movements occur, usually in the transverse colon, only a few times a day. A length of some 20 cm distal to a distended or irritated site in the colon contracts for about 30 s, so emptying that section of the colon. Mass movements may be brought about reflexly through the myenteric plexus as a result of distension of the stomach or duodenum by ingested food (gastrocolic and duodenocolic reflexes). This sudden movement of colonic

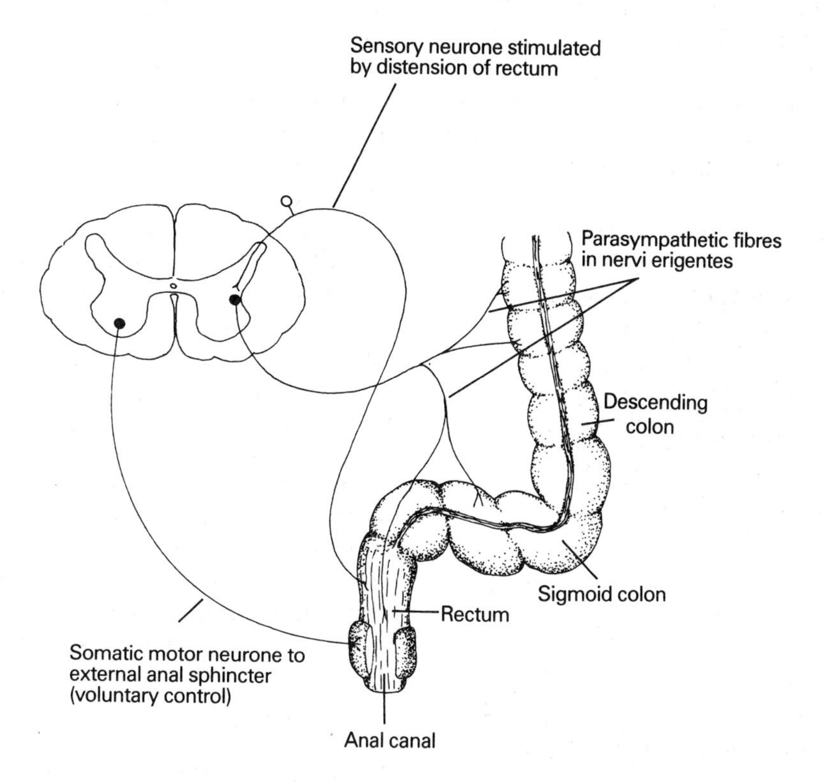

Fig. 6.48 Pathways of the defaecation reflex.

contents can push large amounts of faeces into the rectum and initiate the desire to defaecate (see below).

Movement of chyme through the large intestine is a slow process. Although as much as 75% of the remains of a meal are lost in the faeces within 72 hours, it may take up to a week for the entire meal to pass through the gut.

Defaecation

The removal of salts and water from the intestinal contents renders them increasingly solid, so that by the time they have reached the rectum they are in a semi-solid state and have formed the faeces.

The faeces are composed of the indigestible remains of food, such as: cellulose and other dietary fibres, and substances associated with them; fat; ions and water; digestive enzymes; mucus; cells from the alimentary tract lining; and bile pigments, which give colour to the faeces.

The presence of faeces in the rectum initiates the defaecation reflex as well as a desire to defaecate. The reflex involves the stimulation of sensory neurones which travel from the rectum to the sacral segments of the spinal cord, and the parasympathetic fibres in the nervi erigentes stimulate peristalsis in the descending and sigmoid colon, the rectum and anal canal (see Fig. 6.48). The contraction may be sufficiently powerful to empty the colon from the splenic flexure onwards, with one peristaltic wave. There is also a weaker, intrinsic reflex, mediated by the myenteric plexus.

When the peristaltic wave reaches the internal anal sphincter, the latter relaxes. However, defaecation also depends upon the state of contraction of the external anal sphincter. Voluntary relaxation of this sphincter will enable the faeces to be voided. The afferent signals transmitted via the spinal cord to the cerebral cortex also usually result in voluntary contraction of the abdominal muscles and deep inspiration, to lower the diaphragm and raise intra-abdominal pressure and assist evacuation.

Secretion in the large intestine

The glands of the large intestine secrete mucus, which protects the intestinal wall and acts as a binding agent for the faeces. Large quantities of water and ions may be produced in response to infection, serving to dilute and flush out the irritants (and causing diarrhoea).

Mucus secretion is brought about primarily in response to direct tactile stimulation by the intestinal contents, although local reflexes and extrinsic parasympathetic stimulation can also increase mucus production. The latter route is used in psychogenic diarrhoea, which may accompany stressful activities (such as taking examinations!).

Absorption in the large intestine

The proximal colon is the main site for the absorption of substances from chyme, whilst the distal colon has a storage function.

Sodium ions are actively absorbed by two mechanisms, an electrogenic pump, and in a one-to-one exchange with hydrogen ions. Sodium transport in association with glucose or amino acids, as in the small intestine, does not occur.

Chloride ions are also transported by two methods, one is passive, following sodium, and the other is active in which chloride is exchanged for bicarbonate.

Potassium absorption may be by active transport. Calcium ions are absorbed in a similar fashion as in the small intestine, by a vitamin D-dependent mechanism, and magnesium ions may also be absorbed by this mechanism.

Water is absorbed along the osmotic gradient produced by the absorption of the ions (500 to 1000 ml per day).

Urea is metabolized in the colon wall to ammonia, which is subsequently absorbed and then metabolized in the liver. Some of it may be used to synthesize amino acids.

Although the majority of bile acids are absorbed in the small intestine, bacteria in the colon deconjugate them, making them more lipid-soluble so that they may be absorbed from this site.

Bacterial flora

Although the lumen of the large intestine is sterile at birth, it is quickly invaded by a number of different bacteria, the major ones being the Gram-positive anaerobes *Eubacteria* and *Bifidobacteria*, and *Bacteroides*.

It has been shown that the intestinal lymphoid tissue proliferates as a result of the presence of bacteria in the gut. Immunoglobulins (IgA) are present on the mucosa and help to prevent the passage of harmful organisms into the blood.

The bacteria produce several vitamins, K, B_{12}, thiamin and riboflavin, although only vitamin K is absorbed in sufficient amounts to be physiologically significant.

The bacteria are able to metabolize some cellulose, digestive enzymes and other cellular debris and they produce various gases.

The main gases in flatus (not all produced in the colon) are odourless (nitrogen, carbon dioxide, methane, hydrogen and oxygen). Traces of odoriferous gases are present, including ammonia, hydrogen sulphide, indole, skatole, short-chain fatty acids and volatile amines.

7

The nervous system

The nervous system can be divided into two primary divisions, the central nervous system (CNS), consisting of the brain and spinal cord, and the peripheral nervous system, consisting of nerves which connect the CNS with the tissues.

Chapter 1 includes a section on nervous tissue, in which details of the structures and functions of neurones and nerves are given. This chapter covers general aspects of sensory physiology (special senses are covered in Ch. 8), the autonomic nervous system, temperature regulation, the brain, spinal cord and spinal reflexes.

SENSORY PHYSIOLOGY

SENSORY RECEPTORS

The brain is informed of events occurring both within and outside the body by nerve impulses which originate at a large number of sensory receptors. These receptors may simply be nerve endings, single specialized cells or, alternatively, aggregations of cells forming sense organs. The receptors act as transducers, converting the energy of the stimulus into a pattern of impulses which then pass to specific areas of the brain including the cerebral cortex.

Perception of conscious sensations seems to be caused by the arrival of impulses at specific locations in the cerebral cortex, although it is far from clear how this is sufficient for an individual to distinguish between different types, intensities and quality of sensations, as well as their points of origin in the body.

Classification of sensory receptors

Receptors can be classified in a number of ways, often with regard to the types of stimuli to which they respond. The classic 'special' senses of sight, hearing, taste and smell are part of a much longer list of senses which inform the brain about the body's activities and environment. As a consequence, the classification of receptors has become somewhat complicated. One of the simplest methods of division is into 'exteroceptors' which respond to stimuli originating in the external environment and 'interoceptors' or internal receptors. A further class of 'proprioceptors' monitor body position.

A more detailed classification is given in Table 7.1. This divides sense organs into a number of functional classes: electromagnetic receptors which respond to electromagnetic waves, e.g. light; chemoreceptors, which respond to a variety of chemical stimuli; mechanoreceptors, which respond to any mechanical stimuli; thermoreceptors, which detect heat and cold; and nociceptors, which originate impulses leading to the perception of pain.

It is important to realize that not all receptors give rise to a sensory experience. Many of the deep-lying receptors provide information which assists in the regulation of the internal environment, but as the impulses are not relayed to the cerebral cortex no conscious sensation is perceived.

Modality of sensations

Each of the sensations we are able to experience is said to have a modality. Sight and hearing are examples of sensory modalities.

Each modality depends upon the precise area of the brain where the impulses which are derived from the sensory receptors terminate. The nature of the stimuli and the manner by which the receptors react to those stimuli are specific, but the methods by which the information is carried to the brain are identical. Thus stimulation of, say, a pathway carrying impulses from a pain receptor, by electrical or any other means, will lead to the perception of pain. Although sensations are experienced in the brain, they appear to be localized somewhere in the body. Thus a painful stimulus to the hand will be 'felt' in that hand, because the brain 'projects' the sensation to the point where the stimulus was introduced.

Table 7.1 Classification of sensory receptors

Receptor	Organ/Site	Stimulus	Modality
Electromagnetic receptors			
Rods	Eye	Low intensity light	Vision
Cones	Eye	High intensity light	Colour vision
Chemoreceptors			
Gustatory cells	Taste buds	Chemicals in oral cavity	Taste
Olfactory nerve endings	Olfactory epithelium	Air-borne chemicals	Smell
Nerve endings	Aortic/carotid bodies	Change in blood CO_2 Change in blood H^+	—
		Change in O_2	—
Nerve endings	Surface of medulla	Change in c.s.f. H^+	—
Osmoreceptors	Hypothalamus	Change in blood osmotic pressure	—
Nerve endings	Hypothalamus	Change in blood amino acids, fatty acids, glucose	Hunger
Mechanoreceptors			
Hair cells	Ear (cochlear)	Sound waves	Hearing
Hair cells	Ear (saccule/utricle)	Linear acceleration	Linear acceleration
Hair cells	Ear (semicircular canals)	Rotational acceleration	Rotational acceleration
Nerve endings	Skin	Distortion of skin	Touch/Pressure
Meissner's corpuscles	Skin	Distortion of skin	Touch/Pressure
Krause's corpuscles	Skin	Distortion of skin	Touch/Pressure
Merkel's discs	Skin	Distortion of skin	Touch/Pressure
Pacinian corpuscles	Skin	Distortion of skin	Touch/Pressure
Ruffini's end organs	Skin	Distortion of skin	Touch/Pressure
Nerve endings	Carotid sinus, aortic arch. L. atrium	Change in arterial blood pressure	—
Nerve endings	R. atrium, great veins	Change in venous pressure	—
Stretch receptors	Muscle spindle	Length of muscle	—
Nerve endings	Joints	Movement of limb	Position
Thermoreceptors			
Nerve endings	Skin	Drop in temperature	Cold
Nerve endings	Skin	Rise in temperature	Warmth
Nociceptors			
Nerve endings	Most tissues	Various	Pain

Nature of receptor function

There are basically two types of receptor, or sense organ. The first type employs a particular type of non-neural element, often a modified epithelial cell, which responds to the stimulus and then generates a pattern of action potentials in an associated nerve fibre. The second type consists of a nerve ending which can respond directly to the stimulus.

Stimulation of either type of receptor leads to the development of a graded potential, known as the generator potential (see *Graded potentials* in Ch. 1), whose amplitude is dependent upon the strength of the stimulus. This relationship is not linear, however, since the amplitude rises rapidly at first but then by steadily smaller amounts until a maximum is reached. Thus, above a certain point, increasing the intensity of a stimulus will have no further effect.

The method by which a generator potential is induced varies with the type of receptor. It may be due to physical deformation of the receptor membrane, e.g. touch receptors; combination with a chemical substance, e.g. taste buds; alterations in the chemical balance within the cell, e.g. rods and cones; or change in the rates of certain chemical reactions, e.g. temperature receptors. Once the generator potential rises above the threshold potential, then it will initiate action potentials in the nerve fibre. If the receptor cell is a non-neural type, then the stimulus generates a receptor potential in the receptor cell, and the generator potential is only that part of the receptor potential which is transferred to the nerve endings. In the case of the nerve-ending type of receptors, the terms receptor potential and generator potential are synonymous. In both cases it is the generator potential which initiates action potentials in the nerve fibres.

Generally, the size of the generator potential determines the number of action potentials which arise in the sensory nerve fibre. A large generator potential will give rise to more action potentials than a small one. Thus increase in the size of a stimulus results in an increase in the number (frequency) of impulses going to the brain. A larger stimulus may also increase the number of neurones firing from a given area.

Adaptation of receptors

If a receptor is stimulated continuously for a long or in some cases quite a short period of time, the frequency of impulses in the accompanying nerve fibre reduces and may cease completely. The process is known as adaptation. This may be due to an adaptation of the receptor itself as, for example, in the photoreceptive cells of the retina which show an overall reduction in their internal concentrations of photosensitive chemicals on prolonged exposure to light. Alternatively, it may be due to the change in the responsiveness of the nerve fibres as they accommodate to continual stimulation.

Different receptors adapt at different rates. Pacinian corpuscles, for example, are skin receptors which are sensitive to pressure and they adapt so quickly that they stop firing in a fraction of a second. Such receptors provide information about sudden changes in the body and are therefore known as rate receptors.

Other receptors such as muscle spindles adapt very slowly and some never stop firing as long as a stimulus is maintained. Such tonic receptors include baroreceptors, chemoreceptors, Golgi tendon organs, pain receptors and otolith organs. Such receptors constantly appraise the brain of their local environment.

SOMATIC SENSORY PATHWAYS

Those modalities which are associated with general body sensations, such as touch, pressure, vibration, pain and temperature may be grouped together as the 'somatic senses'. Impulses are carried from the receptors to the spinal cord and/or brain along sensory or afferent nerve fibres. Sensory pathways between the tissues and the cerebral cortex of the brain generally consist of three consecutive neurones, described as first, second and third order. First order neurones originate in the tissues and end within the spinal cord or brain. They are pseudounipolar cells, which have their cell bodies in the dorsal root ganglia. Second order neurones lie within the CNS and cross over from one side to the other. Third order neurones connect the thalamus to the sensory cortex of the same side.

Further details of specific sensory pathways are given below as well as in *Ascending pathways* (pp 253 and 254).

Touch, pressure and vibration

There are many receptors which are able to respond to all three stimuli, although it is generally observed that touch is detected by superficial receptors, pressure by deep-lying receptors and vibration by rapidly adapting receptors.

Sense organs consist of either free nerve endings or encapsulated nerve ending (Fig. 7.1). Free endings are found everywhere in the skin and can detect light touch and pressure. Similar endings are wrapped around the bases of the hairs on the skin so that disturbance of the hair gives rise to the sensation of touch. A rather specialized type of free nerve ending is the Merkel's disc, the expanded end of a

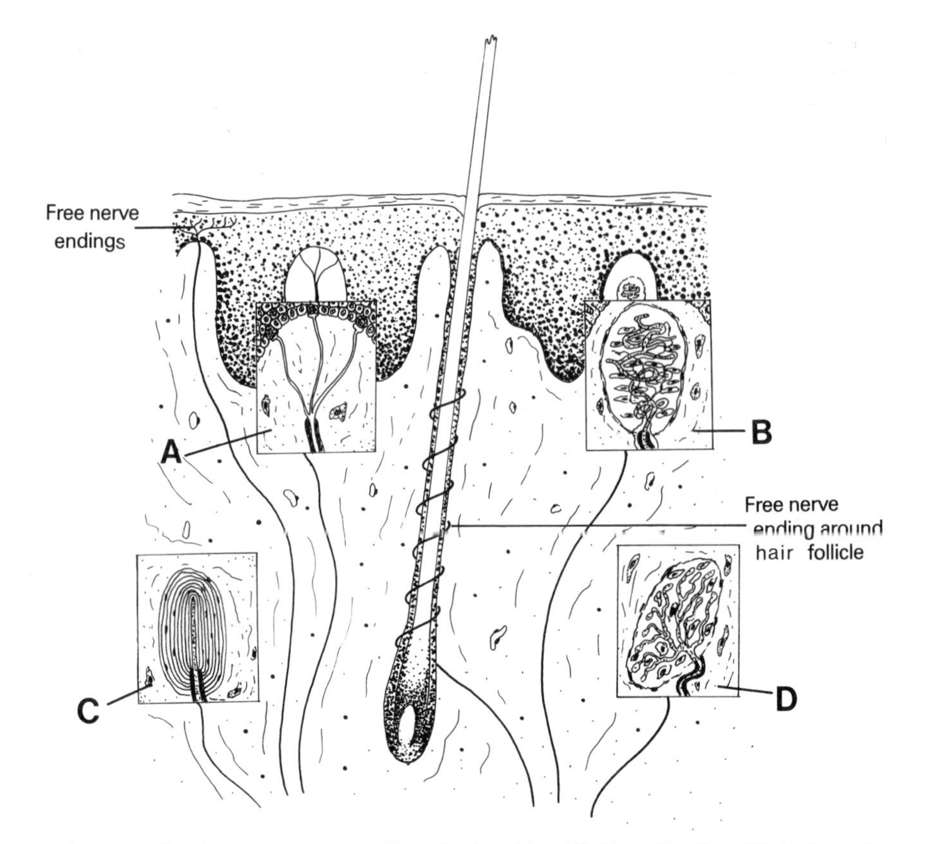

Fig. 7.1 Free and encapsulated sensory nerve endings in the skin: (A) Merkel's discs (B) Meissner's corpuscle; (C) Pacinian corpuscle; (D) Ruffini's organ.

sensory fibre which is closely applied to a specialized (Merkel) cell in the epidermis. Several of these structures may aggregate to form a 'touch-dome'.

Encapsulated endings include Meissner's corpuscles. These are fairly rapidly adapting receptors found in areas of skin where the sense of touch is highly developed, e.g. fingertips, lips. Ruffini's organs consist of branched nerve endings enclosed within a simple capsule of connective tissue. They are found in the deeper layers of the skin and also joint capsules and are slowly adapting receptors. It has been found that they supply a continuous stream of information to the brain.

A group of rapidly adapting receptors, which can detect vibration, are the Pacinian corpuscles. These consist of a nerve ending surrounded by a multilayer of concentrically arranged supporting cells. They are found beneath the skin and also in deeper tissues, as for example the connective tissue of the joints. Their rapid action enables them to provide information to the brain concerning the rates of movement of different parts of the body and therefore enable such complex movements as running to be co-ordinated.

The more elaborate receptors generally give rise to impulses which pass rapidly along large diameter fibres (Fig. 7.2). Free nerve endings, on the other hand, employ smaller diameter fibres which transmit at much slower rates.

All of these receptors respond to some sort of physical pressure, either steady or intermittent. It is probable that they all react in the same way, as their membrane potential is altered as a result of physical deformation of the cell surface.

Impulses associated with light touch, two point discrimination and vibration travel to the

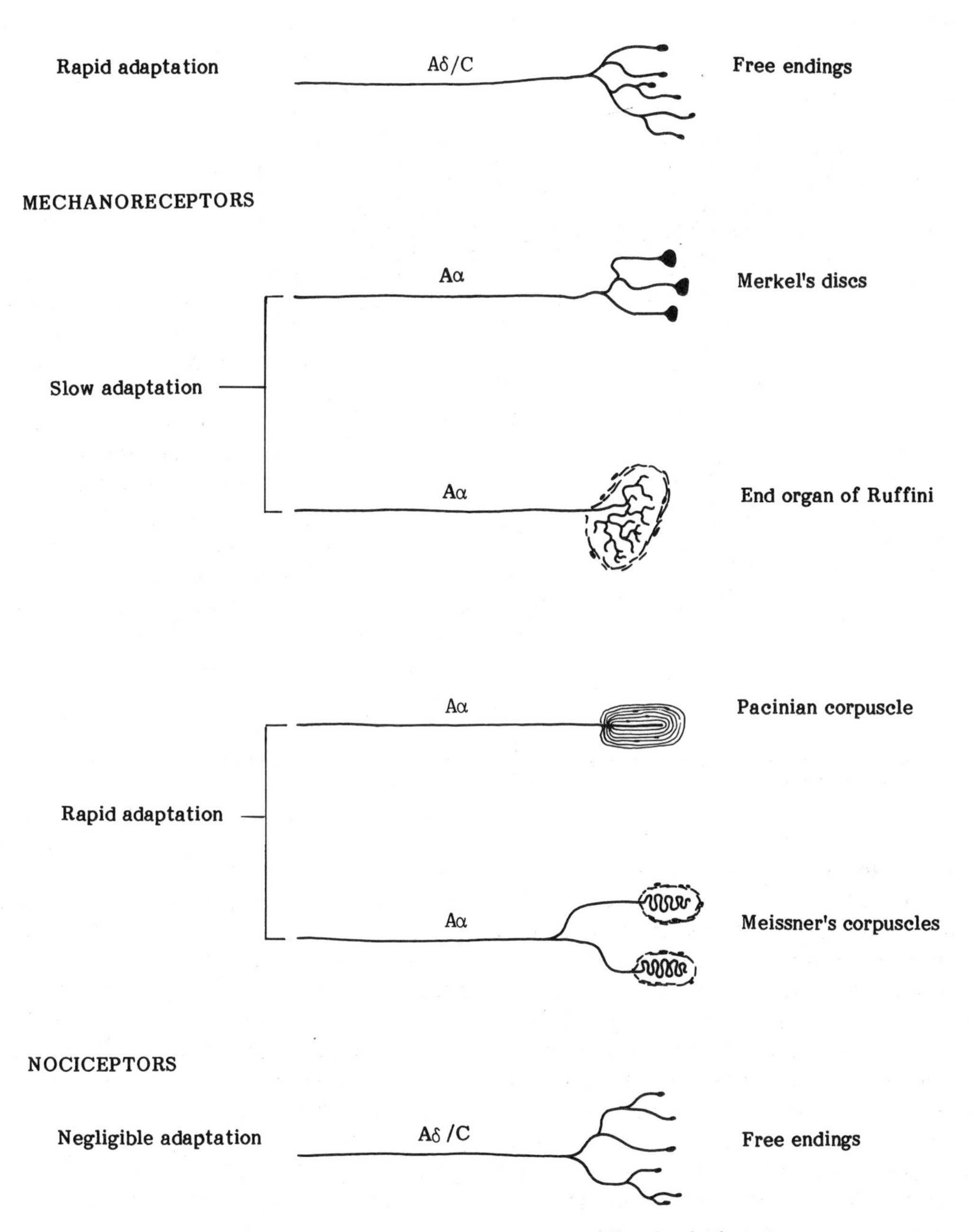

Fig. 7.2 Sensory receptors in the skin, their nerve fibres and adaptation rates.

cord and enter the pathways in the dorsal columns of the white matter. Since these pathways contain large diameter myelinated fibres, impulses are transmitted to the brain very rapidly. These pathways also maintain spatial localization in that the sensations induced in the brain can be pinpointed on the body surface with great accuracy. Crude touch and pressure sensations are transmitted along the ventral spinothalamic pathways. These are slow conducting fibres and there is poor spatial localization.

Temperature and pain

The sensations initiated by changes in temperature and painful stimuli are detected by free nerve endings in, for example, the skin. The receptors employed are histologically indistinguishable from those which initiate the sense of crude touch. They are, however, physiologically specific and an ending which responds to temperature change will not, for example, initiate the sense of touch.

The response to temperature is brought about by a change in the rate of chemical reactions within the fibre ending. Pain responses are usually brought about by any extreme stimulus, such as a change in temperature or mechanical deformation.

Impulses concerned with pain and temperature travel in the lateral spinothalamic pathways in the spinal cord. Impulses travel more slowly and with much less retention of spatial localization than those of the dorsal tracts. Thus temperature and pain are only poorly pinpointed on the body's surface.

THE AUTONOMIC NERVOUS SYSTEM

The autonomic nervous system is usually defined as a motor or efferent system, which connects the CNS with smooth muscle, cardiac muscle and some glands. It is a two-neurone pathway, so that there is a ganglion between the CNS and the effector. The ratio of preganglionic to postganglionic neurones is not one to one, however, and there are more postganglionic than preganglionic neurones. Preganglionic neurones are myelinated whereas postganglionic ones are unmyelinated. The autonomic nervous system is controlled by various parts of the brain, principally the hypothalamus but also the prefrontal cortex, the limbic lobe, the cerebellum and the reticular formation.

The autonomic nervous system is subdivided into two divisions, the parasympathetic and the sympathetic, which have clear anatomical and functional differences.

STRUCTURE OF THE AUTONOMIC NERVOUS SYSTEM

Parasympathetic system

The parasympathetic system has a craniosacral outflow from the CNS. Figure 7.3 shows the distribution of the principal parasympathetic pathways in cranial nerves III (oculomotor), VII (facial), IX (glossopharyngeal) and X (vagus). The only points of emergence from the spinal cord are in the pelvic splanchnic nerves arising from segments S2, S3 and S4. The cell bodies of the preganglionic neurones are found in the lateral columns of the grey matter in this part of the spinal cord.

Parasympathetic ganglia are generally located near to or lie within the innervated organ. In the cranial region of the system there are four ganglia, the ciliary, pterygopalatine, submandibular and otic, forming pathways to the eyes, lacrimal glands, salivary glands and the oral and nasal mucosae. The vagus nerves have a wide distribution, including the heart, respiratory tract and the gastrointestinal tract as far as the proximal colon. The ganglia are located within the organs themselves, e.g. the intramural plexuses in the gastrointestinal tract. The distal colon, rectum, kidney, bladder and sexual organs are supplied by the pelvic splanchic nerves. The ganglia either lie within the organs themselves, or there are very small ones found where the parasympathetic fibres meet branches of the sympathetic pelvic plexuses.

Because of the position of the ganglia, it can be seen that the parasympathetic preganglionic neurones are relatively long compared with the postganglionic neurones.

Sympathetic system

Reference to Figure 7.3 will show that the sympathetic pathways emerge from the thoracic and upper lumbar segments (L1 and L2 and sometimes L3) of the spinal cord. The system therefore has a thoracolumbar outflow.

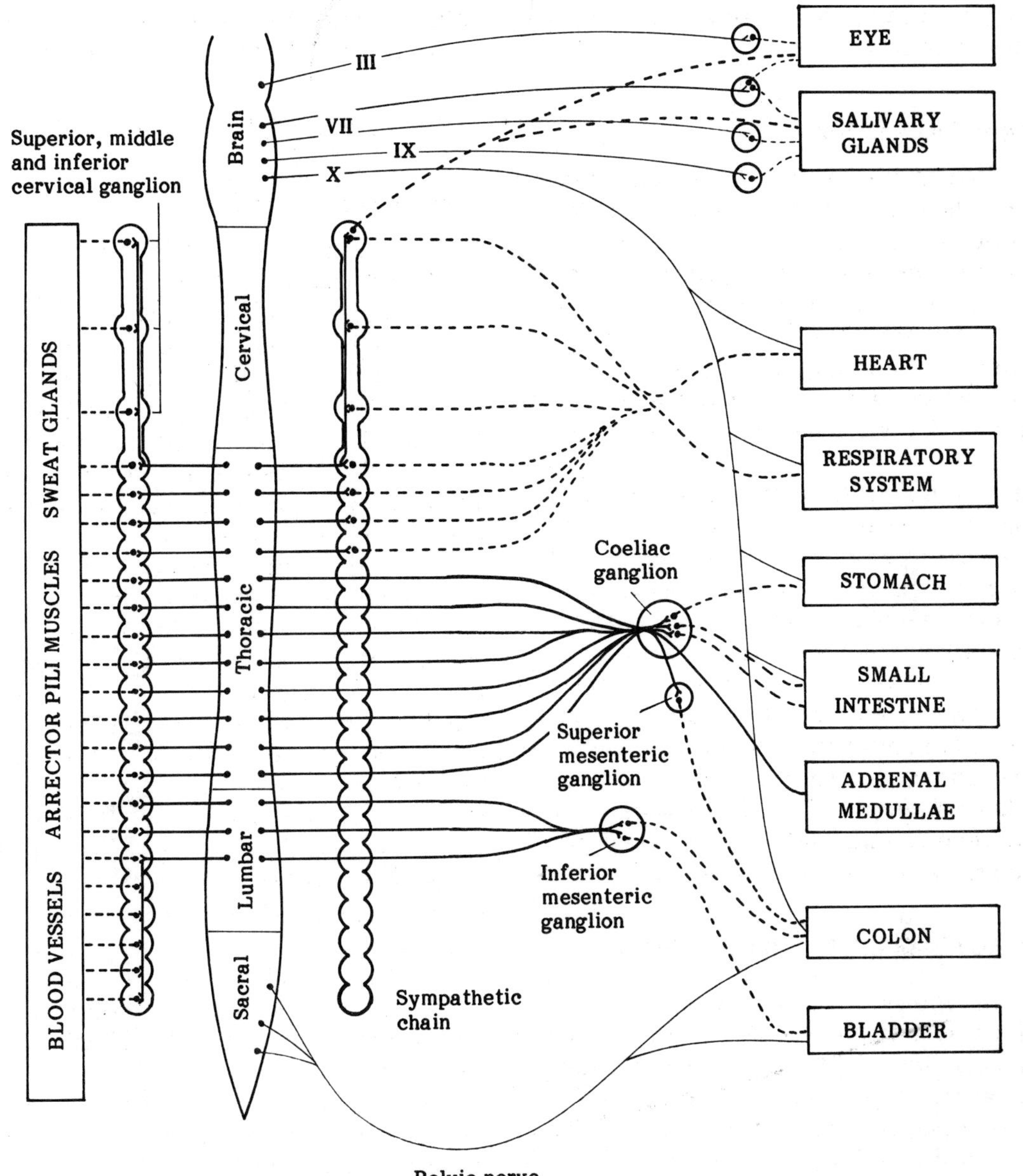

Fig. 7.3 Plan of the autonomic nervous system. The system is bilaterally symmetrical, but the innervation of some parts of the body is shown only on the left-hand side, others only on the right-hand side. Postganglionic neurones are shown by dotted lines.

Sympathetic ganglia are arranged as two interconnected chains immediately outside the spinal cord. Each chain, or trunk, consists of 22 or 23 ganglia extending alongside the whole length of the spinal cord (not just the thoracolumbar sections). Another group, the collateral or prevertebral ganglia, are found at various points along the outside of the abdominal aorta. Their names reflect their positions, so that the coeliac ganglion (strictly two ganglia) is at the junction of the aorta and the coeliac artery, while the superior mesenteric and inferior mesenteric ganglia lie adjacent to their respective arteries.

The preganglionic sympathetic neurones have their cell bodies in the lateral columns of the grey matter in the thoracolumbar region of the spinal cord. The axons leave in the ventral roots of the spinal nerves and then enter the sympathetic chains in the white rami communicantes (Fig. 7.4). The latter are so-named because they contain myelinated axons. Thereafter, the preganglionic sympathetic fibres may:

a. synapse immediately in a ganglion of the sympathetic chain
b. pass up or down the chain before synapsing, or
c. not synapse in the chain, but pass through it into a splanchnic nerve to a collateral ganglion. This is the pathway to the gastrointestinal tract.

The postganglionic neurones also follow a variety of routes. They may:

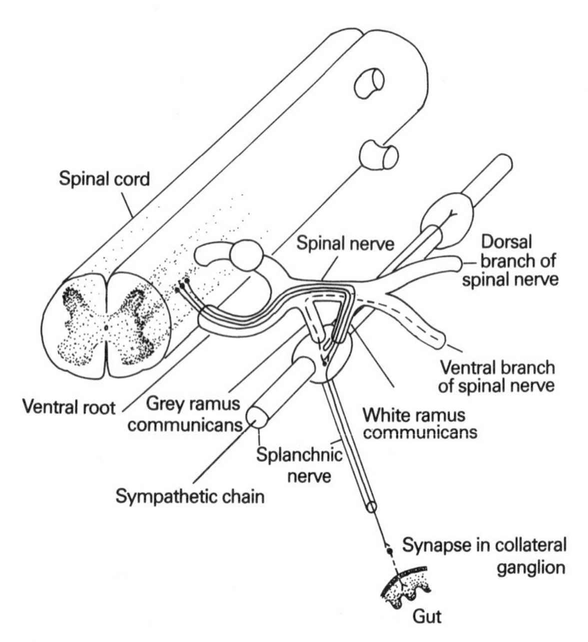

Fig. 7.4 Sympathetic pathways from the spinal cord. Preganglionic fibres leave the cord in the ventral root of the spinal nerve and may then either (a) synapse immediately within a ganglion of the sympathetic chain (in this case the post-ganglionic fibre passes back into the spinal nerve through the grey ramus communicans); (b) pass up or down the chain before synapsing; (c) not synapse in the chain at all, but instead pass through it and out into a splanchnic nerve and synapse in a collateral ganglion.

d. pass back to the spinal nerve in a grey ramus communicans (postganglionic neurones are unmyelinated and so appear grey), and then travel in the dorsal or ventral branches of the spinal nerves; this is the route taken to blood vessels, sweat glands and hairs
e. be distributed alongside blood vessels in plexuses, or
f. ascend or descend the sympathetic trúnk before leaving, as in d or e.

As sympathetic ganglia lie close to the CNS, the preganglionic fibres are relatively short compared with the postganglionic fibres. This contrasts with the relative proportions of the two types of parasympathetic neurones.

It many be noted from Figure 7.3 that the distribution of the sympathetic system is more widespread than the parasympathetic. Sweat glands, arrector pili muscles and most blood vessels receive only sympathetic innervation. This is also true of the adrenal medullae, which are supplied by preganglionic sympathetic fibres. Those organs which receive fibres from both sympathetic and parasympathetic divisions are said to have dual innervation.

FUNCTIONS OF THE AUTONOMIC NERVOUS SYSTEM

Neurotransmitters

The neurotransmitter released by the preganglionic fibres in both parasympathetic and sympathetic nervous systems is acetylcholine. The transmitter–receptor interaction is excitatory, so that autonomic ganglia are collections of excitatory synapses (see Ch. 1). However, at the neuro-effector junctions, the neurotransmitters differ in the two divisions. In the parasympathetic system, the postganglionic fibres release acetylcholine (cholinergic neurones), whereas in the sympathetic system noradrenaline is employed (adrenergic or strictly noradrenergic neurones). Sweat glands and skeletal muscle arterioles are exceptions, however, since they are supplied by sympathetic postganglionic fibres which secrete acetylcholine. As usual at synapses and neuro-effector junctions, the neurotransmitters are rapidly removed by enzymes (see *Neurotransmitters* in Ch 1).

The sympathetic and parasympathetic systems exert opposing infuences upon their target organs, and the fact that the two divisions release different chemicals into the tissue largely emplains the different responses. The nature of the tissue receptors, however, also influences whether individual responses are excitatory or inhibitory.

Parasympathetic system

The effects of the parasympathetic system are generally associated with inactivityy and the building up of food reserves by digestion. Someone asleep after a big meal demonstrates a dominant parasympathetic nervous system.

Although the vagus nerves have extensive innervation and gastrointestinal reflexes often affect many sites at once (e.g. the cephalic phase of the control of gastric secretion also stimulates the flow of pancreatic juice and bile, and gall bladder contraction), other parts of the system have discrete innervation and therefore local reflexes, e.g. lacrimation, pupillary reflexes.

Table 7.2 shows that at sites where the sympathetic system causes changes appropriate for exercise, the parasympathetic system has the opposite effect. On the other hand, the parasympathetic system generally has a stimulating effect on the digestive system, whereas the sympathetic system does not.

It may be noted that most blood vessels are not controlled by the parasympathetic system, so that vasodilation is usually caused by reduced vasomotor tone. Although the sympathetic system is generally associated with fear, there are occasions when the parasympathetic system initiates changes in fearful or anxious situations. Fainting is caused by reduced cerebral blood flow. If there is a massive parasympathetic discharge to the heart via the vagus nerves (a vasovagal attack) then the heart will be slowed and weakened to such a degree that cerebral blood flow is reduced sufficiently to cause fainting. Psychogenic diarrhoea is another example of high parasympathetic activity caused by emotional disturbance. These examples seem to be a reaction to extreme sympathetic stimulation.

Sympathetic system

This system generally causes physiological changes associated with physical exercise such as stimulating the heart, raising blood pressure, releasing adrenaline. It is also stimulated when one is frightened or anxious, causing changes such as dilated pupils, gooseflesh (skin hairs being raised). W.B. Cannon suggested in the 1920s that the sympathetic system was activated in states of emergency in 'fright, fight or flight' and that there was generalized activity in the system, the 'sympathetic discharge'.

It must be appreciated, however, that the autonomic nervous system is important in maintaining basal levels of activity. So that, at rest, the heart is controlled by both divisions of the autonomic nervous system, even though the parasympathetic system is dominant. An increase in heart activity can be brought about, therefore, by a reduction in parasympathetic activity as well as by an increase in sympathetic activity.

Table 7.2 summarizes the effects of sympathetic stimulation at various named sites. It can be seen that the responses are generally associated with fear and/or physical exercise, and that the parasympathetic and sympathetic divisions generally have opposite effects. There are a few anomalies, however. With regard to pupil size, the sympathetic system causes dilation, whereas the parasympathetic system causes constriction. The two effects, however, are not caused by the same set of muscles. Pupillary dilation is caused by contraction of the radial muscles in the iris, whereas pupillary constriction is caused by contraction of the circular muscles in the iris. Both divisions stimulate salivation, albeit with different results. Parasympathetic stimulation causes a copious, watery secretion whilst sympathetic stimulation brings about the

Table 7.2 Effects of sympathetic and parasympathetic activity

Site	Sympathetic activity	Parasympathetic activity
Eye		
Iris radial muscle	Contraction (pupil dilation) (α)	—
Iris circular muscle	—	Contraction (pupil constriction)
Ciliary muscle	Relaxation (far vision) (β)	Contraction (near vision)
Salivary glands	Scant, viscous secretion Vasoconstriction (α)	Copious, watery secretion
Heart	Increased rate (β) Increased force (β)	Reduced rate Reduced force
Arterioles		
General	Vasoconstriction (α)	—
Skeletal muscle	Vasoconstriction (α) Vasodilation (β) Vasodilation (cholinergic)	— —
Coronary	Vasoconstriction (α) Vasodilation (β)	Vasodilation
Veins	Venoconstriction (α)	
Bronchi	Bronchodilation (β)	Bronchoconstriction
Gastrointestinal tract		
Peristalsis	Decreased (α, β)	Increased
Sphincters	Constriction (α)	Relaxation
Secretion	Reduced	Increased
Gall bladder	Relaxation	Contraction
Bladder		
Detrusor	Relaxation	Contraction
Trigone and sphincter	Contraction (α)	Relaxation
Skin		
Sweat glands	Secretion (cholinergic)	—
Arrector pili muscles	Contraction	—
Lacrimal glands	—	Secretion
Penis	Ejaculation	Erection
Adrenal medullae	Secretion (Cholinergic)	—
Pancreas (Exocrine)	Decreased secretion (α)	Increased secretion
Adipose tissue	Lipolysis (β)	—
Liver	Glycogenolysis (β) Bile secretion	— —

production of small amounts of very viscous saliva.

The sympathetic system is dominant during physical exercise and its activity is manifested particularly in the responses of the cardiovascular system. The heart rate is increased as is stroke volume, both of which increase cardiac output and thus systolic arterial blood pressure. There is generalized vasoconstriction, but the blood vessels in skeletal muscle and the coronary blood vessels dilate. Local factors play an important part in vasodilation at these sites too (see Ch. 4). If the vasodilation in skeletal muscles exceeds the vasoconstriction elsewhere, then peripheral resistance will fall and so will diastolic pressure. Sympathetic stimulation also causes venoconstriction and thus increases venous return during exercise.

The sympathetic system has a generally inhibitory effect on the gastrointestinal tract so that peristalsis is inhibited and the sphincters constrict. Activity is further reduced by vasoconstriction which reduces the blood supply.

The sympathetic system innervates the

adrenal medullae. The preganglionic neurones release acetylcholine in the glands, which respond by releasing adrenaline and noradrenaline (in a ratio of about four to one). The glands, therefore, are analagous to postganglionic sympathetic neurones. Once noradrenaline and adrenaline are released into the blood, they potentiate and prolong the effects of sympathetic stimulation. The hormones are said to have a sympathomimetic effect.

The two hormones, although similar chemically, do not have identical effects at particular sites. The differences may be explained in terms of the tissue receptors present. Adrenergic receptors may be subdivided primarily into α-and β- receptors. Noradrenaline stimulates mainly α-receptors, whereas adrenaline stimulates both α- and β-receptors. Table 7.2 shows some of the effects of these receptors. It may be noted that the β-receptors (to which adrenaline attaches) are generally associated with smooth muscle relaxation, e.g. vasodilation in skeletal muscle and coronary arterioles, intestinal relaxation and bronchodilation. β- Receptor stimulation is also associated with metabolic changes such as glycogenolysis and lipolysis both of which increase the nutrients available for active muscles.

REGULATION OF BODY TEMPERATURE

The internal temperature of the body, the 'core' temperature, is maintained within narrow limits, irrespective of atmospheric temperature, at about 37°C. The actual core temperature does not normally vary by more than 1°C, although rectal and oral temperatures which are conventionally measured and taken to approximate to that of the core have a range of about two degrees. Prolonged exercise and/or extremes of atmospheric temperature may cause variations outside these limits.

Heat is continually being produced by the body as a consequence of metabolic activity. As a result, temperature regulation is primarily an operation which employs mechanisms for heat loss.

Physical mechanisms of heat loss

Heat is lost primarily from the surface of the body. About 60% is lost by radiation (infrared electromagnetic waves). A much smaller percentage may be lost by conduction, either to surfaces with which the body surface is in contact or to the layer of air next to the body. As air molecules become excited by the heat energy from the body's surface, convection currents are set up which carry the heat away. Evaporation of water from the skin also carries heat away. This 'insensible' loss of water cannot be controlled and is due to the fact that the skin is not completely waterproof and some water diffuses through it. 'Sensible' loss of water by sweating is, on the other hand, an important mechanism by which heat loss can be regulated.

As long as the ambient temperature is below that of the body temperature, heat loss should present no problems as radiation provides an efficient mechanism. When ambient temperature rises above body temperature, or body temperature rises for some reason, evaporation will be required to cool the body. If humidity is also very high, then adequate heat loss in this way will not be possible and heat stroke may result.

Hypothalamic control of body temperature

The core temperature of the body is controlled by a temperature regulating 'centre' in the hypothalamus. This area receives information concerning the temperature of various parts of the body and sends out impulses to the structures which are able to modify heat loss (or gain). Heat-sensitive receptors in the pre-optic area of the hypothalamus are the prime temperature receptors, but others in skin and many internal structures also play a part. The hypothalamic temperature-regulating centre acts like a thermostat which initiates heat-losing activities should the body temperature start to rise, and heat-gaining and reduced heat-losing activities should it start to fall.

MECHANISMS COUNTERACTING A RISE IN BODY TEMPERATURE

If the temperature of the blood flowing into the hypothalamus starts to rise, impulses are generated which bring about the inhibition of the vasomotor centre. As a result sympathetic vasoconstrictor impulses to the arterioles in the skin are reduced in frequency and vasodilation occurs (see Ch. 4). The extra blood passing through the skin warms it, thereby increasing the temperature gradient between the body and its surroundings and heat is lost by radiation, conduction and convection.

The hypothalamus also initiates sweating via sympathetic nerves to sweat glands on all parts of the skin surface. The stimulated sweat glands then secrete fluid on to the skin surface. It is the evaporation of this fluid which carries heat away. It is estimated that as much as 1.5 litres of fluid can be lost per hour through sweating. This represents 3600 J (870 calories) of heat (1 calorie = 4.2 joules).

It is interesting to note that acclimatization to a hot climate leads to an increase in the amount of fluid which can be lost through sweating. Furthermore, the amount of sodium chloride lost in the sweat reduces, as increased aldosterone output causes sodium retention; aldosterone exerts a similar influence on sweat glands as it exerts on kidney tubules (see also Ch. 3).

MECHANISMS COUNTERACTING A FALL IN BODY TEMPERATURE

If the blood flowing into the hypothalamus is relatively cool, the vasomotor centre becomes excited and there is an increase in sympathetic vasoconstrictor impulses to the skin. The resultant increase in vasoconstriction will cut down heat loss by radiation, conduction and convection.

In certain parts of the body (hands, feet, lips, nose, ears) arteriovenous anastomoses connect the arterial vessels of the skin to deeper-lying venous plexuses. Constriction of these anastomoses prevents blood flow into the plexuses, which then become relatively empty and the skin cools (see also *Skin circulation* in Ch. 4). This cuts down heat loss from those areas. In addition, blood flowing from the extremities is diverted into veins (venae comitantes) which run close to the arteries taking blood out from the core. As a result, the blood entering the core is warmed and that passing to the extremities is cooled. In addition to these cardiovascular responses, sweating is also inhibited.

A minor consequence of cooling the body is that the hairs on the surface stand out. This is pilo-erection and is of importance to animals with a thick covering of hair which is able to trap a layer of warm air, but it has little or no significance in humans.

Most of the mechanisms employed by the body when it is cooled are directed towards increased heat production rather than decreased heat loss. When the hypothalamus is cooled, an inhibitory influence from the thermoregulatory area upon a hypothalamic 'shivering centre' is removed. In addition, impulses from peripheral 'cold receptors' have a stimulatory effect. The result of these activities is that impulses are generated which pass down the spinal cord to motor neurones at various levels and out to the skeletal muscles. These impulses bring about a rise in tension in the stimulated muscles. Following this initial phase, signals become spasmodic and the reflex contractions unco-ordinated and 'shivering' results. Shivering is an important source of heat when the body is cooled. The precise mechanism which governs shivering is not understood, however, although it is likely that the stretch reflex is involved in some way.

The hormones thyroxine and tri-iodothyronine are released when the body is cooled and stimulate metabolism to generate heat, although the effect is slow in onset and is of doubtful physiological importance. The general increase in sympathetic activity which occurs in cool conditions also results in the release of adrenaline and noradrenaline; these hormones also stimulate metabolism, an effect which is much more marked in children than in adults.

Fever (pyrexia)

The hypothalamic thermostat can be reset by the introduction of certain chemicals or pyrogens. These substances may be secreted by bacteria when the body is suffering from an infection. In this case the thermostat is reset at a higher level than normal and since the body temperature will, at that time, be lower than this new level, then the heat-gain mechanisms come into play. Vasoconstriction, inhibition of sweating and shivering occur which raise the bodyy temperature to the new level. If at the higher temperature the pyrogens are destroyed, then the thermostat is reset to a lower value and heat-loss mechanisms come into play. Vasodilation and profuse sweating cause cooling of the body back to the level of the reset thermostat.

Thus in a fever, a period of shivering and feeling cold (chills) is followed by a point at which the fever breaks and the patient becomes flushed and sweats profusely (crisis).

Hypothermia

If the body is exposed to extreme or prolonged cold and the core temperature drops by more than about 2°C, then the regulatory mechanisms are unlikely to be able to restore the normal core temperature. As body temperature falls, metabolic rate is reduced which means that the body cells produce very little heat; respiratory and cardiovascular activities reduce. At around 33°C, muscle rigidity replaces shivering and amnesia and mental confusion occur. By 30°C, tendon reflexes are lost, unconsciousness is likely and respiratory and heart rates become irregular.

THE BRAIN

This brief outline of brain function aims to identify some key structures and the major functions with which they are associated, largely as a reference section for the other chapters in this book.

Traditionally, neuro-anatomists and neurophysiologists have studied the brain with a view to locating specific functions in particular 'centres' or anatomical regions. Such ideas emanated from animal experiments in which parts of the brain were either electrically stimulated or, alternatively, removed or isolated, so that the resultant changes could be related to the physiological role of the area of the brain (e.g. the inspiratory and expiratory 'centres' in the medulla oblongata). Data from human subjects is relatively sparse and largely involves observations of altered functions resulting from either congenital malformations; physical traumas such as as cerebrovascular accident or road accident, or a brain tumour.

The relative crudity of animal experiments and the paucity of data on humans mean that much of the historical view of brain function is questionable. Additionally, the remarkable recovery made by some people who have suffered physical trauma, such as a cerebrovascular accident, suggests that the concept of discrete localization of function in the brain may not be accurate. Future textbooks on brain function will almost certainly be even more complex than those available at the present time.

It must be emphasized that this section of the book is merely an introduction to the brain and that detailed neuro-anatomy is not included. For this reason, the structure and function of each part are covered together.

Embryonic development

The central nervous system develops by invagination of the neural plate. The latter arises when a strip of ectoderm (surface layer) along the midline of the dorsal surface of the embryo increases in thickness. The sides form neural folds and by the end of the third week of embryonic life, the folds start to fuse to form the neural tube (Fig. 7.5), the precursor of the brain and spinal cord. At the head end, the neural folds enlarge prior to closing the neural tube, and the primitive brain thus formed has two transverse constrictions

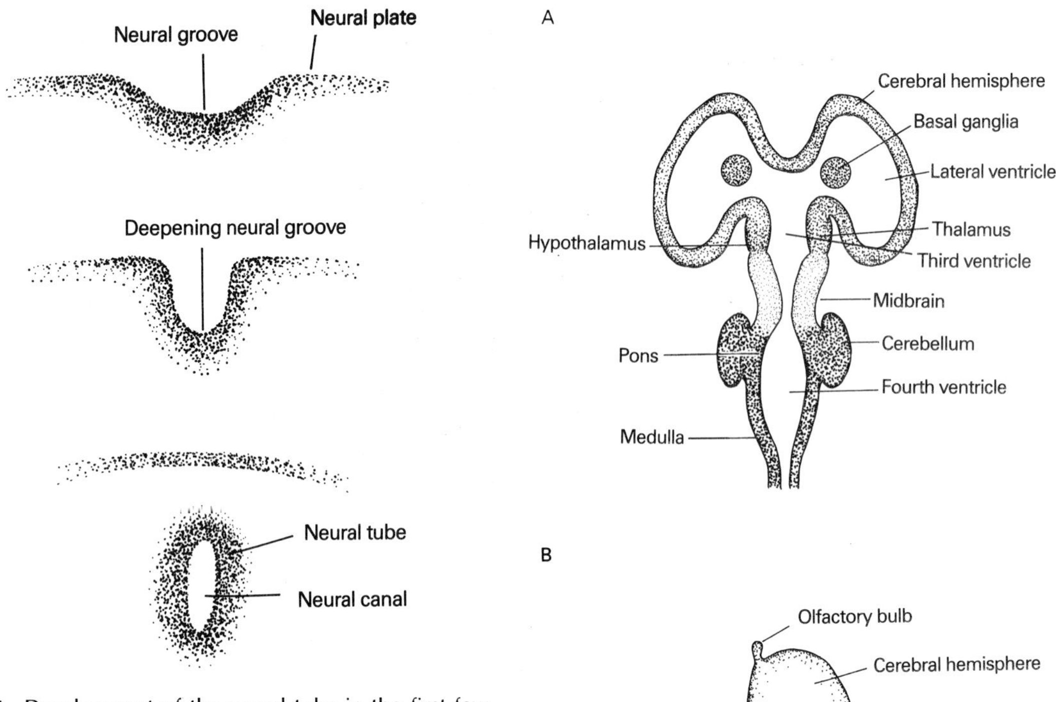

Fig. 7.5 Development of the neural tube in the first few weeks of embryonic life.

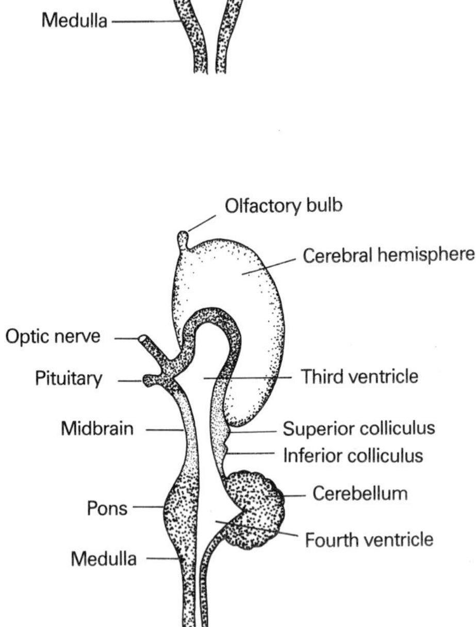

Fig. 7.7 Major areas of the brain in a four-month embryo: (A) horizontal section; (B) midline section.

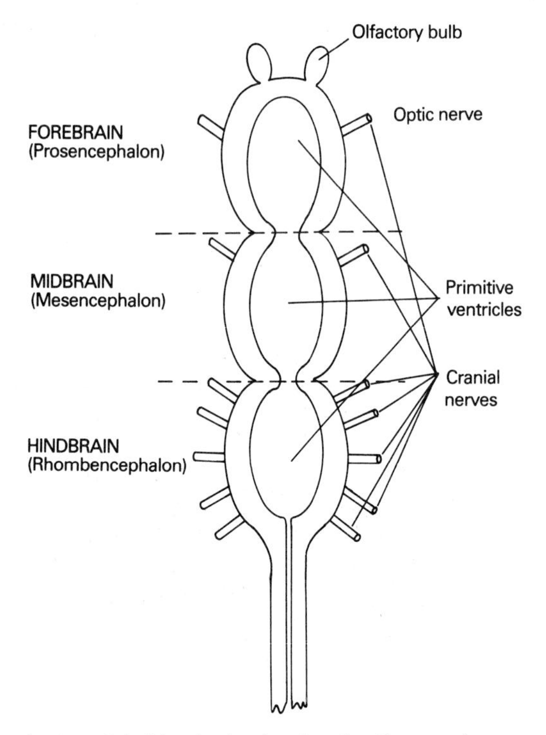

Fig. 7.6 Primitive brain showing the three main areas and the associated cranial nerves.

demarcating the three major divisions of the brain — the forebrain (prosencephalon), midbrain (mesencephalon) and hindbrain (rhombencephalon) (Fig. 7.6). These three areas are also known as cerebral vesicles at this stage. The lumen of the neural tube persists, eventually becoming the four ventricles of the brain. In the spinal cord it is reduced to a narrow cavity, the spinal canal.

The cerebral vesicles enlarge and change shape, although the forebrain develops more than the other two areas, and curls over

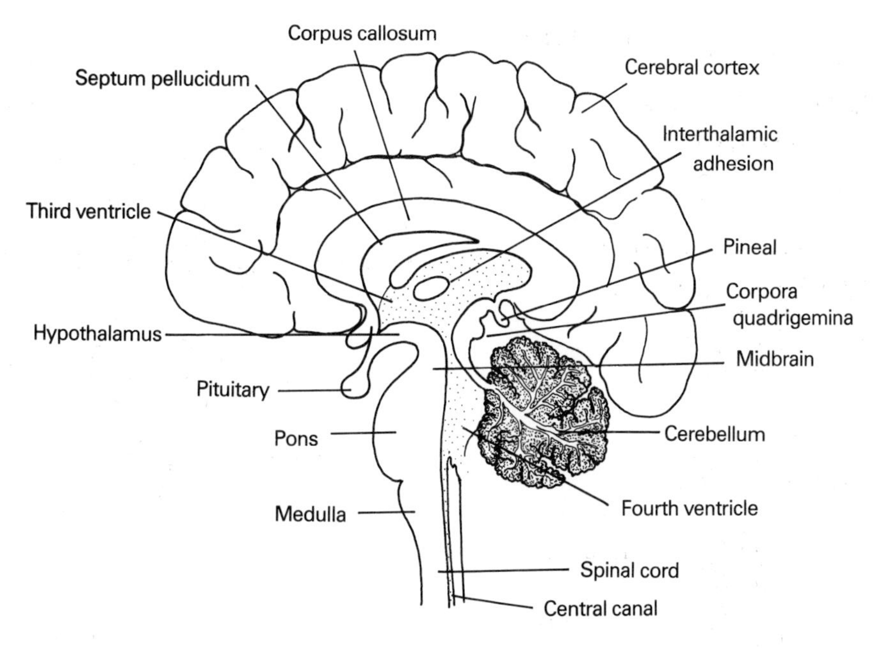

Fig. 7.8 Midline section of the adult brain.

ventrally to meet the hindbrain. By about four months, most of the major brain structures are evident (Fig. 7.7). As each brain structure is mentioned in the text, it may useful for the reader to locate the structure in a diagram of the embryonic brain before moving to the diagram of the adult brain in Figure 7.8.

STRUCTURE AND FUNCTION OF THE ADULT BRAIN

The brain weighs about 1.4 kg in the adult male. It is held in position within the skull by the meninges (see below). The brain is ovoid and slighty wider posteriorly than anteriorly. The major structure visible from above is the cerebrum, consisting of the right and left cerebral hemispheres, which form the bulk of the forebrain. When viewed from below, the base of the brain reveals the hindbrain (medulla, pons and cerebellum) and the midbrain as well as the cerebrum (Fig. 7.9).

Although the brain can be divided into the three principal areas, forebrain, midbrain and hindbrain, the term brain stem is used to describe the stalk-like section of the brain which connects the spinal cord with the cerebrum. It comprises the medulla, pons and midbrain.

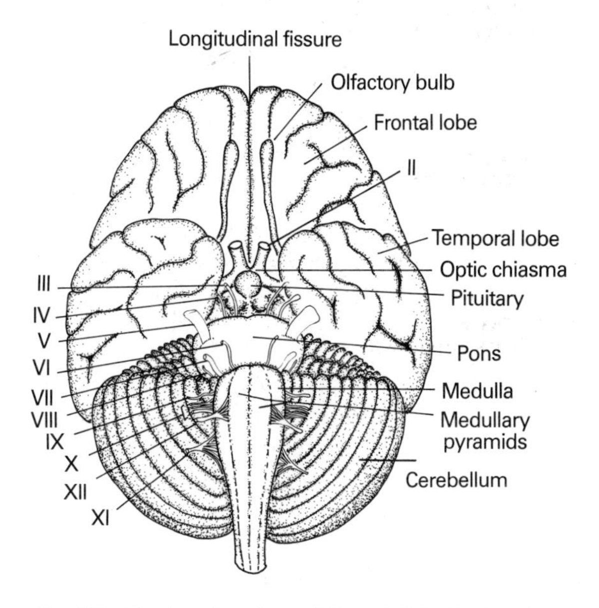

Fig. 7.9 The basal surface of the adult brain showing the major structures and numbered cranial nerves.

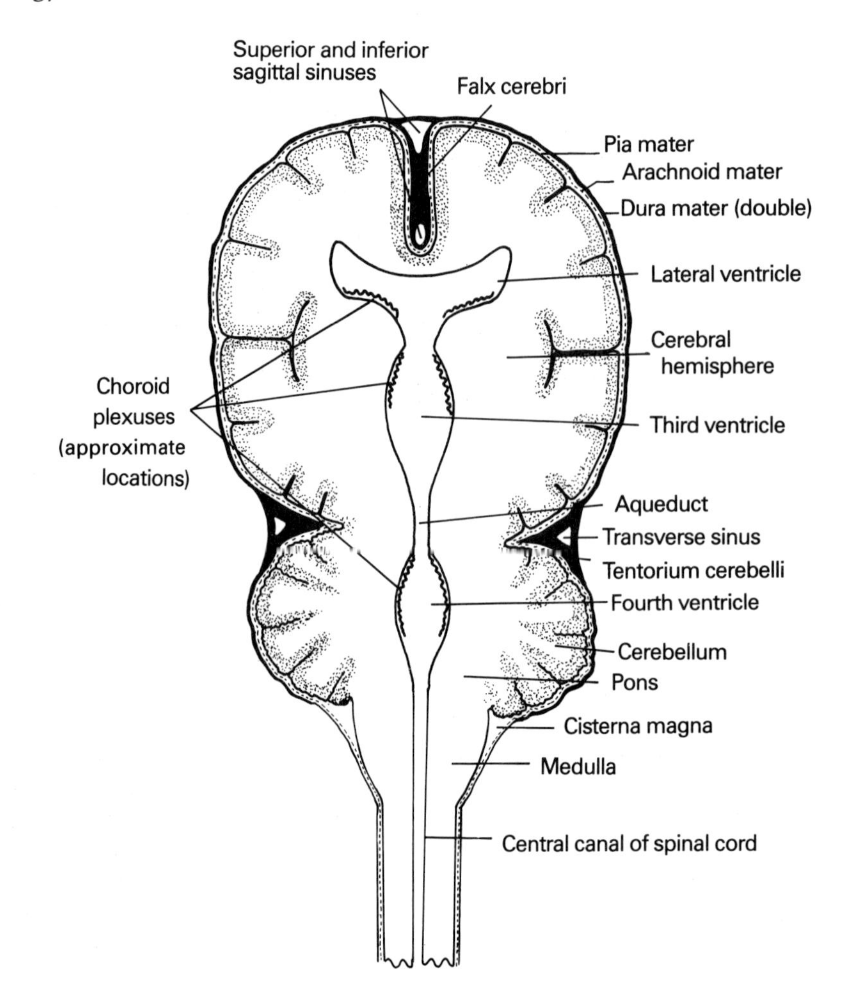

Fig. 7.10 The cerebral and spinal meninges and the locations of the choroid plexuses.

Cerebral meninges

Several membranes cover the brain and line the skull; from outer to inner they are the dura mater, arachnoid mater and pia mater (Fig. 7.10).

The dura mater is a double layer of dense fibrous tissue. The outermost layer is attached to the skull, forming an endosteum, so that there is no epidural space like that surrounding the spinal cord, where the dura mater is a single layer only. The two layers of dura mater separate at certain points where they contain venous sinuses. This gives rise to a structure known as the falx cerebri which runs along the longitudinal fissure between the two cerebral hemispheres and contains the superior and inferior sagittal sinuses.

An equivalent structure, the tentorium cerebelli, lies between the cerebellum and the cerebrum and contains the transverse sinus and the superior petrosal sinus on each side. Other projections of the inner dura mater form septa in the brain, the falx cerebelli and the diaphragma sellae.

The arachnoid mater is a more delicate structure than the dura mater. It is composed of simple squamous epithelium and connective tissue. It is separated from the dura mater by a potential space, the subdural space, and from the pia mater by the subarachnoid space,

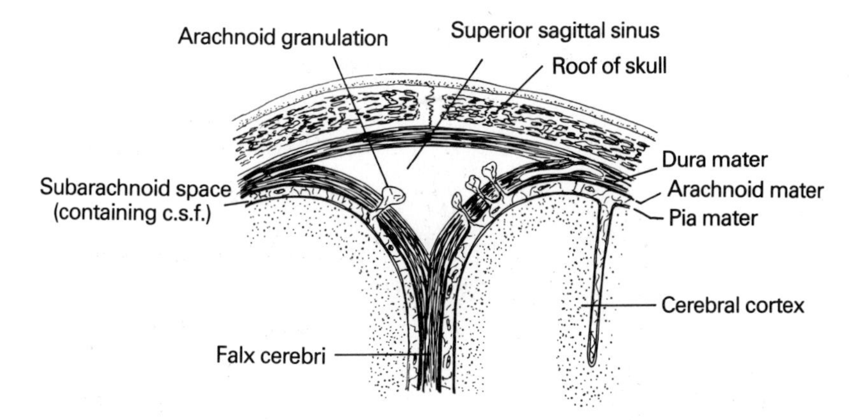

Fig. 7.11 Coronal section through the falx cerebri showing arachnoid granulations projecting into the superior sagittal sinus.

containing cerebrospinal fluid. Connective tissue strands connect the arachnoid and pia mater together. The archnoid mater projects into several of the sinuses as finger-like arachnoid villi, or as the larger arachnoid granulations (Fig. 7.11). Cerebrospinal fluid (c.s.f.) passes through these projections into the blood in the venous sinuses (see *Cerebrospinal fluid*). The cavity in the hindbrain, the IVth ventricle, has median and lateral apertures into the subarachnoid space.

The pia mater is a fine vascular membrane which closely covers the surface of the brain. The membrane invaginates the brain at certain points to join the lining of the ventricles, to form fringe-like structures, projecting into the ventricles. These are the choroid plexuses, highly vascular structures which give rise to c.s.f.

The forebrain

The forebrain (prosencephalon) comprises the telencephalon and the diencephalon. The telencephalon consists of the cerebrum which itself is made up of the two cerebral hemispheres containing the lateral ventricles (imcompletely separated by the septum pellucidum), and the rhinencephalon (olfactory areas) on each side. The diencephalon consists of the thalami (joined by the interthalamic adhesion) and the hypothalamus.

Each cerebral hemisphere has a surface layer of grey matter, the cerebral cortex, with white matter below. This is in contrast to the spinal cord where the white matter encloses a core of grey matter.

The white matter in each cerebral cortex contains tracts of fibres orientated in three principal directions: connecting upper and lower parts of the brain (projection fibres), connecting different parts of the same hemisphere (association fibres), and connecting the right and left hemispheres (commissural fibres).

The surface of the cerebrum is covered with ridges (gyri), separated by shallow furrows (sulci) or deeper fissures. The two cerebral hemispheres are incompletely separated by the longitudinal fissure. In the middle section, the fissure extends down to the corpus callosum, the wide band of commissural fibres which link the hemispheres.

During evolution, the telencephalon has enlarged and developed disproportionately compared with other areas of the brain, a process known as telencephalization, so that man has a proportionally larger telencephalon compared with other animals. The telencephalon is thought to be responsible for 'higher' levels of mental activity such as reasoning, intelligence, memory, many aspects of personality, interpretation of sensations, initiation of voluntary movement and the

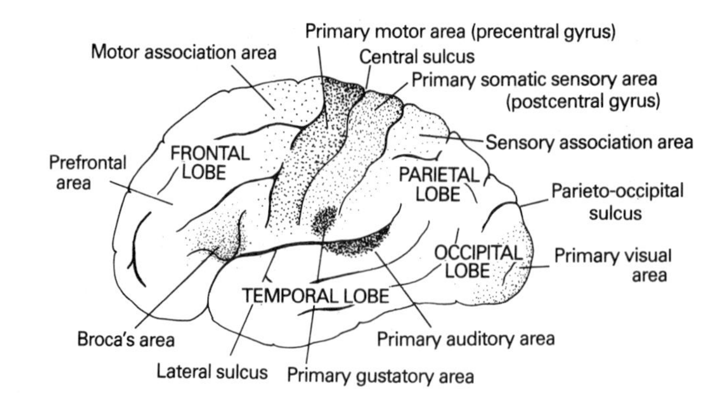

Fig. 7.12 Surface view of the left cerebral cortex showing the lobes and some functional areas.

moderation of reflex movements. Physiologists have often used the rather vague term 'higher centres' when referring to areas of the cerebral cortex.

The cells in the cerebral cortex are arranged in vertical columns (0.5 to 1 mm diameter) and it is usually considered that the columns are arranged in six layers, the relative thickness of each layer varying in different parts of the cortex. It has been suggested that layers I to IV receive sensory fibres, whereas layers V and VI contain the cell bodies of motor neurones.

Each cerebral cortex may be divided into four primary areas, the frontal, parietal, occipital and temporal lobes (Fig. 7.12).

Frontal lobes

Each frontal lobe lies anterior to the central sulcus and above the lateral sulcus. Anterior to the central sulcus is the precentral gyrus, the primary motor area. This controls voluntary movements in the opposite side of the body. Each part of the body is represented by a particular location along the gyrus, and although generally these representations are in approximately the right place, though inverted, they are not in anatomical proportion. Those parts of the body which are capable of complex muscular movements, such as the face and hands, are represented by a relatively large area of motor cortex, whereas areas with only limited movements, such as the trunk, cover a much smaller area of cortex. A body with its parts in proportion to their representation in the motor cortex is known as a motor homunculus (Fig. 7.13).

The principal nerve pathways controlling voluntary movements are the pyramidal tracts (corticonuclear and corticospinal) and they arise in the motor cortex. Some extrapyramidal tracts start in the cortex and then synapse in other areas of the brain before innervating skeletal muscle.

Anterior and parallel to the primary motor cortex is an area known as the motor association area, which appears to be involved in the co-ordination of complex movements (see Fig. 7.12). Broca's area controls the movements producing speech. In right-handed people it is found in the left inferior frontal gyrus, whereas in left-handed people it is in the corresponding position in the right cerebral hemisphere. Above Broca's area, a section of the cortex is concerned with voluntary eye movements; above this is an area concerned with head rotation, and another area is concerned with hand movements.

Anterior to the motor areas of each frontal lobe lie the prefrontal areas. For many years these were associated with higher intellectual functions, but more recently it is thought likely that other parts of the cortex are just as important in this respect.

The frontal lobes have nerve connections with the hypothalamus, thalamus, brain stem and spinal cord. Lesions in the frontal lobes

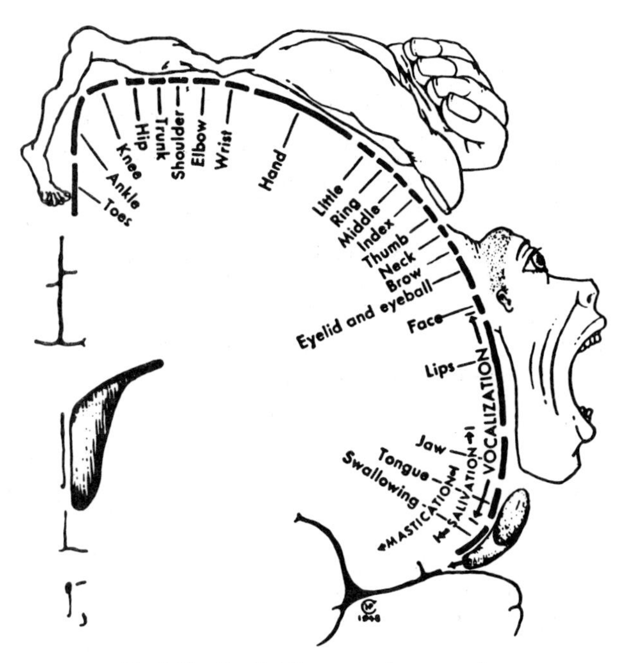

Fig. 7.13 Penfield's motor homunculus.

cause somatic motor and autonomic disturbances, as well as alterations of behaviour and character such as easily being distracted, having a reduced ability to perform complex sequences of thoughts, and altered social behaviour.

Parietal lobes

The parietal lobes lie behind the central sulcus and above the lateral sulcus on each side. The postcentral gyrus on each side is known as the somatic sensory (somaesthetic) area and it is concerned with the perception of somatic sensations. Like the motor cortex, the sensory cortex can be mapped out and the sensory homunculus is similar to the motor one (Fig. 7.14), with the body being represented upside down, and the left hemisphere receives impulses from the right side of the body and vice versa. Just above the lateral sulcus, at the bottom of the somatic sensory area, lies the primary gustatory area, responsible for the perception of taste.

Next to the primary sensory area is a sensory association area where somatic sensations are interpreted, for example by comparing them with a memory of a similar sensation or contrasting them with different sensations. Lesions in this area cause problems of interpreting sensations so that a person may not appreciate where the sensation is coming from.

Occipital lobes

Each occipital lobe lies behind the parieto-occipital sulcus and contains the visual areas. The primary visual cortex is found right at the back of the cortex on the walls of the calcarine sulcus. The sensory pathways ending here form the optic radiation from the lateral geniculate body on each side (see Fig. 8.14). There is spatial separation of parts of the retina in the visual cortex, and interpretation of the stimulus is aided by the visual association area found adjacent to the primary area. Images are represented upside down in the visual area.

Temporal lobes

Each temporal lobe lies below the lateral sulcus. Areas for auditory and olfactory sensations are located in the temporal lobes. The primary auditory area, which receives the

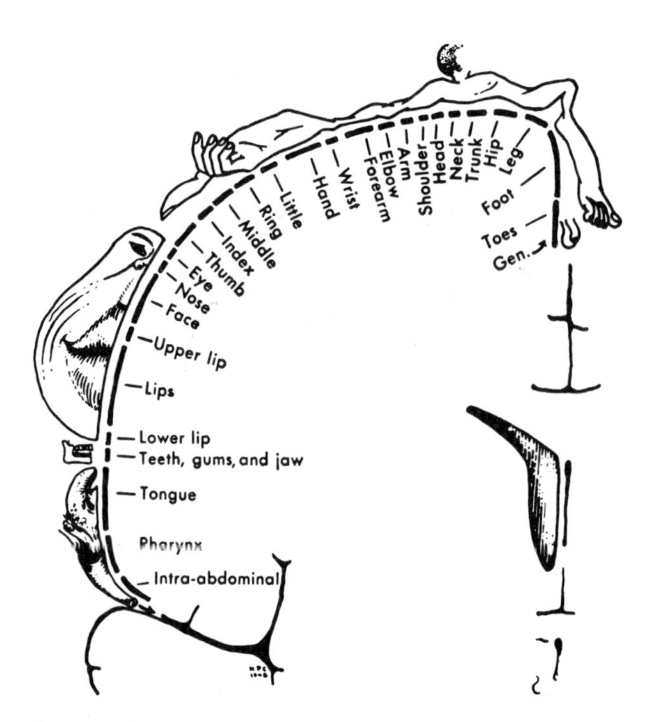

Fig. 7.14 Penfield's sensory homunculus. (Figures 7.13 and 7.14 reprinted with permission of Macmillan Publishing Company from Penfield W, Rasmussen T 1950 The cerebral cortex of man.)

auditory radiations, perceives sound and depends on the adjacent association or auditopsychic area for interpretation. Impulses from the vestibular apparatus are conveyed to areas which lie adjacent to those dealing with auditory information.

The somatic, visual and auditory association areas all converge on an area in the temporal lobe known by a variety of names including general interpretive area and Wernicke's area. This area has been regarded as of prime importance in intellectual functions, and in right-handed people the left side is dominant.

Cerebral asymmetry

Although the right and left cerebral hemispheres may appear to be structurally similar, detailed studies have shown some anatomical differences. Generally, in right-handed people the left occipital and parietal lobes are wider than the right, whereas the right frontal lobe is usually wider than the left.

Functionally too, there tend to be differences between the two hemispheres, but there are some individual exceptions to this. The left hemisphere is usually associated with language, both spoken and written, mathematical skills and reasoning. Damage to the left hemisphere usually causes psychological disturbance and depression. The right hemisphere is generally associated with non-verbal skills such as musical and artistic awareness, the perception of space and pattern and imagination. Damage to the right hemisphere often does not cause the emotional disturbances that are seen in left-sided damage.

Rhinencephalon

The rhinencephalon consists of those parts of the brain concerned with smell. In many animals this is a dominant feature of the brain, and in the early embryo in man, the olfactory bulbs are prominent (Fig. 7.6). In the adult brain, however, the bulbs are relatively small and consequently the sense of smell is relatively unimportant.

The olfactory bulbs are small, ovoid structures on the base of the brain on the orbital surface of the frontal lobes (Fig. 7.9). The olfactory tracts pass back from the bulbs and

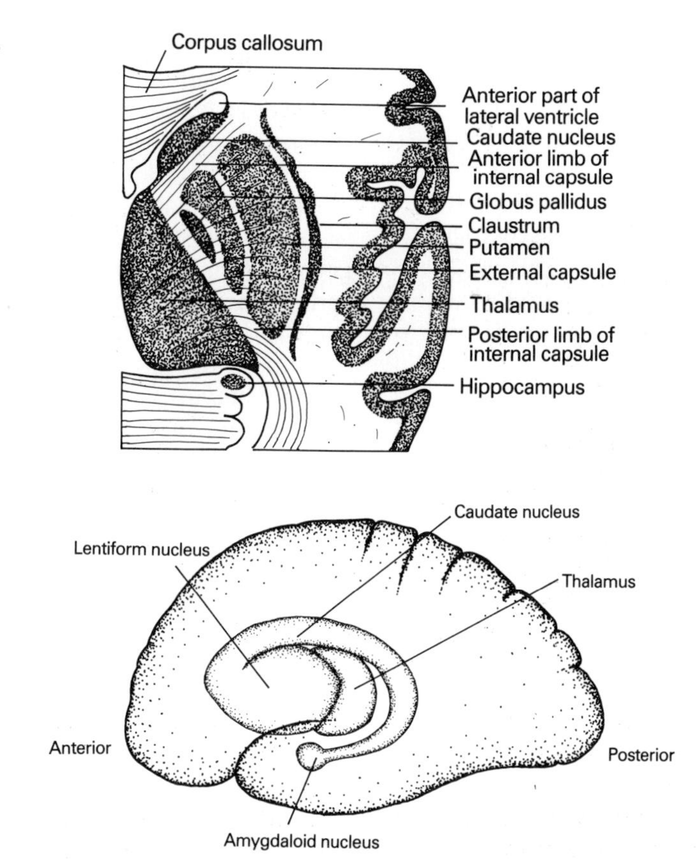

Fig. 7.15 (A) Coronal section through the brain to show the basal ganglia. (B) Three-dimensional representation of the basal ganglia.

then divide into medial and lateral striae (Fig. 8.32B). Other parts of the rhinencephalon include the anterior perforated substance, piriform area, hippocampal formation, fornix and habenular region.

Basal ganglia

The basal ganglia or nuclei are areas of grey matter embedded within the white matter of each cerebral hemisphere, near to the thalamus (Fig. 7.15). The basal ganglia include the corpus striatum which consists of the caudate and lentiform nuclei, the latter being subdivided into the putamen and the globus pallidus. The putamen and the caudate nucleus together constitute the neostriatum or striatum. The nuclei are surrounded or separated by broad bands of white matter, the internal capsule, formed by projection fibres.

The basal ganglia also include the amygdaloid nucleus and the claustrum, and some authorities further include the subthalamic nucleus and two areas in the midbrain, the substantia nigra and the red nucleus.

The basal ganglia play an important role in the control of voluntary motor activity. They have nerve connections with the motor cortical areas as well as the thalamus, which in turn has nerve connections with the cerebellum. All of these parts of the brain are concerned with motor activity. The ganglia are connected to each other, as well as other areas of the brain, and they give rise to some of the extrapyramidal pathways to skeletal muscle (pathways connecting the brain and skeletal muscle apart from the pyramidal tracts, see also *Spinal cord — descending pathways*).

As the motor areas of the brain appear to be interdependent and modify the activities of the other areas, it is difficult to ascribe specific functions to each component of the motor system. The basal ganglia do, however, appear to exert a generally inhibitory influence on muscle tone (via the nerve connections with the motor cortex and the lower brain stem). The corpus striatum has a role in the initiation of gross voluntary movements, which are normally refined by the motor cortex.

Damage to the basal ganglia can give rise to several motor disorders (depending on the precise location of the damage) including chorea, athetosis, hemiballismus and Parkinson's disease. The symptoms of Parkinson's disease, rigidity and akinesia, seem to be due to the loss of dopamine-secreting neurones in the basal ganglia. Since these neurones are inhibitory in nature their destruction results in increased muscle activity.

Thalami

The thalami are two oval masses of grey matter which form the lateral walls of the IIIrd ventricle (Fig. 7.16). Each thalamus is subdivided into a number of nuclei which are arranged in three major groups, anterior, medial and lateral. In addition, the medial and lateral geniculate bodies are sometimes included as part of the thalamus (Fig. 7.18).

All areas of the cerebral cortex have both afferent and efferent connections with the thalamus and the two areas function together as the thalamocortical system. All sensory pathways (except smell) synapse in the thalamus and then project to the sensory cortex. The auditory pathways relay in the medial geniculate bodies, the visual pathways relay in the lateral geniculate bodies and the general somatic senses, and taste, relay in the ventral posterior nuclei. Some sensations, such as pain, may reach consciousness at the thalamic level.

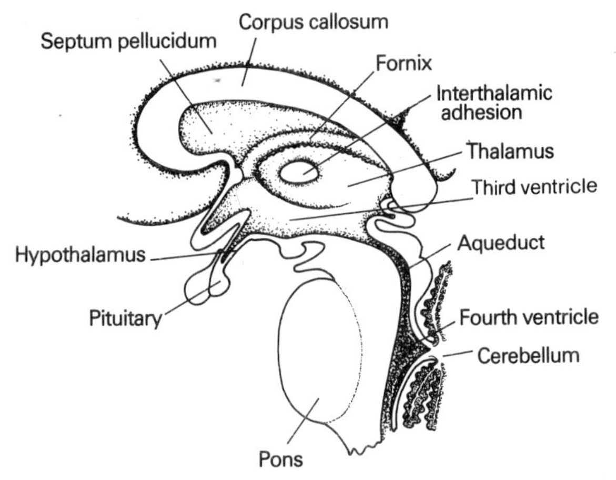

Fig. 7.16 Midline section of the brain in the region of the third ventricle. The surfaces of the thalamus and hypothalamus can be observed in the wall of the third ventricle.

The thalamus also plays a role in the control of somatic motor activity (see also *Basal ganglia*) and has nerve connections with the other motor areas, the motor cortex, cerebellum and basal ganglia.

The thalamus receives afferent neurones from the hypothalamus and brain stem nuclei as well as the spinal cord, basal ganglia, cerebellum and cerebral cortex. The anterior thalamic nucleus is associated with memory and emotions.

Hypothalamus

The hypothalamus, as its name suggests, lies beneath the thalamus, forming the lateral walls of the IIIrd ventricle (Fig. 7.16). Several nuclei have been differentiated within the hypothalamus, but generally they do not appear to have discrete functions. The pituitary gland is closely connected to the hypothalamus and indeed is sometimes regarded as part of it (certainly the neurohypophysis is embryologically derived from the hypothalamus; see *The pituitary gland*, Ch. 9). The mamillary bodies are also part of the hypothalamus.

The hypothalamus controls all the activities of the autonomic nervous system. The latter controls some glands, the heart, and structures containing smooth muscle, e.g. blood vessels, gastrointestinal tract and urinary tract (see *Autonomic nervous system*). The hypothalamus receives sensory inputs from the viscera and initiates appropriate responses by means of its nerve connections with the sympathetic and parasympathetic nervous systems. In many cases the pathway between the hypothalamus and the effectors includes secondary integrating 'centres' in the brain stem or spinal cord, e.g. cardiovascular and much digestive activity is mediated in the medulla oblongata and defaecation and micturition reflexes involve the sacral region of the spinal cord.

The nerve connections between the cerebral cortex and the hypothalamus provide a 'psychosomatic' link whereby emotional feelings may result in diverse bodily changes such as a rise in blood pressure, urgent defaecation and micturition and the release of adrenaline.

By means of its connection with the anterior pituitary gland, the hypothalamus controls the output of many hormones, including those from the gonads, adrenal cortices and thyroid, as well as growth hormone and prolactin from the anterior pituitary itself (see Ch. 9). Nerve cell bodies in the supraoptic and paraventricular nuclei in the hypothalamus synthesize the hormones ADH and oxytocin, respectively, which are subsequently released from the posterior pituitary.

The hypothalamus contains various sensory receptors including osmoreceptors, temperature receptors and glucoreceptors. Furthermore, it contains various 'centres' concerned with temperature regulation, feeding, satiety and thirst. In addition, the hypothalamus also appears to house a pacemaker, or biological clock, which generates circadian rhythms such as variations in hormone levels (e.g. cortisol) and body temperature at different times of the day.

Behavioural manifestations of emotional states such as pleasure, pain, rage, docility, affection, sexual feelings have been elicited in many animals by stimulating parts of the hypothalamus and adjacent structures, which are collectively called the limbic system. This includes the limbic lobe or cortex (the cingulate gyrus and the hippocampal gyrus), beneath which lie the hippocampus, the amygdala, anterior nuclei of the thalamus and the mammillary bodies (Fig. 7.17). The limbic system has been called the 'emotional brain'. By virtue of its nerve connections with the reticular activating system in the brain stem, the limbic system controls the general levels of alertness in the brain. The hippocampus seems to play a particular role in memory. Destruction of the hippocampi in man leads to an inability to learn anything new as well as some deficit in previously learned memories. It may be that the hippocampus is involved in translating short-term memory into long-term memory.

The midbrain

This section of the brain connects the pons with the diencephalon, is some 2.5 cm long and contains the cerebral aqueduct. The midbrain is essentially tubular, with two ventral elevations known as the cerebral

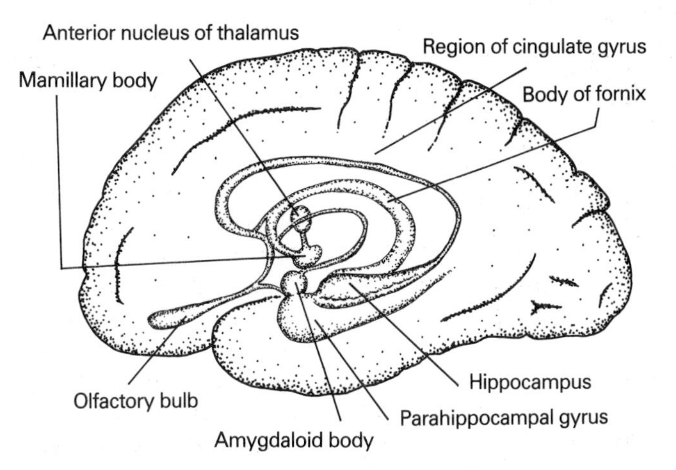

Fig. 7.17 Three-dimensional representation of the major components of the limbic system.

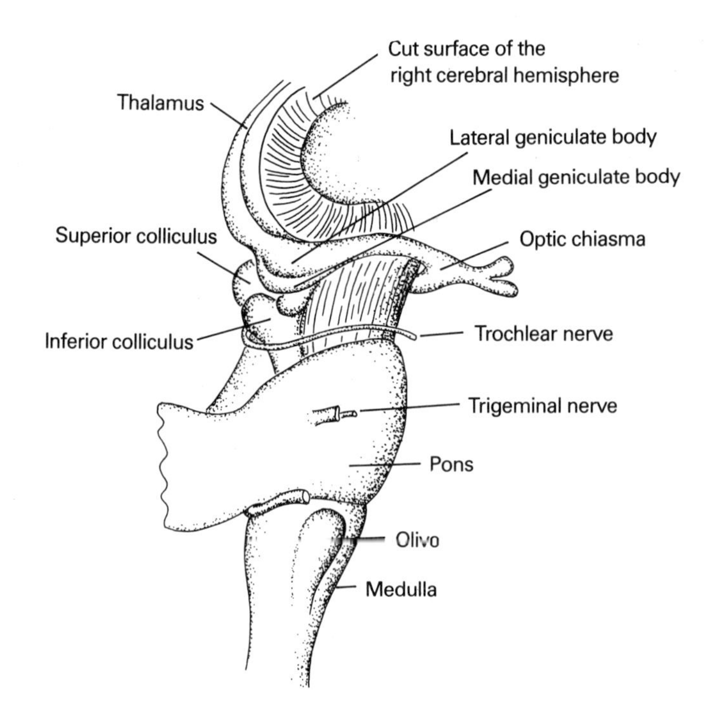

Fig. 7.18 Surface view of the midbrain.

peduncles, and four dorsal elevations comprising the tectum (Fig. 7.18).

The cerebral peduncles consist of ascending and descending fibre tracts. The tectum comprises the corpora quadrigemina made up of the superior and inferior colliculi. The two superior colliculi contain reflex pathways for eyeball movements as well as pupillary reflexes and accommodation. They receive the optic nerves (cranial nerves II), and the oculomotor nerves (III) and trochlea nerves (IV) originate here. Table 7.3 summarizes the locations of the cranial nerves and their principal functions. Figure 7.19 shows the location of cranial nerve nuclei in the brain. The ascending spinotectal and spinothalamic tracts from the spinal cord terminate in, and pass through, respectively, the superior colliculi. The tectospinal tract originates here.

The two inferior colliculi receive auditory stimuli and initiate head and trunk movements. They also receive vestibular stimuli which initiate eye and postural movements.

The midbrain contains the substantia nigra (see *Basal ganglia*) and the red nucleus which receives afferents from the cerebellum and the cerebrum and efferent fibres leave in the rubrospinal tract.

The medial leminiscus is a band of white fibres found in the medulla and pons as well as the midbrain, which carries sensory fibres concerned with touch, proprioception and vibrations to the thalamus.

The hindbrain

The hindbrain comprises the pons varolii, the medulla oblongata and the cerebellum.

Pons varolii

The pons is about 2.5 cm long and lies anterior to the cerebellum and above the medulla (Fig. 7.8). It contains large numbers of fibres orientated both transversely and longitudinally. The transverse fibres connect the pons with the cerebellum (ponticerebellar fibres), whilst the longitudinal fibres consist of sensory and motor tracts between the spinal cord and the higher parts of the brain. The

Table 7.3 Cranial nerves

Nerve		Origin	Termination	Functions
I	Olfactory	Olfactory mucosa	Olfactory cortex	Sensory: Smell
II	Optic	Retina	Visual cortex (synapses in lateral geniculate body)	Sensory: Vision
III	Oculomotor	Midbrain	Upper eyelid muscle	Motor: Eyelid movement
			Extrinsic eye muscles (superior, medial and inferior recti, inferior oblique)	Eyeball movement
			Ciliary muscles	Accommodation of lens
			Sphincter muscle of iris	Pupillary constriction
		Proprioceptors in extrinsic eye muscles	Midbrain	Sensory: Proprioception
IV	Trochlear	Midbrain	Extrinsic eye muscle (superior oblique)	Motor: Eyeball movement
		Proprioceptors in extrinsic eye muscle (superior oblique)	Midbrain	Sensory: Proprioception
V	Trigeminal	Pons	Muscles of mastication	Motor: Chewing
		Ophthalmic branch takes sensory fibres from skin of upper eyelid, eyeball, lacrimal glands, nasal cavity, side of nose, forehead and anterior half of scalp	Midbrain, pons and medulla	Sensory: Touch, pain, temperature, proprioception
		Maxillary branch takes sensory fibres from mucosa of nose, palate, parts of pharynx, upper teeth, upper lip, cheek and lower eyelid		
		Mandibular branch takes sensory fibres from anterior two thirds of tongue, lower teeth, skin over mandible and side of head in front of ear		
VI	Abducens	Pons	Extrinsic eye muscle (lateral rectus)	Motor: Eyeball movement
		Proprioceptors in lateral rectus	Pons	Sensory: Proprioception
VII	Facial	Pons	Facial, scalp and neck muscles. Lacrimal and salivary glands (sublingual and submandibular)	Motor: Facial expression Salivation Lacrimination
		Taste buds on anterior two-thirds of tongue	Gustatory cortex (synapses in pons and thalamus)	Sensory: Taste
		Proprioceptors in muscles of face and scalp		Proprioception

Table 7.3 (*cont'd*)

Nerve		Origin	Termination	Functions
VIII	Vestibulo-cochlear	Cochlear and vestibular portions of the ear	Cochlear nuclei in pons Vestibular nuclei in medulla	Sensory: Hearing Equilibrium
IX	Glossopharyngeal	Medulla	Swallowing muscles in pharynx Parotid gland	Motor: Swallowing Salivation
		Taste buds on posterior one-third of tongue Carotid sinus	Gustatory cortex (synapses in medulla and thalamus)	Sensory: Taste Regulation of blood pressure.
		Proprioceptors in swallowing muscles		Proprioception
X	Vagus	Medulla	Visceral muscles (muscles of pharynx, larynx, respiratory tract, oesophagus, heart, stomach, small intestine, proximal half of large intestine, gall bladder, liver, pancreas)	Motor: Swallowing, digestive movements and secretions
		Receptors in the same structures that the motor portions innervate	Medulla and pons	Sensory: Range of sensory inputs from organs supplied and proprioception from muscle
XI	Accessory	Bulbar portion: medulla	Muscles of pharynx, larynx, soft palate	Motor: Swallowing
		Spinal portion: cervical spinal cord	Sternocleidomastoid and trapezius muscles	Head movements
		Proprioceptors in muscles supplied by motor fibres	Medulla	Sensory: Proprioception
XII	Hypoglossal	Medulla	Muscles of tongue	Motor: Tongue movements
		Proprioceptors in tongue	Medulla	Sensory: Proprioception

corticopontine tracts connect all parts of the cerebral cortex with the pons and then relay with the ponticerebellar fibres.

The pyramidal tracts arise in the motor cortex and comprise the corticonuclear and corticospinal fibres. The former end in the brain stem and relay with somatic motor neurones in cranial nerve nuclei. The latter descend to the spinal cord (see below).

The cranial nerve nuclei in the pons include: V, which is associated with chewing and various sensations from the head; VI, concerned with eyeball movements; VII, facial expression, salivation lacrimation and taste; and VIII, hearing (see Fig. 7.19). The cochlear nuclei, part of the auditory pathway, are also found in the pons; from here, fibres travel to the medial geniculate bodies and the inferior colliculi and thence to the temporal cortex. Two 'respiratory centres', the pneumotaxic centre and the apneustic centre, are found in the pons. These modify the activity of the medullary respiratory centres (see *The respiratory centres* Ch. 5).

Medulla oblongata

The medulla is 2.5 to 3 cm long and widens as it approaches the pons, at which point it contains several nuclei.

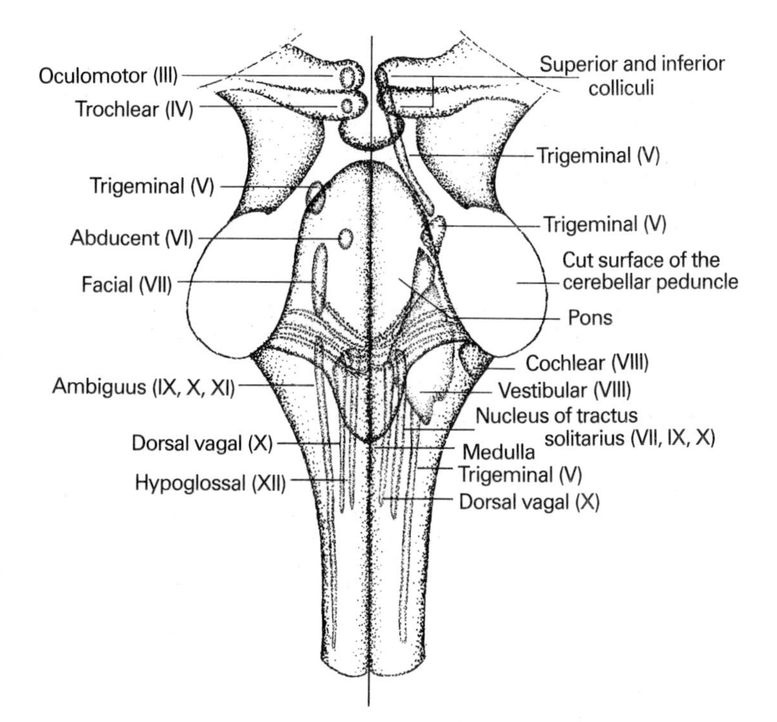

Fig. 7.19 The positions of major motor (left) and sensory (right) cranial nerve nuclei.

The central canal of the spinal cord is continuous with a canal in the medulla and this in turn connects with the IVth ventricle (Fig. 7.8).

Ventrally, either side of the anterior median fissure, there are two rounded columns known as the pyramids (Fig. 7.9). These contain the pyramidal (corticospinal) tract fibres. Most of these fibres cross over in the medulla, the decussation of the pyramids. Fibres then descend in the lateral cortico-spinal tracts of the spinal cord. Fibres in the lateral part of each pyramid, however, remain uncrossed and travel in the ventral cortico-spinal tracts of the spinal cord.

The fasciculus gracilis and cuneatus on each side of the spinal cord contain ascending fibres which continue up in the lower medulla and terminate at the bottom of the IVth ventricle as the nucleus gracilis and nucleus cuneatus on each side. Second order neurones arising from these nuclei cross over in the medulla and ascend in the medial leminiscus to the thalamus.

There are many other fibre tracts passing through the medulla, including the anterior and posterior spinocerebellar, the anterior and lateral spinothalamic, spinotectal and tectospinal, vestibulospinal and rubrospinal.

Cranial nerves IX to XII arise in the medulla (Fig. 7.19): IX is associated with swallowing, salivation and taste; X contains sensory and motor fibres from and to the thoracic and abdominal viscera; XI controls head and shoulder girdle movements; and XII, tongue movements. The vestibular nuclei of cranial nerve VIII, the lateral, medial and inferior, are also found in the medulla.

On each upper lateral surface there is an oval structure, the olive, which contains several olivary nuclei having nerve connections with the cerebellum.

The medulla co-ordinates many autonomic activities and is said to contain several specialized 'centres' which, in turn, are controlled by the hypothalamus. These include the cardiovascular, inspiratory and expiratory, swallowing, vomiting, coughing and sneezing centres.

Cerebellum The cerebellum lies behind the pons and medulla and below the occipital lobes of the cerebrum. The two cerebellar hemispheres are joined by a smaller vermis. The whole structure is divided into three lobes, the anterior, middle and posterior or flocculonodular (Fig. 7.20). Inferiorly, the vermis is divided into the uvula and the pyramid. Also inferiorly, there is a lobule of each cerebellar hemisphere known as the tonsil. The tonsils, appropriately enough, lie near to the uvula.

The surface of the cerebellum is composed of grey matter (like the cerebrum) with parallel curved fissures separating narrow folia. These are analagous to the sulci and gyri of the cerebrum.

The deeper white matter contains fibre tracts orientated like those in the cerebrum, i.e. there are association, commisural and projection fibres. The projection fibres are arranged in three bundles on each side, forming the superior, middle and inferior cerebellar peduncles. The white matter projects into the grey as fine laminae to form a structure known as the arbor vitae (tree of life).

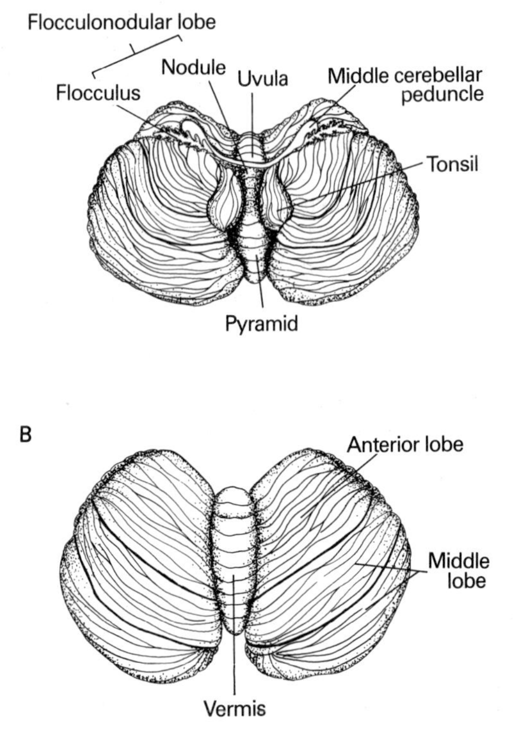

Fig. 7.20 The cerebellum: (A) inferior surface; (B) superior surface.

The cerebellum functions principally in the control of rapid muscular movements such as running. Whereas other parts of the brain and spinal cord may initiate muscle contraction, the cerebellum serves to monitor and make appropriate adjustments to the timing and strength of contraction of groups of muscles. It can be regarded as comparing intended with actual movements.

The cerebellum receives sensory input from the periphery, from the muscle spindles, Golgi tendon organs, touch receptors in the skin and joint receptors. It is appropriate for the rapid adjustments made by the cerebellum that the sensory fibres are among the most rapidly conducting in the body (over 100 m/s). Sensory input to the cerebellum also connects it to other parts of the brain, the motor cortex, olivary nuclei, vestibular apparatus, and the reticular formation via the corticocerebellar, olivocerebellar, vestibulocerebellar and reticulocerebellar tracts, respectively.

The cerebellum also has motor connections with other parts of the brain. The pathways arise in the deep cerebellar nuclei the main ones being the dentate, interpositus and fastigial. Voluntary motor activity is co-ordinated via motor pathways to the motor cortex. Equilibrium and posture involve pathways to the medulla and pons, whereas the co-ordination of conscious control from the motor cortex with subconscious body postural control, involves nerve connections between the cerebellum, thalamus and motor cortex as well as the basal ganglia and reticular formation.

Reticular formation

The reticular formation is a core of grey matter which extends from the spinal cord, through the medulla, pons and midbrain, to the hypothalamus and part of the thalamus. It is analagous to the central grey matter in the spinal cord.

One of the main functions of the reticular formation is that of determining the level of wakefulness or alertness in the brain as a whole. The reticular formation receives a variety of sensory signals including proprioceptive ones from joints and muscles, pain from the skin, visceral signals, and visual and auditory ones. Sensory input of this kind can be relayed via nerve pathways to the thalamus, hypothalamus, basal ganglia, cerebellum and cerebrum, producing an arousal reaction. The motor cortex has reciprocal nerve connections with the reticular activating system, so that the flow of information is two-way

Many specific nuclei lie within the general area covered by the reticular formation such as the red nucleus, substantia nigra, and the vestibular nuclei. Such specific nuclei are, however, often not included as part of the reticular formation as such.

The vestibular nuclei, together with the reticular formation control the muscles that support the body against gravity, i.e. the extensor muscles in the limbs. Sensory stimuli from the vestibular apparatus can result in changes in the degree of contraction, thereby maintaining equilibrium.

Muscle tone is controlled from the reticular formation by means of the sensitivitiy of the spindles, which in turn is controlled by γ-efferents (see *The stretch reflex*). The reticular formation alters the level of activity of these γ-efferents. There are inhibitory influences on the reticular formation from other parts of the brain, and if the nerve connections are damaged, excess muscle tone may result in such conditions as spasticity.

Cerebrospinal fluid

Cerebrospinal fluid (c.s.f.) surrounds the brain and spinal cord, as well as filling the central ventricles and canals. It performs a protective function in that the brain floats in the c.s.f. within the skull, so providing a cushion against a blow to the head.

As the c.s.f. is formed from the blood and is subsequently returned to it, the fluid is an extension of that which flows through the cardiovascular system, and is therefore analagous to interstitial fluid or lymph.

The composition of the fluid is similar, but not identical to plasma (Table 7.4). It is produced by a combination of filtration and active secretory processes by the choroid plexuses found in the inferior horns of the lateral ventricles, the posterior part of the third ventricle and the roof of the fourth ventricle (Fig. 7.10). Some c.s.f. is also produced through the ependymal lining of the ventricles.

The total volume of fluid is about 150 ml and it is produced at a rate of approximately 0.5 ml/min (600–700 ml/day). The c.s.f. is filtered and secreted into the ventricles, with most of it being produced in the lateral ventricles. Fluid from the latter flows through the interventricular foramina into the third ventricle, and thence to the fourth ventricle. There are two lateral and one median aperture through which c.s.f. can flow out of the fourth ventricle. Any obstruction to the flow at this point leads to an accumulation of fluid and a rise in pressure and may result in hydrocephalus. Cerebrospinal fluid flows from the fourth ventricle into a large space, the cisterna magna, which lies behind the medulla and beneath the cerebellum. From here the fluid fills the subarachnoid space, and the main direction of flow is upwards towards the

Table 7.4 Comparison of the major constituents of plasma and cerebrospinal fluid (c.s.f.) (mmol/l). *From Passmore and Robson 1976*

	c.s.f.	plasma
Na^+	140.0	142.0
K^+	2.9	4.5
Mg^{++}	1.0	0.8
HCO_3^-	22.0	25.0
Cl^-	115.0	98.0
glucose	4.5	5.0
urea	5.0	5.0
P_{CO_2}(kPa)	6.6	5.3
pH units	7.33	7.4

cerebrum, through the tentorial opening around the midbrain (another potential point of obstruction). The fluid also fills the central canal and surrounds the spinal cord.

The return of c.s.f. to the blood is achieved in the arachnoid villi and granulations which project through the walls of the venous sinuses (Fig. 7.11). The pressure of c.s.f. is slightly higher (average 10 mmHg when the body is horizontal) than that of the venous blood in the head and the arachnoid villi act like one-way valves, allowing c.s.f. into the blood in the sinus, but not in the reverse direction. The mechanism is similar to that operating in lymphatic capillaries.

Blood-brain barrier

With the exception of the hypothalamus, the capillaries supplying the brain including those in the choroid plexuses, are composed of endothelial cells with very tight junctions, which makes their permeability relatively low. This means that some substances are either completely or partially prevented from gaining access to brain tissue. This blood-brain barrier acts as a protective mechanism, whereby the local environment of the brain is kept more constant than the environment in other parts of the body. Bilirubin and circulating neurotransmitters such as acetylcholine, noradrenaline, dopamine and glycine are prevented from crossing the blood-brain barrier. Any of these substances could cause disturbances of function should they gain access to the brain. Potassium ion concentration is kept relatively low, a condition which favours optimal resting and action potential function; hydrogen ions also are largely excluded from gaining access to c.s.f. by this route.

Water, carbon dioxide, oxygen, alcohol and anaesthetics can, however, gain access to the brain via the blood stream. Epidural administration of drugs (around the spinal cord) will exclude them from access to the brain because of the lack of an epidural space there. If, however, drugs are administered into the c.s.f., then they can readily gain access through the ependyma lining the ventricles, or through the pia mater covering the brain, as these are highly permeable membranes.

THE SPINAL CORD

The spinal cord lies within a canal inside the vertebral column. It extends from the opening on the underside of the skull, the foramen magnum, to the level of the first or second lumbar vertebra and averages 46 cm in length. Below this the vertebral canal is occupied by nerves from the lumbar and sacral segments of the cord which exit well below their point

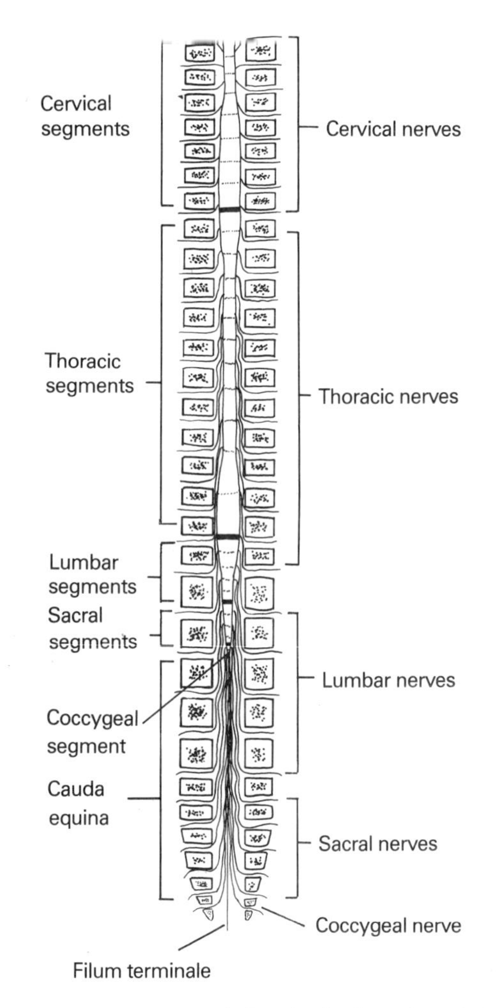

Fig. 7.21 Diagrammatic longitudinal section of the vertebral column to show the segmentation of the spinal cord and the origins of the spinal nerves.

of origin (see Fig. 7.21). These constitute the cauda equina ('horse's tail').

The cord may be considered to be divided into segments and from each segment arises two spinal nerves which emerge between adjacent vertebrae. (Table 7.5 shows the number of segments, spinal nerves and vertebrae associated with each section of the spinal cord.) Each spinal nerve has a dorsal (posterior) and a ventral (anterior) root. The dorsal roots contain sensory fibres which carry impulses from the periphery to the cord and are therefore designated sensory or afferent roots. The cell bodies of the pseudounipolar sensory neurones lie in the dorsal root ganglia. The ventral or efferent roots contain motor fibres which carry impulses away from the cord. With the exception of the nerves which originate in the cervical segments, all spinal nerves travel at least a short distance within the vertebral canal before exiting.

Table 7.5 Comparison of the numbers of vertebrae and spinal cord segments in each region. Each spinal cord segment is connected to one pair of spinal nerves. With the exception of the first pair of cervical nerves, each pair of spinal nerves emerges below the corresponding vertebra

	Vertebrae	Spinal cord segments and pairs of spinal nerves
Cervical	7	8
Thoracic	12	12
Lumbar	5	5
Sacral	5 fused	5
Coccygeal	4 fused	1

The diameter of the cord is not uniform, it narrows from top to bottom. It is enlarged in the cervical and lumbar regions and these enlargements correspond with the origins of the nerves to the upper and lower limbs. The lower extremity of the cord is very narrow and is known as the conus medullaris.

The cord is incompletely divided into right and left halves along the whole of its length by a deep ventral median fissure and a shallower dorsal median sulcus (Fig. 7.22).

Running down the whole length of the cord from the fourth ventricle of the brain is a narrow central canal filled with cerebrospinal fluid (c.s.f.).

Blood enters the cord from a series of anastomotic channels which are derived from branches of the intercostal, vertebral, deep cervical and lumbar arteries and the anterior and posterior spinal arteries. Blood drains out of the cord into six vessels which form a tortuous plexus in the pia mater. They drain into the intervertebral veins.

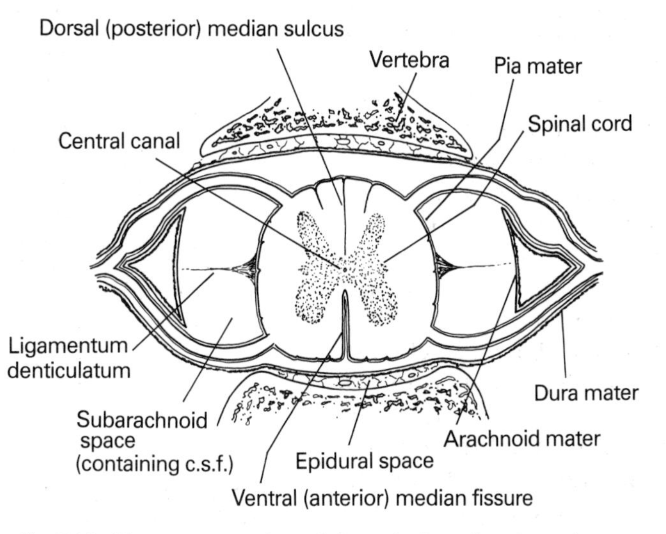

Fig. 7.22 Transverse section of the spinal cord and meninges.

Spinal meninges

The spinal cord is surrounded and protected by three membranes or meninges. Closely applied to the surface of the cord is a thin membrane, the pia mater. This is separated from the thin arachnoid mater by the subarachnoid space containing c.s.f. The arachnoid mater is separated from the outer, tough, fibrous dura mater by a potential, subdural space (see Fig. 7.22).

Outside the dura is the epidural space which is filled by fatty tissue and the epidural veins. These tissues separate the dura from the wall of the vertebral canal.

The pia mater is connected to the dura mater at each thoracic and cervical segment by means of a ligamentum denticulatum on each side. The latter anchor the cord to the dura. The lower end of the cord is anchored to the lower end of the vertebral column by an extension of the pia mater, the filum terminale.

The spinal cord is thus suspended within a fluid-filled cavity which is then enclosed by a layer of spongy tissue; the latter is itself surrounded by bone.

STRUCTURE OF THE SPINAL CORD

In a transverse section the spinal cord has two distinct regions which can be observed with the naked eye (see Fig. 7.23). In the centre is a roughly H-shaped mass of grey matter which is embedded in the surrounding white matter. The dorsal and ventral projections of the grey matter are known as the dorsal and ventral horns respectively. Additionally, in the thoracic, upper lumbar and sacral regions there are short lateral horns or columns. The white matter may be separated into three regions on each side of the cord, the dorsal, ventral and lateral columns. The grey matter consists of large numbers of neurones, their processes and neuroglia. The white matter contains primarily myelinated axons and supporting elements. It is the heavy proportion of myelin which gives this area its white appearance.

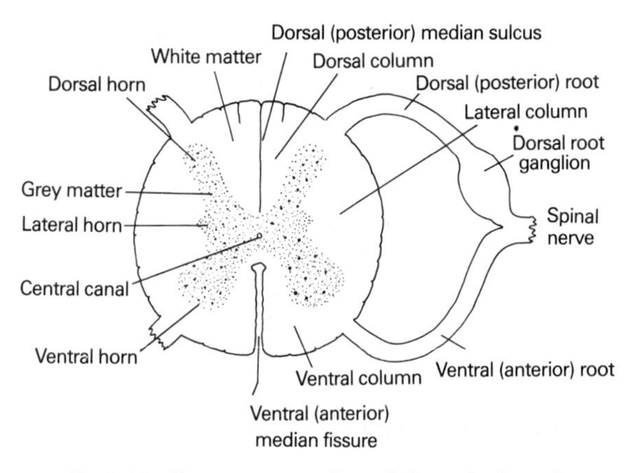

Fig. 7.23 Transverse section of the spinal cord.

Grey matter

In tranverse section the grey matter can be seen to contain large numbers of nerve cell bodies. It has been shown experimentally that

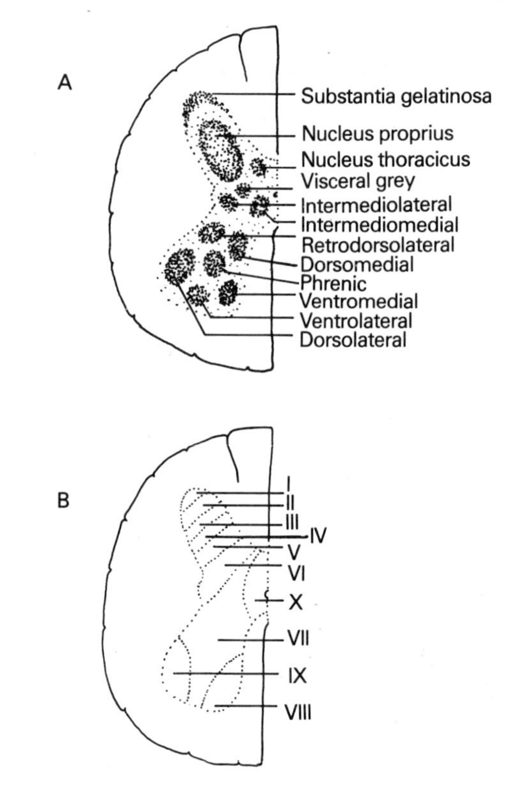

Fig. 7.24 (A) Transverse section of half of the spinal cord to show the distribution of the cell nuclei in the grey matter. (B) Laminar interpretation of the organization of the grey matter.

these nerve cell bodies are grouped according to specific functions. For example, the cell bodies of somatic motor neurones are found in the ventral horns of the grey matter, whereas autonomic (sympathetic and parasympathetic) preganglionic neurons are found in the lateral horns. Groups of cell bodies are referred to as nuclei (Fig. 7.24A). The cells of a nucleus are not only associated in the horizontal plane, however, but also from segment to segment. Some extend through the whole length of the spinal cord, others for shorter lengths. Therefore the nuclei which are distributed within the various columns of the grey matter are themselves often columnar. Table 7.6 lists the spinal nuclei and the functions with which they are associated.

Recent studies have led to alternative concepts of the organization of grey matter. Nine, roughly parallel, laminae have been recognized on each side with a tenth area around the central canal (Fig. 7.24B). These laminae are distinguished by the appearance, size, shape and density of the constituent neurones.

White matter

The white matter of the spinal cord consists primarily of longitudinally orientated nerve axons, functionally grouped together to form tracts (see Fig. 7.25).

Ascending tracts carry sensory information up to the brain, whilst descending tracts transmit motor impulses to appropriate levels in the cord. Both ascending and descending tracts may pass up or down the cord on the same side as their origin or they may cross over to an equivalent position on the opposite side. It should be remembered that the two sides of the cord are mirror images and therefore each ascending and descending tract is duplicated on the opposite side.

Ascending pathways

It may be recalled that ascending (sensory) pathways involve three consecutive neurones, first order, second order and third order. Sensory pathways involving conscious proprioception, fine touch and vibration ascend in the *fasciculus gracilis* and *fasciculus cuneatus*

Table 7.6 The spinal nuclei, their positions and functions

Name	Distribution	Function
Dorsal cell column		*Sensory*
Substantia gelatinosa	All	Pain, temperature
Nucleus proprius (dorsal funicular)	All	Muscle tone/joint position
Nucleus thoracicus (dorsalis)	T1–L2	Proprioception
Visceral grey	T1–L3	Visceral sensation
Lateral cell column		*Autonomic motor*
Intermediolateral	T1–L3	Preganglionic sympathetic
Intermediomedial	T1–L3	Preganglionic sympathetic
Sacral parasympathetic	S2–S4	Preganglionic parasympathetic
Ventral cell column		*Somatic motor*
(*Medial*)		
Ventromedial	All	Innervation of upper and lower limb muscles
Dorsomedial	C1, T1–L3	Innervation of upper and lower limb muscles
(*Lateral*)		
Ventrolateral	C4–C8, L2–S1	Innervation of upper and lower limb muscles
Dorsolateral	C5–T1, L2–S2	Innervation of upper and lower limb muscles
Retrodorsolateral	C8–T1, S1–S3	Innervation of upper and lower limb muscles
(*Central*)		
Phrenic nucleus	C3–C7	Innervation of diaphragm
Lumbosacral	L2–S2	Innervation of lower limb muscles

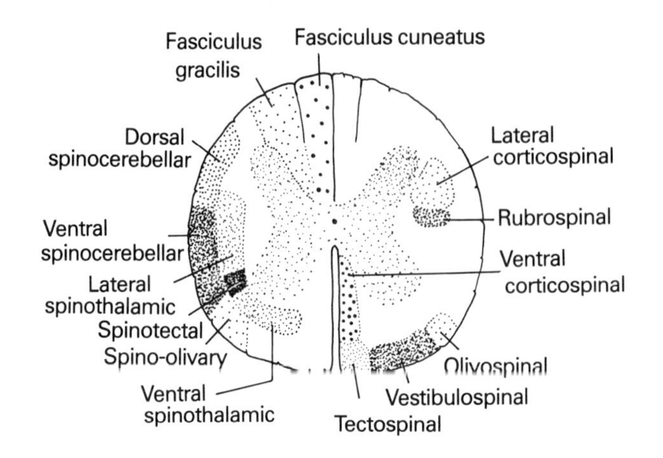

Fig. 7.25 Pathways in the white matter of the spinal cord. Left side: ascending (sensory) tracts. Right side: descending (motor) tracts.

(Fig. 7.25) before synapsing in the medulla oblongata. The fasciculus gracilis runs the whole length of the cord and contains fibres from the lower part of the body. As fibres are added at each level, the old ones are displaced medially (Fig. 7.26). The fasciculus cuneatus arises in the thoracic region and contains fibres which derive from the upper thoracic and cervical nerves. These first order neurones also have collateral branches in the grey matter which relay with reflex spinal pathways. The location of different second order sensory cell bodies in the dorsal horns is given in Table 7.6.

First order sensory neurones of pain and temperature pathways enter the spinal cord, divide into ascending and descending branches which travel one or two segments before entering the dorsal grey matter and synapsing with second order neurones. The second order neurones cross over to the opposite side of the cord and then ascend in the *lateral spinothalamic tract* (Fig. 7.26). These neurones travel through the pons and midbrain in the spinal leminiscus. They synapse with third order neurones in the ventrolateral nucleus of the thalamus. The third order neurones pass through the internal capsule to the postcentral gyrus (somatic sensory cortex).

Crude touch and pressure pathways travel a similar route to pain and temperature, but in the *ventral spinothalamic tracts* on each side of the cord. (Fig. 7.25).

The *dorsal spinocerebellar tract* (Fig. 7.26) begins at the level of the second or third lumbar nerves in the thoracic nuclei of the grey matter. The cell bodies of these fibres receive proprioceptive impulses from collateral branches of the ascending fibres of the fasciculus gracilis and cuneatus and from intermediate ascending fibres in the dorsal columns of the grey matter.

The *ventral spinocerebellar tract* (Fig. 7.25) arises in the lumbar and sacral regions and passes up to the pons before turning back and travelling to the cerebellum. The fibres originate in the grey matter and are concerned with proprioceptive information like the dorsal spinocerebellar tract. The fibres of the ventral spinocerebellar tract cross the cord and pass up on the opposite side from their point of origin.

The *spino-olivary tract* (Fig. 7.25) arises from cells in the grey matter whose axons cross over and ascend on the opposite side. This tract carries touch and proprioceptive infor-

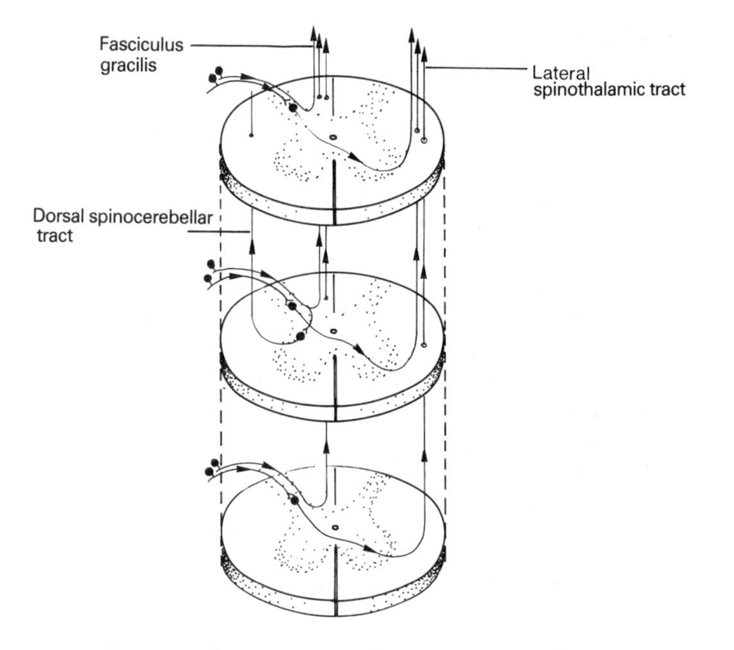

Fig. 7.26 Three sensory pathways in the spinal cord.

mation from receptors in skin, muscles and tendons.

The *spinotectal tract* (Fig. 7.25) arises from cell bodies in the grey matter. Again, axons cross over and ascend on the opposite side, terminating in the superior colliculus of the midbrain. Impulses passing up this tract result in movement of the head and eyes towards a source of visual stimulation.

Descending pathways

Fibres which originate in the motor cortex pass downwards through the internal capsule, the cerebral peduncle and the pons to enter the pyramids of the medulla oblongata. At least two-thirds of the fibres then cross over and enter the cord as the *lateral corticospinal tract* (Fig. 7.27). The remaining fibres descend on the same side as the *ventral corticospinal tract* (Fig. 7.25). These tracts are also known as the pyramidal pathways.

Most corticospinal neurones terminate on interneurones in the basal part of the dorsal grey matter. The ventral tract is present only in the cervical and upper thoracic segments, while the lateral tract extends virtually the whole length of the cord.

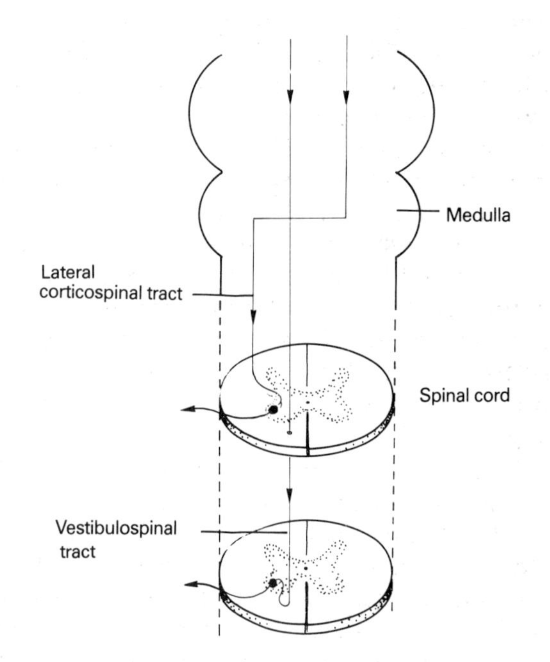

Fig. 7.27 Two motor pathways in the spinal cord.

The pyramidal tract connects the motor cortex (and other parts of the cerebral cortex) with lower motor neurones (somatic motor neurones) which innervate skeletal muscle. It is a pathway, therefore, that is concerned with the regulation of voluntary movements.

The other descending motor pathways to skeletal muscle are interrupted by synaptic junctions in the medulla and cerebrum and have, in the past, been given the collective term 'extrapyramidal pathways'. These can arise in the cerebral cortex and in subcortical areas of the brain.

The *vestibulospinal tract* (Fig. 7.27) arises from cells in the vestibular nucleus of the pons. Fibres descend uncrossed and terminate around motor cells in the ventral grey matter. The tract conveys impulses which are concerned with posture and equilibrium.

The *rubrospinal tract* (Fig. 7.25) originates from nerve cells in the red nucleus of the midbrain. Fibres cross over and then descend on the opposite side and terminate in the dorsal part of the ventral grey matter. Some fibres run the whole length of the cord and, while it is not certain, it is suggested that muscle action in both upper and lower limbs may be associated with this tract.

The *tectospinal tract* (Fig. 7.25) arises from the tectum of the midbrain. Fibres cross over and descend to terminate in the ventral grey matter. It is probable that few fibres actually reach the spinal cord since most appear to terminate within the brain stem.

The *olivospinal tract* (Fig. 7.25) has long been thought to originate in the olivary nucleus, although this is now open to question. Its fibres terminate in the ventral grey matter of the upper cervical segments.

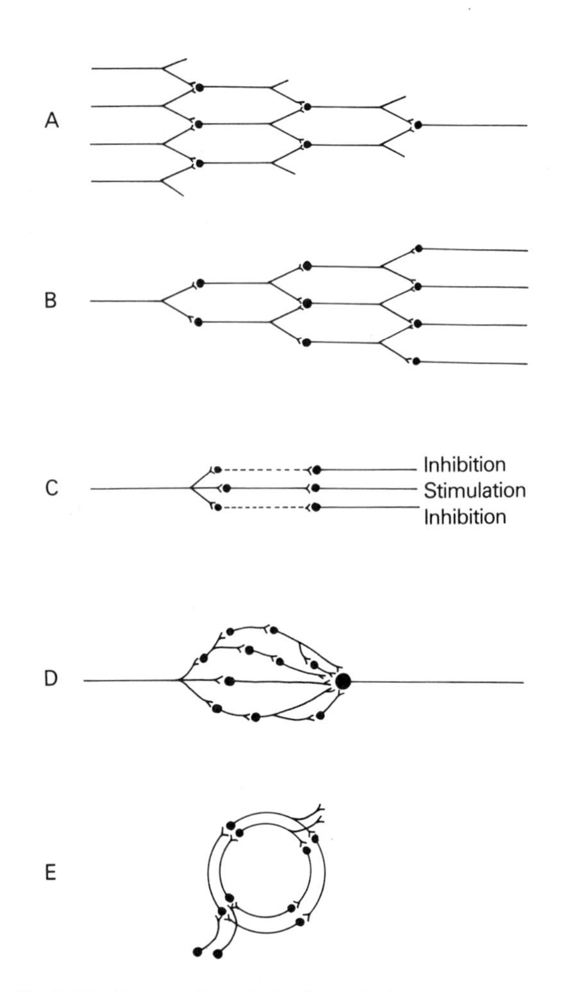

Fig. 7.28 Neuronal pools in the spinal cord. Input neurones are shown on the left, output neurones on the right in each case. (A) a converging pathway; (B) a diverging pathway; (C) inhibition and stimulation of neurones within a pool; (D) a parallel circuit causing after-discharge; (E) a reverberating neuronal circuit.

Neuronal pools

The central nervous system (CNS) may be considered to consist of a large number of variably sized collections of interconnecting groups of neurones described as neuronal pools. These pools may be as small as the individual nuclei in the medulla or as large as the entire cerebral cortex. The ventral and dorsal grey matter columns of the spinal cord are also considered to be longitudinal neuronal pools. Several examples of the organization of neuronal pools are shown in Figure 7.28.

Impulses enter a pool through a number of input (afferent) fibres and leave via output (efferent) fibres. Within the pool, depending upon the anatomical organization and the source and frequency of the impulses entering, a number of different things might happen. Impulses may enter the pool along several fibres simultaneously to stimulate a smaller number of output fibres; this is known as 'convergence' (Fig. 7.28A). In divergence, a smaller number of input fibres stimulate a larger number of exit fibres (Fig. 7.28B). The inclusion of both excitatory and inhibitory synapses allows even greater flexibility

(Fig. 7.28C) (see also *Excitatory synapses* and *Inhibitory synapses* in Ch. 1).

Another phenomenon commonly found in neuronal pools is that of neural after-discharge. A single action potential induces a postsynaptic potential that lasts for several milliseconds during which time impulses may be transmitted in the output fibre. Figure 7.28D shows a parallel circuit involving different numbers of synapses which allows impulses to arrive sequentially at the output fibre, thereby allowing much more prolonged excitation.

Another means of causing after-discharge is by reverberating or oscillating neuronal circuits (Fig. 7.28E). These consist of arrangements of fibres that allow re-excitation within the circuit. Activity within such a circuit may last from milliseconds to several hours.

SPINAL REFLEXES

In addition to its role in the vertical transmission of information the spinal cord also has an integrative function. Much of this integration takes place through the medium of reflexes and the basic unit for this activity is the reflex arc. This consists of a sense organ or receptor, an afferent or sensory neurone, usually one or more connecting or interneurones, an efferent or motor neurone and an effector, usually a muscle (Fig. 7.29).

The sensory neurone enters the spinal cord through the dorsal root, its cell body being located within the dorsal root ganglion. The fibre terminates in the dorsal column of the grey matter, but not necessarily in that particular segment. Branches of the fibre may pass right up to the brain, or sensory information may be relayed through other fibres upon whose cell bodies the afferent fibre terminates. Thus the brain, though not necessarily the higher centres, is informed of the sensory event.

The afferent fibre also terminates on one or more interneurones or directly upon one or more motor neurones which then pass out of the cord through the ventral root on the same side.

A reflex must contain a minimum of one synapse, i.e. the junction between the sensory and motor neurones, but usually it contains more. It is these synaptic gaps which provide the opportunity for integration within the spinal cord.

The stretch reflex (monosynaptic)

A sharp tap on the patellar tendon leads to a contraction of the quadriceps femoris muscle which raises the leg. Tapping the tendon stretches the muscle and receptors within the muscle initiate impulses in the afferent neurones passing to the spinal cord (Fig. 7.30).

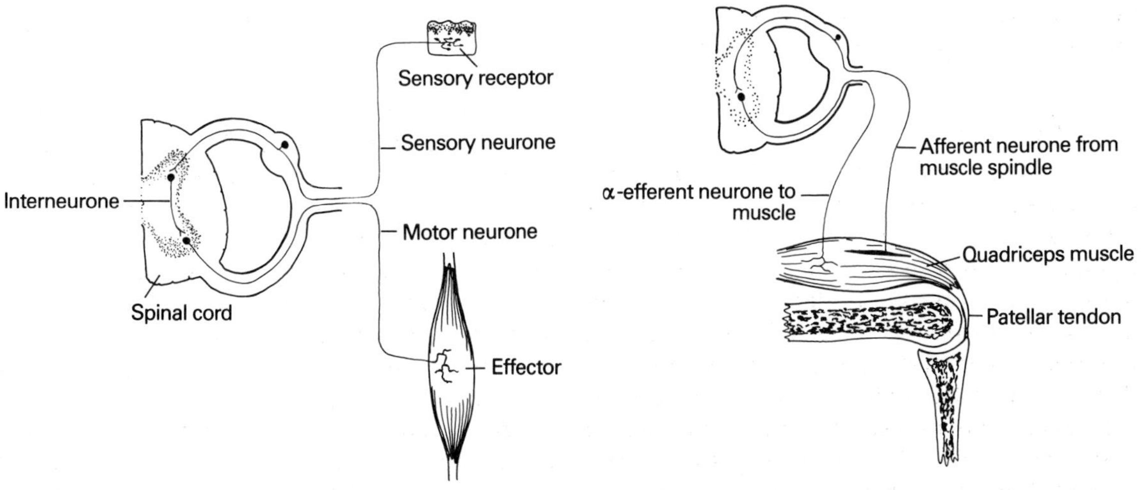

Fig. 7.29 The components of a reflex arc.

Fig. 7.30 The monosynaptic nerve pathway of the stretch reflex.

Within the cord impulses are generated in motor fibres which pass back to the muscle. The reflex is thus initiated by stretching the muscle and it is therefore known as a stretch reflex. Stretch reflexes are the only monosynaptic reflexes in the whole body.

Structure of the muscle spindle

The receptors which detect stretch in muscle are known as muscle spindles. Each muscle spindle consists of a small number (less than 14) of small fibres, between about 1 and 5 mm in length, which are poorly striated in their central regions. They are known as intrafusal fibres to distinguish them from the rest of the fibres of the muscle, the extrafusal fibres. The intrafusal fibres are enclosed in a connective tissue capsule and attached to the connective tissue sheaths of the surrounding muscle fibres and are orientated parallel to them.

Two types of intrafusal fibres are present. The larger fibres have a cluster of nuclei in a swollen central region and hence have been named 'nuclear bag' fibres. The smaller and narrower fibres have their nuclei arranged in a row along the centre of the fibre and are known as 'nuclear chain' fibres (Fig. 7.31).

Both types of fibre are associated with sensory nerve endings of large myelinated (group Ia) neurones. These 'annulospiral' nerve endings wrap around the central, nuclear regions of both types of intrafusal fibre. Smaller (type II) neurones also contact the intrafusal fibres with 'flower-spray' terminals on either side of type I junctions (Fig. 7.31). Type II endings are largely confined to the nuclear chain fibres.

Two types of fine γ-efferent fibres innervate the intrafusal fibres. One type forms typical motor end-plates on the nuclear bag fibres while the other forms unspecialized 'trail-endings' primarily on the nuclear chain fibres.

Function of the muscle spindle

Tapping the patellar tendon causes a sharp increase in length of the quadriceps muscle. As a result the muscle spindle is stretched which distorts the primary afferent nerve endings and generates action potentials which are then transmitted to the cord. The afferent fibres synapse on the cell bodies of the motor fibres which pass to the extrafusal fibres of the muscle (see Fig. 7.30). As a result, in less then one millisecond, the quadriceps muscle begins to contract and the leg is raised. This reaction is known as a dynamic stretch reflex since it is elicited while the intrafusal fibres are increasing in length. Experimental evidence

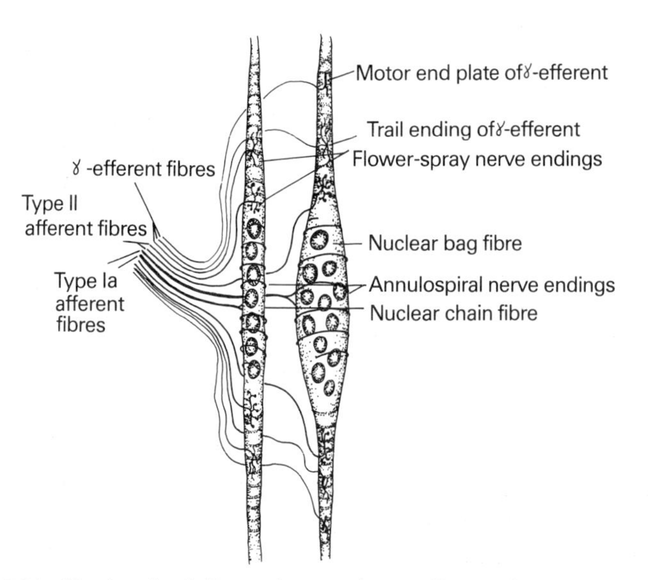

Fig. 7.31 The intrafusal fibres of a muscle spindle and their innervation.

suggests that this reflex relies primarily upon excitation of the nuclear bag fibres.

Stretching a muscle induces excitation of both type Ia and type II afferent nerve endings primarily of the nuclear chain fibres. In both cases the frequency of discharge is proportional to the degree of stretch. The resulting muscle contraction will last as long as the muscle is stretched and this is therefore known as a static stretch reflex.

γ-Efferent stimulation regulates the sensitivity of the spindle so that if the central region is prestretched, only a small amount of further stretching will lead to rapid afferent discharge. Alternatively, if γ-efferent activity is reduced, the central regions of the intrafusal fibres slacken and their sensitivity is correspondingly reduced. As muscles contract, γ-efferent discharge causes concomitant shortening of the spindle fibres so that they maintain their sensitivity. The activity of these γ-efferents is regulated by descending spinal pathways which originate in the reticular formation in the brain stem.

Since muscle spindles are stimulated when muscles are stretched and their effect is to oppose that stretch, then they have a damping function, i.e. they help to smooth out contractions that would otherwise be very jerky. Passive stretching of a muscle induces the stretch reflex which opposes the stretch. This constitutes muscle 'tone' which exerts a slight inhibitory effect upon passive movements of the limbs. Excessive tone is observed in a spastic muscle where such movements are very difficult, but very low tone is seen in flaccid muscles. Muscle spindles also play an important role in the maintenance of posture since they provide constant feedback to the cerebellum through the dorsal spinocerebellar tracts. The cerebellum, in turn, controls γ-efferent activity via the brain stem.

Reciprocal innervation

When a muscle contracts in the stretch reflex, the muscle which would normally oppose the contraction of that muscle relaxes, or more accurately is prevented from contracting. This is brought about by a pathway which includes a collateral branch from the Ia afferent which synapses with an inhibitory interneurone. The latter then synapses with a motor neurone to the antagonistic muscle. As a result, this muscle is prevented from contracting (Fig. 7.33 shows the reciprocal innervation pathway in the flexor and crossed extensor reflex).

Again, the knee jerk serves as an example. Tapping the patellar tendon stretches the muscle spindle within the quadriceps muscle and results in the passage of impulses to the spinal cord. Impulses then pass down to the quadriceps and initiate contraction, but at the same time inhibitory interneurones prevent excitation of the antagonistic ('hamstring') muscles.

Reflexes initiated by the Golgi tendon organ

Golgi tendon organs are found in tendons, near to the junction with the muscle. Each organ is about 500–700 μm long and consists of an encapsulated bundle of tendon fibres which are innervated by Ib sensory fibres with spray-type endings (Fig. 7.32). The Golgi tendon organs are 'in series' with the muscle and are stimulated by stretch (increase in tension) of the tendon. The reflex response is a reduction in contraction (i.e. relaxation) of the muscle. This is mediated by inhibitory synapses in the interneurone pool. The reflex can be initated by either stretch of the muscle or contraction of the muscle.

Passive stretch does not increase the tension in the tendon very much, so the Golgi tendon organ is not stimulated until the muscle is stretched considerably. If this

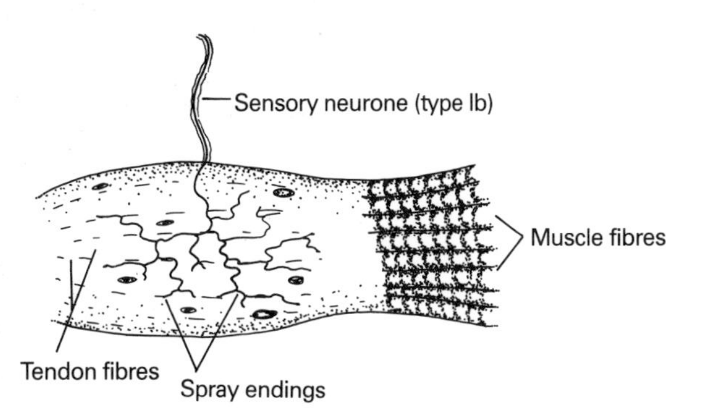

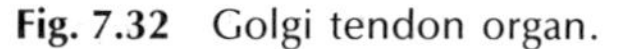
Fig. 7.32 Golgi tendon organ.

happens, the stretch reflex initiated by the muscle spindles is suddenly replaced by relaxation, the inverse stretch reflex, initiated by the Golgi tendon organ.

Whereas the inverse stretch reflex is initiated only when a muscle is considerably stretched, when a muscle contracts only slightly, it initiates the tendon reflex which exerts a degree of restraint on the contraction. Just as the muscle spindle has a dynamic and a static response, so does the Golgi tendon organ, so that while the tension in the tendon is changing, there is a burst of activity from the sensory nerve endings, and once the tension is static, the frequency of impulses from the Golgi tendon organ is proportional to the level of tension present.

As well as initiating reflex muscle relaxation, sensory impulses from the Golgi tendon organ ascend to the cerebellum and to the cerebral cortex.

Flexor or withdrawal reflex

A painful stimulus to the hand or foot initiates reflex contraction of the flexor muscles which causes withdrawal of the arm or leg, respectively. Reciprocal innervation allows the extensor muscles to be inhibited (Fig. 7.33).

Diverging pathways (Fig. 7.28B) cause accessory muscles (e.g. trunk and shoulder muscles) to be stimulated, aiding movement of the limb. Pathways causing after-discharge are also involved (Fig. 7.28D)

Crossed extensor reflex

A strong painful stimulus will not only induce withdrawal of the affected limb but also extension of the other (Fig. 7.33). Afferent impulses to the cord pass through a series of interneurones to the opposite side. Here, the synapses are so arranged that stimulation of the extensor muscle and inhibition of the flexor muscle occur. The presence of a number of synapses in the pathway causes a delay of up to 0.5 seconds in the crossed extensor reaction.

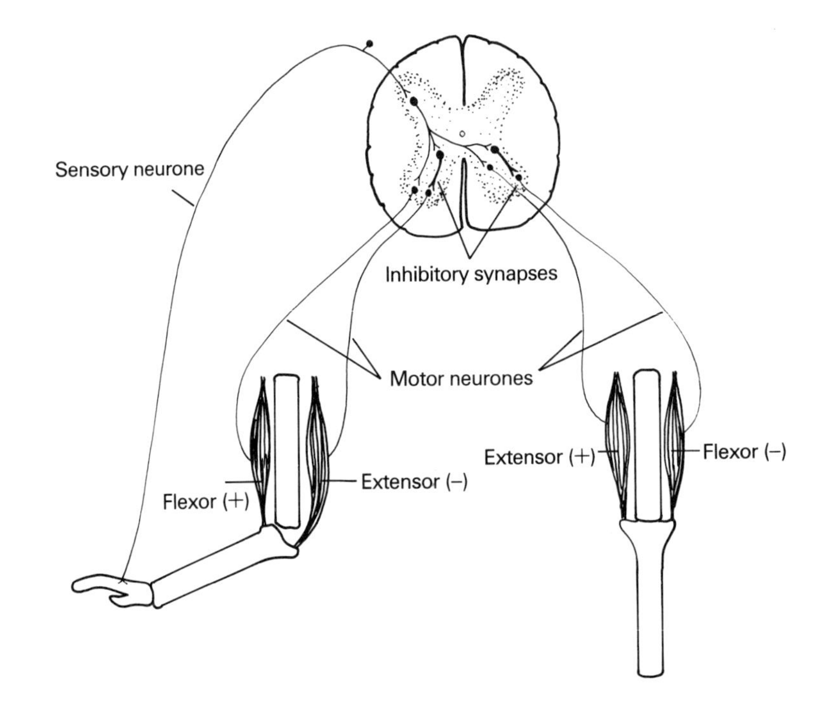

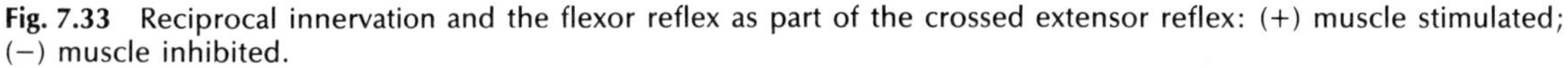

Fig. 7.33 Reciprocal innervation and the flexor reflex as part of the crossed extensor reflex: (+) muscle stimulated; (–) muscle inhibited.

8

The special senses

This chapter is concerned with a more detailed consideration of the special senses, sight, hearing, taste and smell. The relevant anatomy is described in relation to these functions. As the vestibular apparatus is part of the ear, vestibular functions are also covered in this section, even though they are not strictly 'special' senses.

THE EYE

The two eyes are located within the orbits of the skull, one on either side of the nose and, in humans, face forwards. The eyeball is quite a rigid structure but nevertheless receives protection from the bony socket within which it is located and from the eyebrow ridge. The eyeball is attached to the socket by extrinsic muscles and cushioned by pads of fat. Its frontal surface is protected by the eyelids.

Each eye receives light from the environment and forms images on the photosensitive lining, the retina, which converts them into nerve impulses which are passed to the visual centres of the brain.

STRUCTURE OF THE EYE

The eye is a roughly spherical structure about 2.4 cm in diameter, consisting of three principal layers, the outer sclera and cornea, the middle uvea and the inner retina (Fig. 8.1).

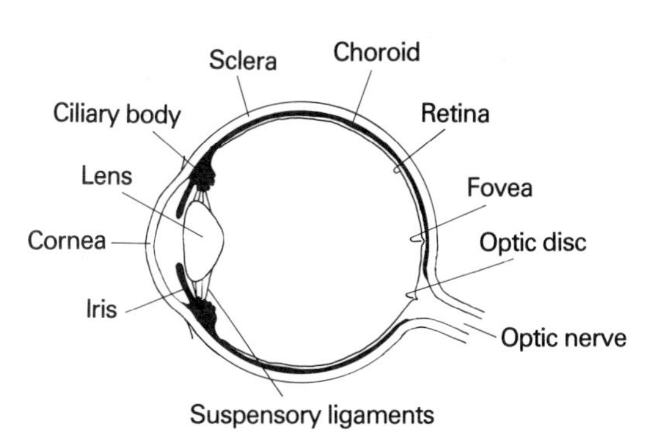

Fig. 8.1 Horizontal section through an eyeball.

Sclera and Cornea

About five-sixths of the outer surface of the eyeball is made up of a tough, white fibrous layer, the sclera. The anterior one-sixth consists of the cornea, a transparent structure which is continuous with the sclera. Both sclera and cornea exhibit similar microscopic structures consisting of layers of collagen fibres. In the cornea these fibres are very regular in size and in arrangement and this leads to the transparency of the layer. The frontal surface of the cornea is covered by stratified squamous non-keratinized epithelium which is continuous with the conjunctiva. The latter covers the sclera anteriorly and also lines the inner surface of the eyelids.

Within the sclera, near the sclerocorneal junction, lies an oval canal which passes around the circumference of the eyeball. This is the canal of Schlemm (sinus venosus sclerae). At the rear the sclera is pierced by the optic nerve which carries away the nerve fibres originating in the retina.

Uvea (choroid, ciliary body and iris)

Beneath the sclera is a thin, highly vascular, pigmented layer, the choroid. This is the main nutritive tissue of the eyeball which supplies the retina with oxygen and foodstuffs. The pigment cells are thought to absorb light and prevent it from being reflected back through the retina.

At the front of the eyeball the choroid thickens to form a circular structure, the ciliary body, from which the lens is suspended by fine ligaments (Fig. 8.2). The ciliary body extends forward from the anterior border of the retina proper (ora serrata) as a thin layer which connects with a ridge surmounted by a number of ciliary processes. The ciliary body contains meridional, radial and oblique smooth muscle fibres and blood vessels. The entire structure is covered by two layers of epithelium separated by a basal lamina. Both of these layers are derived from the retina.

In front of the lens the choroid gives rise to the iris. This structure, which may be variously

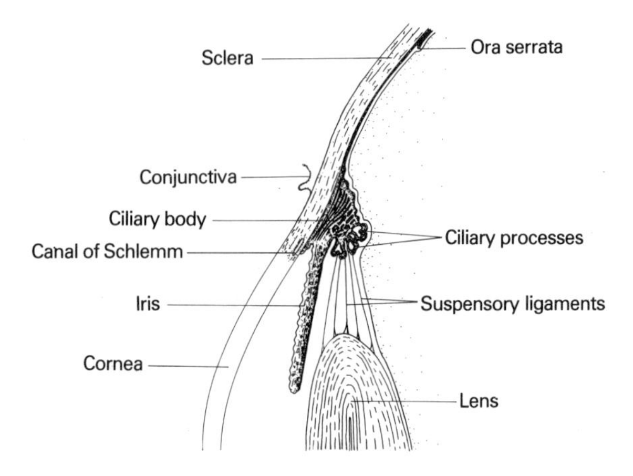

Fig. 8.2 Section through the ciliary body and part of the lens.

pigmented to give colours ranging from pale blue to dark brown, surrounds a small hole, the pupil, through which light enters the eye. The iris has two layers of smooth muscle: the sphincter pupillae which consists of muscle fibres that are arranged in a circular manner so that their contraction brings about constriction of the pupil (meiosis) and the dilatater pupillae which consists of radially arranged fibres and brings about pupillary dilatation (mydriasis). By these means pupil size can be varied from about 1 mm to about 8 mm diameter.

Retina

The retina is the innermost layer of the eyeball, lining the whole of the posterior chamber and terminating in a ragged line (ora serrata) behind the ciliary body. The retina has an extremely complex structure consisting of nerve elements, supporting cells and light-sensitive rods and cones. The histological appearance of the retina has given rise to a collection of terms that are described below (see Fig. 8.3).

Pigment layer. The outermost layer of the retina lying next to the choroid. The cells have microvilli on their free surfaces which extend between the rod processes (see below). They contain melanin. The precise function of this layer is still obscure, but it may be partly nutritive, partly mechanical, in that the cells help to separate and support the rod processes, and partly optical in that it absorbs light. The cells of this layer are also phagocytic and remove material from the apices of the cells of the next layer.

Rod and cone layer. The outer processes of

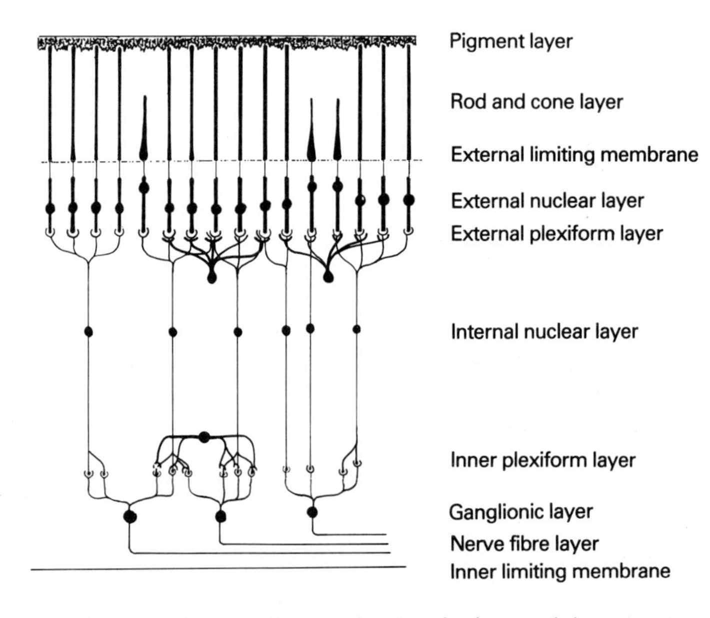

Fig. 8.3 Schematic diagram showing the layers of the retina.

the rods and cones lie next to the pigment layer. Their detailed structure will be described in a later section.

External limiting membrane. Appears optically to be a membrane but is, in fact, formed by the close juxtaposition of the supporting glial cells (Müller cells) and the rod and cone inner segments.

External nuclear layer. Contains the inner segments and nuclei of the rods and cones.

External plexiform layer. Consists of the innermost parts of the cell bodies of the rods and cones. They resemble nerve axons and terminate in bulbous extensions called rod spherules and cone pedicels which synapse with the nerve elements in the next layer.

Internal nuclear layer. Contains the nuclei of the Müller cells, together with those of a number of nerve elements, the horizontal, bipolar and amacrine cells.

The horizontal cells lie close to the basal portions of the rods and cones and may be subdivided into rod and cone types. Rod horizontal cells may synapse with up to 12 rod spherules, while cone horizontal cells synapse with only seven cone pedicles.

Bipolar cells may similarly be subdivided. Rod bipolar cells connect with up to four ganglion cells and perhaps as many as 50 rods. Cone bipolar cells, of which there are two types, are more specific in their relationships. A flat cone bipolar synapses with up to seven cones, while a midget cone bipolar will connect with a single cone cell.

The amacrine cells lie horizontally at the level of the junctions of the bipolar and deep ganglion cells.

Inner plexiform layer. Consists of a mass of interconnecting processes derived from the amacrine, bipolar and ganglionic neurones.

Ganglionic layer. This is the innermost layer of nuclei and is made up of the nuclei of those nerve cells which carry sensory information to the optic nerve. Several different types of ganglion cells have been identified (see *Electrical aspects of retinal function*).

Nerve fibre layer. The axons of the ganglion cells which are converging towards the optic nerve.

Inner limiting membrane. This is the thin basement membrane of the glial cells of the inner retina and effectively seals off the retina from the cavity of the eyeball.

Rods and cones

Rods and cones are distributed as a tightly packed mass throughout the retina, except at the point where the ganglionic fibres converge to form the optic nerve. This area, which is about 1.5 mm in diameter, is termed the 'optic disc'. Since it possesses no photosensitive cells it is also known as the blind spot.

In the centre of the retina is an oval, yellowish area, the macula lutea, at the centre of which lies a small depression, the fovea centralis. In this region the neural layers of the retina become very thin since they radiate out from the fovea at an acute angle. As a result, the photosensitive cells, which in this region are exclusively cones, are more exposed to light than over the rest of the retina. Therefore at this point visual resolution is at its highest.

At the fovea, cone density is about 150 000 per sq. mm falling to about 5000 per sq. mm over the rest of the retina. Rods are absent from the fovea, but they appear at its edge and their density quickly rises to 160 000 per sq. mm and then falls gradually until the ora serrata, where it is still about 30 000 per sq. mm. It has been estimated that the total number of rods in the human retina is of the order of 120 million and the number of cones 6.5 million.

Ultrastructurally, rods and cones have some common features, although the former are much longer than the latter (120 μm, compared with 75 μm).

Each cell type consists of an outer segment, an inner segment containing the nucleus and a dilated basal region which synapses with the neural processes of the adjacent layer (Fig. 8.4).

The inner segments are longer and have greater diameters than the outer segments. At the apex of the inner segment of each cell type there is a projection which has an internal structure characteristic of a cilium, from which the outer segment appears to

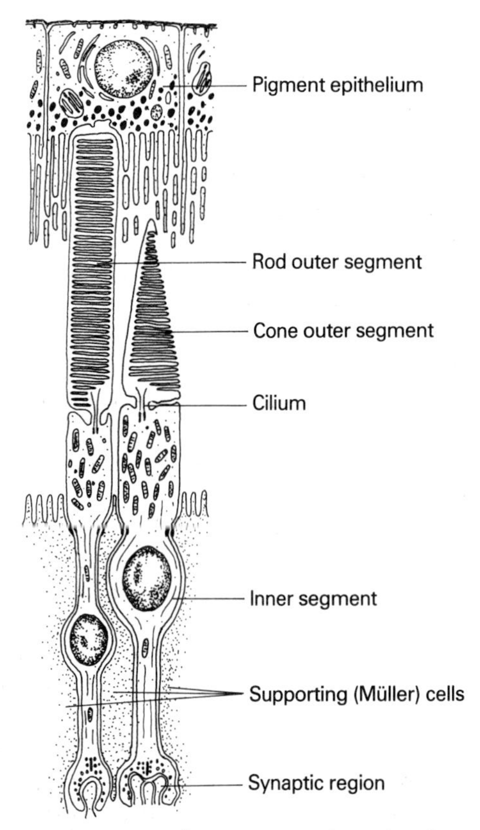

Fig. 8.4 Ultrastructural appearance of a rod and a cone in situ.

develop. This cilium links the inner and outer segments.

The outer segments of the rods and cones resemble one another in that both contain stacks of closely packed flattened membrane-bound sacs. So flattened are these sacs that they are usually described as being disc-like. In rods these discs are totally separated from the plasma membrane surrounding the outer segment but in cones a continuity exists.

The outer segments of the rods are generally longer than those of cones, although at the fovea cones are quite long and very narrow (resembling rods), becoming broader and shorter at the ora serrata.

It has been shown that discs arise at the basal end of the outer segment (by the cilium) and are lost at the apex in rods and probably in cones. The discs which are lost at the apex are phagocytosed by the cells of the pigment epithelium. The turnover of discs is very high and may approach 100 per day from a single outer segment.

In rods the disc membranes are very rich in the photochemical substance rhodopsin. Similar but not identical molecules are found attached to the discs of cone outer segments.

Ocular refractive media

The eyeball is filled with material which is structured and arranged to focus light onto the retina. The bulk of the cavity is filled by the gelatinous vitreous body, in front of which lies the lens. In front of and around the lens is the watery aqueous humor.

Aqueous humor

The aqueous humor is a clear, watery fluid, containing much less protein than plasma, which fills the cavities around the lens (anterior and posterior chambers). It is derived from the plasma in the capillaries of the ciliary processes by means of active secretion and by diffusion through the surface epithelium of the ciliary processes into the posterior chamber. It then passes forward through the pupil to the anterior chamber where it is absorbed into the ciliary veins and the canal of Schlemm. Its function is to nourish the tissues of the cornea and the lens.

Vitreous body

The vitreous body occupies the bulk of the eyeball cavity, being the space behind the lens and ciliary body. It consists of a transparent gel composed mostly of water with some electrolytes, mucoprotein (vitrein), hyaluronic acid and some collagen fibres. The gel is condensed at its edges to form a distinct membrane. It is penetrated by a narrow (hyaloid) canal which passes from the optic disc to the posterior surface of the lens. In the fetus the canal contains the hyaloid artery which nourishes the anterior structures of the eye until about six weeks before birth, when it atrophies.

The vitreous body is attached to the rear surface of the ciliary body and while its presence may help to keep the retina in position, the two are not fused in any way.

Lens

The lens is a transparent, biconvex, encapsulated structure suspended from the ciliary body posterior to the iris. The front surface of the lens consists of a layer of cuboidal epithelial cells, and at the equator the cells become elongated before losing their nuclei and developing into lens fibres which make up the bulk of its structure. The lens is an elastic structure and this allows its shape to change when the eye is focused. This is brought about by contraction of the muscles in the ciliary body which alters the tension in the suspensory ligaments (see *Formation of an image*). It is interesting to note that the lens steadily enlarges throughout life and steadily loses its flexibility. As a result of the latter change the lens loses its ability to accommodate in old age, a condition known as presbyopia.

Accessory apparatus

The eyeball is anchored into position by a number of structures including the extrinsic (extraocular) muscles, the conjunctiva and the eyelids. It is further protected by the bony socket and eyebrow ridge and some fatty tissue. It is lubricated by the lacrimal glands.

Extrinsic muscles

The extrinsic muscles, which bring about rotational movements of the eyeball, comprise four rectus muscles (superior, inferior, medial and lateral) and two oblique muscles (superior and inferior) (Fig. 8.5).

The four recti are attached to the sclera, posterior to the corneal margin and to a fibrous ring (common annular tendon) which surrounds the exit point of the optic nerve from the orbit.

The oblique muscles are arranged such that for part of their length, at least, they lie around the circumference of the eyeball. The superior oblique lies above the eyeball and is attached at the rear of the orbit. However, near the front of the orbit it passes through a fibrocartilage loop (trochlea) and the tendon of the muscle then attaches to the eyeball behind its equator. Thus the superior oblique turns at an angle of more than 90 degrees between its origin and insertion. The inferior oblique muscle is attached to the sclera on its lateral margin, runs around beneath the eyeball and is inserted into the floor of the orbit.

Lacrimal apparatus

The lacrimal gland has two portions, an orbital part and a palpebral part. The former is the larger of the two parts and lies in a depression

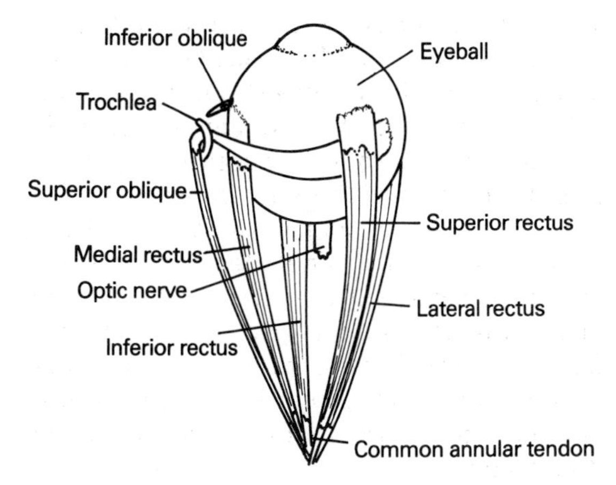

Fig. 8.5 Extrinsic muscles of the right eye viewed from above.

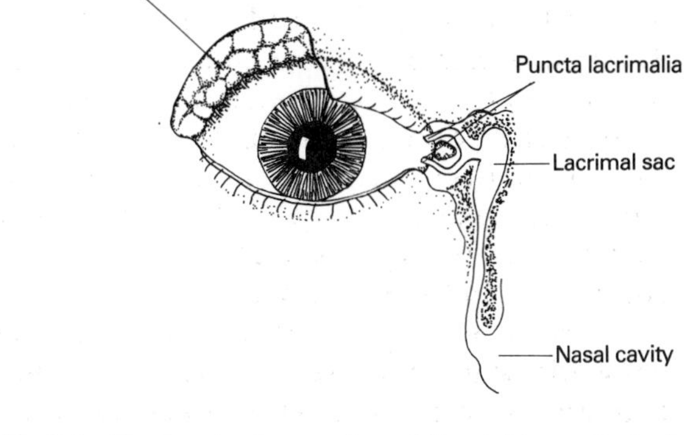

Fig. 8.6 The larcrimal apparatus of the right eye. Part of the upper eyelid and orbit has been removed to show the lacrimal gland, as has the bone on the nasal side to reveal the lacrimal sac etc.

in the frontal bone above the eyeball and towards its outer margin (Fig. 8.6).

The gland has a lobular tubulo-acinar structure which secretes a clear fluid over the front of the eyeball.

A small opening at the inner corner of each eyelid (puncta lacrimalia) leads into canaliculi which drain into the lacrimal sac. The latter is the upper, blind end of the nasolacrimal duct which carries the lacrimal secretions (tears) down to the nasal cavity where it terminates in the anterior part of the inferior meatus.

Tears are secreted on to the frontal surface of the eyeball and are spread over it by the blinking movements of the eyelids. Excess fluid then drains down into the nasal cavity.

The lacrimal glands are stimulated by parasympathetic fibres in response to chemical and mechanical irritants, so producing tears which may wash away the irritants. Pain and emotion can also cause the production of tears.

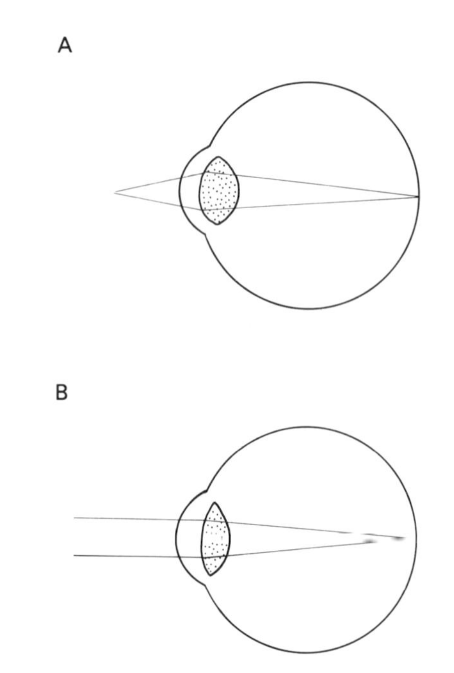

Fig. 8.7 Accommodation. (A) Light from a source close to the eye is focused by a rounded lens. (B) Light from a distant source is focused by a flattened lens.

FUNCTION OF THE EYE

Focusing the image

Accommodation

The eye is able to focus light from objects at different distances by altering the curvature of the lens in a process known as accommodation (Fig. 8.7). As an object moves closer to the eye the curvature of the lens increases, so that the image remains in focus upon the retina.

When the eye is at rest, i.e. it is focused on a distant object, the ligaments supporting the lens are under tension and it is flattened. As the object moves closer, the ciliary muscles contract, causing the ciliary body to decrease in diameter and thereby reducing the tension in the ciliary ligaments. As a consequence, the lens is less restrained and becomes more spherical owing to its own in-built elasticity.

There is a limit to how close an object may be to the eye and its image still be focused upon the retina. The closest point is called the 'near point' and it grows larger throughout life, e.g. at age 20 years it is about 10 cm and at age 60 years is about 83 cm. This effect is mainly due to hardening of the lens with age.

The image which is focused upon the retina is inverted but is interpreted as being the right way up by the brain.

Iris and pupil

The diameter of the pupil also influences image formation. If the pupil is constricted, then a reduced amount of light is able to enter the eye. As a result, the beam of light passing to the lens will only strike the central portion which reduces chromatic and spherical aberrations which tend to occur at the edges of the lens. Secondly, the depth of focus of the eye also improves, since with a narrow pupil the cones of light entering the eye are themselves very narrow and remain in focus over a short distance rather than at a specific point (see Fig. 8.8).

The iris responds reflexly to changes in ambient light intensity. An increase in light intensity initiates pupillary constriction (the pupillary light reflex), whereas a decrease in light intensity causes pupillary dilatation. If only one eye is illuminated, then that eye

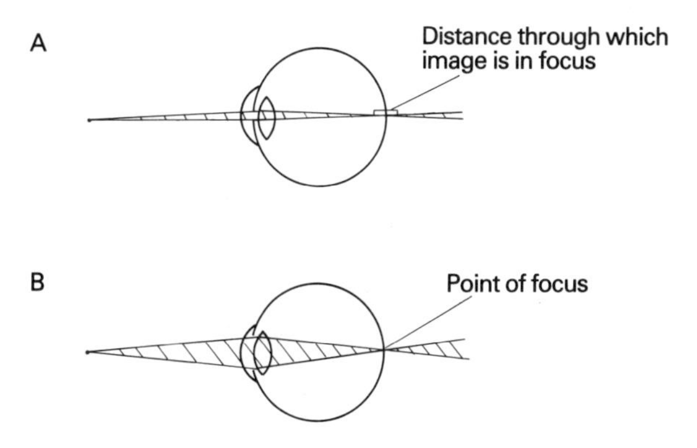

Fig. 8.8 The effect of changes in pupil diameter upon image formation. (A) A constricted pupil produces a narrow cone of light which is in focus over a short distance. (B) A dilated pupil produces a wide cone which is focused at a single point.

shows a direct effect in response to the light change. The other eye, however, also shows an identical response and this is known as the 'consensual light reflex'.

The pupil is capable of about a fivefold change in area, although light intensity with a millionfold variation may be experienced. Generally, pupil diameter changes rapidly in response to changes in light intensity, but longer term adaptation depends upon photochemical changes in the retina.

Eye movements

When the focus of the eye shifts from a distant to a near object, the two eyes move within their sockets so that their optical axes converge. In this way, images can be formed upon the corresponding areas of the retinas of the two eyes. In convergence the medial recti of the two eyes contract simultaneously to bring the two optical axes together.

Focusing upon distant objects requires that the optical axes be parallel to one another and that the corresponding muscles of the two eyes work in concert. This is known as conjugate movement, and such movement may either be quite smooth or jerky, as for example in reading; the latter is known as saccadic movement.

Neural regulation of image formation

In order that clear images may be formed upon the retinae it is evident that the shape of the lenses must be correct, the pupils have an appropriate diameter and the two eyes are positioned so that light falls upon corresponding points. All three of these conditions are regulated by the nervous system.

Parasympathetic fibres from the oculomotor nucleus in the mibrain pass in the third cranial nerve to the ciliary ganglion, which lies just behind the eye. From here they pass through the ciliary nerves into the eyeball, terminating in the ciliary body and on the circular muscle fibres of the iris. Parasympathetic stimulation causes contraction of the ciliary muscles and the lens becomes more convex (see *Accommodation*); stimulation of the circular muscle fibres of the iris causes a reduction in pupil size (as in the pupillary light reflex).

Sympathetic fibres originate at the level of the first thoracic segment of the spinal cord and pass up to the superior cervical ganglion. They then follow a number of arteries to reach the eyeball where they innervate the radial muscle fibres of the iris. Contraction of these fibres causes an increase in pupil size. Sympathetic fibres also exert an inhibitory influence upon the muscles of the ciliary body.

The precise mechanism by which the shape of the lens is altered when the distance of an object from the eye is varied is not certain. It is dependent upon the production of an image that lacks sharpness on the retina. This information is conveyed to the brain which then increases or decreases activity in the parasympathetic pathways to the ciliary body, resulting in compression or stretching of the lens until a sharp image is formed. The main area of uncertainty is the mechanism by which the retina detects an unsharp image.

Alterations in the size of the pupil depend upon light hitting the retina. This then passes impulses to the midbrain, which in turn sends impulses along parasympathetic pathways to bring about constriction, or along sympathetic pathways to bring about dilatation as appropriate.

The extrinsic muscles of the eye receive innervation from the third, fourth and sixth cranial nerves. Each muscle receives reciprocal innervation, so that while one muscle of a pair is active the other is inhibited.

Fixation of the image of a moving object on the retina is achieved in two stages. The first or voluntary movement involves directing the gaze to a particular object. This activity is controlled by an area within the premotor cortex. The second, involuntary movement, which allows the moving object to be followed, is regulated by the occipital cortex. In the latter case the eye is moved in response to the image drifting from the centre of the fovea. A continuously moving panorama will result in sacchidic movements as the eye fixes on successive objects.

As a distant object moves closer to an observer, the eyes gradually converge while still exhibiting conjugate movements. The lens becomes more convex as it accommodates and the pupil constricts, thereby helping to sharpen the image.

Common abnormalities of image formation

There are many commonly encountered problems found with the image-forming mechanism. The eyeball may be short in an anteroposterior direction so that images are formed at a point beyond the retina. This is far-sightednes (hyperopia) and may be compensated by the use of convex lenses.

If the eyeball is too long, an image is formed in front of the retina and the person is said to be short-sighted or near-sighted (myopia). This is compensated by the use of biconcave lenses.

Astigmatism describes a condition where the cornea is not uniformly curved. This often causes images in different places to be focused at different points. It may be compensated by the use of cylindrical lenses.

Photoreceptor function

Both rods and cones contain pigments which alter physically and chemically as a result of the effect of light.

Rod pigments

The rod pigment is known as rhodopsin and it is incorporated into the disc membranes of the outer segments. Rhodopsin is a complex of vitamin A aldehyde, retinal (also known as retinene) and a protein scotopsin (an opsin). Since the intact molecule has a purplish colouration, it is also known as 'visual purple'.

In the intact rhodopsin molecule retinal exists as the 11-cis isomer or folded form and it is so structured that it fits into a pocket on the surface of the much larger scotopsin molecule (Fig. 8.9). Light energy causes the retinal to alter its configuration to the all-trans or straight form which, because of its shape, no longer fits into the scotopsin. As a result, the retinal and the scotopsin separate and the 'visual purple' becomes colourless.

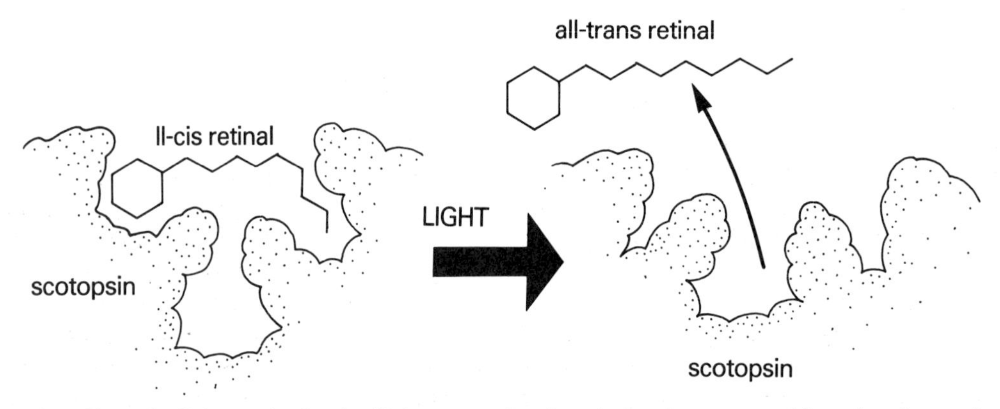

Fig. 8.9 The effect of a light on rhodopsin. Light causes the cis-retinal to be converted into the all-trans form. The change in the molecular configuration of retinal means that it will no longer fit into its slot on the scotopsin molecule. As a result, the retinal is released from the scotopsin. The scotopsin also changes its shape slightly.

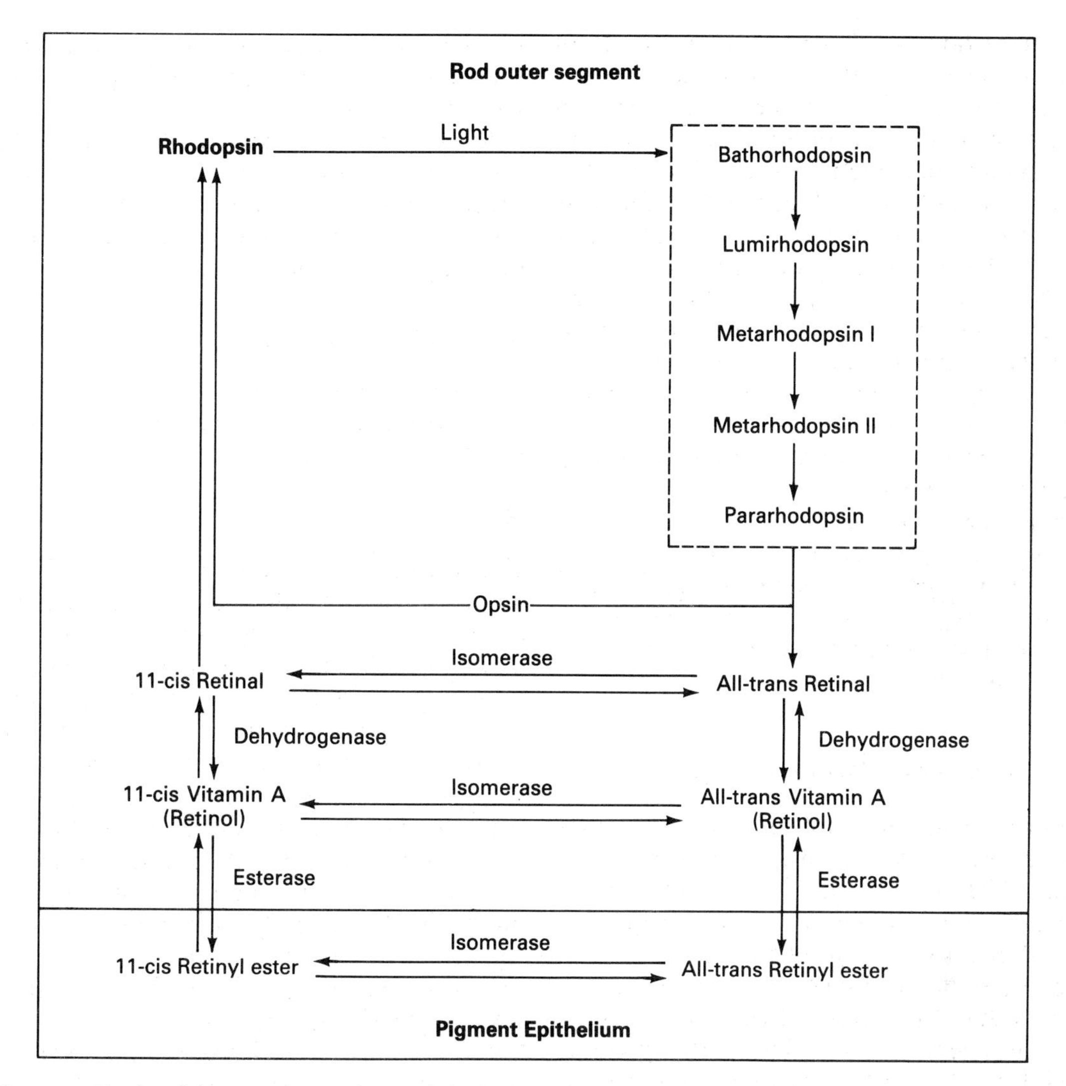

Fig. 8.10 The breakdown and resynthesis of rhodopsin. Intermediate compounds formed in the conversion of bathorhodopsin to pararhodopsin are very short-lived.

The above is a summary of the degradation of rhodopsin. Further details of this complex process are given in Figure 8.10, for the chemically minded. It should be noted however, that only the first stage of the reaction, i.e. the conversion to bathorhodopsin, is light-dependent. All of the remaining reactions occur in the dark.

Rhodopsin is regenerated with the aid of the enzyme retinal isomerase, which converts the all-trans form of retinal back to the 11-cis form, so that it will once again combine with scotopsin.

Cone pigments

The Young-Helmholtz theory of colour vision depends on the existence of three types of cones each with a different pigment, each of which has a different spectral sensitivity (Fig. 8.11). One pigment absorbs light maximally in the blue-violet portion of the spectrum, one in the green portion and one in the yellow-red portion. These are normally referred to as blue, green and red cones, although this is somewhat misleading as it refers to the colour of light to which each responds maxi-

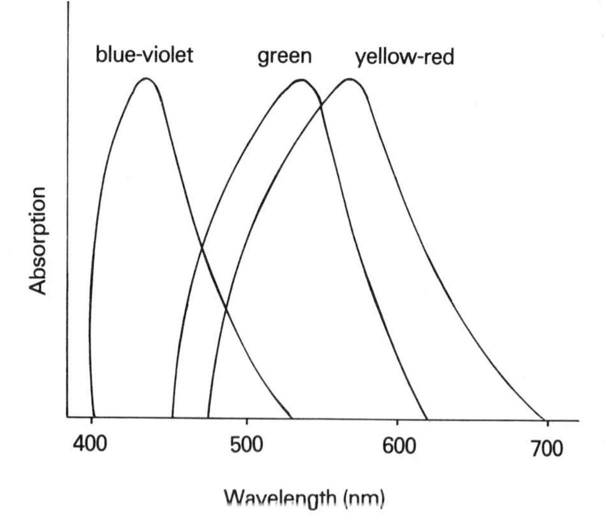

Fig. 8.11 Spectral sensitivities of the cone pigments.

mally and not to the actual colour of the cones. The theory holds that for a particular colour of light, the pattern of response from the three types of cone form a code which the brain interprets as that particular colour. Since blue, green and red are the primary colours then stimulation of the different cones to different levels of excitation will allow the brain to 'see' any colour. For example, if the red cones are excited to 83% of their maximum capability, green cones to 83% and blue cones not at all, the brain will 'see' the colour yellow. If all three cone types were excited to the same level, white light would be perceived.

The eye will react in the same way to a monochromatic light source, e.g. yellow light as to an appropriate mixture of, in this case, red and green light. In both cases the brain will receive an impression of yellow light.

The red-sensitive pigment is called iodopsin and is formed by a combination of retinal (as in rods) and the protein photopsin. The other pigments are also known to contain retinal but differ in their opsins.

If one or more types of cone are congenitally absent from the retina, the person will be 'colour blind'. Most commonly, lack of red- or green-sensitive cells are found and in either case the person will not be able to discriminate colour changes over the red–green portion of the spectrum. He, or more rarely she, will therefore suffer from red-green colour blindness.

Light and dark adaptation of the retina

The sensitivity of the retina is dependent upon the amount of pigments in the rods and cones. If the eye is exposed to bright light for a long period, then the total amount of pigment will be greatly reduced. As a result the eye is said to be light-adapted and its overall sensitivity will be reduced.

If the eye is kept in darkness for a long time, then photopigments are synthesized to their maximum concentration, which is determined by the number of opsin molecules present. In this case the eye is said to be dark-adapted and it exhibits maximum sensitivity.

The rods adapt more slowly than the cones, but they eventually reach a higher level of sensitivity. This means that the rods are used more than the cones in poor lighting conditions, in which case one can only discriminate between shades of black and white (scotopic vision). In bright light, however, when the cones are operating, vision is described as photopic because different colours can be discriminated.

Light and dark adaptation are necessary in order that the eyes can function under widely different levels of lighting since the pupil size and therefore the amount of light entering the eyes can only be varied by relatively small amounts.

Electrical aspects of retinal function

In the dark-adapted retina there is a steady flow of electrical current (sodium ions) from the inner segment of the photoreceptor to the outer segment (Fig. 8.12). This flow relies upon the presence of a sodium pump in the membrane of the inner segment which continually removes Na^+ from the cell and creates a negative membrane potential. The membranes of the outer segments are highly permeable to Na^+ and therefore Na^+ re-enters the cell down its electrochemical gradient.

Under the influence of light the photo-

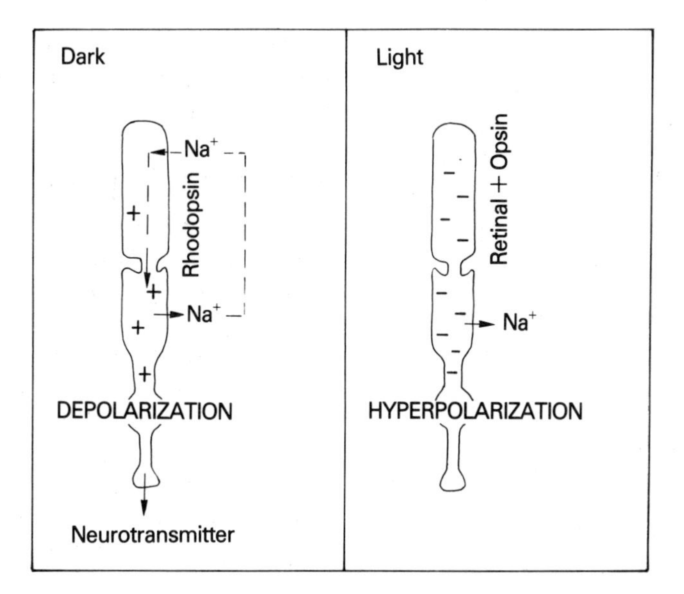

Fig. 8.12 Effect of light on a photoreceptor (see text for details).

chemical changes which take place in the outer segment in some way reduce its permeability to Na^+ so that while the latter continues to be pumped out, fewer ions re-enter the cell. The mechanism by which the Na^+ permeability is reduced is not understood, although several theories have been suggested, e.g. Ca^{++} may be used to block the Na^+ pores. As a result of the reduction in Na^+ permeability the overall negative charge within the cell is increased, i.e. it becomes hyperpolarized, giving rise to the so-called receptor potential. This is very unusual, as receptors usually become depolarized on excitation

The degree of hyperpolarization is proportional to the amount of light striking the retina (strictly to the logarithm of light energy) and its duration depends upon how long the light is switched on.

It is thought that photoreceptors release transmitter substances in the absence of light. These substances maintain the horizontal and bipolar cells in a constant state of depolarization. Light causes the cessation of transmitter release from the photoreceptors and therefore removes inhibition from the other cells. As a result the horizontal cells become hyperpolarized while the bipolar cells become either hyperpolarized or depolarized. It is not understood how these alternative responses can occur in the bipolar cells, but it is evident that there are two different types of cell which respond differently to the same stimulus.

Monitoring the behaviour of individual bipolar cells shows that a spot of light falling on to its attached receptors may therefore cause either depolarization ('on-centre') or hyperpolarization ('off-centre'). An annulus or ring of light falling upon the receptors which surround the central area, but which link to the bipolar cell through horizontal cells produces an opposite response, i.e. if the bipolar cell is depolarized by a spot of light it will be hyperpolarized by an annulus (see Fig.8.13). Thus two types of receptive field ('on-centre' and 'off-centre') can be demonstrated.

Even illumination over the whole receptive field or indeed the whole retina will produce a cancelling-out effect that greatly reduces the amount of information passing to the brain. The presence of a light pattern (i.e. an image) with sharp edges will, however, produce excitation and inhibition of adjacent fields which lead to the generation of a pattern of impulses that can be interpreted by the brain.

The bipolar cells pass information to the ganglion cells whose fibres pass into the optic

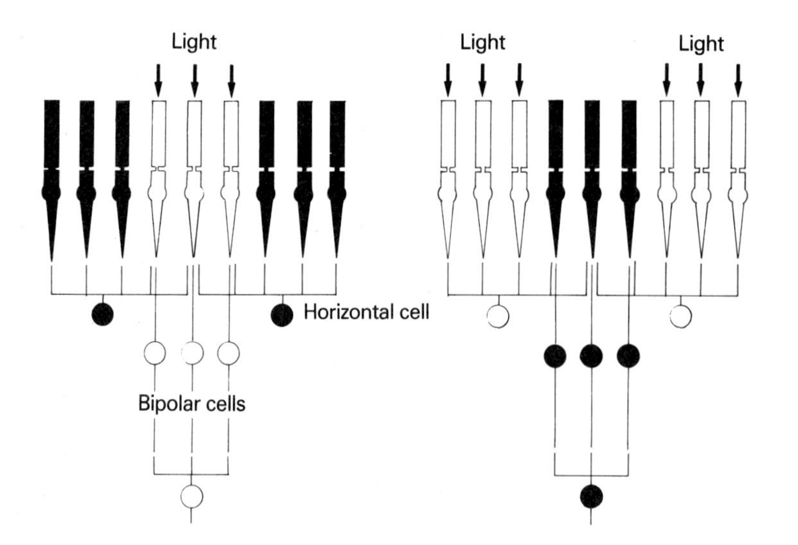

Fig. 8.13 Diagram to show the 'centre-surround' organization of a retinal field. This is an example of an 'on-centre field'. On the left, the centre of the field is stimulated by a spot of light which excites the photoreceptors and leads to excitation of the bipolar and ganglion cells. On the right, an annulus (ring) of light stimulates the peripheral receptors and their horizontal cells, leading to inhibition of the bipolar and ganglion cells.

nerve. There appear to be at least two types of ganglion cells present in the retina. The first shows a tonic response, a constant rate of discharge when the centre of a receptive field is stimulated but which is inhibited when the periphery is excited. The second type of ganglion cell produces a sudden burst of action potentials when a field is illuminated and a second burst when the illumination is switched off.

The first type of ganglion cell seems therefore to provide information on the intensity of illumination whilst the second type signals change in the visual field and would, for example, allow detection of a moving object. The amacrine cells lying at the boundary of the bipolar and ganglion cell layers also exhibit transient responses, showing a burst of activity at the onset of light. Ganglion cells show a certain amount of electrical activity even when the retina is not receiving illumination. Stimulation of the photoreceptors and the subsequent activities of bipolar, horizontal and amacrine cells therefore bring about change (i.e. increases or decreases) in the number of impulses travelling in the optic nerve fibres, rather than simply initiating action potentials.

Although the preceding account has not differentiated between the responses of rods and cones, their responses to light are, in fact, different. Cones respond to light with a sharp hyperpolarization which rapidly dies away (i.e. fast onset and offset). Rods, on the other hand, respond with sharp hyperpolarization which dies away slowly (fast onset and slow offset). The level of cone response changes with regard to the background illumination but that of rods does not. Therefore, cones show a good response to changes in light intensity at high levels of illumination, while rods differentiate between the presence or absence of light.

The perception of colour starts in the photoreceptive layer with the three types of cones and continues through a processing system within the neural layers. Ultimately this results in a pattern of action potentials in the ganglion cells.

It has been shown that individual ganglion cells may be stimulated by one colour but inhibited by a second 'complementary' colour. For example, red light will cause excitation and green light inhibition. It is probable that red light excites a red cone which then transmits its excitation through a bipolar cell to a ganglion cell. Green light excites a green cone which then hyperpolarizes a horizontal cell which is linked to the 'red' bipolar cell and the latter is inhibited.

Yellow and blue are also complementary but in this case the link is between red or green cones (both of which are stimulated by yellow light) and blue cones.

Some ganglion cells are presumably stimulated by all three types of cones and will therefore be responsible for the transmission of information concerning white light.

On average 140 rods and six cones converge on each ganglion cell. The level of convergence varies in different parts of the retina, however, from about 300 rods to one ganglion cell at the periphery, to one cone to one ganglion cell at the fovea where there are no rods and no convergence. The relative sizes of the visual fields in different parts of the retina vary accordingly. This explains why the highest visual acuity is achieved at the fovea.

The processing of information from the retina

Visual pathways

Figure 8.14 shows the nerve pathways from each retina to the lateral geniculate body in the thalamus and thence to the visual cortex. It can be seen that fibres from the lateral retina travel to the thalamus and visual cortex of the same side, whereas fibres from the nasal retina cross over. The lateral geniculate bodies are also the relay points for nerve pathways to the nuclei of cranial nerves III, IV and VI which regulate lens shape, pupil size and eye movement (see *Neural regulation of image formation*).

The cells of the lateral geniculate body are arranged in six distinct layers, or laminae, and it has been shown that the crossed (nasal) fibres of the optic tract terminate in layers 1, 4 and 6 while the uncrossed (lateral) fibres terminate in layers 2, 3 and 5. Each fibre of the optic tract may synapse with up to 30 lateral geniculate cells, but in humans each geniculate cell receives impulses from only one fibre. Thus in each lateral geniculate body there is not only a sorting out of information that has come from the two eyes but also separation of stimuli from individual points on the retina.

In passing to the visual cortex the fibres from the lateral geniculate body are arranged so that those from the six layers come

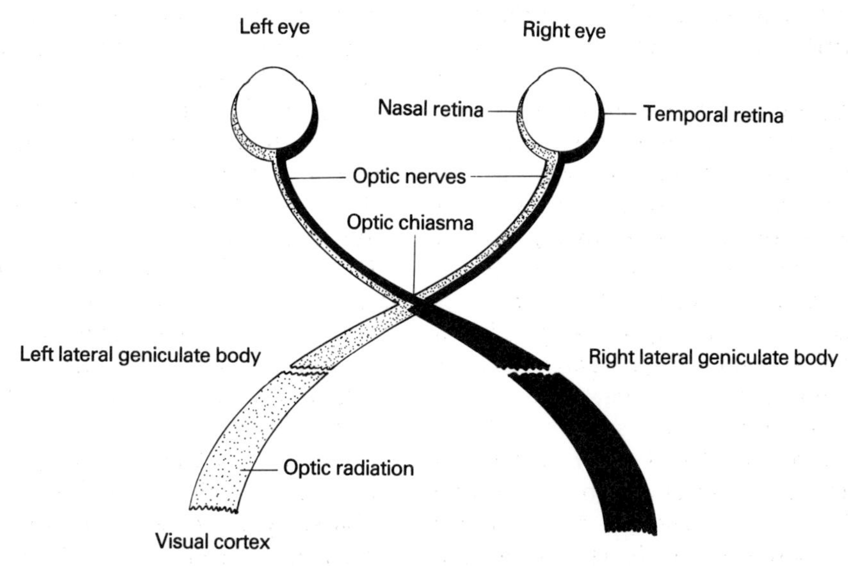

Fig. 8.14 Visual pathways. Nerve impulses travel from the retina along the optic nerve to the optic chiasma. Fibres from the lateral (temporal) retina of each eye travel on the same side to the lateral geniculate body, whereas fibres from the nasal retina of each eye cross over and travel to the lateral geniculate body, of the opposite side. The neurones synapse in the geniculate bodies with neurones which travel in the optic radiation to the visual cortex.

together at the cortex. Thus information from corresponding parts of the two retinae juxtapose in the visual cortex. Moreover studies of the relationship between the retina and the cortex have shown that the former can be represented by a projection onto the surface of the brain. This projection is not uniform, however, since the macula is represented by a disproportionately large area on the surface of the cortex.

Impulses from the visual cortex pass to other areas of the brain including the visual association areas where further processing occurs such that the image is given meaning.

Visual processing

The cells of the visual cortex exhibit responses which can be tied to the receptive fields of the retina. However, the configurations of the retinal fields which elicit responses in cortical cells are not identical to those which elicit such responses in bipolar cells. Cortical responses rely upon stimulation of the retina by lines and edges rather than minute spots of light, i.e. the retinal fields of the cortical cells are linear rather than circular.

It has been demonstrated that there are a number of different types of neurones present in the visual cortex which are arranged hierarchically. Thus it is probable that several lateral geniculate fibres converge on a 'simple' cortical cell, several of which converge on a 'complex' cell. At least two of the latter fibres then converge on a 'hypercomplex' cell. The retinal fields associated with each type of cell show a corresponding increase in complexity.

Binocular vision

Each eye when facing directly forwards is capable of 'seeing' over quite a wide area, or visual field. This field is not circular, as might be expected, but is cut off by the nose and orbital ridge. Careful mapping of the visual fields shows that there is a considerable area of overlap with smaller temporally located areas that can be seen only by one eye. Impulses generated at corresponding points on the two retinae are fused into a single image within the cerebral cortex. Impulses generated within the temporal part of a single retina obviously cannot be fused in a like manner. Therefore images formed on the nasal and central portion of the retinae give rise to fused or binocular images which lead to an appreciation of depth. Peripheral temporal images on the other hand give rise to monocular or flat images within the brain.

It should perhaps be emphasized that the perception of depth depends upon the fact that the two images generated at corresponding points on the two retinae are not absolutely identical. Rather, they are two views of the same object obtained at points which are about 10 cm apart (the distance between the fovea of the two eyes). It is also important to realize that a great deal of information concerning depths of fields is derived from a knowledge of the comparative sizes of objects. This is particularly true when viewing distant objects.

THE EAR

The externally visible portion of each of the two ears, the auricle, is located on the lateral surface of the head on each side. The remaining parts of each ear are embedded in the bone of the skull beneath. The different positions of the two ears enables directional sources of sound to be pinpointed.

Each ear receives sound vibrations from the environment and converts them into nerve impulses which are transmitted to the auditory centres of the brain.

The ears also play an important part in the appreciation of head position and movements and initiate reflex postural adjustments which maintain balance.

STRUCTURE OF THE EAR

Each ear can be divided into three primary areas, the external, middle and internal ear. The external ear comprises the auricle, or

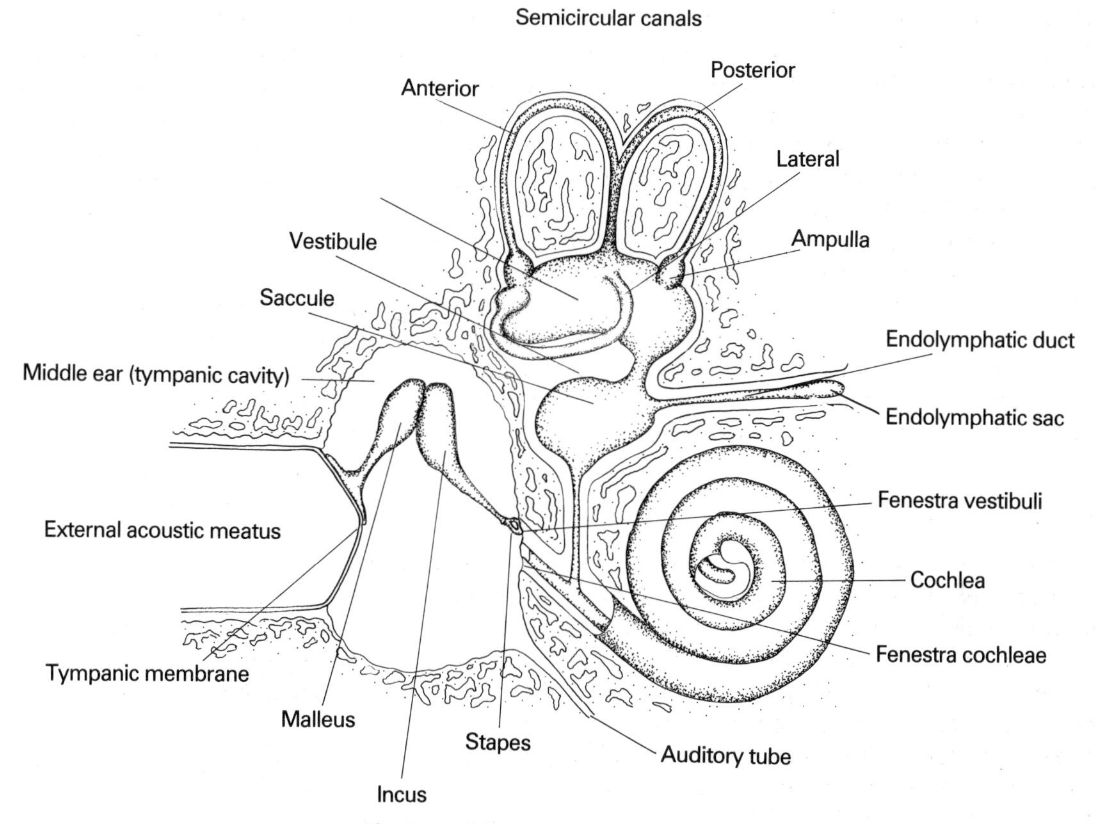

Fig. 8.15 Schematic diagram of the ear.

pinna, together with the external acoustic meatus. The air-filled middle ear contains three bones, or ossicles, which transmit sound from the external ear to the internal ear. The internal ear is composed of an osseous (bony) labyrinth, which is a cavity in the temporal bone of the skull and which contains a membranous labyrinth, consisting of several interconnecting fluid-filled sacs and tubes suspended in fluid (Fig. 8.15).

The external ear

Each auricle consists of a piece of elastic cartilage with a number of associated ligaments and muscles and it is covered with skin. The lobule (ear lobe) does not contain cartilage, but is composed of fibrous and adipose tissue which makes it softer than the rest of the auricle. The cartilage in the auricle is continuous with that forming the lateral (outermost) third of the wall of the external acoustic meatus. The skin of the auricle is covered with fine hairs and is continuous with the skin lining the external acoustic meatus and covering the tympanic membrane (eardrum) at the junction between the external and middle ears.

There are two sets of auricular ligaments; two extrinsic ones, which connect the auricle to the temporal bone; and several intrinsic ones, which link different parts of the cartilage. Similarly, there are two sets of auricular muscles; the extrinsic ones being connected to the skull and scalp and responsible for limited movements of the whole ear, whereas the intrinsic auricular muscles connect different parts of the auricle and are only capable of very limited movements. These muscles are supplied by various branches of the facial nerve.

The external acoustic meatus is an S-shaped

tube, oval in cross-section and about 2.5 cm long. The first third of its length is cartilaginous, the remaining two-thirds osseous. The skin lining the meatus contains ceruminous glands; these produce cerumen (wax), which protects the lining from damage by trapped water and deters attack by insects.

The tympanic membrane at the end of the external acoustic meatus is semitransparent, oval and orientated at an angle of 55 degrees, with the floor of the external acoustic meatus. Its longest diameter is about 10 mm. The membrane has a three-layered structure, the outermost layer consisting of skin (hairless), the middle layer of connective tissue containing collagen and elastic fibres, and the innermost layer is continuous with the mucous membrane lining the middle ear.

The middle ear

The middle ear is also known as the tympanic cavity, although strictly there is a tympanic cavity proper, and an epitympanic recess which lies above the level of the tympanic membrane.

The air-filled middle ear is very irregular in shape and measures some 15 mm vertically and from front to back. It is bounded laterally by the tympanic membrane, medially by the internal ear, posteriorly it connects with the mastoid antrum and then the mastoid air cells, and anteriorly it connects with the auditory tube.

The auditory tube (also known as the pharangotympanic or Eustachian tube) connects the middle ear with air in the nasopharynx. It is usually about 36 mm long and is lined with mucous membrane, the surface of which is composed of ciliated, columnar epithelium. The pharyngeal orifice is normally closed, but opens on swallowing.

The three bones or ossicles lying in the middle ear serve to transmit sound from the tympanic membrane to the fenestra vestibuli (oval window) at the junction of the middle and inner ears. They are named malleus, incus and stapes (mallet, anvil and stirrup) because of their respective shapes. The malleus is attached to the tympanic membrane, the stapes is attached to the circumference of the fenestra vestibuli, and the incus articulates with the other two ossicles by means of synovial joints. The ossicles are attached to the wall of the middle ear by ligaments.

The malleus is some 8–9 mm long and consists of a head, neck and three processes, the manubrium (handle), anterior process and lateral process (see Fig. 8.16). The head articulates with the incus at a cartilage-covered facet. The manubrium is connected to the tympanic membrane.

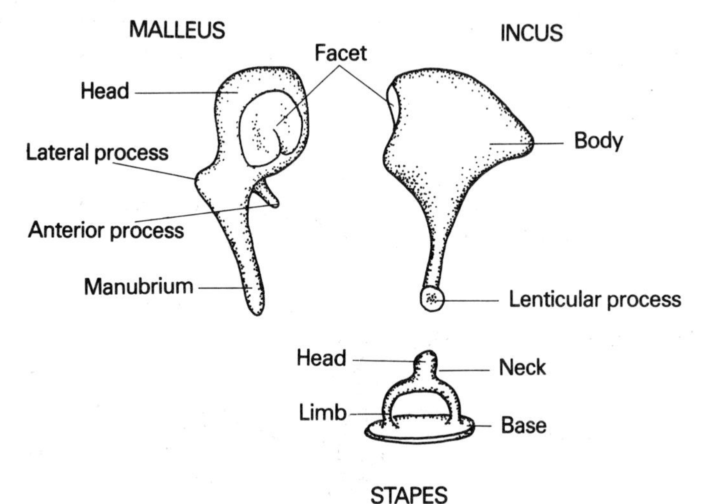

Fig. 8.16 The ossicles.

The incus has a body and two processes. The body articulates with the malleus, whereas the longer of the two processes has a rounded end, called the lenticular process, which articulates with the head of the stapes.

The stapes consists of a head, neck, two limbs and a base. The head articulates with the incus, whereas the base or footplate is held in the fenestra vestibuli by an annular ligament.

The tensor tympani muscle lies in a canal above and parallel to the auditory tube and has its origin on the cartilaginous wall of the tube. The muscle inserts into the neck of the malleus. It is supplied by the trigeminal nerve and pulls the malleus inwards. The stapedius muscle arises from the posterior wall of the middle ear and inserts into the neck of the stapes. This muscle is supplied by the facial nerve and it acts to pull the stapes away from the fenestra vestibuli. These muscles are stimulated to contract in the attenuation reflex (see below).

The internal ear

Osseous labyrinth

The bony cavity within the temporal bone known as the osseous labyrinth is lined with periosteum, which in turn is lined with perilymphatic cells which have a variety of shapes in different parts of the labyrinth. It is composed of three parts, the vestibule, the cochlea and the semicircular canals.

The vestibule lies medial to the tympanic cavity and measures about 5 mm from front to back and 3 mm across. Anteriorly, the cochlea extends from the vestibule and the semicircular canals arise from the posterior aspect. Another extension, the aqueduct of the vestibule, arises just below the semicircular canals.

The cochlea is a helical cavity (like the inside of a snail's shell), consisting of a canal in the bone winding around a central cone-shaped piece of bone, called the modiolus, for two and three quarter turns. The cochlea measures 9 mm across at its base and is 5 mm from base to apex. Projecting from the modiolus, rather like the thread on a screw, is a shelf of bone called the osseous spiral lamina (see Fig. 8.17).

The cochlea is some 35 mm long and its diameter reduces from the base (where it is about 3 mm) to the apex. At the base of the cochlea there are three openings in the bone, the fenestra vestibuli; the fenestra cochleae or round window which lies below the fenestra vestibuli and which is closed by the secondary tympanic membrane; and the aperture of the cochlear canaliculus, which opens out on the inferior surface of the temporal bone and thereby provides a connection between the subarachnoid space and the scala tympani in the cochlea.

Each ear contains three semicircular canals orientated approximately perpendicular to each other. They are the anterior or superior canal, the posterior canal and the lateral canal. Each canal has an expansion, the ampulla, at one end.

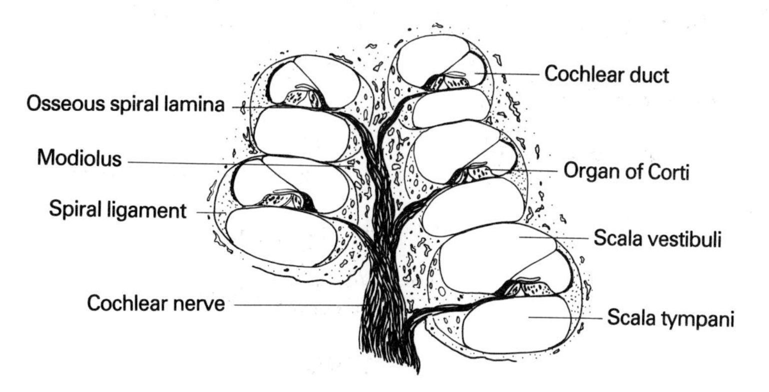

Fig. 8.17 Section through the cochlea.

The anterior canal measures between 15 and 20 mm and is orientated vertically when the head is bent forward at an angle of about 30 degrees. Posteriorly it joins the posterior canal, which is also vertical at right angles to the anterior canal, to form a common limb, the crus commune. The lateral canal is orientated in the horizontal plane, is 12–15 mm long and its ampulla is just above the fenestra vestibuli, while the other end of the canal opens just below the crus commune.

Membranous labyrinth

The membranous labyrinth lies within the osseous labyrinth to which it is attached at certain points, but otherwise it is surrounded by fluid, known as perilymph, with a composition like that of cerebrospinal fluid. The fluid contained within the membranous labyrinth is known as endolymph and resembles intracellular fluid.

The membrane has three layers. The outermost layer consists of fibrous tissue with blood vessels, the middle layer is looser, more highly vascular connective tissue, while the inner layer is simple epithelium, variously squamous, cuboidal or polygonal.

The membranous laybrinth may be subdivided into the vestibular apparatus, which is concerned with proprioceptive information concerning head position and rotation; and the cochlear duct, which is concerned with hearing.

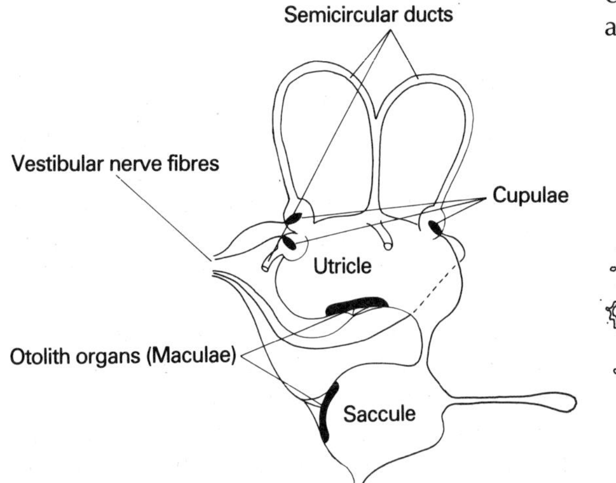

Fig. 8.18 The vestibular apparatus.

Vestibular apparatus

The vestibular apparatus (see Fig. 8.18) comprises the membranous semicircular ducts lying within the bony canals, each one having an ampulla within the bony ampulla; the utricle, which is a sac connecting with the semicircular ducts and with a second sac, the saccule. Arising posteriorly from the saccule is the endolymphatic duct, which ends in a blind expansion, the endolymphatic sac.

The semicircular ducts are only about a quarter of the diameter of the bony canal and are attached to the wall by fibrous tissue (see Fig. 8.19).

Within each ampulla, the membrane is thickened and projects into the cavity as a ridge called the ampullary crest (see Fig. 8.20A). On the surface of the crest are two types of hair cells, together with supporting cells. The type I hair cells are surrounded by a goblet-shaped nerve terminal, the calix, which is very close to, but does not actually touch, the hair cell (see Fig. 8.20B). Type II cells, on the other hand, are not surrounded with a calix but by a number of boutons. Some of these are afferent (sensory) terminals, whereas others are efferent (motor) and may be involved in altering the threshold of the type II cells.

Each sensory fibre innervating the type I cells innervates only a few such cells which are located near to each other. In contrast,

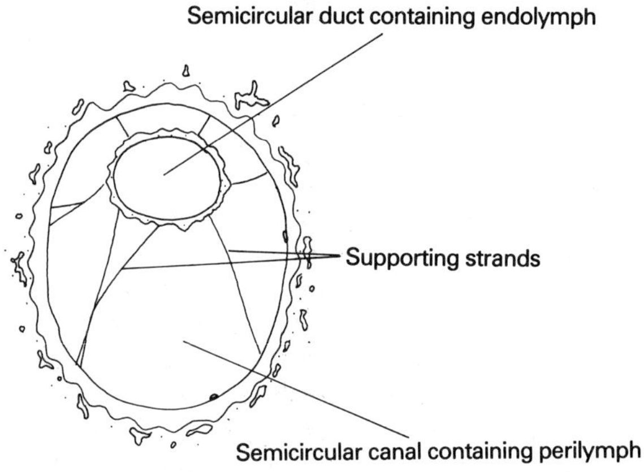

Fig. 8.19 Section through a semicircular canal.

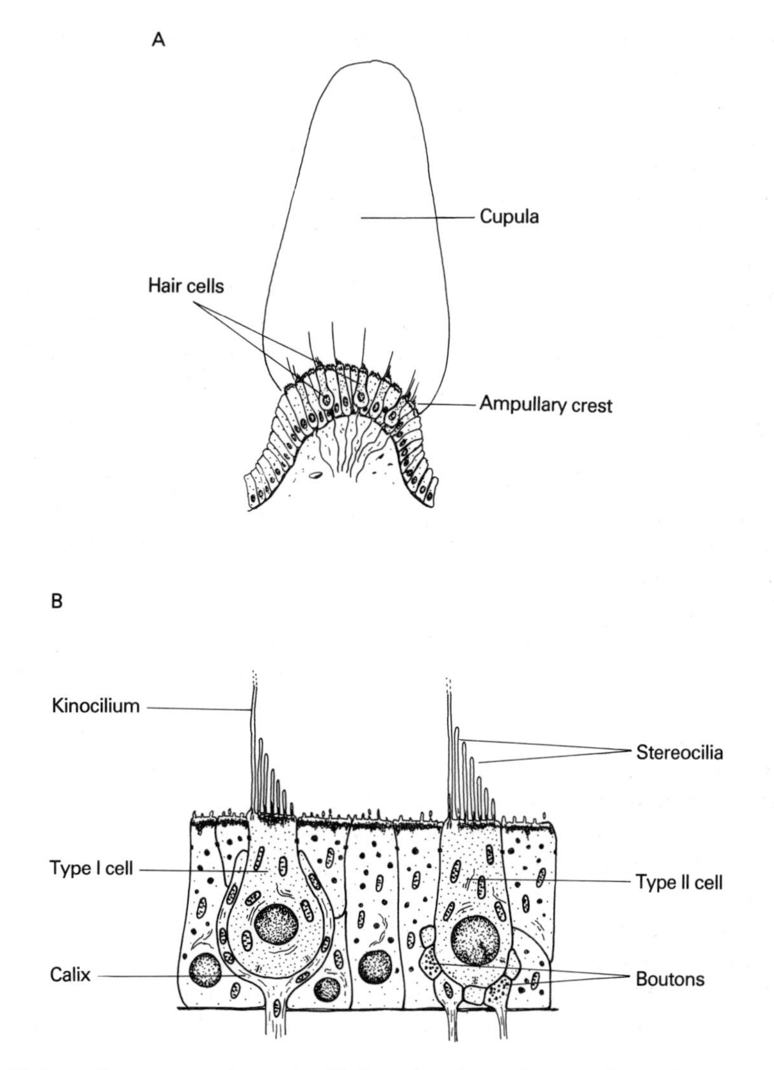

Fig. 8.20 (A) Ampullary crest and cupula. (B) Type I and type II hair cells on the ampullary crest.

type II cells are innervated by a number of different neurones, which themselves innervate a large number of type II cells spread over a large area. This pattern of innervation results in the type I cells being more discriminative.

The hairs of the type I and type II cells are in fact stereocilia, or modified microvilli. There are between 40 and 100 stereocilia per cell, arranged into a hexagonal group. In addition one border of the cell has a kinocilium, which has the usual nine doublets of microtubules around the periphery but the central pair are normally poorly developed or may be absent.

Within each ampulla, the hair cell processes project into a jelly-like, dome-shaped structure, the cupula, which projects across the ampulla and can be deformed by endolymph swirling within the semicircular duct when the head starts, or stops, turning. The jelly is composed of a protein–polysaccharide complex.

Within the utricle and the saccule is a structure known as a macula or otolith organ (see Fig. 8.21). In the utricle the macula measures some 3 by 2 mm and is orientated horizontally on the lateral wall, whereas in the saccule the macula is orientated vertically on the anterior wall, at right angles to the utricular macula.

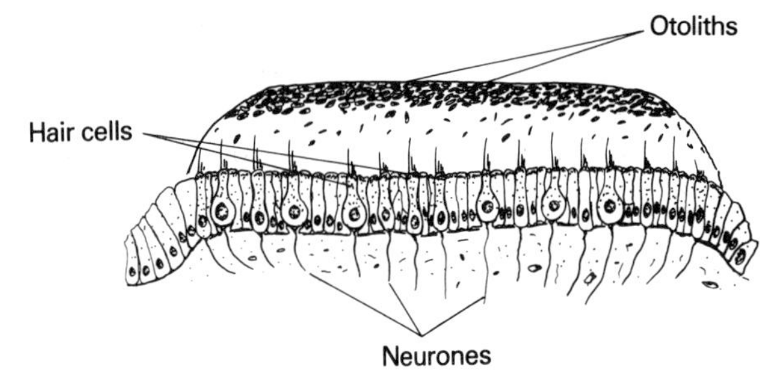

Fig. 8.21 Macula or otolith organ.

The hair cells and supporting cells of the maculae are similar to those on the ampullary crests, but they project into an otolithic membrane composed of jelly containing particles of calcium carbonate and protein, called otoliths or otoconia. The otolithic membrane is flat and kidney-shaped in the case of the utricle, and hook-shaped in the saccule.

Cochlear duct

A cross-section through the cochlea shows a kidney-shaped structure comprising three canals, or scalae (Fig. 8.22). The middle canal (scala media, or cochlear duct) is part of the membranous labyrinth and is filled with endolymph which is continuous with that in the vestibular apparatus.

Either side of the middle canal there are two canals filled with perilymph. The scala vestibuli contains fluid which is continuous with that in the vestibule, while the fluid in the scala tympani is continuous with that which bathes the secondary tympanic membrane.

The outer wall of the cochlear duct is known as the stria vascularis and is composed of stratified epithelium containing capillaries. This structure overlies the spiral ligament which is composed of thickened endosteum lining the bone.

The cochlear duct is separated from the scala vestibuli by the vestibular membrane. This is a thin, delicate structure composed of two layers of squamous epithelium separated by a basal lamina.

The basilar membrane separates the cochlear duct from the scala tympani. The membrane is about 35 mm long, attached to the spiral ligament along its outer edge, and to the osseous spiral lamina along its inner edge. The basilar membrane is much stronger than the vestibular membrane and contains transversely orientated protein filaments, 8 to 10 nm thick, embedded in a matrix.

The structure and physical properties of the basilar membrane vary along its length. At the basal end the membrane is narrow (0.21 mm), the fibres short and the membrane stiff; towards the apex the membrane is much wider (0.36 mm) and more flexible. The membrane ends just short of the apex of the cochlear so that there is continuity between

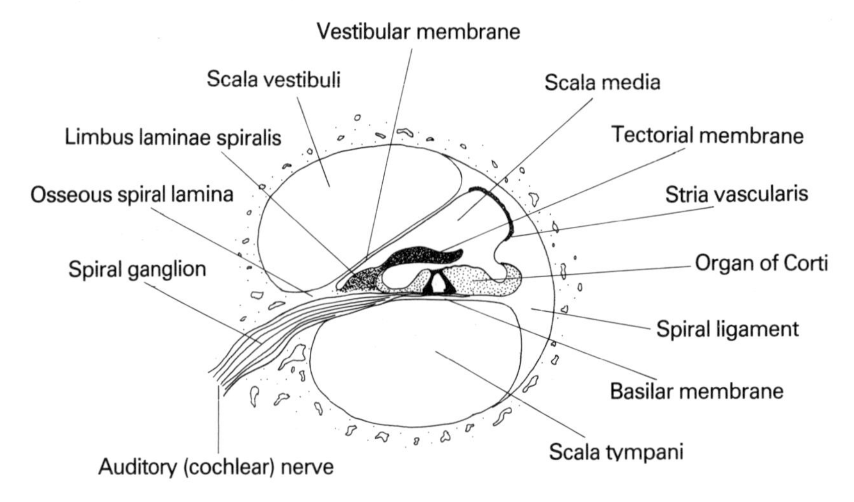

Fig. 8.22 Section through the cochlea.

the scala vestibuli and the scala tympani at the helicotrema (Figs. 8.15 and 8.24).

The sensory receptors associated with hearing are found in the organ of Corti, which lies along the length of the basilar membrane. The receptors are hair cells, which are arranged in two groups (outer and inner) and are surrounded by a variety of packing cells. Perhaps the most distinctive of these supporting cells are the rod cells of Corti, which separate the two groups of hair cells. The rod cells have expanded bases resting on the basilar membrane and microtubules running along the length of the cells. They are orientated in such a way as to create a triangular-shaped tunnel of Corti (Fig. 8.23).

The outer hair cells are arranged in three rows. Each cell has some 50–100 stereocilia projecting from its apex in a V- or W-shaped pattern. The tallest of these cilia are embedded in the overlying tectorial membrane. This is a gelatinous structure, containing fibres, which is attached to the limbus laminae spiralis, a layer of thickened periosteum covering the osseous spiral lamina. The cells comprising the single row of inner hair cells have 50–60 stereocilia arranged in a shallow U-shaped pattern. These stereocilia do not embed in the tectorial membrane.

Both types of hair cells are innervated by sensory neurones, bipolar cells whose cell bodies lie in the spiral ganglion. The inner hair cells are supplied by many more sensory neurones than those in the outer group.

Motor neurones are also present and those which innervate the outer cells are known to have an inhibitory function. The inner hair cells do not have any direct synaptic contact with motor neurones, however, and instead the latter synapse with the sensory neurones whose activities they presumably modify.

FUNCTIONS OF THE EAR

Hearing

Sound waves pass through the external acoustic meatus and cause the tympanic membrane to vibrate. They then pass through the middle ear by vibration of the ossicles, and on to the liquid-filled internal ear.

Movement of the sensory receptors in the cochlea, induced by the sound waves, stimulates their sensory neurones and the information thus collected is conveyed along the auditory pathways to the brain. Here occurs the remarkable process of perception of the great variety of types of sound, its loudness and its source.

Transmission of sound in the middle ear

Sound waves hitting the tympanic membrane cause it to vibrate. These vibrations are then transmitted directly to the malleus, then to the incus and finally the stapes, each bone moving against the next at its joint. The foot-

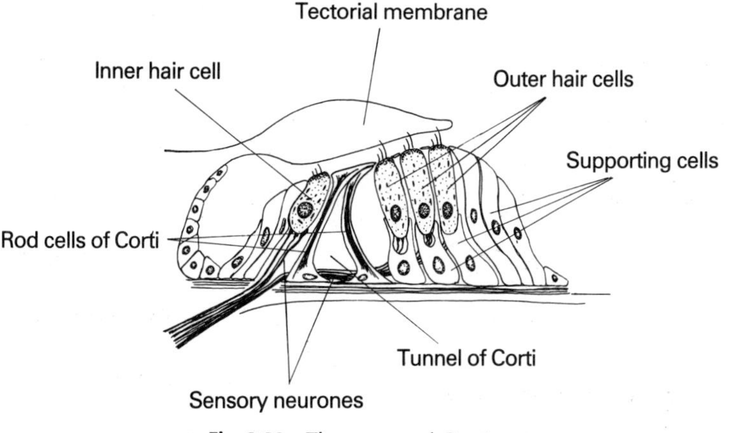

Fig. 8.23 The organ of Corti.

plate of the stapes moves in and out of the fenestra vestibuli employing a rocking motion. The latter results from the uneven thickness of the annular ligament by which the stapes attaches to the margin of the fenestra vestibuli. The force of the sound waves is increased by an estimated 1.3 to 1 as they pass through the middle ear, because the ossicles act as a system of levers. Furthermore, the eardrum is some 15 to 17 times larger than the footplate of the stapes and therefore the force is exerted on a smaller surface area. The pressure (i.e. force exerted per unit area) of the sound waves is thus increased some 20 times as they pass through the middle ear.

Attenuation reflex

A sudden loud noise causes reflex contraction of the tensor tympani and stapedius muscles in the middle ear. This causes the ossicles to move closer together and become more rigid, thereby reducing their ability to transmit sound waves, particularly those of low frequency.

This reflex is a protective mechanism which prevents excessive vibrations from being transmitted to the internal ear. The reflex does not, however, prevent damage to sensory cells in the internal ear caused by prolonged excessive noise such as from rock concerts or loud machinery.

Transmission of sound and sensory processing in the internal ear

The movement of the stapes in and out of the oval window sets up vibrations in the perilymph within the vestibule. The vibrations are then transmitted through the perilymph into the scala vestibuli and from there into the scala media (cochlear duct) across the easily deformed vestibular membrane (Fig. 8.24). The basilar membrane then vibrates and the sound is transmitted across into the scala tympani, where it is dissipated as the secondary tympanic membrane bulges from the round window.

The pitch of a note depends upon the frequency of the sound waves, high notes having a high frequency of vibration compared with low notes. Intrinsic variability in the degree of flexibility of the basilar membrane results in it being maximally displaced (resonating) by high frequency notes at positions relatively near to the base of the cochlea, where it is stiff. Low frequency sound waves, on the other hand, travel further towards the apex of the cochlea where the membrane is much more flexible, before causing it to resonate (Fig. 8.25).

Another factor affecting the position of

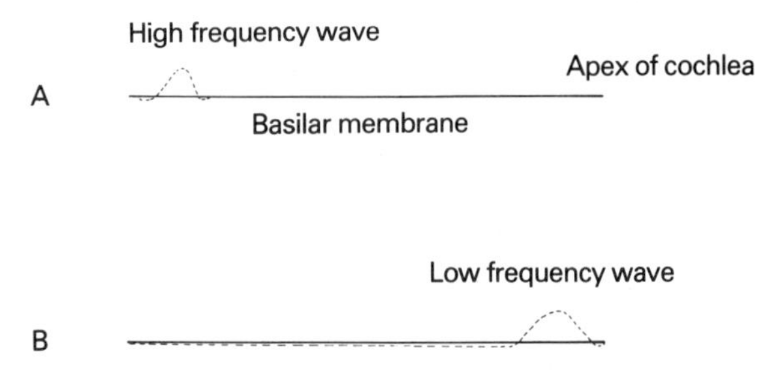

Fig. 8.25 Response of the basilar membrane to (A) high and (B) low frequency sounds.

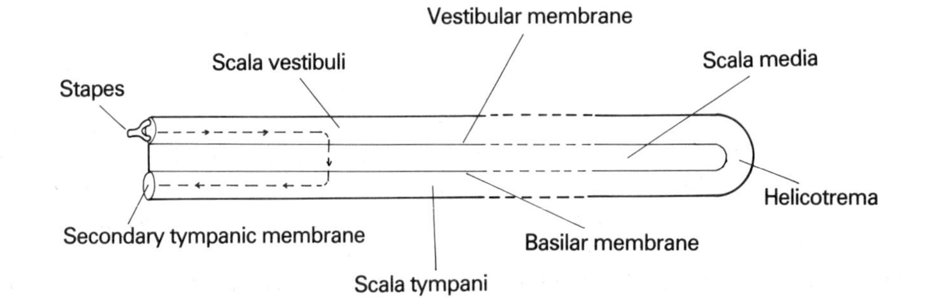

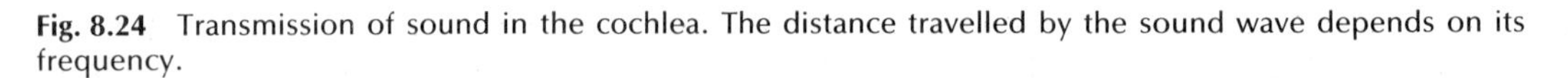
Fig. 8.24 Transmission of sound in the cochlea. The distance travelled by the sound wave depends on its frequency.

resonance of a particular note along the basilar membrane is the mass of fluid involved in the vibration. Vibrations near to the base of the cochlea involve a smaller mass of fluid than vibrations travelling further along the cochlea. This means that the basilar membrane has a smaller fluid 'loading' nearer to its base than its apex, which also affects its resonating properties.

Displacement of the basilar membrane brings about movement of the organ of Corti which lies upon it. The cilia of the outer hair cells of the organ of Corti which are embedded in the tectorial membrane are alternately bent and stretched as they move against this stationary membrane. As the cilia bend and stretch, they are alternately depolarized and hyperpolarized, which leads to the development of alternating receptor potentials within the hair cells. As a result of these excitatory activities, neurotransmitters are released and these stimulate the sensory neurones to which they are attached. It is interesting to note that the alterations in the excitatory state of the hair cells and the accompanying ionic movements across their membranes give rise to alternating changes in the electrical potential of the surrounding fluid. These electrical potentials coincide in frequency with the frequency of the stimulating sound wave and are termed 'cochlear microphonics'. The cilia of the inner hair cells which are not embedded in the tectorial membrane are thought to be stimulated by movements of the endolymph. The importance of this latter group of cells should not be overlooked since they are connected to about 95% of the fibres in the auditory nerve.

Pitch discrimination

There is, then, crude pitch discrimination in the organ of Corti itself. A high-frequency note will stimulate hair cells near to the base of the cochlea and stimulate the sensory neurones supplying them. Low frequency notes will, on the other hand, stimulate neurones nearer to the apex of the cochlea. As the nerve pathways from the different positions along the organ of Corti are kept separate right up to the auditory cortex in the brain, then the same principle is also used in the cortex. It is called the place principle for the discrimination of pitch. Typically, man is able to distinguish pitch over a frequency range of about 30 to about 20 000 cycles per second.

Loudness

The detection of the loudness of a note is achieved in several ways. A louder note causes a greater amplitude of vibration of the basilar membrane and therefore a greater stimulus to the hair cells. This results in an increased frequency of action potentials from the sensory neurones innervating them.

The greater amplitude of vibration of the basilar membrane also means that more neurones on the edge of the vibrating area will be stimulated. Thirdly, some sensory neurones have a higher threshold than others and therefore will only be stimulated by a louder note.

Perilymph and endolymph

Perilymph has a similar composition to c.s.f., although some differences have been reported. The source and circulation of perilymph is, however, still controversial. One view is that it is continuous with the c.s.f. in the subarachnoid space surrounding the brain through the cochlear canaliculus, but although this tube is open in children it is not always patent in adults. Another suggestion is that perilymph is produced as a transudate by blood vessels in the spiral ligament, and that it flows first into the scala vestibuli and then into the scala tympani.

Endolymph is produced by blood vessels lining the membranous labyrinth, both from the stria vascularis in the cochlear duct, and by blood vessels in the vestibular apparatus. After circulating through the membranous labyrinth, the endolymph is returned to the blood in vessels surrounding the endolymphatic sac.

Endolymph has a high K^+ concentration and a low Na^+ concentration, which results in a potential difference (endocochlear potential) between it and the perilymph of some

+80 mV with respect to the former. The total potential difference between the endolymph and the cytoplasm of the hair cells is about −150 mV with respect to the hair cells. It is thought that this relatively large membrane potential sensitizes them to respond to slight movements of the cilia.

Auditory pathways

As impulses are relayed along the nerve pathways in the brain, the spatial separation of neurones stimulated by different sound frequencies is maintained and indeed sharpened. That is, successive neurones in the pathway respond to a narrower frequency band. Figure 8.26 shows the major routes between the sensory neurones from the cochlea to the brain stem and cerebral cortex. It may be noted that impulses from each ear travel to both sides of the cerebral cortex, i.e. some pathways cross over from one side to the other whereas others do not. This is different from sensory pathways in general, where sensory information from the left side of the body is relayed to the right cerebral cortex, and vice versa.

The auditory pathways between the cochlea and the auditory cortex consist of between four and six consecutive neurones. Synapses occur in the medulla (in the cochlear nuclei), in the midbrain (in the superior olivary nuclei, the nuclei of the lateral leminiscus, and inferior colliculi) and in the thalamus (in the medial geniculate nuclei).

There are several points at which crossing over (decussation) from one side of the brain to the other occurs; between superior olivary nuclei, between nuclei of the lateral leminisci and between inferior colliculi.

There are also pathways to the reticular formation and the cerebellum.

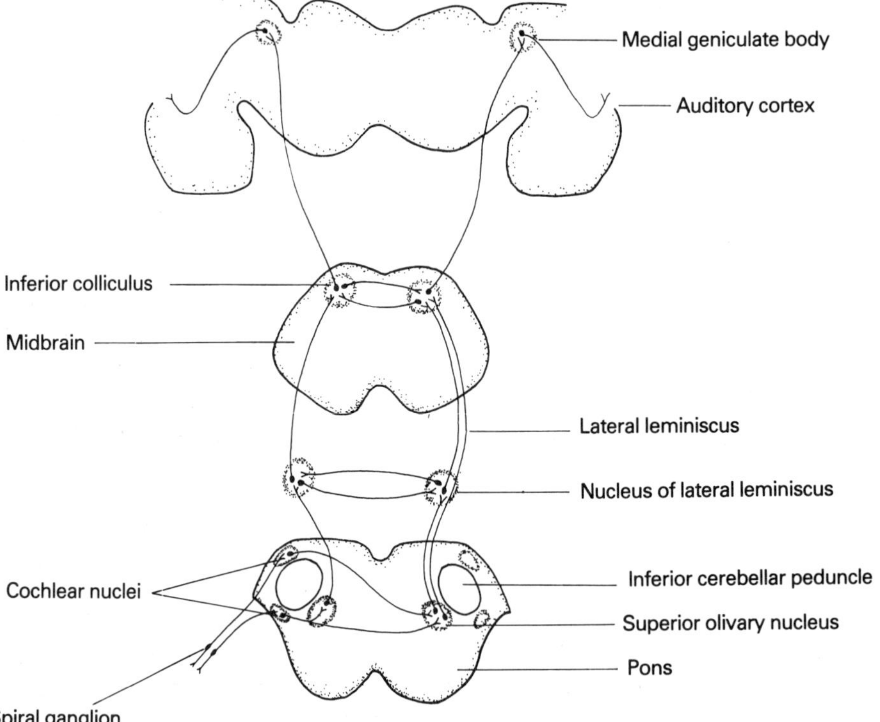

Fig. 8.26 Auditory pathways. Sensory neurones from the cochlea have their cell bodies in the spiral ganglion and enter the upper medulla of the brain and synapse in the cochlear nuclei. The second order neurones mostly cross over through the trapezoid body to the superior olivary nucleus where most of them synapse. The third order neurones ascend through the lateral leminiscus where some of them synapse, others terminate in the inferior colliculus. Some fibres cross over at the level of the lateral leminiscus, others cross over between the inferior colliculus. Neurones ascend to the thalamus and synapse in the medial geniculate nucleus and from there the final neurones in the pathway run to the auditory cortex in the auditory radiation.

Localization of a sound source

If the source of a sound is not exactly equidistant from the two ears, then the sound waves arrive at the two ears at slightly different times and at slightly different intensities. The localization of low frequency sounds (below 3000 cycles per second) is determined by a comparison of the different times the two ears are stimulated by the sound. The information from the two ears is firstly compared in the superior olivary nuclei where the neurones change their firing rate as the difference between the two ears increases. Intact pathways from here to the auditory cortex are required in order to interpret the information.

Higher frequency sounds are localized by the difference in intensity of the sound arriving in each ear. Neurones in the superior olivary nuclei are stimulated by impulses from one ear and inhibited by impulses from the other so that if the two are exactly balanced (sound source equidistant from the two ears) there is no change in activity. Depending on which ear is nearest to the sound, the firing pattern either increases or decreases.

In both of these mechanisms of sound localization, turning the head changes the pattern of response and gives additional information about the direction of the sound.

Functions of the vestibular apparatus

Otolith organs

The otolith organs appraise the brain of the static position of the head and also of linear acceleration. As the head is tipped, the otolithic membrane containing the heavy otoliths moves, causing some of the embedded stereocilia to be bent and changing the basal frequency of impulses emanating from the sensory neurones supplying the hair cells. Movement of the head in a different direction will cause a different pattern of response involving different hair cells.

If there is a sudden movement forward (e.g. if a car accelerates suddenly), then the otolithic membrane, which has greater inertia than the endolymph, tends to drag backwards. The sterocilia that are bent are those which would otherwise be moved if one were falling backwards. This explains why, when one suddenly moves forward, one can have the sensation of falling backwards. The discharge pattern from the otolith organs initiates reflex postural changes which result in the body leaning forwards.

Semicircular ducts

The receptors within the ampullae give information concerning the direction and rate of change (acceleration or deceleration) of rotational movements of the head. The way that the receptors are stimulated explains why they only function when the speed of rotation is changing and not if it is constant.

At the onset of rotation, while the semicircular canals themselves move as the head moves, the fluid within lags behind. Therefore, inside those canals which lie in the plane of rotation, the cupula is dragged in a direction opposite to that of the movement. It should be remembered that in any such movement at least two semicircular canals will be stimulated, one in each ear. Since the two sets of canals are mirror images then the two cupulae are effectively dragged in different directions.

There is a basal discharge rate of impulses from the sensory neurones supplying the cristae, and as a cupula is moved it pulls the stereocilia either towards or away from the kinocilium. Movement of the stereocilia towards the kinocilium leads to depolarization of the cell and an increase in impulse frequency, while movement away from the kinocilium leads to hyperpolarization and a decreased impulse frequency. In any rotational movement of the head, for reasons outlined above, one cupula will generate an increased number of impulses, the corresponding one in the other ear a decreased number.

A similar effect occurs if the head, already turning, suddenly moves faster. If, on the other hand, the movement is at a steady speed, then the endolymph will catch up with the semicircular canal, the cupula will recover its resting position and the basal discharge rate will be resumed.

At the end of movement, the endolymph continues to flow and drags the cupula in the direction of the movement. The sterocilia will bend in the opposite direction to that at the onset of the movement, so that the discharge rate will correspondingly change in the opposite direction.

So, for a particular head movement, impulses from the six cristae are coveyed to the sensory cortex and the pattern of change at the onset or the end of the movement forms a code, which is interpreted as the movement in question.

Such information is valuable when a person is involved in fast, intricate movements. If, for example, an individual is running and suddenly veers off in a different direction, then the change in firing patterns from the cristae will lead to reflex postural adjustments, preventing him or her from falling over. Because the firing patterns from the cristae alter at the start of a movement, the information they generate rapidly becomes sufficient to initiate the appropriate postural changes, so that the new position is anticipated, and equilibrium maintained by the end of the turn.

Nystagmus

If the head is turned suddenly, signals from the semicircular ducts initiate reflex movements of the eyes whereby the gaze moves slowly in a direction opposite to the direction of movement. If the head continues to turn, the eyes 'jump' forwards and fix on an object in view. With further movement of the head the eyes continue their backward gaze remaining fixed upon the object, before jumping forwards again to focus upon a new object.

Nystagmus consists of a slow component (the backward gaze) and a fast component (the jump forward). The effect of these movements is that the eyes have a better chance of focusing on an object if the gaze is maintained on it. If the eyes always looked forwards, everything would be out of focus all the time as the head turned.

Nerve pathways from the vestibular apparatus

Sensory neurones from the otolith organs and the cristae travel in the vestibular branch of the vestibulocochlear nerve. Figure 8.27 shows the nerve connections made by the vestibular nerves with the brain.

The main pathway for reflexes maintaining equilibrium is via the vestibular nuclei, the cerebellum, the reticular formation and thence down the spinal cord to the muscles.

Eye movements are mediated via pathways from the vestibular nerves to the vestibular

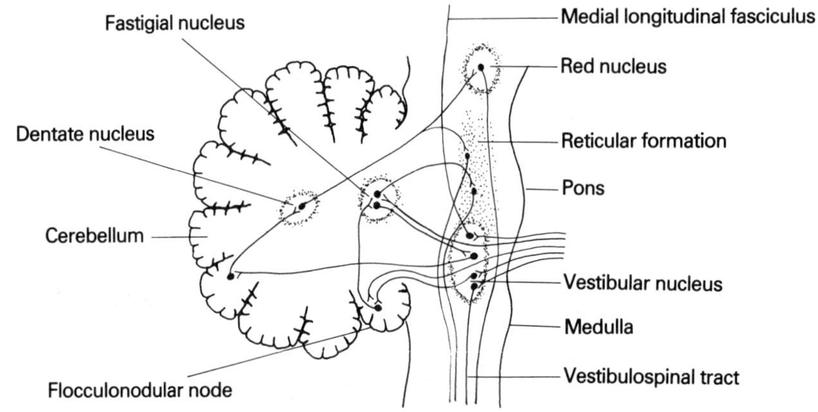

Fig. 8.27 Vestibular pathways. Most vestibular nerve fibres synapse in the vestibular nuclei at the medullary/pontine junction. From here, second order neurones travel to various parts of the cerebellum; to the vestibulospinal tract; into the medial longitudinal fasciculus to the midbrain; and to the reticular formation in the brain stem.

nuclei which then pass into the medial longitudinal fasciculus and on to the ocular nuclei in the midbrain.

Perception of the body's state of equilibrium involves pathways from the vestibular nerves to the cerebral cortex.

TASTE BUDS

The organs of taste or gustation are located within the walls of the buccal and pharyngeal cavities. They are chemoreceptors which respond to substances present in food and generate patterns of nerve impulses which are then passed to the brain for interpretation.

STRUCTURE OF TASTE BUDS

Taste buds are located in the epithelial coverings of the tongue, soft palate, posterior wall of the pharynx and epiglottis. They are most numerous on the upper and lateral surfaces of the tongue, especially on the side walls of the vallate papillae which are located on its rear, dorsal surface. There are approximately 10 000 taste buds in all.

Each taste bud consists of a barrel-like arrangement of about 20 sensory and supporting cells lying deep within the thick epithelium and opening onto the surface through a small gustatory pore (see Fig. 8.28). Running between the cells of the taste bud are the fine non-myelinated endings of afferent nerve fibres derived from any one of a number of cranial nerves (see *Nerve pathways from the taste buds*). The sensory cells have synaptic contacts with the nerve fibres, and are similar in shape and size to many of the supporting elements. Both have microvilli (gustatory hairs) projecting from their apical surfaces.

Two different types of supporting cells may be distinguished; the first type forms a layer around the taste bud, separating it from the surrounding epithelium. The second type supports the individual sensory cells, insulating them from one another and also insulating the basal portions of the nerve fibres. A third type of cell which is neither sensory nor supportive in function, lies towards the base of the taste bud and is capable of dividing to give rise to cells of the two other types. Thus all the cells of the taste bud are constantly being renewed.

FUNCTION OF TASTE BUDS

Four classes of taste have long been recognized; these are sweet, sour, salt and bitter. The application of pure solutions to the surface of the tongue can demonstrate experimentally that it is most sensitive to sweet and salt at its tip, to sour at its edge and to bitter at its back. It should be noted, however, that this distribution is not exclusive since the

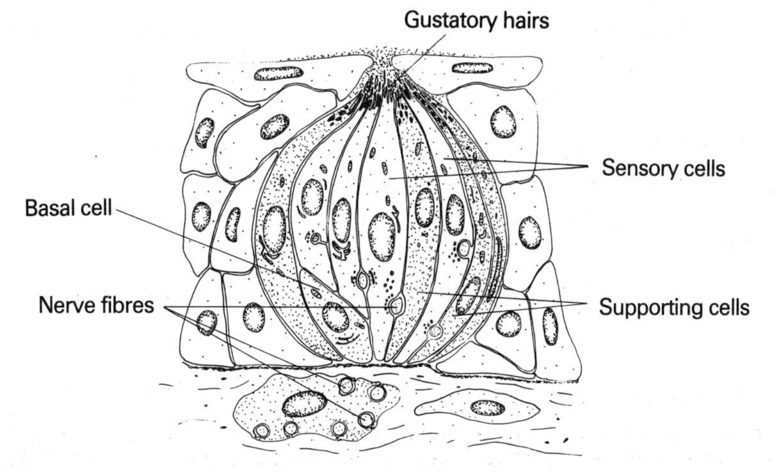

Fig. 8.28 Section through a taste bud.

tongue is, for example, also quite sensitive to bitter substances on its tip and edges. The foregoing might be taken to imply that the tongue possesses taste buds that are specifically sensitive to particular tastes, but microelectrode studies have shown that individual taste buds may be sensitive to one, two, three or even all four primary tastes.

The tongue as a whole does not exhibit identical maximal sensitivities to the four tastes and overall it is generally most sensitive to bitter substances and least sensitive to salt and sweet substances.

Nature of the primary substances

It is not always clear how different substances are able to elicit the same taste response since they do not always form a very cohesive chemical group.

Sweet substances include such organic molecules as sugars, alcohols, aldehydes, ketones, esters and amino acids and the inorganic salts of lead and beryllium. Perhaps the sweetest substance of all is the protein saccharin (600 times sweeter than sucrose).

Sour substances are all acids and generally the taste is proportional to the H^+ concentration. Organic acids, however, tend to have a stronger sour taste than inorganic acids, probably because they penetrate the receptor cells more easily.

Salt taste is elicited by the anions of inorganic salts. Some organic salts can, however, also taste salty.

Bitter substances include a variety of organic compounds such as caffeine, nicotine, urea and quinine and inorganic salts of magnesium, calcium and ammonium.

Mechanism of taste bud function

It is not known exactly how taste buds react with substances to stimulate sensory neurones leading to the sensation of taste. It is believed, however, that molecules pass into solution on the surface of the tongue and then combine in some way with the surface membranes of the receptor cells and/or their processes. This leads to the generation of receptor potentials within the cells and the liberation of transmitter substances which evoke action potentials in the sensory nerve fibres.

While individual taste buds or indeed individual receptor elements may be responsive to more than one class of substance, specificity of sensation can still be achieved. It is believed that this occurs because individual receptors discharge at particular frequencies determined by the type and concentration of

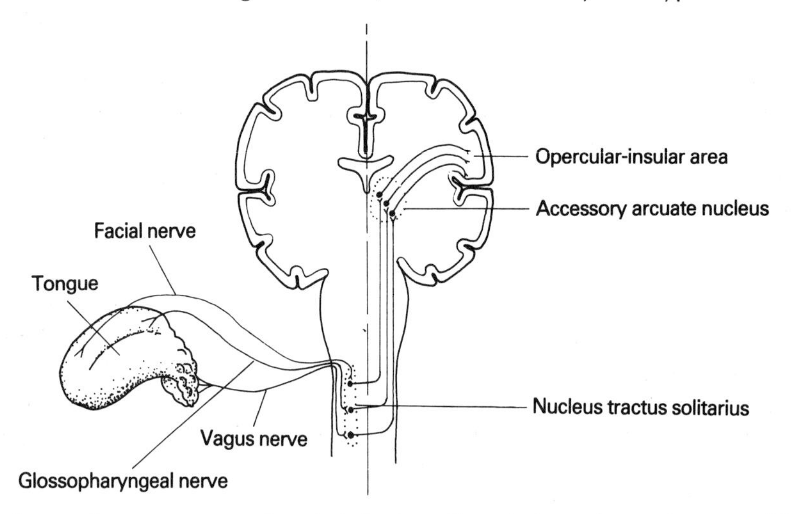

Fig. 8.29 Taste pathways. Sensory neurones from taste buds on the anterior and posterior parts of the tongue and the pharynx are carried by the facial, glossopharyngeal and vagus nerves to the nucleus tractus solitarius in the brain stem. From here, impulses cross over and pass up to the accessory arcuate nucleus in the thalamus. From here, neurones travel to the operacular-insular area of the cerebral cortex.

specific substances. Thus an individual taste sensation is dependent upon a frequency pattern produced by the discharge of several receptor cells.

This theory also illustrates how an appreciation of the four 'primary' taste sensations can lead to an appreciation of the almost infinite variety of tastes that are available to man. Each taste is dependent upon particular discharge patterns derived from a number of taste buds.

Nerve pathways from the taste buds

Taste buds on the anterior two-thirds of the tongue connect with fibres of the facial nerve in the chorda tympani (see Fig. 8.29). They may pass directly to the brain within the facial nerve itself or, in some individuals, go via the otic ganglion to the greater petrosal nerve and then to the geniculate ganglion of the facial nerve.

Taste buds on the posterior one-third of the tongue are associated with fibres of the glossopharyngeal nerve within which they pass to the brain. Pharyngeal taste buds send impulses to the brain in the vagus nerve.

Fibres from all three nerves enter the medulla and combine to form the tractus solitarius, most of which terminate in the nucleus tractus solitarius. Secondary neurones then pass up to the brain stem, to the accessory arcuate nucleus from which fibres travel to the parietal opercular-insular area of the cerebral cortex. This is the area of the brain where taste is experienced, and it lies close to the area where other sensations from the tongue are received.

Importance of taste

Although at first sight the ability to detect different flavours may appear to have little biological value, experiments have shown that, in fact, the reverse is true. It has been demonstrated, for example, that supplying a sodium-deficient diet to a laboratory rat will result in the animal selectively taking salt-rich foods and drink. An adrenalectomized animal will also select a salt-rich diet, while an animal that lacks parathyroid hormone will selectively drink calcium chloride solution. It appears that the animals taste buds provide the mechanism by which it is able to select its dietary requirements.

It is also well known that animals and indeed humans will avoid bitter substances, many of which are poisonous, which demonstrates the role of the taste buds as protective agents.

The sensation of taste, apart from its many psychological benefits and hazards, also initiates the flow of gastric juice in the cephalic phase of digestion (see Ch. 6).

OLFACTORY ORGANS

The organs which are responsible for the sense of smell, the olfactory organs, are located in the nasal passages. They are chemoreceptors which respond to airborne chemicals. Olfactory sensitivity in man is very low compared with that in most animals, but nevertheless some substances can be detected in extremely low concentrations, e.g. mercaptan can be detected at a concentration of 4×10^{-8} mg per litre of air.

STRUCTURE OF THE OLFACTORY MUCOSA

The olfactory mucosa occupies an area of about 2.5 sq. cm on the roof of each nasal cavity. It does not lie in the air stream which enters the nose during normal respiration,

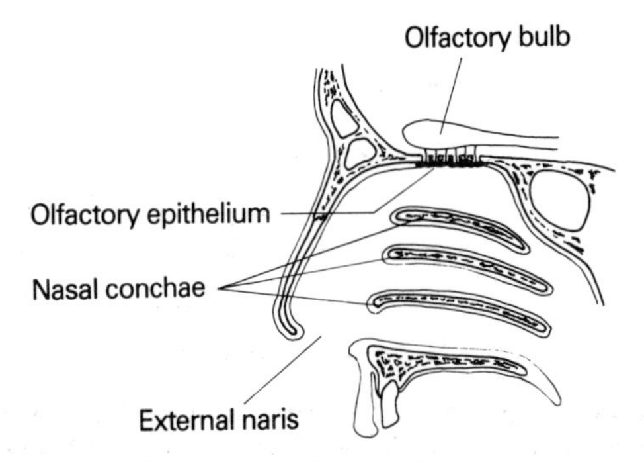

Fig. 8.30 Section through the nasal passage of one side to show the position of the olfactory epithelium.

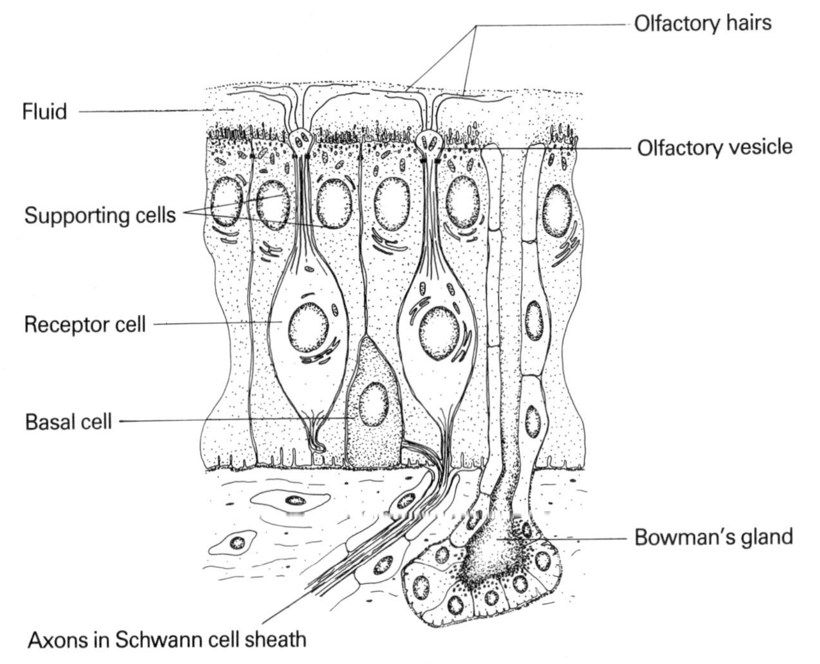

Fig. 8.31 Ultrastructure of the olfactory epithelium.

although sudden sharp inspirations ('sniffing') will carry molecules up to the sensory cells (see Fig. 8.30).

The olfactory epithelium is somewhat thicker than the respiratory epithelium which lines the nasal cavities and is composed of three types of cells. These are the receptor cells, supporting or sustentacular cells and basal cells (see Fig. 8.31).

The receptor cells are actually bipolar sensory neurones with a cell body lying basally within the epithelium and with a process which terminates in a small knob or olfactory vesicle from which a number of long cilia (olfactory hairs) project above the epithelium. Basally, each cell gives rise to a slender non-myelinated axon; this joins with other axons to form small bundles which penetrate the bone. Each bundle of axons (not each axon) then receives a Schwann cell sheath. Bundles of axons pass through holes in the ethmoid bone and enter the olfactory bulb, where they synapse with a variety of sensory fibres.

Lying close to the basal lamina are two cell are the sustentacular cells, irregular columnar cells with nuclei lying close to their apical surfaces which have numerous microvilli.

Lving close to the basal lamina are two cell types, one of which appears to have a supporting role, the other being capable of division and giving rise to new receptor cells. The olfactory epithelium is the only known structure in the human body which is capable of routine replacement of nerve cells.

The olfactory epithelium also contains the openings of the Bowman's glands whose acini lie beneath the basal lamina. Bowman's glands secrete an enzyme-containing fluid onto the surface of the olfactory epithelium. The precise function of this fluid is not known but its occurrence in this situation leads to the supposition that it provides a medium for the passage of molecules from the air to the sensory cells.

FUNCTION OF OLFACTORY RECEPTORS

It is not known how olfactory receptors react with substances and generate action potentials. It is presumed that sensory cells 'recognize' substances in some way, perhaps by means of a chemical reaction or possibly a physical 'lock and key' type of association. It has also been hypothesized that there are a number of primary odours, similar to the primary tastes. One such classification recognizes seven odours; camphoraceous, floral, musky, peppermint, ethereal, pungent and putrid. More recent work has indicated that a much longer list of primary odours, if indeed they do exist, might be more appropriate.

The receptor cells respond to the presence of molecules with a slow negative shift in potential. This has been named the electro-olfactogram (e.o.g.) and is believed to represent the summated generator potentials from a number of different receptor cells. These generator potentials promote the formation of action potentials which pass from the receptor cells to the olfactory bulb.

It is believed that individual receptor cells have at least some specificity and are able to respond more strongly to one odour than to

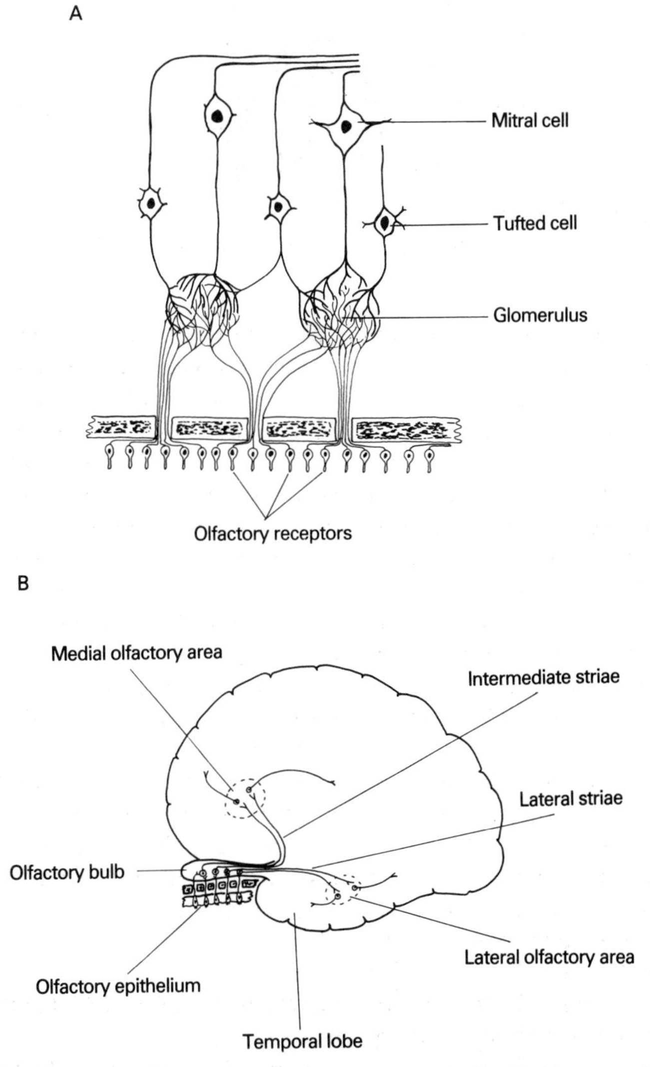

Fig. 8.32 (A) The principal neural connections within the olfactory bulb. (B) Neurones from the olfactory bulb pass through either the intermediate or lateral olfactory striae to the medial or lateral olfactory areas. From these areas, impulses pass to the hypothalamus, thalamus and various nuclei in the brain stem.

another. It is presumed that the discrimination of up to 4000 substances of which humans are capable depends upon a pattern of impulses generated over a wide area of the olfactory epithelium.

Olfactory pathways

There are around five million receptor cells in each olfactory epithelium (in man), all of which give rise to axons which travel in about 20 nerve bundles passing to the olfactory bulbs. Within the latter, these axons become interspersed with the dendrites of the mitral and tufted (nerve) cells, forming complex globular synapses known as olfactory glomeruli. It is estimated that about 26 000 axons terminate in each glomerulus, which also contains processes from about 24 mitral and 68 tufted cells (Fig. 8.32A).

From the olfactory bulb, nerve fibres pass through the intermediate and lateral olfactory striae which lead ultimately to the medial and and lateral olfactory areas (Fig. 8.32B). The medial olfactory area lies centrally above and in front of the hypothalamus. The two lateral areas lie in the temporal lobes.

Importance of olfaction

Many animals use odours to attract or repel others. Humans also use artificial odours such as perfumes, air fresheners and deodorants to attract or repel others. The sensations of taste and smell are frequently associated together and are not always easily distinguishable. Much of the enjoyment that we obtain from food, for example, is derived as much from its odour as from its taste.

The sense of smell, like that of taste, also has a protective function, e.g. food that is 'going bad' has a repellant odour.

9

Endocrine physiology

The endocrine organs do not constitute an anatomical system but consist of a number of small glands (Fig. 9.1), some of which are structurally independent, and some located within other organs. The glands of the endocrine 'system' do not have any obvious structural features in common and their only physical link is through the circulatory system. An endocrine gland is most clearly identified by its function, as one which secretes minute quantities of chemical messengers, known as hormones, directly into the blood rather than into a duct.

HORMONES

Hormones themselves form no special chemical group; they include amino acid derivatives, small peptides, large polypeptides, glycoproteins and steroids. After having entered the blood, some hormones (steroids and thyroid hormones) become bound to plasma proteins while others are transported in an unbound state. Their concentration in the blood is always very small. Most hormones exert their effects upon tissues which are remote from their site of origin. A hormone may act on a specific 'target' organ or it may exert its influence on most or all tissues. Hormones do not actually initiate reactions, but merely modify the rates of those which are already taking place.

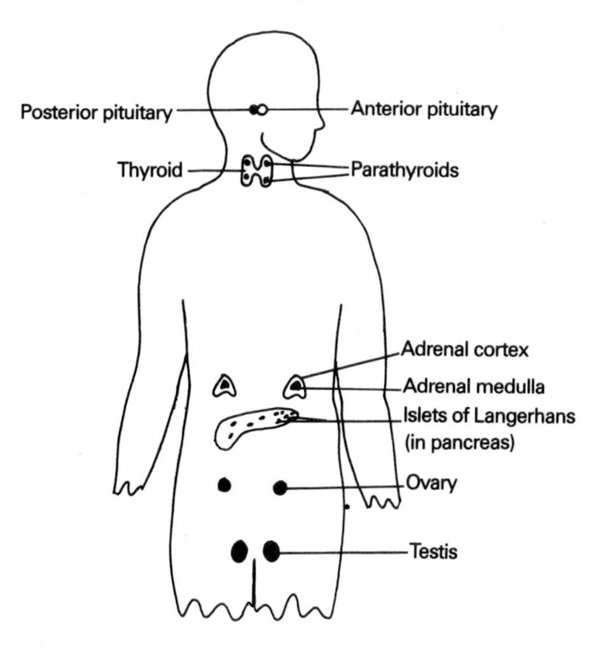

Fig. 9.1 The major endocrine glands.

Most hormones are continuously secreted, but at rates which are modified as a result of stimuli to the glands from which they are released. Thus basal secretion rates can be either increased or decreased.

There may be a delay between receipt of a stimulus by an endocrine gland and the final manifestation of hormone action. In the case of thyroxine for example, it may be two or three days before a rise in metabolic rate is observed following its introduction into the blood (i.e. there is a long latent period). It may then take another eight or so days before there is maximum effect. Other hormones act almost instantaneously and exert an effect that is comparatively short-lived, e.g. adrenaline and noradrenaline which are released from the adrenal medulla.

Table 9.1 is a list of the major known hormones, their sites of origin and their chemical natures. In addition to these hormones, there are those which are produced by and act upon the gastrointestinal tract and its associated glands (see Ch. 6); erythropoietin, which is activated by a kidney factor (see Ch. 2); hypothalamic regulatory hormones (see *The pituitary gland*); as well as lesser known hormones such as thymosin and thymin produced by the thymus gland, and melatonin produced by the pineal gland.

Much of our knowledge of hormone function in man has been derived from studies of hypo- or hyperfunction of endocrine glands. For this reason, a short section on the effects of hormone deficiency and excess is included in the account of each gland.

CELLULAR ACTIONS OF HORMONES

It is still not fully understood how a hormone induces a physiological response in a target

Table 9.1 The major hormones, their sites of origin and chemical nature

Name (with alternatives and abbrevations)	Origin	Chemical nature
Trophic hormones		
Adrenocorticotrophic Hormone (ACTH) (Corticotrophin)	Anterior pituitary	Polypeptide (39 aa)
Thyroid Stimulating Hormone (TSH) (Thyrotrophin)		Glycoprotein
Gonadotrophic hormones		
Follicle-stimulating hormone (FSH)		Glycoprotein
Luteinizing hormone (LH)(female) Interstitial-cell-stimulating hormone (ICSH)(male)		Glycoprotein
Hormones		
Prolactin (PRL) (Luteotrophic hormone) (LTH) (Lactogenic hormone)		Polypeptide (198 aa)
Growth hormone (GH) (Somatotrophic hormone) (STH) (Somatotrophin)		Polypeptide (191 aa)
Antidiuretic hormone (ADH) (Arginine vasopressin) (AVP)	Posterior pituitary	Peptide (9 aa)
Oxytocin		Peptide (9 aa)
Thyroxine (T_4)	Thyroid	Iodinated tyrosine derivative
Tri-iodothyronine (T_3)		Iodinated tyrosine derivative
Calcitonin		Polypeptide (32 aa)
Parathyroid hormone (PTH) (Parathormone)	Parathyroids	Polypeptide (84 aa)
Insulin	Pancreas	Polypeptide (51 aa)
Glucagon		Polypeptide (29 aa)
Adrenaline	Adrenal medullae	Catecholamine
Noradrenaline		Catecholamine
Mineralocorticoids		
Aldosterone	Adrenal cortices	Steroid
Deoxycorticosterone		Steroid
Glucocorticoids		
Cortisol		Steroid
Corticoserone		Steroid
Oestradiol	Ovaries	Steroid
Progesterone		Steroid
Testosterone	Testes	Steroid
Androstenedione		Steroid
Human chorionic gonadotrophin (HCG)	Placenta	Glycoprotein
Human chorionic somatomammotrophin (HCS) Human placental lactogen (HPL)		Polypeptide (191 aa)
Oestriol		Steroid
Progesterone		Steroid

cell, although theories which suggest certain common mechanisms have emerged in recent years.

It has been shown that some cell membranes contain hormone-specific receptor proteins. These receptors may be likened to a lock, while the hormone is the key which fits that lock. This interpretation helps to explain why hormones with similar molecular structures may produce overlapping physiological responses. Furthermore the presence or absence of receptor molecules in a cell membrane (or in some cases within the cell) determines whether or not that cell is sensitive to that particular hormone.

This 'recognition' process is the first stage of hormone action and may be followed by one or more of the following responses.

Second messenger mechanism. Many hormones exert their influence through the formation of a 'second messenger' within the target cell (see Fig. 9.2). In this case a hormone attaches to its specific receptor molecule on the surface of a target cell. This receptor is itself linked to a molecule of the enzyme adenyl cyclase and the interaction of the hormone and receptor serves to activate the adenyl cyclase. The latter then brings about the conversion of cytoplasmic ATP to 3′5′-adenosine monophosphate (cAMP). cAMP is the 'second messenger' and it is this which brings about the physiological response of the target cell. The nature of this response is dependent upon the nature of the target cell since the same second messenger is employed in many different cells and yet can bring about

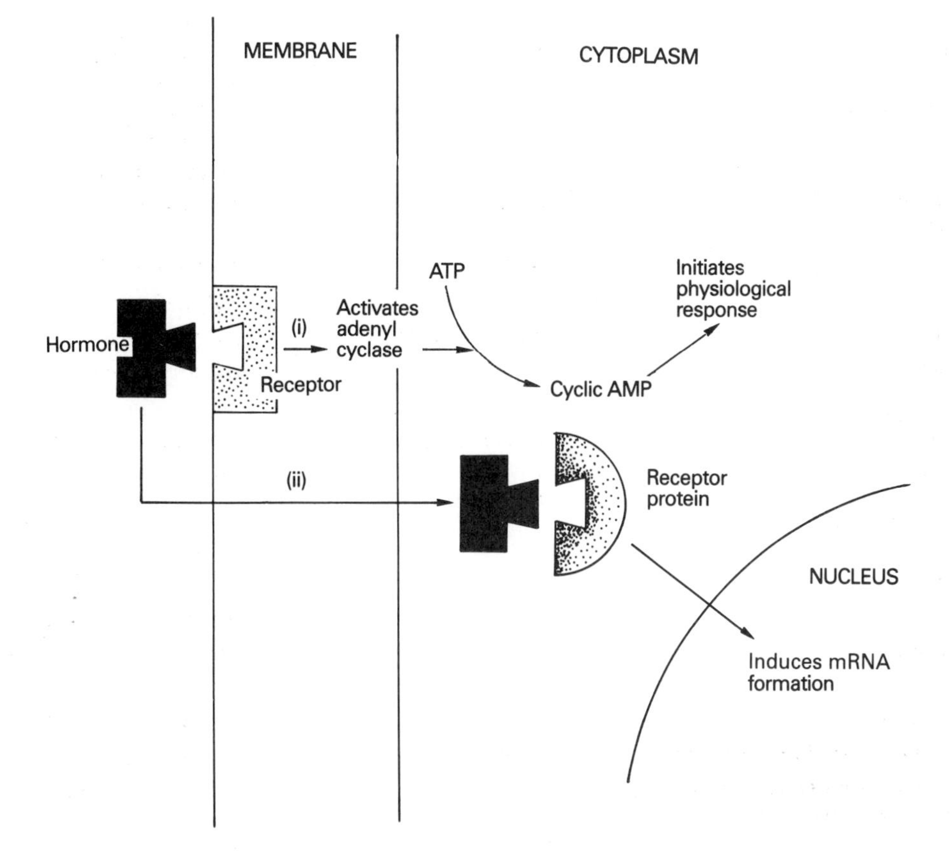

Fig. 9.2 Cellular actions of hormones. Some hormones react with a receptor on the cell membrane and (i) activate a second messenger (cyclic AMP). Other hormones enter the cell and (ii) combine with a receptor protein, which then enters the nucleus and increases mRNA synthesis.

many different responses. Recent evidence suggests that some hormonal effects are mediated via a decrease in cAMP activity (e.g. the effect of insulin on reduced glycogenolysis). Other nucleotides may also act as second messengers.

Hormonal influences on protein synthesis. Steroid hormones do not react with membrane receptors, but instead enter the target cell cytoplasm where they combine with receptor proteins. The resultant complex then diffuses into the nucleus, where it induces the formation of mRNA. The mRNA then passes out into the cytoplasm and promotes protein synthesis (usually of one or more enzymes).

Thyroxine and tri-iodothyronine also enter the cells but there is evidence to suggest that at least some of the molecules pass directly into the nucleus, unattached to receptors.

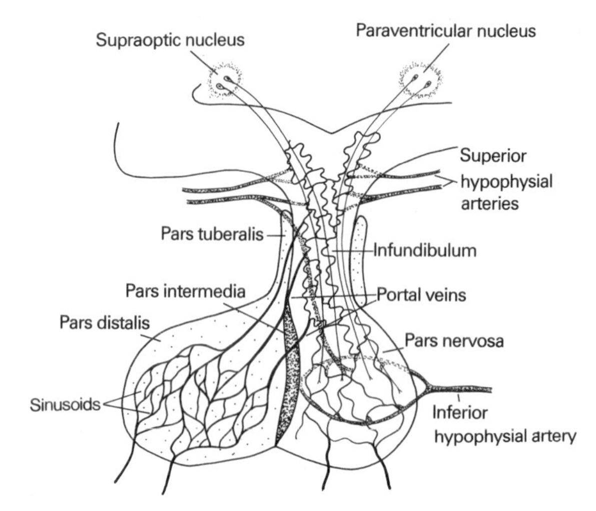

Fig. 9.3 The pituitary gland and its blood supply.

THE PITUITARY GLAND (HYPOPHYSIS)

STRUCTURE OF THE PITUITARY GLAND

The human pituitary weighs about 500 mg and lies at the base of the brain in a depression (the sella turcica) in the sphenoid bone.

The pituitary is, in reality, two glands which are structurally but not functionally related. The anterior portion, the adenohypophysis is derived from the roof of the embryonic mouth cavity and is composed of epithelial tissue. It may be subdivided into three areas, the pars distalis, pars intermedia and pars tuberalis. The posterior portion, the neurohypophysis, is formed by a downgrowth of the floor of the hypothalamus and therefore consists of nervous tissue. It is divided into two regions, the pars nervosa and the infundibulum.

The gland receives blood via the superior and inferior hypophysial arteries. The superior vessels give rise to a number of capillary plexuses which drain into several long and short portal veins on the surface of the pituitary stalk; these constitute the hypophysial portal system and supply the sinusoids of the pars distalis. The inferior vessels primarily supply the neurohypophysis (Fig. 9.3).

Adenohypophysis

The pars distalis forms the bulk of the anterior lobe or adenohypophysis (Fig. 9.3). It is a bulbous structure composed of clumps and cords of cells separated by blood-filled sinusoids. Two principal cell types are found; those which take up stains (chromophils) and those which do not (chromophobes). It seems likely that the latter are an immature form of the former.

Chromophils are classified according to their abilities to react with particular stains, giving rise to the terms acidophils (α-cells) and basophils (β-cells). Both cell types can then be further divided.

Acidophils subdivide into two types: somatotrophs, which secrete growth hormone (somatotrophin); and mammotrophs, which secrete prolactin.

The basophils may also be subdivided into a number of types. These include the thyrotrophs, which secrete thyroid-stimulating hormone; gonadotrophs (of which there may be two types), which secrete follicle-stimulating hormone and luteinizing hormone; and corticotrophs, which synthesize adrenocorticotrophic hormone.

The pars tuberalis consists of a collar of cells wrapped around the infundibular stalk. The cells, whose function is unknown, appear quite different from those of the pars distalis.

The pars intermedia lies between the pars distalis and the pars nervosa. It contains basophilic cells (melanotrophs), which synthesize melanocyte-stimulating hormone. In man, the pars intermedia is rudimentary, and melanotrophs may be found in the pars distalis and even in the pars nervosa.

Neurohypophysis

The pars nervosa, which forms the bulk of the neurohypophysis, contains a rich network of unmyelinated nerve fibres. The cell bodies of these fibres lie in the hypothalamus in the supraoptic and paraventricular nuclei and the fibres themselves pass down to the pituitary via the hypothalamo-hypophysial tract. Within the pars nervosa the nerve endings lie close to a network of vascular sinusoids, and lying between the neural elements are numerous pituicytes which resemble neuroglial cells.

Linking the pars nervosa with the hypothalamus is the infundibulum, a stalk which contains the nerve fibres of the hypothalamo-hypophysial tract. It is largely enclosed by the pars tuberalis.

FUNCTIONS OF THE ADENOHYPOPHYSIS

Four of the hormones released by the anterior pituitary are responsible for normal growth and development of a number of other endocrine glands as well as their maintenance in adulthood. It is for this reason that they are known as *trophic hormones* (the term 'trophic' derives from a Greek word meaning nutrition). The rate of secretion of hormones from the target endocrine glands is also dependent upon the output of the trophic hormones. These hormones are:

1. Thyroid-stimulating hormone (TSH), or thyrotrophin, which stimulates the thyroid to liberate thyroxine and tri-idothyronine.
2. Adrenocorticotrophic hormone (ACTH), or corticotrophin, which stimulates the adrenal cortices to release cortisol.
3. The gonadotrophic hormones: follicle-stimulating hormone (FSH) and luteinizing hormone (LH), which regulate the release of oestrogen and progesterone from the ovaries in the female. In the male, LH, which is also known as interstitial-cell-stimulating hormone (ICSH), together with FSH, regulates testicular function. The testes produce testosterone, the principal male sex hormone.

In addition to the above, the adenophypophysis also secretes two other hormones; prolactin (PRL), which promotes the production of milk in the female mammary glands; and growth hormone (GH), which exerts a number of metabolic effects upon many tissues of the body.

The roles of the various trophic hormones will be considered within the sections dealing with the individual target glands.

A further group of compounds has been discovered in recent years which are derived from the same precursor as ACTH and MSH ('big ACTH'). β-Endorphin (endogenous morphine-like agent) is a fragment of big ACTH, which itself gives rise to smaller enkephalin molecules. Both β-endorphin and the enkephalins have been found to bind to opiate receptors in the brain and to have a role in pain reduction. Their significance and the mechanisms by which they are released have yet to be elucidated.

Hypothalamic control of adenohypophysial hormone secretion

The release of hormones by the pars distalis is regulated directly by the hypothalamus. Specialized neurones in the latter liberate releasing hormones (or factors) into the capillaries on the pituitary stalk which lead into the hypophysial portal vessels. The releasing hormones then pass down to the pars distalis where they stimulate the cells to release their hormones into the blood. Discrete groups of neurones appear to produce specific releasing hormones which, in turn, promote the release of specific pituitary hormones. Table 9.2 lists

Table 9.2 Hypothalamic regulatory hormones

Name	Pituitary hormone
Growth hormone-releasing hormone (GHRH)	Growth hormone
Growth hormone-inhibiting hormone (somatostatin) (GHIH)	
Thyrotrophin-releasing hormone (TRH)	Thyrotrophin
Gonadotrophin-releasing hormone (GnRH)	Luteinizing hormone *and* Follicle-stimulating hormone
or	
Luteinizing hormone-releasing hormone (LHRH)	Luteinizing hormone
and	
Follicle-stimulating hormone-releasing hormone (FSHRH)	Follicle-stimulating hormone
Corticotrophin-releasing hormone (CRH)	Corticotrophin
Prolactin-releasing hormone (PRH)	Prolactin
Prolactin-inhibiting hormone (PIH)	Prolactin

the hypothalamic regulatory substances. It can be seen that the release of growth hormone, thyrotrophin, corticotrophin and prolactin are controlled by their respective releasing hormones, and that growth hormone and prolactin are additionally controlled by inhibiting hormones. The control of secretion of the gonadotrophic hormones FSH and LH is somewhat controversial. Some workers believe that each has its own releasing hormone, others that there is a single gonadotrophin releasing hormone (GnRH).

Figure 9.4 shows the relationships between the hypothalamus, the anterior pituitary and the target endocrine glands. The hormones released by the target glands exert a negative feedback effect (i.e. reduce the rate of secretion of hormones) on either or both the pituitary and the hypothalamus.

Growth hormone

Growth hormone, also known as somatotrophic hormone (STH) or somatotrophin, is a large polypeptide consisting of 191 amino acids and with a molecular weight of 21 500. It is structurally similar to prolactin and, indeed, exhibits lactogenic properties. Growth hormone is required for normal growth and development and its absence results in dwarfism.

The hormone exhibits several metabolic effects:

1. It increases the rate of protein synthesis in all cells, as well as stimulating amino acid uptake by cells. It therefore decreases the concentration of amino acids in the plasma.
2. It increases the rate of fat breakdown in adipose tissue (lipolysis) and the subsequent release of free fatty acids into the circulation. As a result, it increases the concentration of free fatty acids in the plasma.
3. It decreases the rate of glucose utilization by the tissues and increases the output of glucose from the liver. It therefore raises plasma glucose levels.

The release of growth hormone is influenced on an hourly basis by feeding and physical exercise. At its lowest level there may be less than 1 ng/ml in the blood, but at its highest it may reach 60 ng/ml. Pituitary output of growth hormone is regulated by two hypothalamic factors, growth hormone-releasing hormone (GHRH) and growth hormone-

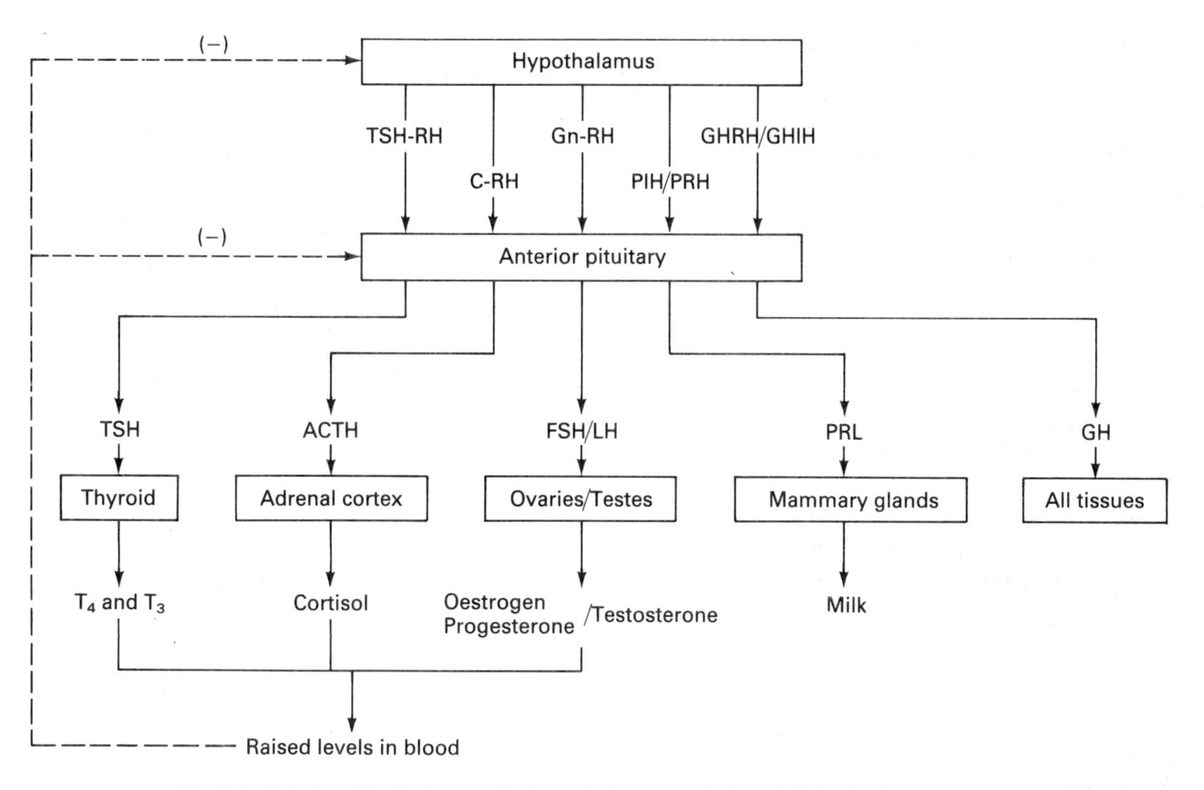

Fig. 9.4 Relationships between the hypothalamus, anterior pituitary and target organs.

inhibiting hormone (GHIH) also known as somatostatin.

A decrease in plasma glucose levels leads to stimulation of the hypothalamic neurones which release GHRH. The latter is secreted into the portal vessels and passes down to the anterior pituitary where it promotes the release of stored growth hormone into the blood. The metabolic consequences of this hormone release is a raised blood sugar which tends to reverse the stimulus. A rise in the levels of amino acids in the plasma will exert a similar effect to a drop in plasma glucose, i.e. it promotes growth hormone release. In this case, the result of hormone action will be to decrease the levels of amino acids in the plasma.

Unfortunately the regulation of plasma glucose and amino acid levels is not a simple matter which only depends upon growth hormone; adrenaline, glucagon and insulin all play important roles (see *Regulation of blood glucose concentration*).

The growth-promoting function of growth hormone, while now recognized as not being its sole function is nevertheless very important. Growth hormone causes growth in all tissues, increasing both cell size and number. The growth and development of children is at least partially dependent upon the presence of growth hormone.

Growth hormone itself does not appear to promote growth in cartilage and bone. This effect is mediated through a number of somatomedins, small proteins formed in the liver, and probably elsewhere, under the influence of growth hormone. It is thought by some physiologists, that many of the actions of growth hormone are actually brought about by the somatomedins.

Growth hormone and/or somatomedin exert a negative feedback effect upon the hypothalamus which ultimately leads to the reduction of growth hormone output by the pituitary. Whether this 'short loop' feedback involves the reduction in GHRH output or an increase in somatostatin release (or both) by the hypothalamus is unknown.

Prolactin

Also known as luteotrophic hormone and lactogenic hormone, prolactin is a protein consisting of 198 amino acids. Its principal function is to initiate and sustain lactation in female mammary glands which have been previously primed by oestrogen and progesterone from the placenta, growth hormone from the pituitary, and steroid hormones from the adrenal cortex.

Prolactin acts directly upon the acini of the breasts to promote the synthesis of milk. The release of this milk is, however, dependent upon the action of oxytocin from the posterior pituitary.

The release of prolactin is largely dependent upon the supression of prolactin inhibitory hormone (PIH), which is in fact dopamine, and to a lesser extent upon prolactin releasing hormone (PRH). PIH is released tonically by the hypothalamus and in its absence, and presumably in the presence of PRH, prolactin is released by the pituitary.

It has been shown that during pregnancy there is a steady rise in the levels of prolactin in the plasma, but that these fall to basal non-pregnant levels about three to four weeks after delivery in the absence of suckling. Suckling of the breast maintains raised prolactin levels. This process involves a neuroendocrine pathway whereby nerve impulses pass from the nipple to the hypothalamus. The latter then responds by decreasing PIH output (and possibly increasing PRH output) and the pituitary releases prolactin (Fig. 9.24). There is strong evidence that prolactin itself can exert a negative feedback influence upon the hypothalamus and pituitary and thus inhibit its own output.

Functions of the pars intermedia

The pars intermedia synthesizes and releases melanocyte-stimulating hormone (MSH). MSH is a small peptide which corresponds in its structure to a fragment of the ACTH molecule. It is therefore not surprising that ACTH itself exhibits MSH activity.

MSH causes skin darkening in amphibians and reptiles by dispersing the pigment granules within the melanophores (pigment cells) of the skin. Human skin is different, but nevertheless prolonged exposure to MSH will promote melanin synthesis and bring about skin darkening in this way.

MSH output is apparently regulated by the hypothalamus through the actions of both releasing (MSHRH) and inhibiting hormones (MSHIH) and also through nerve fibres which normally inhibit MSH output.

MSH does not itself appear to have any functional significance in humans and it should be noted that ACTH, which is present in much larger quantities, probably exerts a greater skin-darkening effect.

Effects of adenohypophysial hormone deficiency

Although the anterior pituitary secretes six different hormones, only rarely do disturbances in the output of individual hormones occur. More commonly, reduced secretion of several hormones at once are found to occur, often as the result of a tumour. It is worth noting that quite a large part of the pituitary structure must be destroyed before the effects of hypofunction are observed.

Destruction of the anterior pituitary, by whatever cause will result in the atrophy and/or reduction in function of dependent glands or tissues. The adrenal cortex atrophies and the output of the glucocorticoids and sex hormones (but not aldosterone) reduces. Thyroid function reduces, lowering the metabolic rate and the ovaries (or testes) atrophy. As a result of the latter, in the female the reproductive cycle ceases; lack of prolactin results in the inability to lactate.

Lack of growth hormone in children leads to the development of a 'pituitary dwarf'. Such an individual might also exhibit hypofunction of the other pituitary hormones or, more rarely, be otherwise normal, since lack of growth hormone alone sometimes occurs.

Effects of adenohypophysial hormone excess

Increases in the circulating levels of the anterior pituitary hormones produce a number of effects which can be related to the individual hormones. Excess gonadotrophin, for example, induces sexual precocity, whilst ACTH stimulates increased activity of the adrenal cortex and induces Cushing's disease (see *Effects of cortisol excess*). Excess TSH, perhaps surprisingly, does not always bring about an increase in thyroid activity. The effect of excess prolactin is equally unpredictable, in that rather than simply increasing milk secretion it actually causes infertility due to a repressive action upon the ovaries.

An increased output of growth hormone causes somewhat different effects depending upon whether it occurs in a child or an adult. In children there is a surge in growth so that gigantism occurs. This is because the epiphysial discs are still present and there is an increase in length of the long bones. In adults the hormone produces a clinical picture known as acromegaly (this also occurs in adult 'giants'). Acromegaly is characterized by thickened features on a long face, large hands and feet and enlargement of many internal organs.

FUNCTIONS OF THE NEUROHYPOPHYSIS

It will be recalled that the posterior pituitary is a downgrowth of the hypothalamus containing nerve fibres whose cell bodies lie in the supraoptic and paraventricular nuclei. The supraoptic nucleus is the principal site of production of antidiuretic hormone (ADH), also known as arginine vasopressin (AVP), whereas the paraventricular nucleus produces mainly oxytocin. Both hormones are synthesized within their respective cell bodies in the hypothalamus; enclosed within vesicles which migrate slowly down the insides of the nerve axons to the nerve endings within the pars nervosa, where they are stored. The transit time is about 10 hours. When a nerve impulse passes down the hypothalamo-hypophysial tract it causes the release of the contents of a number of vesicles into the interstitial fluid and then to the blood by exocytosis. The mechanism is calcium dependent and closely resembles neurotransmitter release. Synthesis and release of the hormone are, therefore, two distinct processes, the former occurring in the hypothalamus, the latter from the posterior pituitary. Figure 9.5 summarizes the relationships between the hypothalamus, posterior pituitary and target organs.

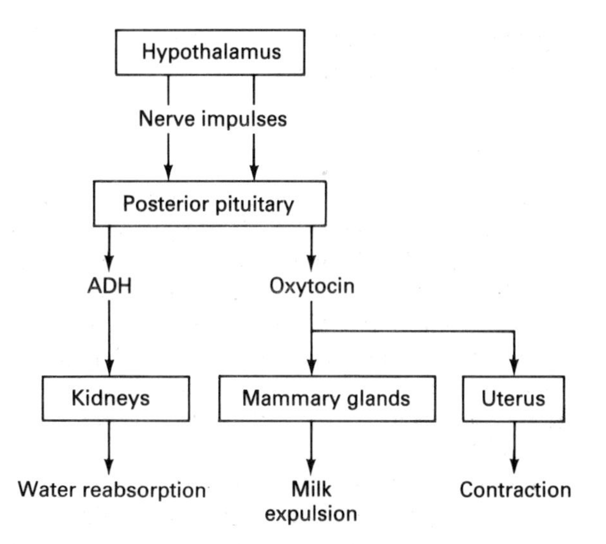

Fig. 9.5 Relationships between the hypothalamus, posterior pituitary and target organs.

Antidiuretic hormone

ADH is a small peptide containing nine amino acids. It exerts a mild stimulatory influence upon smooth muscle, and its action upon this component of the walls of arterioles induces a rise in blood pressure. This particular effect resulted in the alternative name vasopressin (now arginine vasopressin) for the hormone.

At physiological concentrations this pressor effect appears to have little or no significance and the primary role of ADH is to increase the permeability to water of the collecting ducts of the nephrons in the kidneys. As a result of this, water is withdrawn from the tubules down the medullary osmotic gradient into the peritubular capillaries. The net effect of ADH action is, therefore, to reduce urine output (see Ch. 3).

ADH is released in response to changes in the osmotic pressure of extracellular fluid (Fig. 9.6). When the osmotic pressure of the

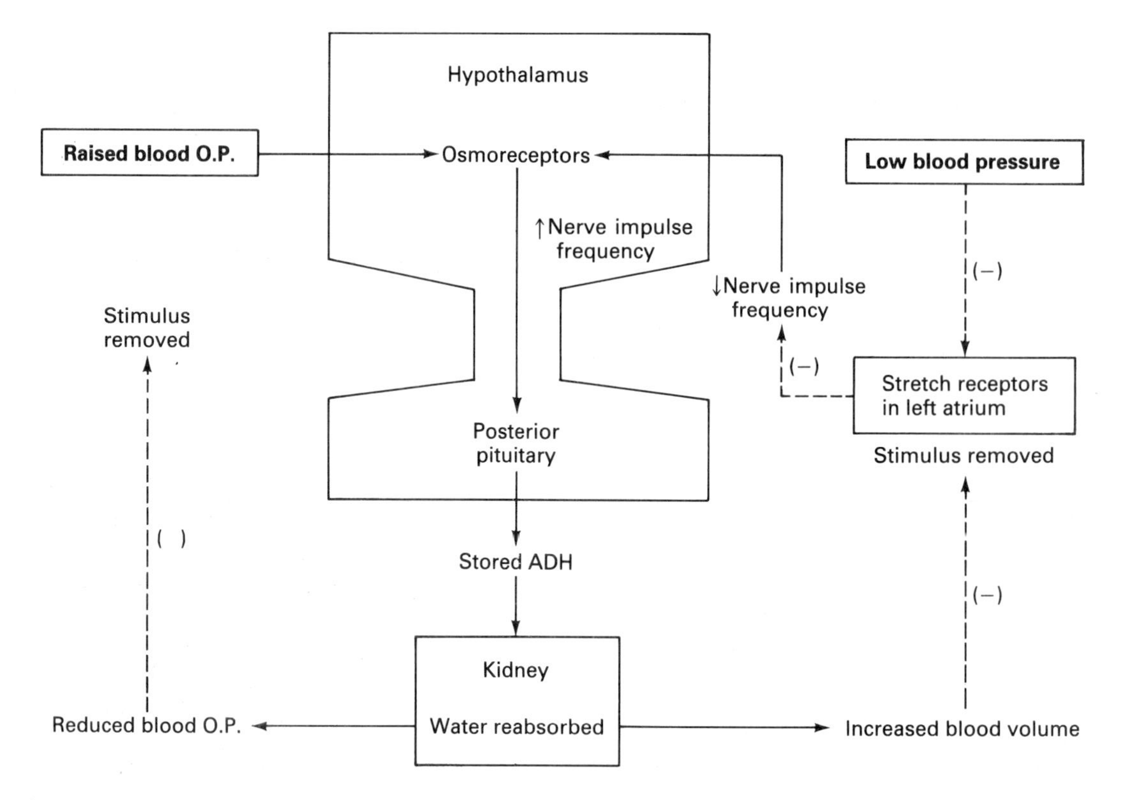

Fig. 9.6 The release and actions of ADH in response to raised blood osmotic pressure (left-hand side) and low blood pressure (right-hand side).

blood flowing to the brain increases, a group of cells in the hypothalamus are stimulated. The cells are known as osmoreceptors and they may be the cell bodies of those neurones which terminate in the neurohypophysis. As a result of this excitation, the basal frequency of nerve impulses which pass down to the pars nervosa is increased and this causes an increased rate of release of stored ADH into the blood.

Once in the blood, ADH travels through the circulation to the kidneys where it increases the permeability of the collecting ducts to water. Water is attracted from the tubular filtrate back into the blood, thereby diluting it. This reduces the stimulus for ADH output and brings about a reduction in its release.

Changes in blood volume, unaccompanied by alterations in osmotic pressure, can also influence ADH output. This effect is mediated by stretch receptors in the left atrium. For this reason, haemorrhage and its accompanying reduction in blood volume (but in the absence of any change in osmotic pressure) is a potent stimulus for ADH release. The resultant water retention will bring about at least a partial reconstitution of blood volume.

Other stimuli have been shown to influence ADH output. For example, alcohol inhibits ADH release while nicotine and morphine exert stimulatory effects.

Effects of ADH deficiency

Lack of ADH release by the posterior pituitary results in a condition known as diabetes insipidus. The most obvious feature in this disease is that there is an overproduction of dilute urine (polyuria) which is accompanied by an insatiable thirst.

Effects of ADH excess

Neurohypophysial tumours are rare, but excessive release of ADH can occur accompanying a number of other clinical conditions. Excess ADH release (termed inappropriate secretion of ADH) results in the production of reduced volumes of urine with a high osmotic pressure. This is accompanied by a reduction in the blood sodium level (hyponatraemia).

Oxytocin

Oxytocin, like ADH, is a small peptide composed of nine amino acids. The two hormones differ by only two amino acids.

Oxytocin exerts two primary physiological effects. Firstly, it promotes milk ejection from the breasts of lactating mothers. Suckling of the nipple by the infant results in the production of nerve impulses which pass to the hypothalamus. The latter stimulates the release of stored oxytocin into the blood in which it travels to the breast. Here the oxytocin initiates contraction of the myoepithelial cells which lie outside the milk-secreting alveoli. Thus milk is ejected (Fig. 9.25).

This is another example of a neuroendocrine reflex similar to that by which prolactin is released.

Oxytocin also has a role in parturition. At the end of pregnancy during the birth process, oxytocin is released and assists in the induction of uterine contractions. It is probable that dilation of the cervix is the primary stimulus for its release via a neuroendocrine pathway. Although oxytocin certainly assists in parturition, it is probably not responsible for its initiation.

It is also suggested that oxytocin is released in response to genital stimulation during intercourse. The hormone then brings about uterine contractions which assist spermatozoa in their journey towards the ova which are passing down the Fallopian tubes. Thus fertilization normally occurs in the fallopian tubes rather than within the uterine cavity.

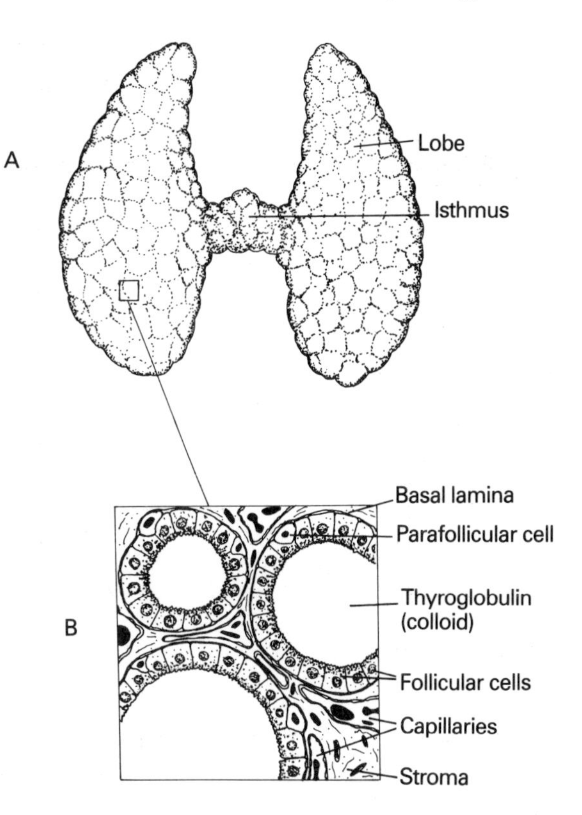

Fig. 9.7 (A) The external appearance of the thyroid gland. (B) A small section of thyroid tissue showing the follicles.

THE THYROID GLAND

STRUCTURE OF THE THYROID GLAND

The thyroid gland lies just below the larynx, is composed of two lateral lobes connected by an isthmus and in adults usually weighs less than 30 g (Fig. 9.7). The gland is covered by a connective tissue capsule and from this septa or trabeculae pass inwards to divide it into lobules. Each lobule is then subdivided into between 20 and 40 roughly spherical follicles which vary in size between 0.02 and 0.9 mm diameter. Lying between and separating the follicles is a connective tissue stroma containing a rich network of capillaries, lymphatics and nerves.

Each follicle consists of a single layer of epithelial cells resting on a basal lamina and encloses a central cavity which contains a variable amount of viscous material known as colloid. The appearance of the follicular cells varies according to the level of activity in the gland, being flattened when it is inactive and columnar in the active state. The colloid is in fact a protein, thyroglobulin, which contains the principal thyroid hormones, thyroxine and tri-iodothyronine.

Scattered between the epithelial cells are a number of cells with a clear cytoplasm which do not penetrate as far as the colloid. These are the parafollicular or C-cells which secrete calcitonin (thyrocalcitonin).

FUNCTIONS OF THE THYROID GLAND

Thyroxine and tri-iodothyronine

Both thyroxine and tri-iodothyronine are synthesized within the follicular cells in a process that involves the iodination of the amino acid tyrosine and the subsequent joining of two such molecules. The follicular cells are able to take up iodide from the blood by active transport; as a result, the thyroid contains the highest level of this element of any tissue in the body.

A tyrosine molecule has either one iodide ion added to it, forming monoiodotyrosine, or two ions in which case di-iodotyrosine is formed. These molecules can then be joined together in various ways. If two di-iodotyrosine molecules are conjugated, then tetra-iodothyronine or thyroxine (T_4) is formed. Alternatively, if one di-iodotyrosine and a monoiodotyrosine are joined then a tri-iodothyronine (T_3) molecule is formed.

Once formed, the hormones are conjugated with a large glycoprotein, also synthesized within the follicular cells, to form thyroglobulin (MW 660 000). This is then passed by exocytosis into the colloid for storage (Fig. 9.8).

The follicular cells also remove colloid from the stored mass by pinocytosis. The membrane-bound colloid vesicles within the cell are then fused with lysosomes to form phagosomes. Within the phagosomes, enzyme hydrolysis of thyroglobulin releases free hormone molecules which then pass into the blood by diffusion.

About 90% of the total quantity of hormone released by the thyroid is in the form of T_4. However, in the blood some of this T_4 is converted to T_3, with further conversion occurring in the tissues. Some scientists

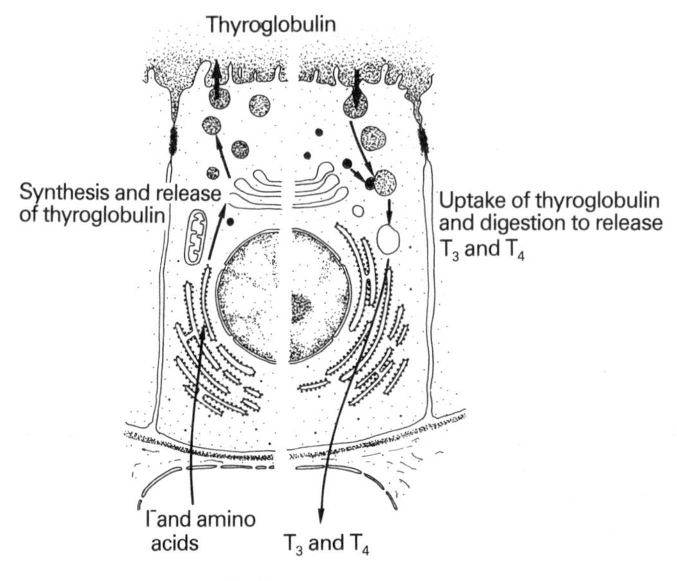

Fig. 9.8 Activities of thyroid cells. Left-hand side: iodide and amino acids are removed from the blood and used to synthesize thyroglobulin, vesicles of which are expelled into the follicle. Right-hand side: thyroglobulin is removed from the follicle and digested to release T_3 and T_4, which are passed into the blood.

consider that T_3 is the only active hormone and that T_4 merely acts as its precursor.

On entering the blood virtually all of the molecules of both hormones are bound to a number of plasma proteins, the principal one being an α-globulin (thyroxine binding globulin). This serves to inactivate the hormones, since only the free form can enter the tissues, as well as preventing their excretion by glomerular filtration. The affinity of T_4 for the binding globulin is about three times that of T_3 so that there is a higher proportion of free T_3 (0.3%) than free T_4 (0.03%). In each case, the free hormone is in equilibrium with the bound form in the blood.

The primary physiological effect of both thyroid hormones is to increase the metabolic rates of many cells and tissues. Some tissues, however, e.g. brain, testes, lungs and spleen, are not affected by either hormone.

Thyroid hormones enter cells and subsequently stimulate protein synthesis and therefore enzyme synthesis. Normal levels of hormones appear to be necessary for protein metabolism and therefore growth, but when they are present in excess, protein breakdown predominates and growth is retarded.

Following the introduction of a single large dose of T_4 there may be a time-lag of two or three days before a rise in metabolic rate is noticed. After this latent period there is a steady rise in BMR, which reaches a peak at about 10 to 12 days after the initial stimulus and then the drop is so slow that an appreciably raised BMR may still be observed after 60 days.

The two hormones also affect carbohydrate metabolism in a number of ways. They stimulate glucose absorption by the gut and enhance adrenaline-stimulated glycogen breakdown thereby raising blood glucose levels. In addition, they increase insulin-promoted uptake of glucose by adipose tissue and glycogen synthesis in muscle.

T_3 and T_4 promote lipolysis in adipose tissue, increased oxidation of free fatty acids and a lowering of plasma cholesterol by increasing its uptake by the adrenal cortex, gonads and liver.

Hypothalamo-hypophysial control of secretion of thyroxine and tri-iodothyronine

The release of T_4 and T_3 is regulated by thyroid-stimulating hormone (TSH) which comes from the adenophypophysis. TSH exerts numerous effects on the thyroid gland, promoting the uptake of iodide into the follicular cells, the iodination of the thyroid hormones and both the storage and breakdown of thyroglobulin.

TSH itself is under the control of the releasing hormone TSH-RH a tripeptide which is synthesized within the hypothalamus. When the concentration of thyroid hormones in the blood is reduced there is an increase in TSH-RH output which promotes the release of TSH. As a result, thyroid activity increases and the levels of thyroid hormones in the blood rise. These higher levels of thyroid hormones then exert an inhibitory influence upon the output of both TSH and TSH-RH. Thus TSH

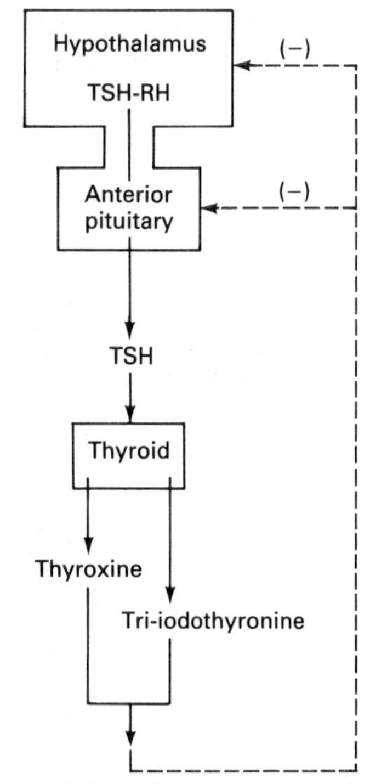

Fig. 9.9 Hypothalamo-hypophysial control of secretion of thyroid hormones. The principal site of negative feedback by the thyroid hormones is the anterior pituitary, although negative feedback on the hypothalamus may be important in long-term regulation of thyroid function.

and thyroid hormone levels are kept in babalance by a negative feedback mechanism (Fig. 9.9).

TSH output has been shown to increase in infants when there is a reduction in body temperature. Temperature receptors in the hypothalamus are activated by a reduction in the temperature of the blood and this results in the release of TSH-RH. The effect, however, is short-lived. In addition, the long latent period of thyroid hormones makes them unsuitable for rapid temperature control.

Chronic stress seems to reduce TSH output, probably by an inhibitory effect of cortisol on TSH-RH release.

Effects of thyroxine and tri-iodothyronine deficiency

Hypothyroidism may be due to any one of a number of causes and its effects are generally predictable from a knowledge of thyroid function. One cause, which used to be particularly common, is lack of iodine in the diet. As a result of this the thyroid gland cannot form enough hormones and thyroidal suppression of TSH is removed. TSH stimulates the thyroid which enlarges, but still cannot produce enough hormones. The end result is a greatly enlarged thyroid gland which produces a swollen neck or goitre. It should be noted that a goitre may also represent an overactive gland (see below).

Hypothyroidism is characterized by a reduction in the BMR observed as mental and physical sluggishness, increased weight, low heart rate, failure of hair growth etc. In serious cases there is also a deposition of mucopolysaccharide in the interstitial spaces producing a swollen appearance. This is known as myxoedema.

In young children, hypothyroidism leads to a retardation of both physical and mental development and produces cretinism.

Effects of thyroxine and tri-iodothyronine excess

A goitre is also often characteristic of hyperthyroidism, but in this case the enlarged gland is actively synthesizing hormones. The disease (Graves' disease) is most commonly produced as a result of the gland's response to certain antibodies in the blood (e.g. long-acting thyroid stimulator, LATS). TSH levels are normally depressed owing to the high levels of T_3 and T_4 in the blood.

The raised levels of T_3 and T_4 increase the metabolic rate and a number of symptoms are produced. These include weight loss, diarrhoea, nervousness, fatigue, sweating, increased heart rate. There is also usually protrusion of the eyeballs (exophthalmos), which is due to the deposition of mucopolysaccharides and oedema in the tissues lining the orbit. The precise cause of this phenomenon is still not understood.

Calcitonin

Calcitonin (thyrocalcitonin) is a peptide consisting of 32 amino acids. It is synthesized and released by the C-cells of the follicular epithelium and contributes to the regulation of blood calcium levels (see Fig. 9.11). A rise in calcitonin reduces blood calcium and phosphate levels.

A rise in blood calcium appears to exert a stimulatory effect directly upon the C-cells resulting in the release of calcitonin. The latter then reduces the level of calcium by acting at a number of sites. It acts on bone by decreasing the breakdown of calcium salts by the bone-dissolving osteoclasts. It also decreases the formation of osteoclasts and increases the rate of formation of osteoblasts from osteoclasts. In the kidneys, the hormone promotes phosphate and calcium excretion.

In adults the physiological effects of calcitonin are probably of little significance since at normal levels they are extremely weak. This is due to the fact that in adults osteoclast activity is very small so that its reduction has very little overall effect on calcium metabolism. Much more important is parathormone from the parathyroid glands.

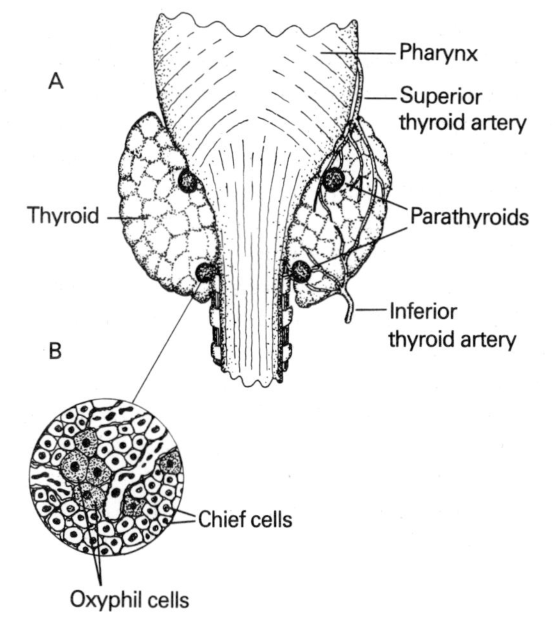

Fig. 9.10 (A) In situ appearance of the parathyroid glands. (B) Histological appearance of parathyroid tissue.

THE PARATHYROID GLANDS

STRUCTURE OF THE PARATHYROID GLANDS

In man there are usually four small glands, which vary in their distribution but which are most commonly attached to the rear surface and within the capsule of the thyroid, one at each of the upper and lower poles of the two lobes. The parathyroids have a combined weight of between 20 and 40 mg.

Each gland is enclosed by a fibrous capsule and contains two cell types; the hormone-secreting chief cells, and the oxyphil cells which are believed to represent an immature stage of the former (see Fig. 9.10).

FUNCTIONS OF THE PARATHYROID GLANDS

The parathyroid glands secrete parathyroid hormone (parathormone, or PTH), a polypeptide consisting of 84 amino acids. It is the hormone which is primarily responsible for the regulation of blood calcium and phosphate levels (see Fig. 9.11).

The parathyroids are stimulated directly by a lowering of blood calcium levels, while a raised blood calcium exerts a depressing effect on PTH output. Magnesium appears to exert similar effects.

PTH appears to raise blood calcium levels in two stages. The first, rapid, stage is brought about when PTH stimulates osteocytes and osteoblasts to remove calcium from their immediate environment and pass it into the extracellular fluid. The lowering of calcium in the bone fluid then allows calcium and phosphate to dissolve out of the bone matrix.

The second, slower, stage occurs when PTH stimulates the formation of osteoclasts and stimulates their bone-dissolving activities.

PTH promotes the reabsorption of calcium and inhibits the reabsorption of phosphate in the renal tubules. Thus, while blood levels of calcium rise, phosphate levels fall.

Lastly, PTH stimulates the conversion of 2,5-hydroxycholecalciferol to 1,25-dihydroxycholecalciferol in the kidneys. The latter is a vitamin D derivative which promotes the intestinal absorption of calcium.

Effects of parathyroid hormone deficiency

Hypoparathyroidism may result from a number of causes, including autoimmune disease, congenital absence of the gland or perhaps surgical damage. As a result of the lack of parathyroid hormone in the circulation, the blood calcium level is depressed (hypocalcaemia) and this then leads to a number of clinical manifestations. Most common among these is increased neuromuscular excitability leading to tetany (spasmodic contractions of muscle).

Effects of parathyroid hormone excess

Hyperparathyroidism is usually caused by a benign tumour of the gland. High levels of parathyroid hormone in the circulation raises blood calcium levels (hypercalcaemia) and lowers blood phosphate levels (hypophosphataemia). Since the increased amount of calcium in the blood has derived from the

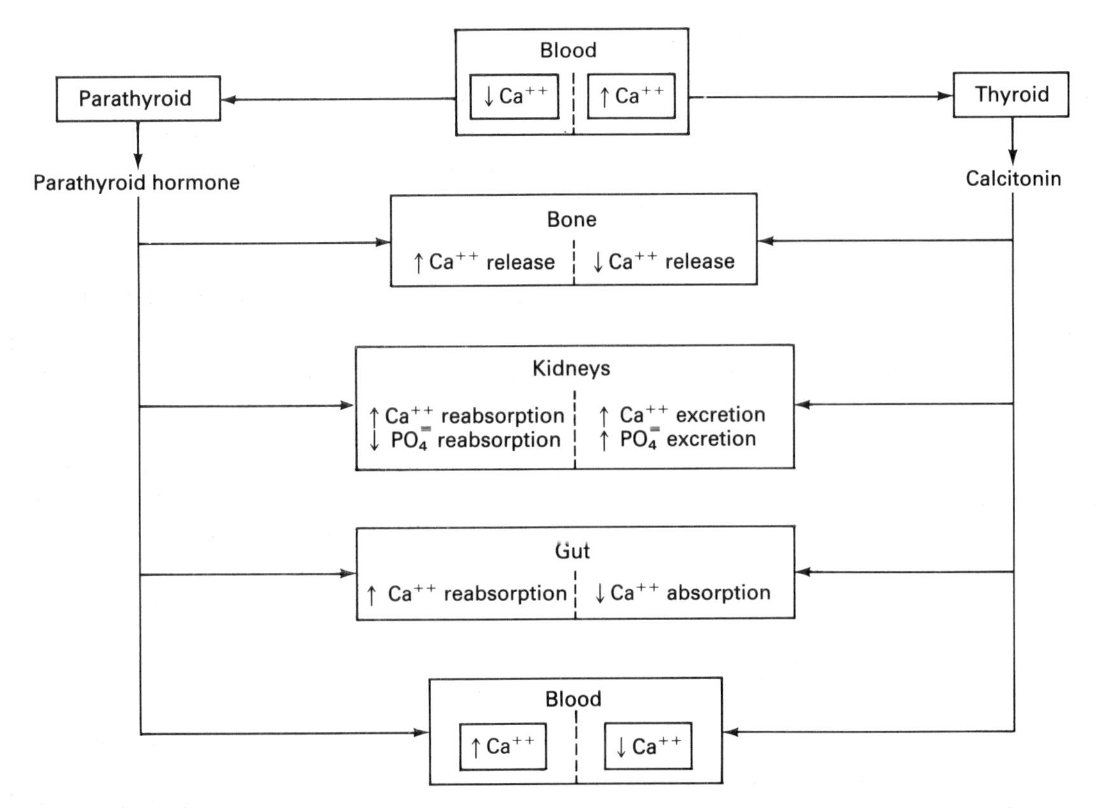

Fig. 9.11 Regulation of calcium metabolism by parathyroid hormone and calcitonin in response to a change in blood calcium concentration (top).

bones due to osteoclastic resorption, the bones themselves may become seriously weakened. Calcium is lost in the urine, which makes it difficult for the kidneys to concentrate the urine effectively and in turn leads to a disturbance of water balance. The high levels of calcium in the kidneys also favours the formation of kidney stones, which are found in about half of the patients suffering from this disease.

REGULATION OF BLOOD CALCIUM CONCENTRATION

Calcium has several functions in the body in addition to its role in bone, where its presence in bone salt confers hardness to the tissue. It stabilizes cell membranes and if calcium levels fall, membranes become overexcitable.

Calcium is involved in: the release of neurotransmitters at nerve endings; in muscle contraction; in modulation of cell metabolism in connection with hormone actions; and in various secretory activities, e.g. salivary glands, pancreas, gastric glands. Calcium ions are also necessary for many of the activation reactions involved in clotting.

Plasma calcium

Plasma calcium levels normally lie between about 9.0 and 10.0 mg% (about 2.5 mmol/l). Approximately half of the calcium is complexed with other molecules (mostly protein) and it cannot leave the bloodstream in this form. The non-conjugated calcium is in the form of ions which can pass into the tissue fluids and into cells quite easily. Nearly 1.2 mmol/l of ionized calcium is found in the blood. The relative proportions of bound and free calcium are pH-dependent. If blood pH rises (i.e. if blood becomes more alkaline) then some of the free calcium becomes bound so that the effective concentration of calcium is lowered. This situation can arise, for example, following hyperventilation and can cause tetany.

Bone calcium

Most of the calcium found in the body is located within the skeleton and about 70% of the weight of compact bone consists of calcium salts in the form of hydroxyapatites ($Ca_{10}(PO_4)_6OH_2$). Of the salts present in bone, as much as 1% is present in a more accessible form such as $CaHPO_4$. This is known as exchangeable bone and it is easily broken down to release calcium into the blood. Equally, if there is a sudden rise in the calcium levels of the blood, this will be rapidly deposited within the bone. Thus exchangeable calcium provides a mechanism whereby rapid changes in blood calcium can be dealt with and calcium levels maintained.

It has already been described in Chapter 1 how bone is continually being resorbed by osteoclastic activity and deposited by the osteoblasts. This resorption and accretion of bone is normally in equilibrium unless particular stress is placed upon particular bones in which case net deposition of calcium occurs and the bones are strengthened.

Role of vitamin D

Calcium is poorly absorbed in the gut, although phosphate is absorbed very easily. In order that calcium may be absorbed at all, it is important that vitamin D is available. This is itself absorbed from the diet and more importantly formed in the skin as a result of the action of sunlight. Vitamin D is converted, through a series of reactions, into 1,25-dihydroxycholecalciferol. The latter promotes calcium absorption in the small intestine by increasing the synthesis of calcium-binding protein and enzymes which pump calcium through the intestinal cell membranes. The 1,25-dihydroxy-form also mobilizes calcium from bone.

Vitamin D is actually converted into 25-hydroxycholecalciferol in the liver; this is a precursor of the 1,25-dihydroxy- form and is independently regulated by a negative feedback influence on its own formation. The final reaction occurs in the kidneys to liberate the active form in a reaction that is regulated by PTH (see *Functions of the parathyroid glands*) and plasma levels of calcium. Low levels of calcium accelerate the reaction, as do raised levels of PTH.

Roles of parathyroid hormone and calcitonin

Parathyroid hormone tends to raise blood calcium level, whereas calcitonin has the opposite effect and tends to lower it (see *Calcitonin* and *Functions of the parathyroid glands* for details of the mechanisms involved).

The role of calcitonin in calcium homeostasis is probably minor in adults; parathyroid hormone is much more important, although its effects take some hours to develop.

THE PANCREAS

STRUCTURE OF THE ISLETS OF LANGERHANS

The pancreas is a compact organ situated behind and below the stomach in the fold of the duodenum. It consists of clusters of cells (acini) which secrete a series of digestive enzymes through ducts into the duodenum (see Ch. 6).

Scattered among the exocrine acini are small groups of cells embedded in well-developed networks of blood capillaries. These groups of cells constitute the hormone-secreting islets of Langerhans, which represent less than 2% of the total pancreatic mass (Fig. 9.12).

The islets contain a number of cells which are characterized by their staining reactions and ultrastructural appearance. The three major cell types are identified as α-, β- and δ-cells (alternatively as A-, B- and D-cells). Most common are the insulin-secreting β-cells, which make up 60–80% of the total number. The α-cells, which secrete glucagon, constitute about 20–30%, while the δ-cells, which secrete somatostatin, only represent a maximum of about 8% of the total cell mass.

Although the islets of Langerhans are scattered throughout the pancreas, they are

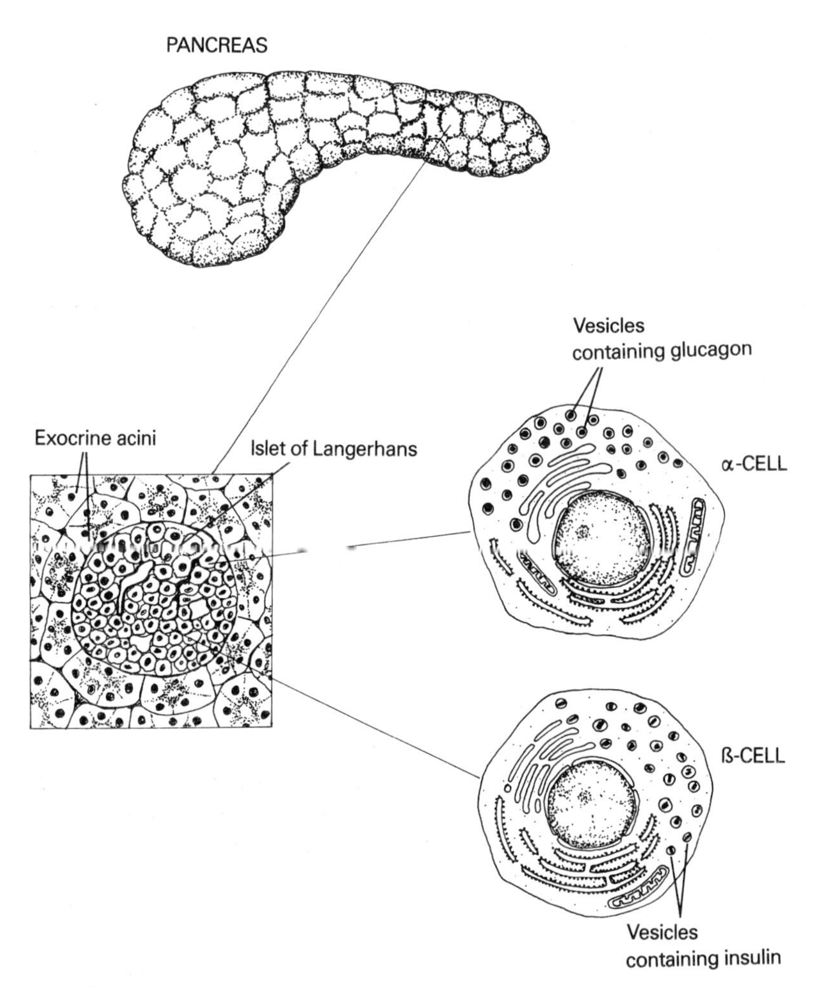

Fig. 9.12 α- and β-cells in the islets of Langerhans of the pancreas.

particularly numerous in the tail region. Like all endocrine tissues, they release their hormones directly into the blood stream and their function proceeds independently of the exocrine pancreas.

FUNCTIONS OF THE ISLETS OF LANGERHANS

Insulin

Insulin is a small polypeptide derived from a long-chain precursor, preproinsulin. The latter loses a terminal fragment and becomes folded with the formation of two disulphide bridges to form proinsulin. A further fragment is then lost leaving insulin which has two chains, the A-chain containing 21 amino acids, the B-chain 30 amino acids. Large quantities of insulin are stored within the granules of the β-cells.

Insulin is released by a process of exocytosis in the presence of high blood sugar. Some proinsulin is also released, but it exhibits low physiological activity.

Although variations in blood glucose may appear to be the primary regulator of insulin release, a number of other factors also play their part.

Some amino acids, such as arginine, increase insulin release and fatty acids have a similar, but much smaller effect. The nervous system also exerts an influence upon insulin secretion, the sympathetic system generally having an inhibitory effect while the vagus is stimulatory. Insulin output is also affected by some other hormones including somatostatin (inhibitory), secretin and cholecystokinin

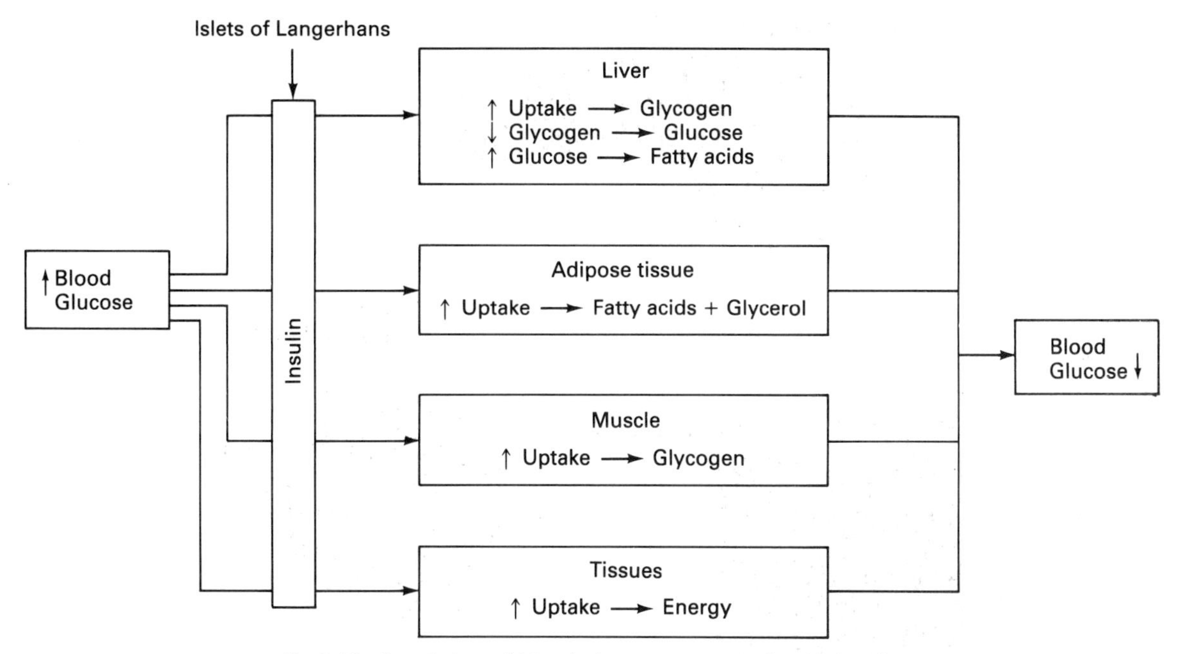

Fig. 9.13 Regulation of blood glucose concentration by insulin.

(stimulatory) and glucagon (also stimulatory).

Insulin molecules are not absorbed by the target cells but combine with receptors in the cell membrane. The precise mechanism by which the insulin then exerts its effect on the cell is not known. At least some of the effects of insulin are associated with reduced levels of cAMP.

Insulin increases the uptake of glucose by many tissues and reduces its release from the liver and thereby decreases the levels of glucose in the blood (Fig. 9.13). It also promotes the formation and storage of both fat and protein. Some tissues, including brain, liver, kidney, intestinal epithelium and the pancreatic islets do not require the presence of insulin for glucose uptake.

The hormone increases the uptake of glucose by the liver by increasing its phosphorylation within the cells, thereby maintaining a steep gradient for glucose absorption. The glucose is effectively trapped within the cell and is then converted into glycogen. Insulin increases the activity of the enzymes which promote glycogen synthesis and at the same time glycogen breakdown is inhibited. The hormone also promotes glucose uptake by muscle but by a different mechanism since it actually stimulates facilitated diffusion. Again, phosphorylation of the glucose occurs within the cells and glycogen is laid down.

Within the liver, insulin promotes the conversion of glucose to fatty acids. The latter escape into the blood and are taken up by adipose tissue. Furthermore, insulin promotes the uptake of glucose by adipose cells, leading to the formation of both fatty acids and glycerol. There is also inhibition of fat hydrolysis in adipose tissue.

Insulin acts generally to increase the uptake of many amino acids by the tissues, to stimulate protein synthesis and inhibit its breakdown. In the liver it depresses gluconeogenesis and thereby conserves amino acids.

Insulin is required in order that glucose can be used as a substrate for the production of energy. It promotes the uptake of glucose by most tissues and allows its subsequent breakdown to release energy. Overall it swings the body's metabolic processes away from protein and fat as energy sources and towards glucose.

Effects of insulin deficiency

Diabetes mellitus is the most common of all

endocrine disorders and involves a reduction or loss in function of the β-cells of the islets of Langerhans. This disease may be observed in 1–2% of the population in European countries.

Although reduction in insulin release by the pancreas leads to the development of diabetes mellitus, it can also be produced as a result of alterations in the circulating levels of a number of other hormones, e.g. excess growth hormone, excess thyroxine.

There are two types of (pancreatic) diabetes: juvenile-onset diabetes and maturity-onset diabetes. The former type begins suddenly in early life and is often very severe. In most affected children, antibodies against islet cells are found in the blood. These then decrease with age. Maturity-onset diabetes develops gradually in later life, usually in individuals who are rather obese.

A reduction in insulin secretion by the pancreas leads to a rise in blood sugar level (hyperglycaemia). This is due to the reduced ability of cells to take up and utilize glucose, and increased glycogenolysis and glucose release by the liver.

The high level of glucose in the blood results in the kidneys being unable to reabsorb it completely and so it appears in the urine (glycosuria). Since the glucose is osmotically active, it reduces reabsorption of water from the tubules and the urine volume increases (diuresis). As a result, the body becomes dehydrated so that thirst becomes a common symptom of diabetes.

Fat is mobilized from adipose tissue and is metabolized instead of glucose. The liver converts fatty acids into ketone bodies (acetone, OH-butyric acid, aceto-acetic acid). Some acetone is eliminated in the lungs which confers a distinctive odour upon the breath.

The presence of excessive amounts of ketoacids causes metabolic acidosis and this in turn stimulates ventilation. Some H^+ replace K^+ in the cells and excessive amounts of potassium may then be lost because of the diuresis.

Protein in skeletal muscle is broken down, leading to muscle wasting. Some of the amino acids are converted into glucose in the liver.

If no insulin is given, all of these changes can happen in a few hours and coma and even death might result. The dehydration can lead to a low blood volume and consequent reduced cerebral blood flow; the presence of ketoacids in the blood and electrolyte depletion (from acidosis and vomiting which can occur) also contribute to the onset of coma.

Effects of insulin excess

Excess insulin may be caused by the presence of an insulin-secreting tumour (insulinoma), but more commonly results from an overdose of insulin. This will lead to a reduction in the level of glucose in the blood (hypoglycaemia). This might be expected to lead to an increase in the desire for food, but such an effect is often not found. The usual symptoms of this condition are behavioural and may resemble drunkeness or mental aberration. The low level of blood sugar stimulates increased sympathetic activity and so the heart rate increases, sweating occurs and there is vasoconstriction in the skin. If the blood sugar level remains low, then brain function may be impaired to such an extent that hypoglycaemic coma may result.

Glucagon

Glucagon comes from the α-cells of the islets of Langerhans. It is a single chain peptide of 29 amino acids. The hormone is released by a process of exocytosis in response to a reduced blood glucose level. Its output is reduced in the presence of raised blood glucose and insulin. Sympathetic stimulation of the pancreas increases glucagon output, as does a rise in the quantity of amino acids in the blood and a rise in CCK–PZ, gastrin and gastric-inhibitory peptide. Secretin, on the other hand, has an inhibitory effect.

Glucagon raises blood sugar levels by stimulating breakdown of glycogen in the liver and inhibiting glycogen synthesis. It also increases the formation of glycogen from non-carbohydrate sources (gluconeogenesis). Glucagon also promotes lipolysis in adipose tissue.

It was the study of glucagon and its action

in increasing intracellular cAMP levels that led to the development of the 'second messenger' concept by Sutherland et al in 1961.

Effects of glucagon excess

Glucagon-secreting tumours are occasionally found, mostly in post-menopausal women. Mild diabetes occurs but is not accompanied by ketoacidosis. It has been shown, however, that glucagon levels are raised in the ketoacidotic state of diabetes mellitus and it has been suggested that the hormone is at least partially responsible for the acidosis. High levels of glucagon may also contribute to the hyperglycaemia observed in uncontrolled diabetes, implying that high blood sugar is not solely the result of low levels of insulin.

THE ADRENAL GLANDS

STRUCTURE OF THE ADRENAL GLANDS

There are two adrenal glands, each weighing about 10 g, one lying on top of each kidney. By virtue of their position, they are also known as the suprarenal glands.

Each adrenal is bounded by a capsule and contains two easily distinguishable areas, the outer cortex and the inner medulla (see Fig. 9.14). These two areas have different embryological origins and functionally may be considered as separate glands.

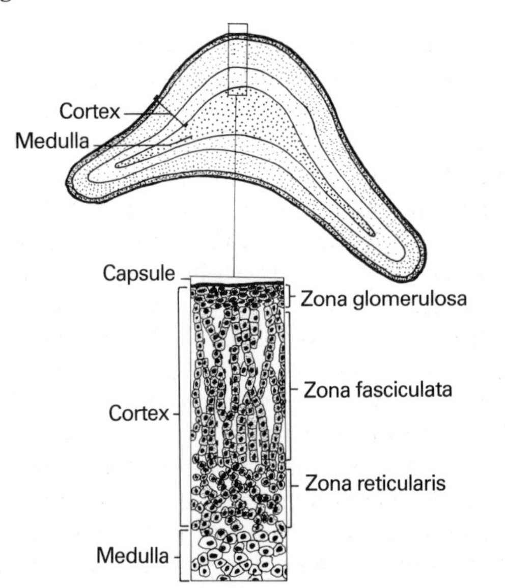

Fig. 9.14 Section through the adrenal gland. (Inset) Histological section of adrenal tissue.

The medulla is regulated by the sympathetic nervous system and its activities parallel those of that system. The cortex, on the other hand, produces a series of steroid hormones which are important in the regulation of both organic and mineral metabolism.

The cortex contains three layers or zones (see Fig. 9.14). The outermost layer, the zona glomerulosa, is a thin band of cells located immediately beneath the outer capsule. The cells are small and arranged in irregular clusters. In the middle of the cortex lies the zona fasciculata consisting of large polyhedral cells arranged in columns. The innermost layer, the zona reticularis, consists of irregular, branching columns of cells. All cortical cells are characterized by the presence of large numbers of mitochondria and numerous small lipid droplets.

The medulla contains irregular clusters of chromaffin cells (containing granules which stain with chrome salts) and are easily distinguished from those of the cortex.

FUNCTIONS OF THE ADRENAL CORTICES

The hormones of the adrenal cortex closely resemble the cholesterol from which they derive. A large number of different steroid molecules have been extracted from the adrenal cortex, although many are intermediate compounds rather than physiologically active hormones. In man, the two major cortical hormones are cortisol and aldosterone, although small amount of oestrogens and androgens are also produced.

The zona fasciculata synthesizes cortisol and a much smaller amount of corticosterone. These hormones are primarily concerned with the regulation of carbohydrate, fat and protein metabolism and are known as glucocorticoids. Their output is regulated by ACTH from the pituitary.

The zona glomerulosa produces aldosterone and a small amount of deoxycorticosterone. These hormones are important in the regulation of sodium balance and are described as mineralocorticoids. Aldosterone secretion rate is influenced by ACTH, but it is mainly regulated by sodium and potassium concentrations in blood, as well as by blood pressure and volume.

The structural similarities of the adrenal steroids lead to an overlap of function. Cortisol, for example, exerts a mild mineralocorticoid effect which is about 2000 times less potent than that of aldosterone. The latter hormone is about 100 times less potent than cortisol as a glucocorticoid. Oestrogens and androgens are secreted by the zona reticularis in very small amounts.

Cortisol

About 90% of the cortisol in the blood is carried bound to plasma protein. The principal carrier is an α-globulin called corticosteroid-binding globulin (CBG). Cortisol controls a wide range of physiological functions (Fig. 9.15).

Cortisol promotes the deposition of glycogen in the liver through a number of mechanisms. Amino acids are removed from the extrahepatic tissues (mainly muscle) thereby raising the level of free amino acids in the plasma. Amino acids are then transported into the cells of the liver and converted into glucose (gluconeogenesis) and then to glycogen. All of these processes are enhanced by the presence of cortisol.

Cortisol also influences carbohydrate metabolism by reducing the uptake of glucose by the cells of the body and reducing its usage within the cells. As a result, the level of glucose in the blood rises.

It is evident from the above that the concentration of free amino acids in the plasma will rise in the presence of cortisol. In addition, some of the amino acids which are transported into the liver will be incorporated into protein. Thus, while there is protein removal from most tissues, there is protein deposition in the cells of the liver.

Cortisol has a lesser, though still important, influence upon fat metabolism. It promotes the mobilization of fatty acids from adipose tissue, which raises the concentration of free fatty acids in the plasma. It also enhances the breakdown of fatty acids within the tissues to release energy. Neither of these two effects is particularly strong, but nevertheless they can, in certain circumstances, shift the overall metabolic function of the cells from carbohydrate breakdown to fat breakdown.

Cortisol generally inhibits the inflammatory response, in a number of specific ways. It decreases vascular permeability which reduces fluid leakage and the migration of blood cells into the damaged areas. The hormone also stabilizes the lysosomes of damaged cells, greatly reducing the release of inflammatory substances. In addition it has a generally suppressive influence upon the immune system, reducing both the numbers of cells

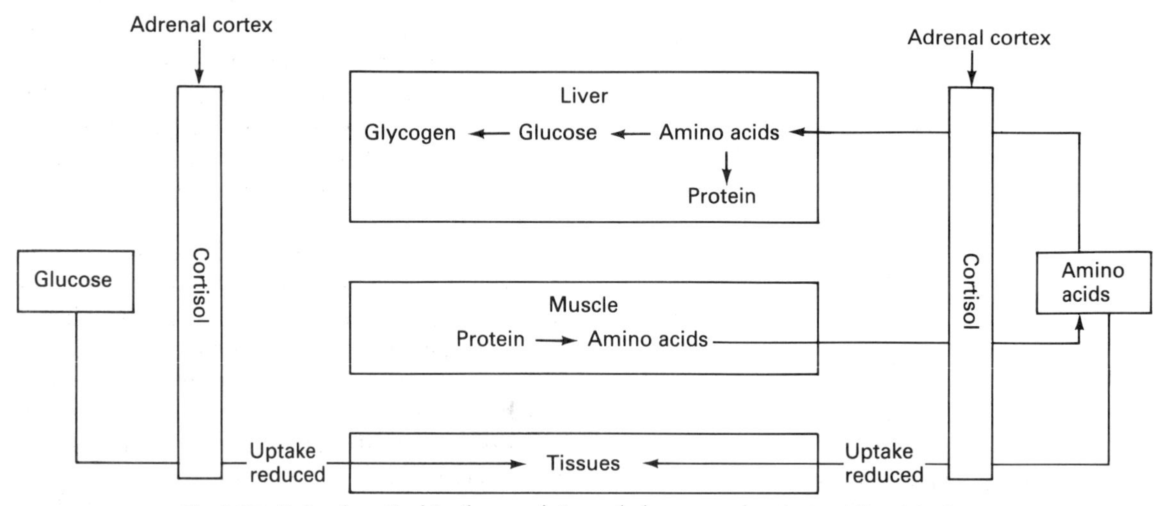

Fig. 9.15 Role of cortisol in the regulation of glucose and amino acid metabolism.

and antibodies which would otherwise be raised. Cortisol delays wound healing.

The anti-inflammatory effects of cortisol are exploited clinically to treat conditions such as acute hypersensitivity states where the inflammatory response is counter-productive.

In addition to the above, it is known that an increase in the level of cortisol above that which is usual for the time of day appears to be necessary in order that the body be able to withstand stressors, e.g. cold, heat, infection, trauma, prolonged exercise. Unfortunately, it is not understood precisely how cortisol is able to provide this protection.

Cortisol appears to exert a number of 'permissive' effects which only really become apparent when the hormone is absent. It appears to be necessary, for example, in order for the smooth muscle in the walls of blood vessels to be able to react normally to adrenaline and noradrenaline. Cortisol does, then, seem to have a role in the maintenance of blood pressure.

The output of cortisol by the adrenal cortex is regulated by adrenocorticotrophic hormone (ACTH) from the anterior pituitary (Fig. 9.16). This, in turn, is regulated by corticotrophin-releasing hormone (CRH), which originates in the basal medial area of the hypothalamus. CRH is a polypeptide which is released into the portal tracts linking the hypothalamus with the adenohypophysis. It passes down into the pars distalis and causes the corticotrophs to release ACTH into the circulation.

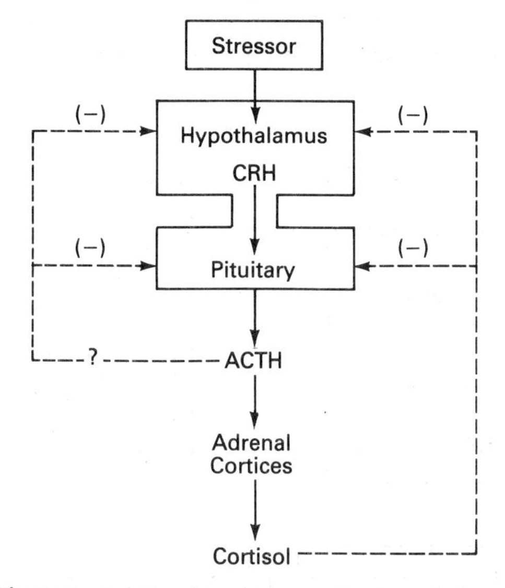

Fig. 9.16 Relationships between the hypothalamus, anterior pituitary and adrenal cortices.

ACTH travels to the adrenal cortex, where it stimulates the cells of the zona fasciculata to synthesize and release cortisol. The latter exerts a negative feedback effect upon both the hypothalamus and the pituitary, thereby reducing the output of both CRH and ACTH.

Physical and mental stressors stimulate CRH release. In addition, CRH brings about a diurnal variation in ACTH output, presumably induced by neural inputs into the anterior hypothalamus. Normally, a high level of ACTH activity is observed in the early morning, often prior to waking up. A change in the length of the day and/or in sleeping habits will induce a corresponding change in hormone levels.

It has been demonstrated that very high levels of antidiuretic hormone and angiotensin II also exert a stimulatory effect upon both the hypothalamus and the adenohypophysis resulting in increased output of ACTH. The significance of these responses at normal physiological levels is, however, not understood.

Aldosterone

Aldosterone promotes the reabsorption of sodium ions and the elimination of potassium ions in the distal convoluted tubules and collecting ducts of the kidneys. The hormone stimulates the synthesis of RNA within these cells to produce protein carrier molecules which assist the carriage of Na^+ from the filtrate into the blood. K^+ is attracted into the filtrate from the cells in exchange for Na^+, a mechanism which ultimately leads to a reduction in the amount of K^+ in the extracellular fluid.

Although a primary effect of aldosterone is to increase Na^+ reabsorption, in reality the Na^+ concentration may increase very little. This is because the raised levels of Na^+ in the extracellular fluid create an osmotic gradient which results in the retention of some water. The net effect, therefore, is that there is an increased volume of extracellular fluid which has only a moderately high level of Na^+. It is interesting to note that aldosterone exerts a similar effect on sweat, salivary and intestinal glands.

Aldosterone is synthesized and released by the zona glomerulosa of the adrenal cortex. There are a number of mechanisms which promote its release and, although they are individually quite well understood, there is some discussion as to which are the more important in man. The stimuli for aldosterone release include a high level of K^+ and/or a reduced Na^+ level in the extracellular fluid, a drop in blood volume or blood pressure, and the presence of ACTH in the blood.

A drop in the concentration of Na^+ in the blood or a rise in K^+ will directly stimulate the cells of the adrenal cortex to liberate aldosterone (Fig. 9.17). The latter will then promote the reabsorption of Na^+ and the elimination of K^+ through the distal convoluted tubules and collecting ducts of the kidneys; thus either or both stimuli will be removed. A drop in blood K^+ levels will normally lead to a decrease in the extracellular fluid volume, and this may act as a stimulus for aldosterone release through a different route.

A decrease in blood volume leads to an increase in the activity of the renal nerves and a decreased pressure in the renal arteries. As a result, the juxaglomerular cells which lie in the walls of the afferent arterioles leading into the glomeruli are activated to release the enzyme renin into the blood. Renin converts circulating angiotensinogen, which is formed in the liver, into a decapeptide, angiotensin I. The latter is then converted into the octapeptide, angiotensin II in the presence of Cl^- and a converting enzyme. Angiotensin II acts upon the adrenal which liberates aldosterone which, in turn, promotes Na^+ retention. In the presence of ADH from the posterior pituitary, water is retained to expand the extracellular fluid volume. Blood flow to the kidney is thus increased and the stimulus removed. It seems likely that the primary physiological role of aldosterone lies in its regulation of body fluid volume.

ACTH induces a transient elevation of aldosterone output presumably by a direct stimulatory mechanism. ACTH also exerts a long-term permissive influence, however, and in its absence the zona glomerulosa will regress so that other stimuli will not function with their normal degree of effectiveness.

Effects of aldosterone and cortisol deficiency

Two types of adrenal insufficiency are found.

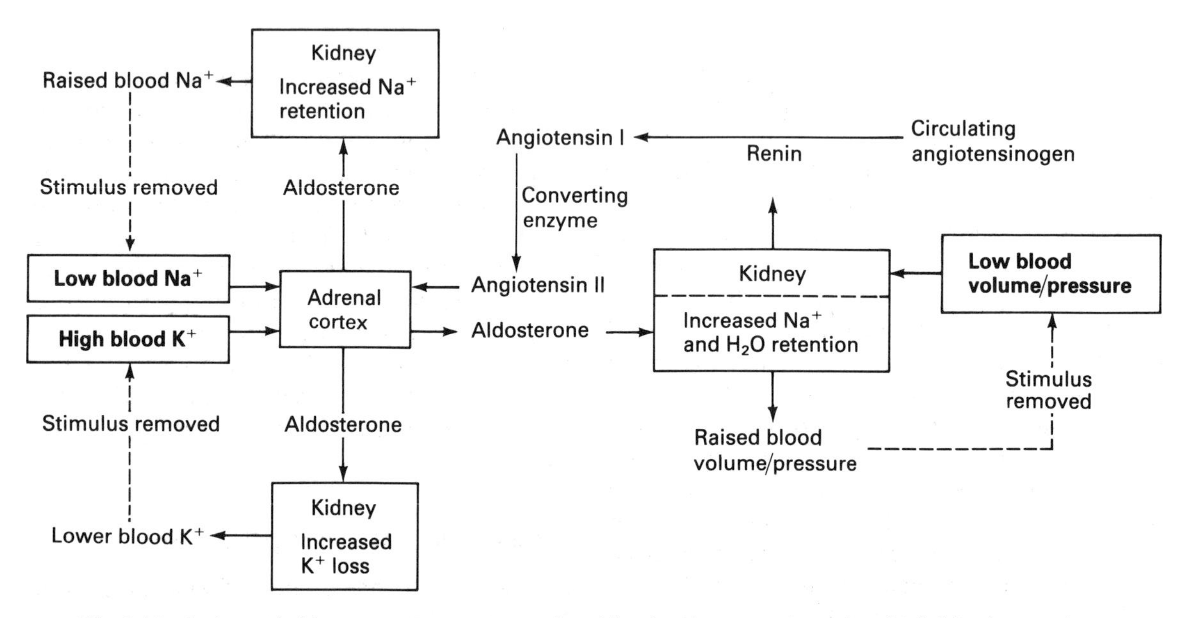

Fig. 9.17 Actions of aldosterone in response to low blood sodium concentration, high blood potassium concentration (left-hand side), and low blood volume/pressure (right-hand side).

Primary insufficiency is due to a direct decrease in adrenal function and involves both cortisol and aldosterone (Addison's disease). Secondary insufficiency is due to a reduction in ACTH and is therefore dependent upon the hypothalamus and/or pituitary. In this case cortisol alone is affected.

Lack of aldosterone secretion decreases sodium and water retention, leading to a reduced blood volume and dehydration. Since sodium is not reasorbed by the kidneys, potassium and hydrogen cannot be secreted in exchange so that hyperkalaemia and acidosis occur.

The reduction in the level of cortisol in the blood reduces the amount of glucose in the circulation and also depresses protein and fat breakdown. Thus weakness, dizzy spells etc. commonly occur. There is also an inability to withstand stress.

Lack of cortisol in the blood removes inhibition from the adenohypophysis so that ACTH is released in large quantities. The latter has some MSH-like properties and since it is present in such large quantities they become significant. As a result there is increased pigmentation in parts of the body's surface.

Effects of cortisol excess

An increase in the rate of secretion of cortisol may either be due to the presence of a tumour in the adrenal cortex (Cushing's syndrome) or by increased ACTH release from the pituitary (Cushing's disease).

Cortisol promotes a rise in blood sugar by reducing its uptake by the tissues, leading to 'adrenal diabetes'. Protein metabolism is also affected since cortisol encourages the removal of amino acids from the tissues and their deposition in the liver. As a result, there is muscle wasting, particularly in the limbs, and a breakdown in collagen, especially in the skin, so that the latter is easily damaged (it may tear subcutaneously, giving rise to purplish striae). The protein portion of bone is also depleted, which renders it fragile.

Cortisol causes the mobilization of fat from the adipose tissues and its deposition in the face, upper back (producing a 'buffalo hump') and the abdominal wall.

Since cortisol is present in large quantities, its mineralocorticoid function becomes significant and there is sodium retention and potassium loss. The latter may contribute to the muscle weakness.

Sodium retention leads to water retention and contributes to the hypertension which is often characteristic of excess cortisol.

The immune-suppressing influence of cortisol renders patients susceptible to infection.

Effects of aldosterone excess

Excessive secretion of aldosterone by a tumour of the adrenal cortex (Conn's syndrome) occurs rarely. Secondary aldosteronism may be due to a decreased rate of removal as in cirrhosis of the liver, or as a compensatory mechanism in other disorders such as cardiac failure. There is usually salt and water retention, inducing hypertension which may then produce a number of secondary effects. Excess potassium is lost by the kidneys; hypokalaemia occurs, which gives rise to muscle weakness or even paralysis, alkalosis and kidney damage.

Androgens

The zona reticularis of the adrenal cortex secretes minute quantities of male sex hormones, the principal one being dehydroepiandrosterone. These androgens may be converted into the primary male hormone, testosterone, outside the adrenal and it is probably this which confers most of the androgenic function ascribed to the adrenal cortex. Adrenal androgens are secreted in both males and females, but their activity in the latter is slight. Small amounts of testosterone may be converted to oestrogen in the circulation.

Effects of adrenal androgen excess

Excessive amounts of the adrenal androgens may result from the presence of a tumour or an enzyme deficiency. In young males this hastens the onset of puberty (precocious puberty).

In young and adult females there is masculinization, with a reduction in breast size, an increase in the growth of body and facial hair and increased muscular development.

FUNCTIONS OF THE ADRENAL MEDULLAE

Adrenaline and noradrenaline

The chromaffin cells of the adrenal medulla synthesize and store the catecholamines adrenaline (epinephrine) and noradrenaline (norepinephrine). Both molecules are derivatives of the amino acid tyrosine. The hormones are stored within secretory granules and liberated into the blood when the gland is stimulated by acetylcholine from preganglionic sympathetic nerve fibres leading from the spinal cord. The chromaffin cells may be likened to postganglionic sympathetic nerve fibres which liberate their neurotransmitter substances into the blood instead of the effector organs. Approximately 80% of the liberated hormone is in the form of adrenaline, the other 20% as noradrenaline.

The physiological activities of the medullary hormones mirror those of the sympathetic nervous system and in many ways the medulla may be considered to be an extension of that system. The primary difference is that the effects of the medullary hormones exhibit a duration about ten times longer than those of the sympathetic system. There are also some differences in the specific functions of adrenaline and noradrenaline.

Both hormones increase the rate and strength of contraction of cardiac muscle but noradrenaline is mainly responsible for vasoconstriction in the skin, gut, kidneys and mucous membranes. In skeletal muscle noradrenaline again brings about vasoconstriction but adrenaline induces dilation. Since four-fifths of the medullary output is adrenaline, however, then the overall effect of medullary activity is vasodilation and not vasoconstriction.

In the respiratory system both hormones bring about bronchodilation, with adrenaline exerting the stronger effect.

Both hormones promote relaxation of the smooth muscle component of the gut wall, but induce constriction of the pyloric and ileocaecal sphincters. In addition, both hormones bring about dilation of the pupil of the eye by stimulating contraction of the radial muscle fibres of the iris.

The apparently contradictory responses of different tissues to the two hormones is due to the presence of two alternative types of receptor present in these tissues. α-Receptors normally bring about vasoconstriction and contraction of smooth muscle generally and are responsive mainly to noradrenaline. β-Receptors, on the other hand, bring about relaxation of smooth muscle, vasodilation and an increased heart rate and are stimulated by adrenaline (see also *Sympathetic nervous system* in Ch. 7).

Adrenaline also has a role in the regulation of organic metabolism, specifically in the maintenance of blood glucose levels. It promotes gluconeogenesis and glycogenolysis in the liver and the glucose which is formed passes into the blood, thereby raising the level of blood sugar. Adrenaline also induces glycogenolysis in skeletal muscle, but since muscle cells lack the appropriate enzymes to convert the resulting glucose-6-phosphate into glucose, the end product is lactic acid. The latter passes into the blood and may be converted into glucose by the liver.

Both adrenaline and noradrenaline stimulate lipolysis in adipose tissue, thereby raising the levels of free fatty acids in the blood.

Adrenaline, like the thyroid hormones, increases oxygen consumption in the tissues and raises the metabolic rate of the body.

Effects of adrenaline and noradrenaline excess

Tumours of the adrenal medullae (phaeochromocytoma) are quite rare. They result in persistent secretion of adrenaline and noradrenaline, which leads to the development of a raised blood pressure (hypertension). This in turn may produce headaches, pain, nausea etc. Sweating is also a common symptom.

REGULATION OF BLOOD GLUCOSE CONCENTRATION

Glucose is only one of a number of energy sources in most tissues, but is normally the sole source of energy for the nervous system. It is imperative that blood sugar levels must not vary too widely. Hyper- or hypoglycaemia can each result in coma (see *Effects of insulin excess* and *Effects of insulin deficiency*). The maintenance of reasonably stable blood sugar concentrations is brought about by the action of a number of hormones whose individual activities are covered in other sections.

Blood glucose level rises after a meal (absorptive state) but falls during the post-absorptive state. The physiological mechanisms tending to reverse these changes are therefore different in the two states.

Absorptive state

After a meal, when digested food is being absorbed, glucose is added to the blood from the small intestine and blood sugar level therefore tends to rise, perhaps reaching a blood concentration of 8 mmol/l. The absorptive state may last some four hours and is a time of relative carbohydrate excess. During this period, various physiological mechanisms serve to reduce the carbohydrate level.

Glucose is converted into glycogen and stored in the liver. The latter can also convert glucose into triglycerides which are subsequently stored in adipose tissue. Some glucose is converted to triglyceride in adipose tissue itself. Finally, the general body tissues oxidize the glucose as a source of energy for cellular activities.

The rise in blood sugar directly stimulates the release of insulin from the pancreatic islets. This hormone then promotes the uptake of glucose by most tissues and the deposition of glycogen in the liver. At the same time, insulin also inhibits triglyceride breakdown thereby reducing the supply of fat to the tissues.

Postabsorptive state

When absorption from the intestine is no longer occurring, there is a relative lack of glucose and a fall in blood sugar level. The level may be as low as 4 mmol/l in fasting. The physiological mechanisms which operate to conteract this drop include the breakdown of liver glycogen to glucose and its subsequent release into the blood; the breakdown of fat and protein and their subsequent conversion to glucose in the liver (gluconeogenesis); and the use by tissues of a higher proportion of fat for oxidation.

Many hormones affect blood sugar level but there are only two which seem to be particularly important in controlling it, glucagon and adrenaline.

A fall in blood glucose stimulates glucagon release from the α-cells of the islets of Langerhans. Glucagon promotes the breakdown of glycogen in the liver to release glucose, thereby helping to elevate blood glucose. It also promotes the conversion of amino acids to glucose in the liver, as well as stimulating fat mobilization from adipose tissue.

Adrenaline from the adrenal medullae exerts some of the same effects as glucagon. It stimulates the breakdown of glycogen in the liver, and promotes fat mobilization. The release of adrenaline is stimulated by the effect of low blood sugar on the hypothalamus, which in turn stimulates the sympathetic nervous pathway to the adrenal medullae. Sympathetic stimulation of adipose tissue also causes fat mobilization.

Growth hormone and cortisol can be important in the control of blood sugar in more extreme circumstances. A rapid fall in blood glucose or prolonged fasting stimulates receptors in the hypothalamus which lead to the release of growth hormone by the anterior pituitary. Growth hormone depresses glucose utilization in the tissues and promotes the deposition of glycogen. It therefore reduces glucose removal from the blood stream and helps to maintain a stable level. The hormone also transfers the emphasis of energy production within the tissues from glucose as a primary source to fatty acids, by increasing triglyceride breakdown in adipose tissue.

Lastly, long-term fasting initiates the production of ACTH by the anterior pituitary which leads to the release of cortisol by the adrenal cortex. Cortisol promotes the formation of glucose from amino acids within the cells of the liver and decreases its usage by the tissues.

THE FEMALE REPRODUCTIVE SYSTEM

The female reproductive system has two functions. Firstly, it is concerned with the processes of reproduction and, secondly, with the production of hormones. There is considerable overlap between these two functions.

The system contains two ovaries whose function is to produce the ova or germinal cells and also secrete sex hormones. They lie on the rear wall of the abdominal cavity and are supported by ligaments. Lying close to each ovary is the dilated open end of a Fallopian tube, thus there are two of these short coiled ducts which lead into a hollow and thick-walled muscular organ, the uterus. The latter is pear-shaped and projects into a short muscular tube, the vagina, which leads to the exterior (see Fig. 9.18).

The uterus is the organ within which, should fertilization occur, a fetus will develop. It is therefore capable of growth and has sufficient elasticity to accommodate the developing child. The innermost layer of the uterine wall, the endometrium is a highly vascular, glandular structure which is able to help support the developing child. During reproductive life, in the non-pregnant female, this layer cyclically increases in thickness and is then shed in a hormone-dependent series of events known as the menstrual cycle.

STRUCTURE AND FUNCTIONS OF THE OVARIES

Each ovary is a small ovoid body, 3–4 cm long and weighing 2–8 g. It is held in place by ligaments but has no peritoneal covering. The external surface of an ovary is made up of a layer of germinal epithelium beneath which lies a cortex containing developing ova within follicles. At the centre of the gland is a medulla of fibrous tissue containing blood vessels and nerve fibres (see Fig. 9.19).

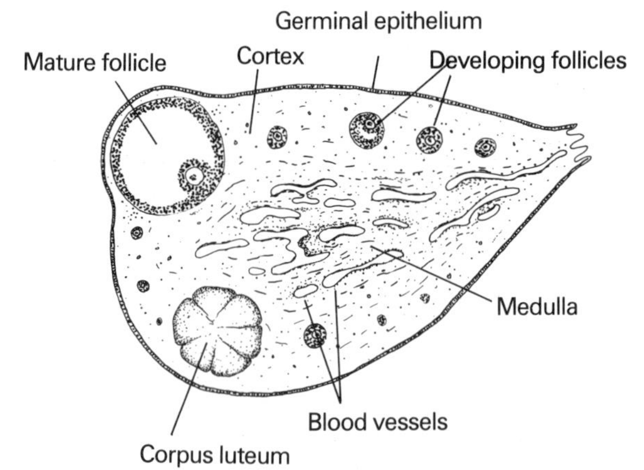

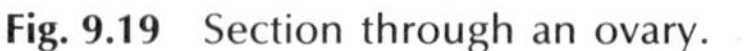
Fig. 9.19 Section through an ovary.

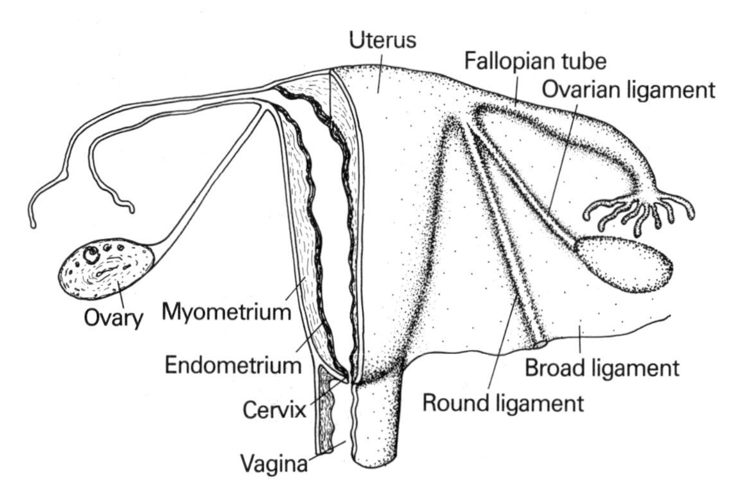

Fig. 9.18 The female reproductive system viewed from the front.

At birth each ovary contains several hundred thousand oocytes (developing germ cells). However, this number reduces, so that at puberty a smaller number, estimated to be of the order of 150 000 in each ovary, are still present. Meiotic division of the germ cells occurs before birth but is arrested at first prophase, giving rise to oocytes. At ovulation first division is completed, but second division and full maturation of the ova is not completed until after fertilization.

Normally only one oocyte matures every 28 days. Therefore, assuming that a female ovulates 13 times a year and that her reproductive life is 35 years, then only about 455 oocytes will ever actually mature. Non-maturing cells degenerate progressively so that by the end of reproductive life (menopause) few, if any, are left.

Each oocyte is surrounded by a single layer of flattened cells, the whole constituting a follicle. When a follicle develops, the oocyte enlarges and becomes separated from the surrounding cells (stratum granulosum) by a thick membrane, the zona pellucida. While these changes are taking place, the connective tissue stroma surrounding the follicle differentiates, giving rise to a layer known as the theca. The latter subsequently separates into a theca interna and a theca externa. With further enlargement of the follicle a fluid-filled cavity, the antrum, appears within the stratum granulosum (Fig. 9.20).

The oocyte will, by now, have reached full size, so that further follicular development will depend largely upon expansion of the antrum. The mature (graafian) follicle measures 1–1.5 cm in diameter and forms a bulge on the surface of the ovary. This eventually ruptures to release the oocyte and its surrounding zona pellucida into the abdominal cavity. This process is known as ovulation.

In the human ovary there may be several follicles which are sufficiently developed to contain an antrum present at any one time. However, usually all but one degenerate.

Following ovulation and rupturing of the follicle the antrum fills with a partially clotted fluid, while the follicular cells enlarge and become filled with a yellow pigment called lutein. The whole structure is then known as the corpus luteum.

It has already been stated that in addition to their role as reproductive organs the ovaries also have an important function as endocrine organs. They secrete three types of steroid hormones, oestrogens, progesterone and androgens. The release of these hormones is controlled by the hypothalamus and adenohypophysis.

During development of the follicle, the theca interna synthesizes oestrogen. Once the corpus luteum develops, it synthesizes progesterone as well. If fertilization does not occur, the corpus luteum degenerates after about 12 days and becomes a corpus albicans.

Oestrogens

The human ovaries secrete two oestrogens, oestradiol and oestrone. The latter is then metabolized to oestriol, which therefore appears in the circulation. The hormones are secreted mainly by the theca interna of the

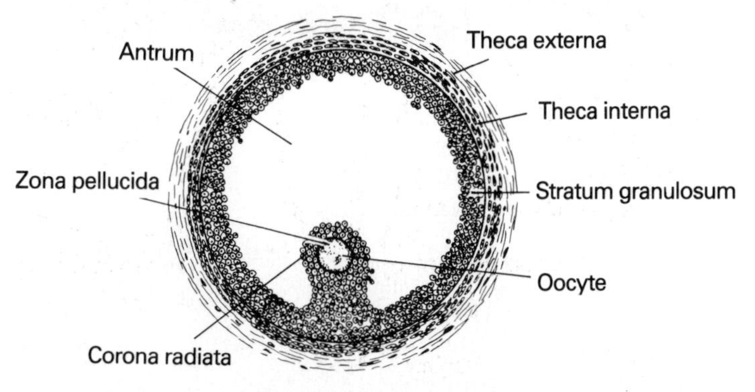

Fig. 9.20 Mature (graafian) follicle from an ovary (see Fig. 9.19).

developing follicle and the cells of the corpus luteum (although very small amounts are produced in the adrenals). In addition, large amounts of oestrogens are synthesized by the placenta of the pregnant female (see below). Once released into the circulation, about 70% of the oestrogen molecules become bound to sex hormone-binding globulin.

Oestrogens are required for the development and maintenance of the female secondary sexual characteristics, including breast development, the redistribution of fat which occurs at puberty, development of the pelvic girdle and body hair. They also promote growth in the genital tract at puberty and continue to influence its activities for the rest of the reproductive life. For example, oestrogens increase the motility of the Fallopian tubes and excitability of uterine muscle and play a part in the regulation of the development of the wall of the uterus in the menstrual cycle.

The hormones also promote salt and water retention in all tissues by the kidneys. They have a mildly anabolic action and promote growth during adolescence and stimulate closure of the epiphyseal discs.

Progesterone

Progesterone is secreted primarily by the corpus luteum in the non-pregnant female and by the placenta. It is released into the blood and is probably bound to a plasma protein. The hormone is rapidly converted into pregnanediol in the liver and has a half-life of only about five minutes in the circulation.

Progesterone promotes the development of the uterine lining, converting it into a secretory structure ready for implantation, should fertilization occur. It then reduces contractility in uterine muscle, thereby reducing the risk of spontaneous abortion. It has a thermogenic effect and is probably responsible for the slight rise in body temperature which occurs at the time of ovulation.

Female puberty

In the young child, the levels of both the ovarian hormones and the trophic hormones which regulate their output are very low. Therefore, the developmental changes, both physical and psychological, which are dependent upon these hormones cannot occur.

It is believed that, at this stage, the hypothalamus is very sensitive to the presence of circulating oestradiol and progesterone so that minute quantities of these hormones effectively inhibit the output of the trophic hormones. Since the latter are therefore only released in tiny amounts ovarian activity is restrained.

As early as the age of eight or nine years, the hypothalamus begins to lose its sensitivity to circulating ovarian hormones. This results in the production of increasing amounts of the trophic hormones LH and FSH. These then stimulate the ovaries to secrete greater quantities of oestrogens and the hormone-dependent changes of puberty begin to take place. These changes include physical growth, fat deposition, breast development, growth of body hair, growth of the reproductive tract and the appearance of the first menstrual period (menarche). The woman is thereafter capable of child-bearing. The whole process of puberty normally takes several years.

The menstrual cycle

The menstrual cycle and its accompanying rhythmical alteration in the production of ovarian oestrogens and progesterone is regulated by the hypothalamus through the anterior pituitary.

The average cycle lasts for approximately 28 days and may be divided into four stages as follows:

1st Stage — menstruation, when the endometrial lining of the uterus sloughs off and menstrual bleeding occurs. Menstruation lasts for about four to six days.

2nd Stage — proliferative phase, when the endometrium builds up. This occurs between six and 13 days after the commencement of the cycle.

3rd Stage — ovulation, which occurs around day 14.

4th Stage — secretory phase. From the fifteenth to the twenty-eighth day the glands in the endometrium are active so that, should fertilization occur, the endometrium forms a suitable environment for implantation.

The adenohypophysis releases two hormones, follicle-stimulating hormone (FSH) and luteinizing hormone (LH), which regulate the cycle and interact with ovarian oestrogens and progesterone in a complex manner. While much is known about the output of hormones from the glands at various stages in the cycle, the underlying regulatory mechanisms are still currently being researched.

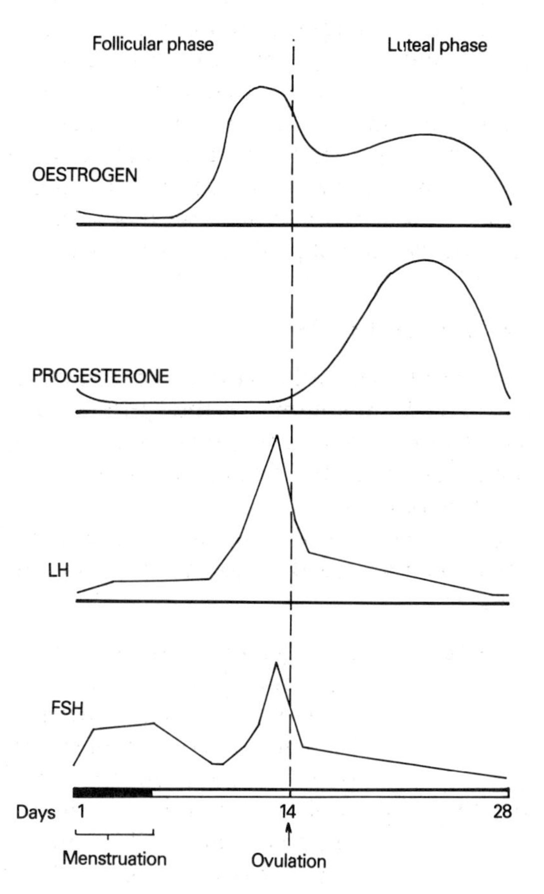

Fig. 9.21 Changes in the plasma hormone levels during the menstrual cycle.

Following menstruation, the levels of oestrogens and progesterone in the blood are low (Fig. 9.21). Under the influence of a releasing hormone about whose nature there is some controversy, but which is identified here as gonadotrophin-releasing hormone (GnRH), the anterior pituitary releases FSH

Negative feedback regulation of FSH and LH secretion

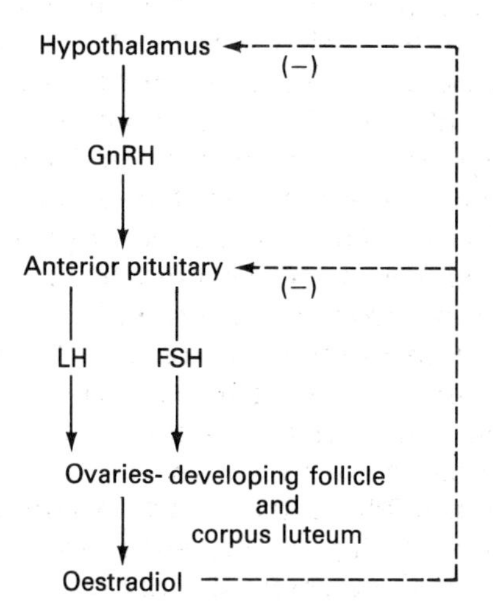

Positive feedback regulation of FSH and LH secretion

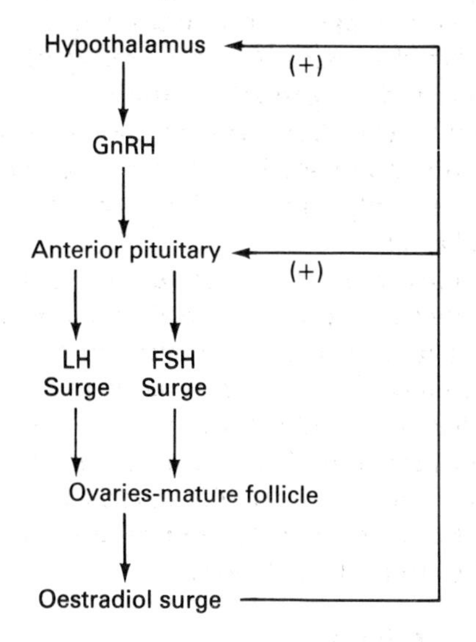

Fig. 9.22 Relationships between the hypothalamus, posterior pituitary and the ovaries during the menstrual cycle.

and some LH which promote the development of an ovarian follicle (follicular phase). As the latter grows, it secretes increasing amounts of oestrogens and under their influence there is a rapid growth of the endometrial lining of the uterus. In the early stages of this part of the cycle they exert a negative feedback influence upon FSH and LH (Fig. 9.22). However, towards the end of the phase there is an abrupt reversal and the oestrogens suddenly effect a positive stimulation upon FSH and probably more importantly LH release from the pituitary. As a result, around days 12–13 there is a surge in the levels of LH and FSH in the circulation. The LH surge is apparently the trigger for ovulation.

The high levels of circulating LH facilitate development of the corpus luteum within the ovary (luteal phase). The corpus luteum produces progesterone and small quantities of oestrogens which support the secretory phase of the cycle. Under the influence of the two hormones the endometrial glands increase in size and secretory activity.

During the latter stage of the cycle the increasing amounts of progesterone and oestrogens secreted by the ovaries exert a negative feedback on both hypothalamus and pituitary. Thus the levels of circulatory LH and FSH fall, causing the corpus luteum to involute and cutting off the production of progesterone and oestrogens. Since the inner endometrium is dependent upon the presence of these last two hormones, then it can no longer be maintained and it sloughs off, i.e. menstruation occurs. The hypothalamus is now released from the suppressive influence of the ovarian hormones and begins to release GnRH, thereby initiating a new cycle.

Pregnancy

Fertilization of the oocyte occurs in the Fallopian tube and is followed by a second meiotic division, after which the male and one of the female nuclei unite and give rise to a zygote. The zygote is carried down to the uterus and is already dividing rapidly to form the beginnings of the embryo. By about the seventh day after ovulation the developing embryo has embedded in the uterine endometrium.

During pregnancy the mother's body becomes adapted to support the developing infant. These adaptations are brought about and are regulated by hormones secreted by the pituitary, ovaries and also by the developing embryonic structure itself.

After about two weeks after fertilization and after the embryo has embedded in the uterine wall, its outer layer, the trophoblast, releases chorionic gonadotrophin (HCG), a glycoprotein similar in structure to LH, into the mother's blood stream. HCG performs an equivalent function to LH in that it maintains the integrity of the corpus luteum. The corpus luteum therefore continues to secrete progesterone and oestrogen which, in turn, maintain the lining of the uterus and prevent menstruation from occuring. This ensures that the embryo has a stable environment for its further development.

HCG reaches a peak 35 to 45 days after fertilization (i.e. 50 to 60 days after the end of the last menstrual period). It then drops to a very low level which is maintained until the end of pregnancy (Fig. 9.23).

In the early stages of development the embryonic tissues differentiate to give rise to two distinct structures. One of these gives rise to the fetus, while the other gives rise to the placenta which links the former to the uterine wall and the mother's circulation. All of the nourishment for the developing infant is derived through the placenta and so too are many of the hormones required for the maintenance of pregnancy. The placenta functions as a fully fledged endocrine organ.

Even in the early stages of its growth, the placenta secretes steadily increasing amounts of oestrogen and progesterone, and by the time the HCG level falls off there is sufficient placental progesterone in the mother's circulation to maintain pregnancy. The levels of circulating progesterone and oestrogens then continue to rise until parturition (see Fig. 9.23).

It is interesting to note that the principal oestrogen synthesized by the placenta is

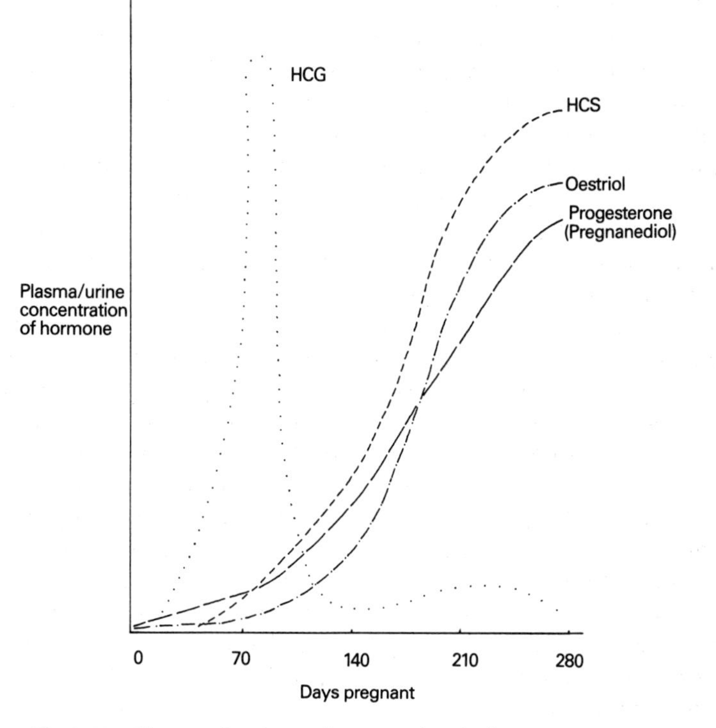

Fig. 9.23 Changes in plasma hormone levels during pregnancy.

oestriol, rather than oestradiol which is released from the ovaries.

The placenta also synthesizes chorionic somatomammotrophin (HCS) previously known as human placental lactogen, which appears in the maternal circulation about six weeks after conception (see Fig. 9.23) The blood levels of this hormone rise steadily throughout pregnancy and flatten out towards parturition. It is structurally similar to growth hormone and exhibits properties characteristic of both growth hormone and prolactin, i.e. it has growth-promoting, lactogenic and lipogenic properties. The precise physiological significance of the hormone is not, however, understood.

In addition to the substances described above, the placenta may also secrete other substances and several have been described at different times in the past. Renin and a polypeptide described as relaxin have been identified, as has a substance with TSH-like activity, which may, in fact, actually be HCG.

Most of the mother's (and some of the fetus's) other endocrine organs are also in a state of elevated activity, with the pituitary, adrenals, thyroid and parathyroid glands exhibiting particularly marked increases.

Parturition

It is not known how parturition (labour) is initiated in women, although there is a considerable amount of information available concerning the processes which influence the event in sheep and goats. Parturition can be considered to consist of two primary events, the co-ordinated contractions of the myometrium (assisted by involuntary contractions of skeletal muscles such as those in the abdominal wall); and the softening of the cervix so that the fetus may pass through.

Uterine contraction is triggered by a rise in intracellular Ca^{++} concentration. The hormone oxytocin stimulates Ca^{++} influx into myometrial cells, as well as lowering their threshold of stimulation, thereby causing contraction. Prostaglandins produced by the myometrium raise intracellular Ca^{++} by liberating it from intracellular binding sites. The prostaglandins PGE_2 and PGE_{2a} promote cervical 'ripening' and in some circumstances are given to induce birth.

The levels of progesterone and oestrogen in the mother's circulation rise steadily throughout pregnancy. Progesterone, which inhibits uterine contractions, is the hormone present in the largest quantities up until about the seventh month. Oestrogen, which promotes uterine contractions, continues to increase in concentration after the seventh month when progesterone output levels off. Therefore, between the seventh and ninth month of pregnancy there is an alteration in the balance of the two hormones, with oestrogen becoming dominant. It is likely that this change in the oestrogen/progesterone ratio renders the uterine muscle more liable to contraction. In many mammalian species it seems that the trigger for this change in the oestrogen/progesterone ratio is in some way related to a rise in fetal cortisol concentration.

It has been established that cervical stretching initiates a neuroendocrine reflex which brings about the release of oxytocin from the mother's neurohypophysis. Furthermore, the uterine muscle is found to be particularly responsive to oxytocin at the time of labour.

It should be noted, however, that while oxytocin normally appears to play a part in the process of birth, by promoting uterine contraction, it is not essential. The process will continue in its absence, although labour may be prolonged.

Parturition normally occurs at 40 ± 2 weeks after the onset of the last menstrual period.

Lactation

Following birth of the infant, milk is secreted by the mother's two mammary glands. During pregnancy the breasts enlarge under the influences of the oestrogens and progesterone, together with the adrenocorticosteroids, growth hormone, insulin, prolactin and chorionic somatomammotrophin. These hormones support the development and growth of the milk-secreting tissues and the duct systems and also promote the laying down of fat.

During pregnancy the output of prolactin from the mother's pituitary rises steadily. It does not promote milk secretion, however, owing to the suppressive actions of oestrogens and progesterone. After birth, the concentrations of these hormones, which originate in the placenta, reduce greatly and this is the stimulus for milk secretion.

At first, a clear fluid, colostrum, which has largely the same composition as milk and which contains IgG antibodies but which has almost no fats, is produced. Milk is secreted after a short period and in much greater quantities.

In addition to prolactin, many other hormones and especially the adrenocortical hormones, growth hormone and parathyroid hormone are required for milk production.

Prolactin output is maintained above non-pregnant levels by suckling. A neuroendocrine reflex (Fig. 9.24) is set up whereby nerve impulses pass to the hypothalamus to inhibit the release of prolactin inhibitory hormone (PIH), which normally prevents a basal release of prolactin from the pituitary. The reflex may also promote the release of a prolactin releasing-hormone (PRH). As a result, prolactin

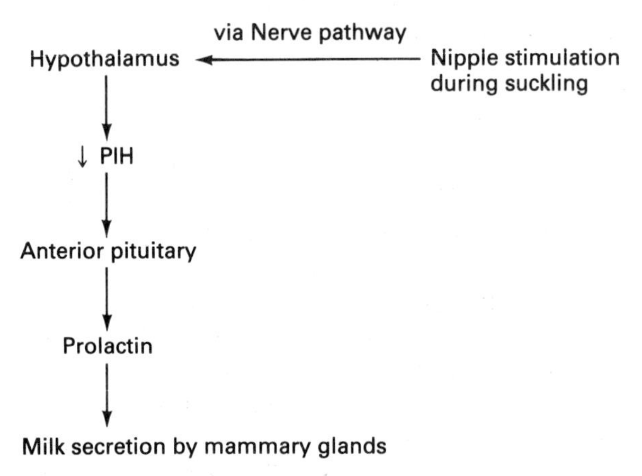

Fig. 9.24 Suckling-induced reflex release of prolactin.

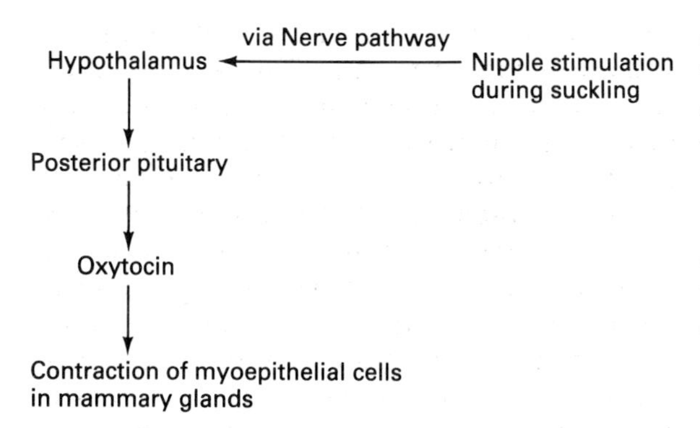

Fig. 9.25 Suckling-induced reflex release of oxytocin.

passes into the mother's blood stream, passes to the breasts and promotes the synthesis of milk. Prolactin output, then, rises and falls sharply depending upon the times of feeding. If breast feeding is not carried out, milk production rapidly ceases.

The ejection of milk from the breasts (let-down) depends upon the presence of a separate hormone, oxytocin. Suckling initiates a second neuroendocrine reflex (Fig. 9.25), which results in the release of oxytocin from the neurohypophysis. Oxytocin passes to the breast and stimulates contraction of the myo-epithelial cells around the milk-producing alveoli and the ducts. Thus milk is expelled into the baby's mouth.

It can be seen from the above that milk secretion and milk expulsion are regulated by two discrete mechanisms, both of which involve the pituitary gland.

Menopause

Menopause is the term applied to the period of time during which ovarian function reduces to a low level of activity. It may last from a few months to a few years and usually falls between the ages of 45 and 55 years.

The menopause is almost certainly due to an ageing process in the ovaries and is linked to a reduction in the number of follicles which remain in the ovaries. As a result oestrogen and progesterone output diminish to the level where they can no longer suppress the output of pituitary FSH and LH and the levels of these two hormones in the blood rise rapidly. The combined effects of low steroids and raised trophic hormones can induce a number of physical and mental changes which may persist for quite long periods. These include, anxiety, irritability, fatigue, dyspnoea and hot flushes (blushing). The menstrual periods and ovulation both eventually cease.

Following the menopause, the ovaries reduce in size and are found to contain few, if any, primary oocytes. Oestrogen output virtually ceases.

Effects of oestrogen deficiency

Hypogonadism may be due to poorly formed or absent ovaries. If there is a total lack of ovarian hormones, then a female eunuch is produced in whom the female secondary sexual characteristics do not appear. The bones of the arms and legs become abnormally long; this is due to the failure of the epiphyses to close.

If small quantities of gonadal hormones are produced there will be very irregular menstrual periods and usually a failure to ovulate.

Effects of oestrogen excess

This occurs rarely and is due to the presence of a hormone-secreting tumour and promotes excessive build-up of the endometrium with irregular bleeding.

THE MALE REPRODUCTIVE SYSTEM

STRUCTURE OF THE MALE REPRODUCTIVE SYSTEM

The male reproductive tract consists of the testes, the penis, a system of tubules which transport seminal fluid from one to the other and a number of small accessory glands.

The testes are suspended between the legs, within a wrinkled skin-covered bag, the scrotum. Each testis is surmounted by an epididymis, a closely packed system of minute tubules leading into the vas deferens, which carries seminal fluid to the penis hanging in

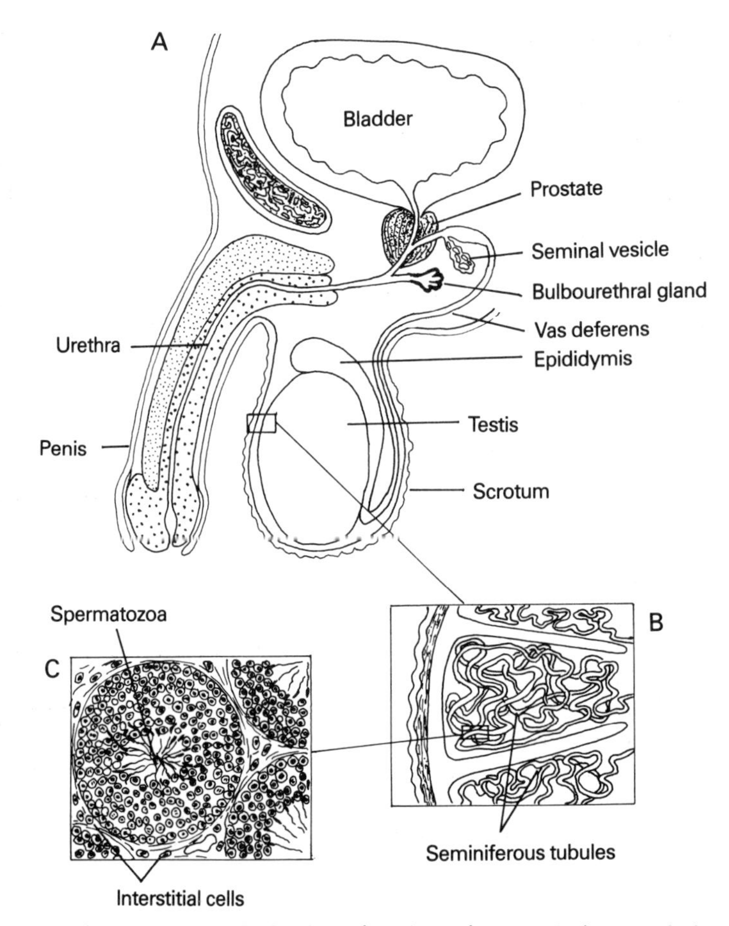

Fig. 9.26 (A) Male reproductive system. (B) Section of testis to show seminiferous tubules. (C) Histological appearance of a seminiferous tubule.

front of the scrotum. In addition, there are the seminal vesicles, prostate and bulbourethral glands (see Fig. 9.26), all of which contribute to the seminal fluid.

STRUCTURE AND FUNCTIONS OF THE TESTES

Each testis is an ovoid structure approximately 2 cm by 4 cm in the adult. It has a connective tissue capsule from which project septae which incompletely divide the gland into 200 or 300 lobules. Each lobule contains one or more highly convoluted seminiferous tubules lined by spermatozoa-producing cells. Lying in the spaces between the tubules are the hormone-producing cells of Leydig, also called interstitial cells.

Testosterone is the main male sex hormone or androgen, androstenedione is also produced in small amounts. In addition, trace amounts of oestrogens are also produced.

Testosterone

Testosterone is a steroid which is structurally similar to the female sex hormones. The testes secrete about 95% of the total testosterone output of the male whilst the adrenals secrete the other 5%. After its release, virtually all of the testosterone is bound to albumin or β-globulin in the plasma. Once it has entered the tissues, in some cases at least, testosterone is converted to dihydrotestosterone and it is this which is actually responsible for the physiological activities normally ascribed to the male sex hormone. Like the other steroids, testosterone and dihydrotestosterone appear to exert their influences through direct actions on protein synthesis within the target cells.

The male hormones are responsible for the development and maintenance of the bodily characteristics of the male. In the embryo the presence of testosterone facilitates the development of the male reproductive organs. A few weeks after birth, however, hormone levels fall and their output is almost non-existent until the onset of puberty which normally occurs between the ages of 10 and 13 years.

Testosterone secretion at puberty causes a considerable enlargement in the penis, testes and scrotum. In addition it has an anabolic effect and thereby promotes physical growth and development, including growth of bone and muscle, enlargement of the larynx with an accompanying deepening of the voice, growth of facial and body hair, but often a decrease in the growth of hair on top of the head.

The hormone also exerts an influence upon the mental state and is associated with development of the libido.

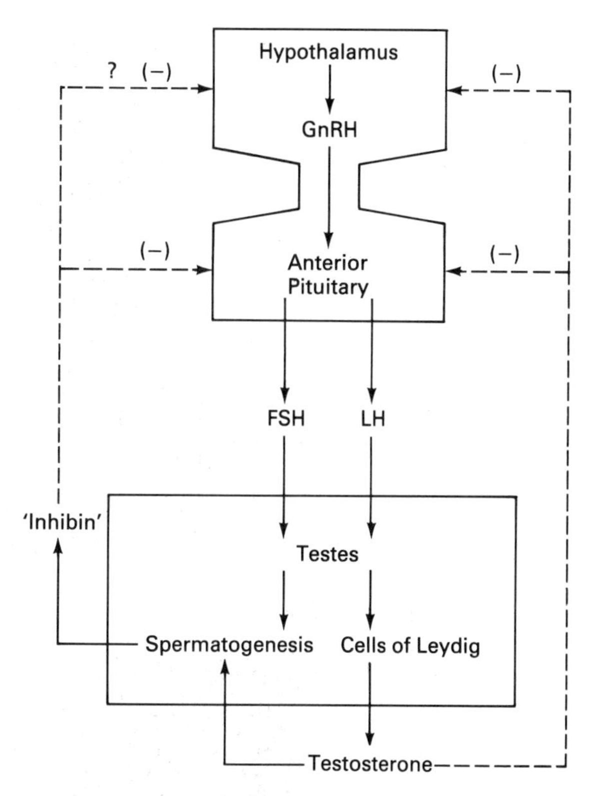

Fig. 9.27 Relationships between the hypothalamus, anterior pituitary and testes.

Control of secretion of testosterone

Release of testosterone by the testes is dependent upon the presence of luteinizing hormone (LH), sometimes known as interstitial-cell-stimulating hormone (ICSH) in the male, from the anterior pituitary. A rise in the level of LH in the circulation stimulates the cells of Leydig within the testes to liberate testosterone which then exerts a suppressive influence upon the output of the hypothalamic releasing hormone and LH itself (Fig. 9.27). It is likely that at puberty, as in the female, there is a change in the sensitivity of the hypothalamus to the suppressive influences of the sex hormone. The minute levels of testosterone in the blood of the young boy are no longer able to inhibit the output of releasing hormone, so that the latter is secreted in increasing amounts. This leads eventually to quantities of testosterone being released by the testes, sufficient to bring about the onset of puberty.

Testosterone is required in order that spermatogenesis may occur within the testes. LH itself is not required directly for this process but FSH is. FSH is apparently required for the early stages of spermatogenesis, whilst testosterone is necessary for final maturation of the spermatozoa.

FSH output is not affected by testosterone, but it is inhibited by a separate substance which has been named 'inhibin'. The latter is secreted by the Sertoli cells, which are situated within the seminiferous tubules and which also supply nourishment for the developing spermatozoa.

It is interesting to note that, in the male, gametogenesis is dependent upon both sex hormones and pituitary trophic hormones from the pituitary. In the female the process is regulated only by the trophic hormones.

Effects of testosterone deficiency

If the testes are reduced or absent, the ensuing lack of testosterone will not allow development of the secondary sexual characteristics or the accessory sex organs. If this occurs before puberty a male eunuch will be produced. He is characterized by long arms and legs; greatly reduced body and facial hair; the larynx is underdeveloped so that he has a high-pitched voice and the genitals are immature.

Effects of testosterone excess

Excessive amounts of testosterone are produced rarely by testicular tumours. In young boys they induce precocious pseudopuberty where the physical signs of puberty develop without sexual maturation of the testes.

PROSTAGLANDINS

The term 'prostaglandin' was first used in the 1930s by van Euler, who discovered that a substance in semen caused uterine smooth muscle to contract. As he thought that the substance was produced by the prostate gland, he named it 'prostaglandin'. Later work by Bergström showed that van Euler's prostaglandin was, in fact, a mixture of substances originating in the seminal vesicles and not in the prostate gland at all.

Since the 1930s some 16 prostaglandins have been chemically characterized, and the Nobel prize for medicine in 1982 was awarded to the scientists Bergström, Samuelson and Vane for their work in this area.

Prostaglandins are lipids, derived from unsaturated fatty acids, in many cases from arachidonic acid. There are three groups of prostaglandins: prostacyclin; prostaglandins (PG) D_2, E_2 and F_2; and thromboxanes A_2 and B_2. Arachidonic acid is a constituent of cell membranes and therefore found all over the body and indeed prostaglandins have been demonstrated in virtually every tissue.

Prostaglandins have been implicated in a great variety of physiological processes including inflammation and hypersensitivity states, pain and fever. The anti-inflammatory, pain-killing and temperature-lowering actions of aspirin are mediated by its inhibition of prostaglandin synthesis.

Some prostaglandins (thromboxanes) have been shown to cause vasoconstriction and platelet clumping, while prostacyclin has the opposite effect, vasodilation and the inhibition of platelet clumping. Since thromboxanes are produced by the platelets themselves, and prostacyclin is produced by blood vessel walls, clotting is evidently controlled by these two opposing systems.

PGF_2 is known to stimulate uterine contraction during parturition in sheep and possibly has the same effect in humans.

Prostaglandins are similar to hormones in many ways in that they are involved in the control of physiological processes, and, furthermore, many of them appear to act via the cyclic AMP system in cells. They are, however, metabolized extremely quickly, and are not dispatched to a site distant from the site of production like conventional hormones.

At present, then, it seems that prostaglandins are 'local hormones', which in some way interact with the actions of the better established control substances in the body, hormones and neurotransmitters.

Recommended further reading

The books and articles listed below are a small selection of the available literature which we have found to be particularly useful to us and to our students.

Ambrose E J, Easty D M 1977 Cell biology, 2nd edn. Nelson, London

American Physiological Society 1959 onwards Handbook of physiology. Williams and Wilkins, Baltimore

Amos W H G 1981 Basic immunology. Butterworths, London

Åstrand P-O, Rodahl K 1977 Textbook of work physiology, 2nd edn. McGraw-Hill, London

Avers C 1982 Basic cell biology, 2nd edn. Van Nostrand, New York

Bendall J R 1969 Muscles, molecules and movement. Heinemann, London

Bloom W, Fawcett D W 1975 A textbook of histology, 10th edn. Saunders, Philadelphia

Bradbury S (ed) 1975 Hewer's textbook of histology for medical students, 9th edn. Heinemann, London

Cassens R G (ed) 1972 Muscle biology. Dekker, New York

Cohen C 1975 The protein switch of muscle contraction. Scientific American 233: 36–45

Comroe J H 1974 Physiology of respiration, 2nd edn. Year Book Medical Publishers, Chicago

Davenport H W 1977 Physiology of the digestive tract, 4th edn. Year Book Medical Publishers, Chicago.

Eyzaguirre C, Fidone S J 1975 Physiology of the nervous system, 2nd edn. Year Book Medical Publishers, Chicago.

Ganong W F 1983 Review of medical physiology, 11th edn. Lange, Los Altos

Giese A C 1979 Cell physiology, 5th edn. Saunders, Philadelphia

Guyton A C 1981 Textbook of medical physiology, 6th edn. Saunders, Philadelphia

Ham A W, Cormack D H 1979 Histology, 8th edn. Lippincott, Philadelphia

Hardy R N 1981 Endocrine physiology. Arnold, London

Hirsh J, Brain E A 1983 Haemostasis and thrombosis. A conceptual approach. Churchill Livingstone, Edinburgh

Hoffbrand A V, Pettit J E 1984 Essential haematology. Blackwell, Oxford

Hopkins C 1978 Structure and function of cells. Saunders, Philadelphia

Huxley H E 1969 The mechanism of muscular contraction. Science 164: 1356–1366

Johnson M, Everitt B 1980 Essential reproduction. Blackwell, Oxford

Junqueira L C, Carneiro J 1983 Basic histology, 4th edn. Lange, Los Altos

Kirkwood E, Lewis C 1983 Understanding medical immunology. Wiley, Chichester

Lee J, Laycock J 1983 Essential endocrinology, 2nd edn. Oxford University Press

Mitchell G A G, Mayor D 1977 The essentials of neuroanatomy, 3rd edn. Churchill Livingstone, Edinburgh

Moffat D B 1975 Biological structure and function 5. The mammalian kidney. Cambridge University Press

Mountcastle V B (ed) 1980 Medical physiology, 14th edn. Mosby, London

Passmore R, Robson J S (eds) 1976 A companion to medical studies, vol 1, 2nd edn. Blackwell, Oxford

Pitts R E 1974 Physiology of the kidney and body fluids, 3rd edn. Year Book Medical Publishers, Chicago

Roitt I 1984 Essential immunology, 4th edn. Blackwell, Oxford

Romanes G J 1976–79 Cunningham's manual of practical anatomy (three volumes), 14th edn. Oxford University Press

Sanford P A 1982 Digestive system physiology. Arnold, London

Selkurt E E 1984 Physiology, 5th edn. Little, Brown, Boston

Snell R S 1980 Clinical neuroanatomy for medical students. Little, Brown, Boston

Vander A J, Sherman J H, Luciano D S 1986 Human physiology, 4th edn. McGraw-Hill, New York

Weir D M 1983 Immunology. Churchill Livingstone, Edinburgh

West J B (ed) 1985 Best and Taylor's physiological basis of medical practice, 11th edn. Williams and Wilkins, Baltimore

Widdicombe J, Davies A 1983 Respiratory physiology. Arnold, London

Williams P L, Warwick R 1980 Gray's anatomy, 36th edn. Churchill Livingstone, Edinburgh

Williams P L, Wendell-Smith C P 1969 Basic human embryology, 2nd edn. Pitman Medical, London

Wilkie D R 1976 Muscle, 2nd edn. Arnold, London

Index